Dissipative Ordered Fluids

André M. Sonnet • Epifanio G. Virga

Dissipative Ordered Fluids

Theories for Liquid Crystals

Springer

André M. Sonnet
Department of Mathematics and Statistics
University of Strathclyde
Glasgow, United Kingdom

Epifanio G. Virga
Dipartimento di Matematica
Università di Pavia
Pavia, Italy

ISBN 978-1-4899-9143-0 ISBN 978-0-387-87815-7 (eBook)
DOI 10.1007/978-0-387-87815-7
Springer New York Dordrecht Heidelberg London

Mathematics Subject Classification (2010): 76A15, 82D30

Softcover reprint of the hardcover 1st edition 2012

Printed on acid-free paper

Springer is part of Springer Science+Business Media (www.springer.com)

Preface

This book envisages liquid crystals as particular examples of dissipative ordered fluids. While it may be unique in taking this special perspective, it is not the only mathematical book on liquid crystals, and so one should have more than one good reason to read it. We can only give the reasons that made us write it: the reader will decide whether they suffice.

First, we felt the need to formulate a unified mathematical framework within which dynamical theories for liquid crystals can be phrased, a framework that is general enough also to incorporate dynamical theories for other ordered fluids. Our general topic is the evolution of order in fluids and its interaction with flow. Liquid crystals are the ideal arena for testing such a general theory for dissipative ordered fluids, because they are perhaps the best understood incarnation of these fluids. The established dynamical theories for liquid crystals have passed the tests of time and experimental scrutiny. Although we chose to concentrate on this special class of ordered fluids, we also highlight the opportunities that our general method offers in other closely related fields.

Since liquid crystals are here only examples of a wider family of ordered fluids, they are not treated in the full generality of all their condensed phases. Although our study is not limited to the traditional *uniaxial* nematics, since it also embraces the newly discovered (and still disputed) *biaxial* phases, it does not cover *smectic* liquid crystals. This large class of fluids, closer indeed to solids, is too complex to be included in an introductory book such as this. However, we interpret *nematics* in a broad sense, incorporating *chiral* nematics, often also called *cholesterics*.

Our narrative starts from a molecular description of the order that gives rise to the condensed phases of liquid crystals, and it moves on to the construction of continuum theories capable of describing their evolution. We sought secure guidance in such an endeavor and found it in a *dissipation principle*, which can be traced back to both the work and vision of RAYLEIGH. We interpret this principle in precise, mathematical terms and phrase it within a thermodynamic context, though most of the theories we review are purely mechanical in nature.

We have deliberately chosen to talk about theories in the plural. Order in liquid crystals appears in various guises and can be described in different ways, each more appropriate than others for certain purposes or in certain contexts. Theories broadly fall into two classes, depending on how the molecular order is described on larger length scales: there are *director* theories and *tensor* theories. The way in which theories in these large classes are established and how they are related is the leitmotif of the core of this book.

We do not limit our scope to harmonizing in a unified setting existing theories, but we also venture into hitherto unexplored territory. In doing so, we derive a new theory for the acoustic actions in nematic liquid crystals that is capable of explaining quantitatively experiments performed almost half a century ago that cannot be completely understood within the classical dynamical theories.

Since this is a mathematical book, we strive for rigor and precision. However, though we use the languages of analysis, algebra, and geometry, this is not a book in any of these mathematical disciplines. This is a book on *mechanics*, the mathematical science of *motion*, which is the archetype of all dynamical processes.

Although we tried to be as comprehensive as the scope of an introductory book allowed us to be, we could not cover all aspects of nematic order evolution. In particular, defect dynamics and dynamics of thin nematic films on surfaces remain untreated. Given the body of theoretical results available in the literature and the interest in their practical applications, these related subjects would actually deserve to fill a whole book by themselves.

This is a book on theories and their conceptual interplay. We have therefore, apart from rare exceptions, not included excercises or assignments. It is our hope that the reader will learn from this book how to phrase a continuum theory for the dissipative dynamics of ordered fluids that could stand the scrutiny of experimental physics, as did the celebrated theories of ERICKSEN–LESLIE and LANDAU–DE GENNES.

Glasgow, Pavia
September 2011

André M. Sonnet
Epifanio G. Virga

Contents

1 Molecular Theories

Any expository account of liquid crystals would invariably commence by saying that they constitute a state of matter that is intermediate between crystals and liquids. Their apparently contradictory appellation conveys well their being *mesophases*, that is, their participating in properties of two worlds. Macroscopically, they exhibit optical birefringence, a property typical of crystals, while retaining their ability to flow, which characterizes fluids. Microscopically, the tendency of liquid crystals to mediate between diversities is ascribed to the anisotropy of their molecules. In this introductory chapter, we present the microscopic basis for our development, which will mostly be macroscopic. We explore the microscopic origin of the ordering transition that gives rise to nematic liquid crystals. This discussion will ultimately serve to identify the most appropriate macroscopic order parameters for nematic liquid crystals, both in the uniaxial and biaxial phases.

1.1 Molecular Interactions

A picture often drawn describes liquid crystal molecules as rods or ribbons subject to interactions that tend to make them align alike. Whenever such a tendency prevails over disorganizing causes, an ordered phase is established from the isotropic, disordered phase. This ordering phase transition, which is usually first-order,[1] induces a local common molecular orientation that may vary from place to place. In the nematic ordered phase, which is still fluid, molecules move freely; they do not exhibit any spatial order, but even if mobile, they reveal a long-range orientational order in their organization, which at the same place can possibly involve different molecules at different times. What persists in the dynamical evolution of the system is its ability to be self-organized, though the average result of the molecular organization may vary in time.

[1] A *first-order* transition occurs abruptly, with a discontinuity in the order parameters that characterize the phase. By contrast, a *second-order* transition is smooth and takes place with gradual changes in the order parameters.

Liquid crystals are of two types, *thermotropic* and *lyotropic*, according to whether the temperature or the density drives molecules towards alignment, respectively. For thermotropic liquid crystals, there is a critical value of the temperature below which the disorder-to-order transition takes place. Correspondingly, for lyotropic liquid crystals, there is a critical value of the density above which the disorder-to-order transition takes place. Here we shall be concerned only with thermotropic nematic liquid crystals.

The core of this chapter is the celebrated theory of MAIER & SAUPE [208] presented in Section 1.3. This theory rests upon a properly formulated mean-field approximation to the HELMHOLTZ free energy of an ensemble of interacting rodlike molecules.

More generally, mean-field theories play a central role in all of condensed matter physics. In treating an ensemble of N mutually interacting molecules, deriving the HELMHOLTZ free energy $\mathscr{F}$ of the ensemble, from which all thermodynamic properties would follow, is commonly a prohibitive task. In the mean-field approximation, the interactions between molecules are replaced by an effective interaction of every molecule in the ensemble with a mean field $\mathbf{Q}$, which binds all molecules together. Within this approximation, the free energy $\mathscr{F}_0$ can often be computed explicitly, albeit constrained by appropriate self-consistency conditions on the mean field $\mathbf{Q}$, which enters $\mathscr{F}_0$ as a parameter. A modern account of mean-field theories and their ample role in condensed matter physics can be found in [263].

In soft-matter physics, $\mathbf{Q}$ generally has the meaning of an *order tensor*, or a collection of such tensors, which describes at a macroscopic scale the molecular organization underlying an ordered phase. The equilibrium values of $\mathbf{Q}$ that describe the condensed phases are customarily identified by minimizing $\mathscr{F}_0$. Thus, minimizers of $\mathscr{F}_0$ branching off a given phase witness second-order phase transitions, while minimizers of $\mathscr{F}_0$ jumping from one phase to another at the same temperature witness first-order phase transitions. In general, the study of both global and local minimizers of $\mathscr{F}_0$ reveals both the stable equilibrium phases and the phases eligible to become so.

For the mechanician, mean-field theories are also attractive for another reason: they can be seen as precursors of the continuum, phenomenological theories for complex materials. In this perspective, the statistical MAIER–SAUPE theory heralds the continuum theory of DE GENNES [59] outlined in Chapter 4. Thus, mean-field theories stand, as it were, at the border between statistical physics and continuum physics.

Here we formulate in a rigorous manner the mean-field approximation; we phrase it in a language so general as to be applicable to a wide class of molecular interactions, those that can be expressed through a diagonal, bilinear Hamiltonian involving appropriately defined collective molecular tensors $\mathbf{q}$. The mean-field approximation $\mathscr{F}_0$ to the equilibrium ensemble free energy $\mathscr{F}$ will be characterized by a minimum principle. This is the principle of global *least free energy*, asserting that the "best" mean-field approximation to the "true" free energy is achieved by the order tensor collection $\mathbf{Q}$ for which $\mathscr{F}_0$ attains its least value among all critical points. For a large class of interactions, which also includes the one at the basis of the MAIER–SAUPE theory , the mean-field free energy $\mathscr{F}_0$ attains its global minimum, and so the above minimum principle makes all local minimizers of $\mathscr{F}_0$ eligible to become the global minimizer describing the condensed phase. It is shown in [116], on which most of our account in this chapter relies, that the least-free-energy principle also holds when $\mathscr{F}_0$ possesses only saddles as critical points, and so it is indefinite. To keep our development simple, we shall often refer the reader to [116] for a more detailed analysis of indefinite mean-field free energies, though, as shown below in this section, they arise quite naturally in liquid crystals.

1.1.1 Two-Particle Hamiltonian

Molecular interactions are ultimately responsible for the mesogenic[2] behavior of some molecules that, unlike others, tend to form ordered phases. In general, the interaction between two molecules depends on the states of the interacting molecules and their relative position in space. In the past, several theories were developed that derive the interaction energy of two molecules from quantum-mechanical computations of charge distributions [215, 320]. These theories, notable among which is LONDON's *dispersion forces* theory [193], produce a two-particle Hamiltonian that depends on the charge distribution in both molecules and the vector joining their charge centers.[3] The mean-field approximation, which will be described in Section 1.2 below, can only bear Hamiltonians much simpler than this.

The mean-field approximation has a long history; it has often proved useful in describing phase transitions in soft matter systems: perhaps the MAIER–SAUPE theory of uniaxial nematic liquid crystals is its most successful application in this area. Crucial to the success of this theory is the replacement of the space-dependent two-particle Hamiltonian with a space-independent one [70]. This is achieved by assuming that molecules sharing one and the same state—purely orientational, in the MAIER–SAUPE theory—are isotropically distributed in space around any given probe molecule and by computing the average interaction energy between the probe and all other molecules. Such a strategy rests upon the intuitive representation of a fluid bulk as a molecular assembly in which a probe molecule in a given state can be approached in all directions with equal probability by any other molecule in another given state, freely wandering in space among all other molecules. Though the interaction energy for the probe and the wandering molecules depends on both their distance and the direction of relative approach, the energy binding the probe to the average field produced by the system of all possible wandering molecules in one and the same state depends only on this state and that of the probe. In the classical MAIER–SAUPE theory , such an effective energy is simply a function of the relative orientation between the interacting molecules.

---

[2] *Mesophase* is the name often given to an *intermediate* phase that is characterized by some partial degree of molecular order.

[3] The center of a system of positive (or negative) charges is defined as the electrostatic analogue of the center of mass for a system of mass points. For neutral, nonpolar molecules, the only ones considered here, the charge center is the point where the centers of positive and negative charges coincide.

The averaging process that conceptually leads us here to a space-independent Hamiltionan is not to be confused with the one that will lead us in Section 1.2 to the mean field experienced by all molecules in an ensemble. The former is the logic antecedent of the latter. We may say that the mean-field approximation is indeed based on two independent successive averaging processes, of which here we describe only the first, deferring the second to Section 1.2 below, where such a derivation is illustrated in detail. In Section 1.4 we shall also show how the space-independent Hamiltonian can embody subtle structural molecular aspects, such as anisotropy of shape and short-range repulsion.

State Space

Following [116], we make the space-independent two-particle Hamiltonian H the basis of our development. Often H is also called the *pair-potential* of the molecular interaction; the terms "two-particle Hamiltonian" and "pair-potential" will be used as synonyms throughout this chapter, as will also be "molecule" and "particle." In mathematical terms, H is a real-valued mapping defined over $\Omega \times \Omega$, where Ω is the *state space* of the molecules:

$$(\omega, \omega') \mapsto H(\omega, \omega').$$

Here ω and ω' describe the states of two interacting molecules. For rigid molecules, Ω represents all possible orientational states; for flexible molecules, it also embodies the conformational states.

H must satisfy certain general conditions. Since all molecules are indistinguishable particles, H must be invariant under particle exchange,

$$H(\omega', \omega) = H(\omega, \omega'), \quad \forall\, \omega, \omega' \in \Omega.$$

H must also be frame-indifferent, that is, it must be invariant under all state transformations that merely amount to a change of frame or, said differently, that can be reduced to a rigid rotation of both interacting molecules. To formalize this property in general, we introduce more structure in Ω. We shall represent by R the action on Ω of the rotation group $\mathsf{SO}(3)$ in three-dimensional Euclidean space. For a given rotation $\mathbf{R} \in \mathsf{SO}(3)$,

$$R_{\mathbf{R}} : \Omega \to \Omega$$

is the mapping such that for any $\omega \in \Omega$, $R_{\mathbf{R}}(\omega) \in \Omega$ describes the state of the molecule that differs from ω only by the effect of the rotation $\mathbf{R}$. The invariance under rotations of H is then expressed by requiring that

$$H(R_{\mathbf{R}}(\omega), R_{\mathbf{R}}(\omega')) = H(\omega, \omega'), \quad \forall\, \omega, \omega' \in \Omega, \ \forall\, \mathbf{R} \in \mathsf{SO}(3).$$

In a similar way, H must be invariant under the point symmetry transformations appropriate for the specific species of molecules under consideration. More precisely, a symmetry transformation is a mapping $G : \Omega \to \Omega$ in the local frame of the molecule (such as a reflection across a molecular plane or a rotation around a molecular axis) that changes a state of the molecule into an equivalent state, possibly represented by another point of Ω. H is required to have the property

$$H(G(\omega),\omega') = H(\omega, G(\omega')) = H(\omega,\omega'), \quad \forall\, \omega,\omega' \in \Omega,\ \forall\, G \in \mathcal{G}, \tag{1.1}$$

where $\mathcal{G}$ is the *symmetry group* of the molecules.

Envisioning Ω as a compact measurable space, we endow it with an appropriate measure μ, which assigns a precise meaning to the integral over every measurable subset $\Sigma \subset \Omega$ of an integrable real-valued function g. This will be denoted by

$$\int_{\Sigma} g(\omega)\, d\mu(\omega).$$

Rigid Molecular Architectures

We discuss here different representations of a specific state space apt to describe purely orientational states. This space applies to all model rigid molecules, to which our development will henceforth be restricted.

The main building block of our theory is a molecule (or an idealization of a molecule) of a certain architecture and symmetry. We think of this in general as a geometric object, such as a cylinder or a rectangular platelet (both in principle capable of modeling uniaxial and biaxial macroscopic phases), just to limit our attention to two notable examples. Different rigid models can serve the same purpose, provided they share the same symmetry.

The symmetry is characterized by the point-symmetry group of the molecule—here we use the SCHOENFLIES notation (see, for example, [217]). For each molec-

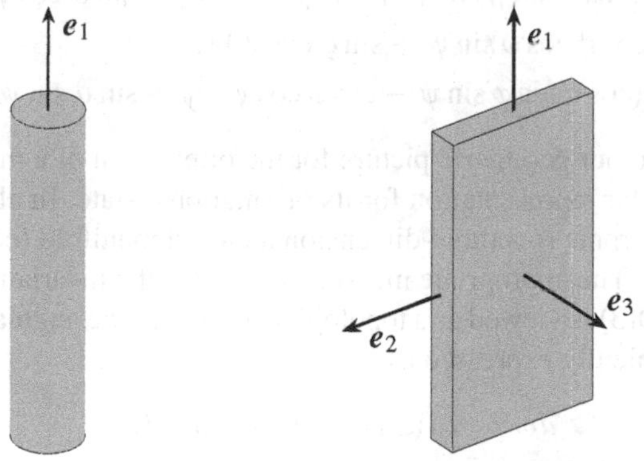

Fig. 1.1. Two molecular architectures and their symmetries: rod (left, $\mathfrak{D}_{\infty h}$), platelet (right, $\mathfrak{D}_{2h}$).

ular species, we imagine that there is a local molecular frame (e_1, e_2, e_3), typically coincident with symmetry molecular axes, contrasted against a reference frame

$(\boldsymbol{e}_x, \boldsymbol{e}_y, \boldsymbol{e}_z)$, both identified with orthonormal triads. Thus the pair-potential is a function only of the relative orientation of the frames of the two molecules, and the form of this function depends on the point-symmetry group of the species. Examples that we shall use are pictured in Figure 1.1.

Here Ω is the set of possible orientations of an individual molecule, i.e., the possible orientations of its frame with respect to a fixed ("lab") frame. Depending on the system being modeled, Ω could be finite or discrete; however, for the applications that we have in mind, we shall take it to be the continuum of all possible relative orientations of a local frame with respect to a fixed frame. The collection of orientational states Ω, then, can be viewed as a manifold, isomorphic to $\mathbb{S}^2 \times \mathbb{S}^1$ (where $\mathbb{S}^n$ denotes the unit sphere in $\mathbb{R}^{n+1}$) or, equivalently, to the rotation group SO(3). The $\mathbb{S}^2$ component of an orientational state $\omega \in \Omega$ orients a designated major (primary) axis of the frame, while the $\mathbb{S}^1$ component corresponds to the orientation of a minor (secondary) axis. The state ω can also be uniquely identified with the proper orthogonal transformation that maps the reference frame into the frame of the rotated molecule.

Orientational states can be represented in terms of several different types of coordinates. For example, a point $\omega \in \Omega$ can be represented by a triple of angles $(\vartheta, \varphi, \psi)$; these may be the familiar EULER angles, which range in the intervals

$$0 \leqq \vartheta \leqq \pi, \quad 0 \leqq \varphi \leqq 2\pi, \quad 0 \leqq \psi \leqq 2\pi.$$

In a notation where the triple $(0, 0, 0)$ corresponds to the coincidence of frames $(\boldsymbol{e}_1, \boldsymbol{e}_2, \boldsymbol{e}_3) = (\boldsymbol{e}_z, \boldsymbol{e}_x, \boldsymbol{e}_y)$, the following is one possible representation of Ω:

$$\boldsymbol{e}_1(\omega) = \sin\vartheta \cos\varphi\, \boldsymbol{e}_x + \sin\vartheta \sin\varphi\, \boldsymbol{e}_y + \cos\vartheta\, \boldsymbol{e}_z, \tag{1.2a}$$

$$\begin{aligned} \boldsymbol{e}_2(\omega) = {} & (\cos\vartheta \cos\varphi \cos\psi - \sin\varphi \sin\psi)\boldsymbol{e}_x \\ & + (\cos\vartheta \sin\varphi \cos\psi + \cos\varphi \sin\psi)\boldsymbol{e}_y - \sin\vartheta \cos\psi\, \boldsymbol{e}_z, \end{aligned} \tag{1.2b}$$

$$\begin{aligned} \boldsymbol{e}_3(\omega) = {} & -(\cos\vartheta \cos\varphi \sin\psi + \sin\varphi \cos\psi)\boldsymbol{e}_x \\ & - (\cos\vartheta \sin\varphi \sin\psi - \cos\varphi \cos\psi)\boldsymbol{e}_y + \sin\vartheta \sin\psi\, \boldsymbol{e}_z. \end{aligned} \tag{1.2c}$$

Thus, in addition to our geometric picture for the orientation of a molecule, we have an equivalent angular representation for its orientational state. In all its realizations, such an Ω is isomorphic to a three-dimensional closed manifold (compact manifold without boundary). The appropriate measure μ on Ω is the invariant HAAR measure on SO(3) when SO(3) is viewed as a topological group. In the angular representation, this is most conveniently expressed as

$$d\omega = d\mu(\omega) = \sin\vartheta\, d\vartheta\, d\varphi\, d\psi,$$

whence it readily follows that the total measure $|\Omega|$ of Ω is

$$|\Omega| = \int_0^{2\pi} d\psi \int_0^{2\pi} d\varphi \int_0^{\pi} \sin\vartheta\, d\vartheta = 8\pi^2. \tag{1.3}$$

We are guilty of some abuses of notation. We have used the same symbol Ω to denote the collection of orientations of a species of molecule (a geometric object) as well as the other mathematical objects to which it is isomorphic, such as $\mathbb{S}^2 \times \mathbb{S}^1$ and $\mathsf{SO}(3)$. In a similar way, we have dilated the meaning of the symbol ω also to include the angular representation of the molecular orientation.

Prototypical Hamiltonian

The molecular states represented by the space Ω can be expressed in different ways in the two-particle Hamiltonian H, depending on the particular theory being adopted. To differentiate the molecular state—which pertains only to the model representing the molecule—from the expression of that state—which pertains to a specific interaction theory and to the way this is reflected onto the two-particle Hamiltonian—we introduce the space $\mathbb{Q}$. This space, which is the image of Ω under an appropriate mapping $\mathfrak{q}$, expresses the ingredients $\mathbf{q}$ that constitute H—typically a collection of molecular tensors, possibly with different ranks. Thus, H can be written as

$$H(\omega, \omega') = \widehat{H}(\mathbf{q}, \mathbf{q}'), \tag{1.4}$$

where

$$\mathbf{q} = \mathfrak{q}(\omega) \quad \text{and} \quad \mathbf{q}' = \mathfrak{q}(\omega').$$

We assume that $\mathfrak{q}$ is a continuous mapping, so that $\mathbb{Q}$ is also compact. We further think of $\mathbb{Q}$ as immersed in a finite-dimensional inner-product space $\mathcal{T}$.

A rather general family of functions $\widehat{H}$ is discussed in [116]; it is also shown there how they can be subsumed in the following diagonal bilinear form

$$\widehat{H}(\mathbf{q}, \mathbf{q}') = -U_0 \left(\alpha_+ \mathbf{q}_+ \cdot \mathbf{q}'_+ - \alpha_- \mathbf{q}_- \cdot \mathbf{q}'_-\right), \tag{1.5}$$

where $U_0 > 0$ is a characteristic interaction energy, $\mathbf{q}_\pm$ are called, for short, the *collective* molecular tensors, and

$$\alpha_\pm \in \{0, 1\}.$$

Formally, $\mathbf{q} = (\mathbf{q}_+, \mathbf{q}_-) \in \mathbb{Q}_+ \times \mathbb{Q}_-$, while $\mathcal{T}$ splits into the Cartesian product $\mathcal{T} = \mathcal{T}_+ \times \mathcal{T}_-$ and $\mathbb{Q}_\pm \subset \mathcal{T}_\pm$. In equation (1.5), $\mathbf{q}_+ \cdot \mathbf{q}'_+$ and $\mathbf{q}_- \cdot \mathbf{q}'_-$ are inner products in the spaces $\mathcal{T}_+$ and $\mathcal{T}_-$, respectively, which may also have different dimensions. Our development, which is here rather formal, will be further illuminated by the specific examples presented in the following subsection. In the prototypical Hamiltonian (1.5), the variables $\mathbf{q}_+$ are said to be *attractive*, while the variables $\mathbf{q}_-$ are said to be *repulsive*: the former tend to be equal in both interacting molecules, $\mathbf{q}_+ = \mathbf{q}'_+$, to reduce their interaction energy, while the latter tend to be "orthogonal," $\mathbf{q}_- \cdot \mathbf{q}'_- = 0$. Thus, if $\alpha_+ = 1$ and $\alpha_- = 0$, the Hamiltonian $\widehat{H}$ in equation (1.5) is called *fully attractive*, whereas if $\alpha_+ = 0$ and $\alpha_- = 1$, it is called *fully repulsive*; finally, if $\alpha_+ = \alpha_- = 1$, then $\widehat{H}$ is called *partly repulsive*. For special spin systems, partly repulsive Hamiltonians were already considered by BOGOLIUBOV JR. [27, 28]; in [116] his approach was systematically extended to a wide class of Hamiltonians appropriate for soft-matter systems.[4]

We present now two specific Hamiltonians in the form (1.5), both relevant to the molecular theory of liquid-crystal phases.

---

[4] See also [260].

Uniaxial and Biaxial Interactions

We consider the examples represented by the molecular architectures in Figure 1.1. The classical molecular model for the uniaxial nematic phase is associated with the description of nematogenic molecules as cylindrical rods (see Figure 1.1 left). In our setting, this model, put forward by MAIER & SAUPE [208], can be derived from a pair-potential written as

$$H_{\mathrm{u}}(\omega,\omega') := -U_0\mathbf{q}(\omega)\cdot\mathbf{q}(\omega'), \tag{1.6}$$

where U_0 is the same interaction energy as in (1.5) and $\mathbf{q}$ is the symmetric, traceless, second-rank tensor defined as

$$\mathbf{q}(\omega) := \boldsymbol{e}_1(\omega)\otimes\boldsymbol{e}_1(\omega) - \frac{1}{3}\mathbf{I}, \tag{1.7}$$

where $\mathbf{I}$ is the second-rank identity tensor in three space dimensions. In (1.6), the inner product between two second-rank tensors $\mathbf{A}$ and $\mathbf{B}$ is defined as

$$\mathbf{A}\cdot\mathbf{B} := \mathrm{tr}(\mathbf{A}\mathbf{B}^{\mathsf{T}}), \tag{1.8}$$

where ^{T} denotes transposition.[5] Inserting (1.7) into (1.6), by use of the following tensor algebra identities[6]

$$(\boldsymbol{a}\otimes\boldsymbol{b})\cdot(\boldsymbol{c}\otimes\boldsymbol{d}) = (\boldsymbol{a}\cdot\boldsymbol{c})(\boldsymbol{b}\cdot\boldsymbol{d}),$$
$$(\boldsymbol{a}\otimes\boldsymbol{b})\cdot\mathbf{I} = \mathrm{tr}(\boldsymbol{a}\otimes\boldsymbol{b}) = \boldsymbol{a}\cdot\boldsymbol{b},$$

valid for all vectors $\boldsymbol{a}$, $\boldsymbol{b}$, $\boldsymbol{c}$, and $\boldsymbol{d}$, we also express H_{u} as

$$H_{\mathrm{u}} = -U_0\left[(\boldsymbol{e}_1\cdot\boldsymbol{e}_1')^2 - \frac{1}{3}\right] = -\frac{2}{3}U_0P_2(\cos\vartheta), \tag{1.9}$$

where ϑ is the angle between the long axes of the molecules in orientations ω and ω', and P_2 is the second LEGENDRE polynomial,

$$\cos\vartheta = \boldsymbol{e}_1\cdot\boldsymbol{e}_1', \qquad P_2(x) := \frac{3}{2}x^2 - \frac{1}{2}.$$

In Section 1.3, we shall build upon the molecular interaction (1.6) the MAIER–SAUPE mean-field theory on uniaxial nematic liquid crystals.

Most liquid crystal molecules are far from possessing the cylindrical symmetry with which they are credited in most of the accounts on their mesogenic nature. Thus, if real liquid crystal molecules resemble more laths than rods, it is natural to imagine that at sufficiently low temperatures, where random rotations about the longer molecular axis may cease to render effectively uniaxial a truly biaxial molecule, an

[5] See also Appendix A.1.
[6] See also Appendix A.2.

ordered phase condenses with the symmetry molecular axes aligned on average parallel to one another, while the molecules retain the spatial mobility characteristic of liquids. Essentially this intuition, made quantitative, formed the basis of FREISER's prediction of the biaxial nematic phase [110, 111] . A long history, paved with both enthusiasm and delusion, followed FREISER's work. YU & SAUPE [365] showed unmistakable evidence of phase biaxiality in certain lyotropic liquid crystals, while the quest for thermotropic biaxial liquid crystals has only recently issued claims that have so far resisted the criticism of the skeptic [197, 198, 199].[7]

A molecular model for the biaxial nematic phase, based upon the platelet geometry (see Figure 1.1 right), was put forward in [321] and reconsidered in [315]. The pair-potential for this model is expressed in [315] as

$$\widehat{H}_b := -U_0\big[\mathbf{q}\cdot\mathbf{q}' + \gamma(\mathbf{q}\cdot\mathbf{b}' + \mathbf{b}\cdot\mathbf{q}') + \lambda\mathbf{b}\cdot\mathbf{b}'\big], \tag{1.10}$$

where $\mathbf{b}$ is the symmetric, traceless second-rank tensor defined by

$$\mathbf{b} := e_2 \otimes e_2 - e_3 \otimes e_3. \tag{1.11}$$

As shown in [24], the Hamiltonian in equation (1.10) can be given the diagonal form

$$\widehat{H}_b = -U_0\left(\alpha_1\mathbf{q}_1\cdot\mathbf{q}_1' + \alpha_2\mathbf{q}_2\cdot\mathbf{q}_2'\right), \tag{1.12}$$

with

$$\mathbf{q}_{1,2} = \mathbf{q} + \gamma_{1,2}\mathbf{b}. \tag{1.13}$$

In (1.12), the tensors $\mathbf{q}_{1,2}$ have different expressions for different values of γ. Precisely, for $\gamma \neq 0$,

$$\gamma_{1,2} = \frac{3\lambda - 1 \pm \sqrt{(3\lambda-1)^2 + 12\gamma^2}}{6\gamma}$$

and

$$\alpha_1 = \frac{\gamma_2 - \gamma}{\gamma_2 - \gamma_1}, \qquad \alpha_2 = \frac{\gamma - \gamma_1}{\gamma_2 - \gamma_1},$$

while for $\gamma = 0$, $\mathbf{q}_1 = \mathbf{q}$, $\mathbf{q}_2 = \mathbf{b}$, $\alpha_1 = 1$, and $\alpha_2 = \lambda$. Moreover, for $\lambda > \gamma^2$, both α_1 and α_2 are positive, and equation (1.12) can be given the form (1.5) with a single attractive term ($\alpha_- = 0$) by appropriately defining $\mathbf{q}_+$:

$$\mathbf{q}_+ = \left(\sqrt{\alpha_1}\mathbf{q}_1, \sqrt{\alpha_2}\mathbf{q}_2\right). \tag{1.14}$$

For $\lambda = \gamma^2$, either α_1 or α_2 vanishes; the expression in (1.12) still reduces to (1.5), with a single attractive term, but now $\mathbf{q}_+$ includes a single tensor:

$$\mathbf{q} := \sqrt{\alpha_h}\mathbf{q}_h, \quad \text{with } \alpha_h := \max\{\alpha_1, \alpha_2\}. \tag{1.15}$$

---

[7] Two more recent contributions [295, 296] have concluded that some allegedly biaxial nematic compounds exhibit rather subtle surface effects. These effects may well be mistaken for signs of bulk biaxiality but are in fact compatible with bulk uniaxiality. This shows that the quest for phase biaxiality in thermotropic nematic liquid crystals is far from over.

For $\lambda < \gamma^2$, either α_1 or α_2 is negative, and equation (1.12) reduces to a partly repulsive Hamiltonian in the form (1.5), with appropriately defined $\mathbf{q}_+$ and $\mathbf{q}_-$, each including a single second-rank tensor:

$$\mathbf{q}_+ := \sqrt{\alpha_h}\mathbf{q}_h, \quad \mathbf{q}_- := \sqrt{-\alpha_k}\mathbf{q}_k, \quad \text{with} \quad \alpha_h > \alpha_k. \tag{1.16}$$

1.1.2 Ensemble Potentials

In this section, we construct the thermodynamic potentials appropriate to describe an *ensemble* of N identical particles, which we index $i = 1, \ldots, N$. These particles interact with each other through the pair-potential H. We can indifferently think of them as occupying the sites of a lattice, for example, or arranged randomly in some region. We can assume that each particle interacts with every other particle or that it just interacts with some proper subset of the rest of the ensemble (such as "nearest neighbors"). We denote by $\mathscr{I}$ the *interaction set*. It is composed of all ordered pairs of interacting particles:

$$\mathscr{I} = \{(i,j) \,|\, i,j \in \{1,\ldots,N\},\ i < j,\ \text{particle } i \text{ interacts with particle } j\}.$$

We assume that each particle in the ensemble interacts with at least one other particle:

$$i \in \{1,\ldots,N\} \Rightarrow \exists\, j \in \{1,\ldots,N\} : (i,j) \in \mathscr{I}.$$

Some examples of ensembles with minimal, nearest-neighbor, and maximal interactions are illustrated in Figure 1.2. We let n denote the cardinality of the set $\mathscr{I}$,

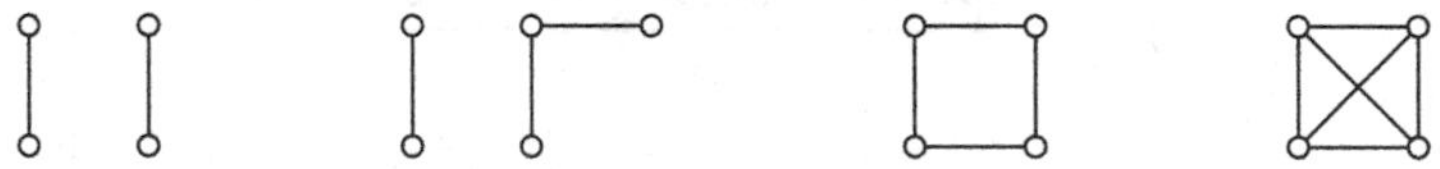

Fig. 1.2. Ensembles with minimal interactions (left, left center), nearest-neighbor interactions (right center), and maximal interactions (right).

$n := |\mathscr{I}|$, that is, the number of particle–particle interactions in the ensemble. One can determine n as a function of N for the cases of a minimal number of interactions, nearest-neighbor interactions, and a maximal number of interactions—all particles interact with each other. These are summarized in Table 1.1 for a simple lattice: equally spaced points in 1-D, square and cubic lattices in 2-D and 3-D, respectively. It should be noticed that we always have

$$n \geqq \frac{N}{2}. \tag{1.17}$$

In fact, the case $n = N/2$ is quite exceptional, associated with the situation in which N is even and each particle is involved in precisely one interaction. This situation is depicted in Figure 1.2 left for the case $N = 4$.

interactions	$n = \lvert\mathcal{I}\rvert$
minimal	$\left\lceil \frac{N}{2} \right\rceil$
nearest neighbor (1-D)	$N-1$
nearest neighbor (2-D)	$2N - 2N^{1/2}$
nearest neighbor (3-D)	$3N - 3N^{2/3}$
maximal	$\frac{N(N-1)}{2}$

Table 1.1. Cardinalities of the interaction set $\mathcal{I}$ for minimal, nearest-neighbor, and maximal interactions. Here $\lceil . \rceil$ denotes the *ceiling* function, the least integer greater than or equal to its argument. Thus, in particular, $\lceil N/2 \rceil = N/2$ (if N is even), $(N+1)/2$ (if N is odd).

A state of the ensemble corresponds to a set of states of each of its particles. An *ensemble state* is represented by

$$\omega = (\omega_1, \ldots, \omega_N) \in \Omega^N .$$

The ensemble state space is thus Ω^N. The total internal energy associated with such an ensemble state is given by the *ensemble Hamiltonian*

$$\mathscr{H}(\omega) := \sum_{(i,j)\in\mathcal{I}} H(\omega_i, \omega_j), \tag{1.18}$$

the sum of the two-particle Hamiltonians associated with all of the interactions. The probability of finding the ensemble in a given state is assumed to follow a BOLTZMANN distribution. The corresponding probability density ρ is

$$\rho(\omega; \beta) := \frac{1}{\mathscr{Z}(\beta)} e^{-\beta\mathscr{H}(\omega)}, \quad \beta := \frac{1}{k_B\theta}, \tag{1.19}$$

where k_B is BOLTZMANN's constant, $\theta > 0$ is absolute temperature, and $\mathscr{Z}$ is the *ensemble partition function*

$$\mathscr{Z}(\beta) := \int_{\Omega^N} e^{-\beta\mathscr{H}(\omega)} d\omega . \tag{1.20}$$

Here and in what follows, the integral over the ensemble state space Ω^N of an integrable function $g : \Omega^N \to \mathbb{R}$ is to be expanded as

$$\int_{\Omega^N} g(\omega)\, d\omega = \int_{\Omega} \cdots \int_{\Omega} g(\omega_1, \ldots, \omega_N)\, d\omega_1 \cdots d\omega_N .$$

Accordingly, the *ensemble average* $\langle g \rangle_\rho$ of g is defined as

$$\langle g \rangle_\rho := \int_{\Omega^N} g(\omega)\rho(\omega; \beta)\, d\omega . \tag{1.21}$$

All relevant thermodynamic potentials are readily derived from the partition function $\mathscr{Z}$. The *ensemble free energy* (or HELMHOLTZ potential) is[8]

$$\mathscr{F}(\beta) := -\frac{1}{\beta}\ln \mathscr{Z}(\beta), \tag{1.22}$$

and the *ensemble internal energy* $\mathscr{U}$ is

$$\mathscr{U}(\beta) := \langle \mathscr{H} \rangle_\rho = -\frac{\partial}{\partial \beta}\ln \mathscr{Z}(\beta). \tag{1.23}$$

The *entropy* $\mathscr{S}$ can be deduced from equations (1.22) and (1.23) with the aid of the thermodynamic relationship

$$\mathscr{F} = \mathscr{U} - \theta \mathscr{S},$$

from which we obtain

$$\mathscr{S}(\beta) = -k_B \langle \ln \rho \rangle_\rho.$$

It is convenient to introduce for $\mathscr{F}$ the corresponding per-particle potential

$$F := \frac{1}{N}\mathscr{F}. \tag{1.24}$$

Analogous expressions could be introduced corresponding to $\mathscr{U}$ and $\mathscr{S}$.

We close this section by heeding a scaling property of $\mathscr{F}$ that is often useful. Suppose that the ensemble Hamiltonian $\mathscr{H}$ is altered by the addition of a constant C, so that

$$\mathscr{H} \mapsto \mathscr{H} + C. \tag{1.25}$$

Then $\mathscr{F}$ is accordingly transformed as

$$\mathscr{F} \mapsto \mathscr{F} + C. \tag{1.26}$$

To prove this, we need only remark that, by (1.20),

$$\mathscr{Z} \mapsto e^{-\beta C}\int_{\Omega^N} e^{-\beta \mathscr{H}(\omega)}d\omega = e^{-\beta C}\mathscr{Z}. \tag{1.27}$$

By applying (1.22), (1.26) readily follows from (1.27). Clearly, the interaction is not affected by (1.25), nor are the thermodynamics of the ensemble affected by (1.26). Moreover, by (1.19), the probability density ρ remains unchanged. Often the scaling constant C is chosen so as to make either $\mathscr{F}$ or $\mathscr{Z}$ attain a desired value at a particular reference phase. We shall customarily scale $\mathscr{F}$ through (1.26) so as to make it vanish in the most disordered phase.

---

[8] For the motivation given in statistical thermodynamics to both definitions (1.22) and (1.23), we refer the reader to the terse booklet by SCHRÖDINGER [290] (see, in particular, Chapter II). The definition of internal energy, entropy, and free energy given in continuum thermodynamics will be recalled in Section 2.2.2 below.

1.2 Mean-Field Approximation

Computing the ensemble free energy $\mathscr{F}$ for a general pair-potential H, even in the diagonalized bilinear form (1.5), in most cases reveals itself to be a formidable task. Much easier is the task of computing $\mathscr{F}$ if the particles in the ensemble interact only with an external field. In the mean-field approximation, the interaction binding the particles together is wisely replaced by an effective internal field that mimics it. We devote this section to laying the theoretical basis of this approximation in our general context.

1.2.1 One-Particle Hamiltonian

Conceptually, the mean-field approximation replaces, as it were, the interaction between particles with the action exerted by a *mean field*, at the same time produced and felt by all particles. Mathematically, the mean-field approximation replaces the two-particle Hamiltonian H with a *one-particle* Hamiltonian H_0, which also depends on the mean field. There are several ways to formulate these ideas. In molding ours, we were influenced by the treatments in both [28] and [263], though they differ to a degree. Still other points of view are exposed, for example, in Chapter 20 of [39] and in [161].

To build the mean-field theory appropriate to a bilinear Hamiltonian in the form (1.4), with $\widehat{H}$ as in (1.5), we start by considering the idealized case in which the ensemble comprises only *two* molecules: $N = 2$. In this case, we define the one-particle Hamiltonian H_0 to be

$$H_0(\omega; \mathbf{Q}) := \langle H(\omega, \cdot)\rangle_{\rho_0} - \frac{1}{2}\langle H(\cdot, \cdot)\rangle_{\rho_0^2}, \tag{1.28}$$

where ρ_0 is the BOLTZMANN distribution function associated with H_0,

$$\rho_0(\omega; \beta, \mathbf{Q}) := \frac{1}{Z_0(\beta, \mathbf{Q})} e^{-\beta H_0(\omega; \mathbf{Q})}, \tag{1.29}$$

Z_0 is the one-particle partition function,

$$Z_0(\beta, \mathbf{Q}) := \int_{\Omega} e^{-\beta H_0(\omega; \mathbf{Q})} d\omega, \tag{1.30}$$

and $\mathbf{Q}$ is the *mean field*, defined by

$$\mathbf{Q} := \langle \mathbf{q}\rangle_{\rho_0}, \tag{1.31}$$

with $\mathbf{q}$ the collective molecular tensor. Often, mean-field theory is referred to as *self-consistent field theory*, a descriptive definition that owes its name to equation (1.31), which is indeed a self-consistency requirement, as will become clearer below.

In (1.28), $\langle H(\omega, \cdot)\rangle_{\rho_0}$ is the ensemble mean-field average of H in only one argument; since H is invariant under particle exchange, there is no ambiguity, since we have

$$\langle H(\omega,\cdot)\rangle_{\rho_0} = \langle H(\cdot,\omega)\rangle_{\rho_0} \quad \forall\, \omega.$$

$\langle H(\cdot,\cdot)\rangle_{\rho_0^2}$ is the ensemble mean-field average in both arguments of H:

$$\langle H(\cdot,\cdot)\rangle_{\rho_0^2} := \int_{\Omega^2} H(\omega,\omega')\rho_0(\omega;\beta,\mathbf{Q})\rho_0(\omega';\beta,\mathbf{Q})\,d\omega\,d\omega'.$$

In general, for a function $g : \Omega^k \to \mathbb{R}$, the average $\langle g\rangle_{\rho_0^k}$ is defined by

$$\langle g\rangle_{\rho_0^k} := \int_{\Omega^k} g(\omega_1,\dots,\omega_k)\prod_{h=1}^{k}\rho_0(\omega_h;\beta,\mathbf{Q})\,d\omega_h.$$

Care is required in interpreting the definition for H_0 in (1.28), since, by (1.29), the probability density ρ_0 also depends on H_0, so that (1.28) is properly an *implicit* definition of H_0. Moreover, the mean field $\mathbf{Q}$, which appears as a parameter in both equations (1.28) and (1.29), is subject to the self-consistency condition (1.31), which may or may not be satisfied. In general, the definition (1.28) is meaningful at most for a finite number of compatible mean fields $\mathbf{Q}$. We now further dwell on the physical interpretation of this definition.

Often, the rationale behind subtracting one-half of the double average $\langle H(\cdot,\cdot)\rangle_{\rho_0^2}$ from the single average $\langle H(\omega,\cdot)\rangle_{\rho_0}$ is explained by the need to avoid double counting of the energy [263]. To make this idea more precise, we compute the ensemble Hamiltonians $\mathscr{H}$ and $\mathscr{H}_0$ corresponding to H and H_0, still in the case of an ensemble of only two particles. By applying (1.18) to H_0, we arrive at

$$\mathscr{H}_0(\omega_1,\omega_2;\mathbf{Q}) := H_0(\omega_1;\mathbf{Q}) + H_0(\omega_2;\mathbf{Q}), \tag{1.32}$$

while $\mathscr{H}$ is simply

$$\mathscr{H}(\omega_1,\omega_2) = H(\omega_1,\omega_2).$$

By computing the mean-field ensemble averages $\langle\mathscr{H}_0\rangle_{\rho_0^2}$ and $\langle\mathscr{H}\rangle_{\rho_0^2}$, we readily obtain that

$$\langle\mathscr{H}_0\rangle_{\rho_0^2} = 2\langle H_0\rangle_{\rho_0} \quad\text{and}\quad \langle\mathscr{H}\rangle_{\rho_0^2} = \langle H\rangle_{\rho_0^2}. \tag{1.33}$$

Since, by (1.28),

$$\langle H_0\rangle_{\rho_0} = \frac{1}{2}\langle H\rangle_{\rho_0^2},$$

it follows from (1.33) that

$$\langle\mathscr{H}_0\rangle_{\rho_0^2} = \langle\mathscr{H}\rangle_{\rho_0^2}. \tag{1.34}$$

This is the formal justification of (1.28): the one-particle approximating Hamiltonian H_0 must be such that the averages in ρ_0^2 of both $\mathscr{H}_0$ and $\mathscr{H}$ coincide at all temperatures. This ensures that the ensembles governed by H and H_0 have the same average internal energy, a natural requirement that makes H and H_0 "consistent" in the language of [263].

We adopt this requirement on the equal scaling of the average internal energy as a guiding criterion in extending the definition (1.28) to a general ensemble with more than two particles. For such an ensemble, we replace (1.28) by

$$H_0(\omega; \mathbf{Q}) := \mathfrak{z}\left(\langle H(\omega, \cdot)\rangle_{\rho_0} - \frac{1}{2}\langle H(\cdot, \cdot)\rangle_{\rho_0^2}\right). \tag{1.35}$$

Here $\mathfrak{z}$ is a parameter to be determined so as to enforce the validity of the analogue of (1.34); we refer to it as the *coordination parameter*. According to (1.32), the mean-field ensemble Hamiltonian $\mathscr{H}_0$ is now

$$\mathscr{H}_0(\boldsymbol{\omega}; \mathbf{Q}) := \sum_{i=1}^{N} H_0(\omega_i; \mathbf{Q}),$$

while $\mathscr{H}$ is given by (1.18). Computing the ensemble averages of $\mathscr{H}_0$ and $\mathscr{H}$ in ρ_0^N, with the aid of (1.35) we arrive at

$$\langle \mathscr{H}_0 \rangle_{\rho_0^N} = N \langle H_0 \rangle_{\rho_0} \tag{1.36}$$

and

$$\langle \mathscr{H} \rangle_{\rho_0^N} = \int_{\Omega^N} \sum_{(i,j)\in\mathfrak{I}} H(\omega_i, \omega_j) \prod_{h=1}^{N} \rho_0(\omega_h; \beta, \mathbf{Q})\, d\omega_h = n \langle H \rangle_{\rho_0^2}, \tag{1.37}$$

where n is the cardinality of the interaction set $\mathfrak{I}$. Equations (1.36) and (1.37) clearly extend (1.33) to the general case in which $N > 2$ and $n \geqq 1$. Since, by (1.35),

$$\langle H_0 \rangle_{\rho_0} = \frac{\mathfrak{z}}{2} \langle H \rangle_{\rho_0^2},$$

the energy scaling

$$\langle \mathscr{H}_0 \rangle_{\rho_0^N} = \langle \mathscr{H} \rangle_{\rho_0^N} \tag{1.38}$$

is guaranteed, provided that

$$\mathfrak{z} = \frac{2n}{N}. \tag{1.39}$$

By (1.17) and the maximal estimate for n in Table 1.1, the coordination parameter $\mathfrak{z}$ obeys the inequalities

$$1 \leqq \mathfrak{z} \leqq N - 1.$$

For N even, a simple combinatoric calculation shows that $\mathfrak{z}$ can be interpreted as the number of molecules that interact with any given molecule in an ensemble with n interacting molecular pairs. The one-particle Hamiltonian (1.35) is then seen to be scaled so as to ascribe the proper amount of potential energy to an individual particle based upon the number of interactions that particle has with other molecules in the ensemble, on average. Proper attention to this scaling and to the value of the coordination parameter $\mathfrak{z}$ also helps in relating mean-field calculations to the results of Monte Carlo simulations based upon the same particle–interaction potential.

Equation (1.35), with $\mathfrak{z}$ as in (1.39), accompanied by equations (1.29)–(1.31), forms the basis of our mean-field framework. For the prototypical Hamiltonian in (1.5),

$$\langle H(\omega,\cdot)\rangle_{\rho_0} = -U_0\left(\alpha_+\mathbf{q}_+\cdot\mathbf{Q}_+ - \alpha_-\mathbf{q}_-\cdot\mathbf{Q}_-\right),$$

where

$$\mathbf{Q}_\pm := \langle\mathbf{q}_\pm\rangle_{\rho_0}, \tag{1.40}$$

while

$$\langle H(\cdot,\cdot)\rangle_{\rho_0^2} = -U_0\left(\alpha_+\mathbf{Q}_+\cdot\mathbf{Q}_+ - \alpha_-\mathbf{Q}_-\cdot\mathbf{Q}_-\right),$$

so that, by (1.35),

$$\begin{aligned} &H_0(\omega;\mathbf{Q}) \\ &= U_{\mathfrak{z}}\left[\alpha_+\left(\frac{1}{2}\mathbf{Q}_+ - \mathbf{q}_+(\omega)\right)\cdot\mathbf{Q}_+ - \alpha_-\left(\frac{1}{2}\mathbf{Q}_- - \mathbf{q}_-(\omega)\right)\cdot\mathbf{Q}_-\right], \end{aligned} \tag{1.41}$$

where

$$U_{\mathfrak{z}} := \mathfrak{z}U_0 \tag{1.42}$$

and $\mathbf{Q} := (\mathbf{Q}_+,\mathbf{Q}_-)$. In general, $\mathbf{Q}$, which we call the *order tensor collection*—for in most applications of this theory it collects average molecular tensors—is a member of the linear space $\mathcal{T} = \mathcal{T}_+ \times \mathcal{T}_-$, the environment of the compact space $\mathbb{Q} = \mathbb{Q}_+ \times \mathbb{Q}_-$, where all molecular states are expressed. $\mathbf{Q}$ must comply with the self-consistency condition (1.31) (more precisely, its components must separately comply with conditions (1.40)). However, we can also regard (1.41) as defining a function H_0 on the whole of $\Omega \times \mathcal{T}$. Such an *extended* function H_0 will be the one-particle Hamiltonian in the literal sense only when the associated self-consistency condition (1.31) is satisfied. In the following subsection, starting from the extended H_0, we shall construct a similarly extended free energy $\mathscr{F}_0$, the critical points of which correspond to the legitimate order tensors that obey (1.31).

1.2.2 Mean-Field Free Energy

To derive the ensemble mean-field free energy $\mathscr{F}_0$, we first compute the ensemble mean-field partition function $\mathscr{Z}_0$. By (1.20),

$$\mathscr{Z}_0(\beta,\mathbf{Q}) := \int_{\Omega^N}\prod_{i=1}^{N} e^{-\beta H_0(\omega_i;\mathbf{Q})}\rho_0(\omega_i;\beta,\mathbf{Q})\,d\omega_i = Z_0(\beta,\mathbf{Q})^N,$$

where Z_0 is as in (1.30). Thus, by (1.22),

$$\mathscr{F}_0(\beta,\mathbf{Q}) := -\frac{1}{\beta}N\ln Z_0(\beta,\mathbf{Q}) = NF_0(\beta,\mathbf{Q}), \tag{1.43}$$

where, also in accordance with (1.24),

$$F_0(\beta, \mathbf{Q}) := -\frac{1}{\beta} \ln Z_0(\beta, \mathbf{Q}) \tag{1.44}$$

is the mean-field free energy per particle. As is the case with H_0, also Z_0 and F_0 can be regarded as extended functions of $\mathbf{Q}$ in $\mathcal{T}$. By (1.41), F_0 is differentiable in $\mathbf{Q}$ (see Section 1.2.4 below for a formal calculus in $\mathcal{T}$), and (1.44) combined with (1.29) and (1.30) implies that

$$\frac{\partial F_0}{\partial \mathbf{Q}} = \frac{1}{Z_0} \int_\Omega \frac{\partial H_0}{\partial \mathbf{Q}} e^{-\beta H_0(\omega;\mathbf{Q})} d\omega = \left\langle \frac{\partial H_0}{\partial \mathbf{Q}} \right\rangle_{\rho_0} . \tag{1.45}$$

On the other hand, equation (1.41) implies

$$\frac{\partial H_0}{\partial \mathbf{Q}} = \left(\frac{\partial H_0}{\partial \mathbf{Q}_+}, \frac{\partial H_0}{\partial \mathbf{Q}_-} \right),$$

where

$$\frac{\partial H_0}{\partial \mathbf{Q}_\pm} = \pm U_3 \alpha_\pm (\mathbf{Q}_\pm - \mathbf{q}_\pm) . \tag{1.46}$$

By comparing equations (1.45) and (1.46), we conclude that the order tensor collections $\mathbf{Q}$ that satisfy the self-consistency condition (1.31) are precisely the points where the function $\mathscr{F}_0(\beta, \cdot) = N F_0(\beta, \cdot)$, extended over the whole space $\mathcal{T}$, is stationary. Thus, only at its critical points does the extended function $\mathscr{F}_0$ acquire the meaning of ensemble free energy for the approximating mean-field Hamiltonian H_0.

1.2.3 Minimum Principle

We develop the principle that characterizes the globally stable phase of the mean-field model even when the mean-field free energy does not have a global minimum. An inequality, which is familiar in this area, plays a key role.

GIBBS–BOGOLIUBOV Inequality

We require the use of a form of what is generally referred to as the GIBBS–BOGOLIUBOV inequality [128].

Theorem 1.1 (GIBBS–BOGOLIUBOV Inequality). *Let $\mathscr{H}$ and $\mathscr{H}'$ be two sufficiently regular ensemble Hamiltonians (on the same ensemble) with associated distribution functions ρ, ρ' and* HELMHOLTZ *free energies $\mathscr{F}$, $\mathscr{F}'$ defined as in* (1.19) *and* (1.22)*. Then the following inequality must be true for every $\beta > 0$:*

$$\langle \mathscr{H} - \mathscr{H}' \rangle_\rho \leqq \mathscr{F} - \mathscr{F}' \leqq \langle \mathscr{H} - \mathscr{H}' \rangle_{\rho'} . \tag{1.47}$$

Here $\langle \cdot \rangle_\rho$ and $\langle \cdot \rangle_{\rho'}$ are the associated averages, defined as in (1.21).

This inequality can be viewed as a consequence of certain convexity properties of the free energy. Several proofs of this and related variants exist in the literature. It was proved in [116] how these inequalities follow directly from JENSEN's inequality.

The inequality (1.47) is valid under quite general circumstances; in particular, it is valid when $\mathscr{H}'$, ρ', and $\mathscr{F}'$ are taken to be the mean-field model functions $\mathscr{H}_0$, ρ_0, and $\mathscr{F}_0$ in their extended interpretations (with the tensor order parameters in $\mathbf{Q}$ free). If we restrict ourselves to self-consistent $\mathbf{Q}$'s, then we obtain the following result as a consequence.

Corollary 1.1. *Let* $\mathscr{F} = \mathscr{F}(\beta)$ *be a target ensemble* HELMHOLTZ *free energy derived from a pair-potential* H *as in Section 1.1.2, and let* $\mathscr{F}_0 = \mathscr{F}_0(\beta, \mathbf{Q})$ *be the mean-field approximation to it constructed as in* (1.43). *For a given* $\beta > 0$, *if the order tensors collected in* $\mathbf{Q}$ *all satisfy the self-consistency conditions for this temperature, then the free energies must necessarily satisfy*

$$\mathscr{F}(\beta) \leqq \mathscr{F}_0(\beta, \mathbf{Q}). \tag{1.48}$$

Proof. The right half of the GIBBS–BOGOLIUBOV inequality (1.47) gives

$$\mathscr{F} - \mathscr{F}_0 \leqq \langle \mathscr{H} - \mathscr{H}_0 \rangle_{\rho_0^N} = \langle \mathscr{H} \rangle_{\rho_0^N} - \langle \mathscr{H}_0 \rangle_{\rho_0^N}.$$

Since $\mathscr{H}_0$ is scaled to satisfy the condition (1.38), we are assured that

$$\langle \mathscr{H}_0(\cdot, \mathbf{Q}) \rangle_{\rho_0^N} = \langle \mathscr{H}(\cdot) \rangle_{\rho_0^N}, \tag{1.49}$$

provided that $\mathbf{Q}$ is self-consistent, and the result follows. □

We emphasize that the general GIBBS–BOGOLIUBOV inequality (1.47) is valid for the extended functions $\mathscr{H}_0$ and $\mathscr{F}_0$ for *any* $\mathbf{Q} \in \mathfrak{T}$, while the inequality (1.48) is valid only when the tensors collected in $\mathbf{Q}$ all satisfy self-consistency. We remind the reader that $\mathscr{F}_0(\beta, \mathbf{Q})$ represents an approximation to $\mathscr{F}(\beta)$ only when $\mathbf{Q}$ satisfies self-consistency, which is equivalent to stationarity of $\mathscr{F}_0$ whenever (1.5) applies. The values of $\mathscr{F}_0(\beta, \mathbf{Q})$ when $\mathbf{Q}$ is far from this critical-point set have no provable relationship to $\mathscr{F}(\beta)$.

It is worthwhile to illustrate this with an example. Recall from (1.6) the two-particle Hamiltonian H_{u} for the uniaxial interaction that is at the basis of the MAIER–SAUPE mean-field theory,

$$H_{\mathrm{u}}(\omega, \omega') = -U_0 \, \mathbf{q}(\omega) \cdot \mathbf{q}(\omega'),$$

where $\mathbf{q}(\omega)$ is the symmetric, traceless, second-rank tensor in (1.7). For this, the single-particle potential Hamiltonian H_0 is given by

$$H_0(\omega, \mathbf{Q}) = U_{\mathfrak{z}} \left(\frac{1}{2}\mathbf{Q} - \mathbf{q}(\omega) \right) \cdot \mathbf{Q}, \tag{1.50}$$

where

$$U_{\mathfrak{z}} = \frac{2n}{N} U_0$$

as in (1.42), and $\mathbf{Q} = \langle \mathbf{q} \rangle_{\rho_0}$ when self-consistency is satisfied. Here, as before, n is the total number of interactions in an ensemble of N particles. Without the self-consistency restriction on the tensor order parameter $\mathbf{Q}$, the right-hand side of (1.47) evaluates to

$$\langle \mathscr{H} - \mathscr{H}_0 \rangle_{\rho_0^N} = -\frac{N}{2} U_{\mathfrak{z}} \langle \mathbf{q} \rangle_{\rho_0} \cdot \langle \mathbf{q} \rangle_{\rho_0} + N U_{\mathfrak{z}} \left(\langle \mathbf{q} \rangle_{\rho_0} - \frac{1}{2} \mathbf{Q} \right) \cdot \mathbf{Q}.$$

At points where $\mathbf{Q} = \langle \mathbf{q} \rangle_{\rho_0}$, this then becomes

$$\langle \mathscr{H} - \mathscr{H}_0 \rangle_{\rho_0^N} = 0,$$

which is consistent with Corollary 1.1. In the absence of self-consistency, the above equality fails to hold.

Thus, in the wide class of interaction Hamiltonians representable in the form (1.5), any self-consistent mean-field free-energy approximation $\mathscr{F}_0(\beta, \mathbf{Q})$ necessarily has value greater than or equal to the true free energy of the system at that temperature, $\mathscr{F}(\beta)$, and so the equilibrium solution with the least value of $\mathscr{F}_0$ gives the best approximation to $\mathscr{F}$ and determines the phase in the mean-field approximate phase diagram. This is true in general, even when the extended $\mathscr{F}_0$ does not possess a global minimum.

Global Least-Free-Energy Principle

We summarize our observations in the following statement.

Theorem 1.2 (Least-Free-Energy Principle [116]). *Let $H : \Omega \times \Omega \to \mathbb{R}$ be a bilinear two-particle Hamiltonian of the form* (1.4) *with $\widehat{H}$ as in* (1.5) *on a state space Ω, and let $\mathscr{F}$ be its associated ensemble* HELMHOLTZ *free energy, as defined in Section 1.1.2. Let $\mathscr{H}_0$ and $\mathscr{F}_0$ be the mean-field approximate ensemble Hamiltonian and free energy, as constructed in Section 1.2. If $\mathscr{H}_0$ is properly scaled so that the condition* (1.38) *is satisfied, then for every $\beta > 0$, all stationary points $\mathbf{Q}^*$ of $\mathscr{F}_0$ provide self-consistent mean-field approximate free energies $\mathscr{F}_0(\beta, \mathbf{Q}^*)$ to the true free energy of the ensemble at that temperature, $\mathscr{F}(\beta)$, and we are guaranteed that*

$$\mathscr{F}(\beta) \leqq \mathscr{F}_0(\beta, \mathbf{Q}^*).$$

We conclude that

> *A globally stable phase of the mean-field model is given by any stationary point $\mathbf{Q}^*$ having the least value of $\mathscr{F}_0(\beta, \mathbf{Q}^*)$, which must give the best mean-field approximation to the true ensemble free energy $\mathscr{F}(\beta)$.*

We observe that the essential ingredients here are simply a mean-field Hamiltonian that gives a consistent approximation to the average internal energy of the ensemble in the sense (1.38) and (1.49) combined with the right-hand side of the GIBBS–BOGOLIUBOV inequality. Also, if the (extended) mean-field free energy $\mathscr{F}_0$ were to have a global minimum (attained at a critical point), then such a point would necessarily coincide with a least-free-energy point. We shall see in the next section that in the cases in which $\mathscr{F}_0$ is *not* bounded from below, all of the candidates for a globally stable phase point necessarily have a saddle-like nature. We also observe that these points (both global minimizing points and least-free-energy saddle points) need not be unique; for example, at a *first-order* phase transition two or more distinct equilibrium phases have precisely the same free energy.

1.2.4 Minimax Principle

The bilinear two-particle Hamiltonian H in (1.4) may fail to possess the diagonal form of $\widehat{H}$ in (1.5), though it can always be reduced to that form through an appropriate change of variables, as already shown for the biaxial interactions described at the end of Section 1.1.1. The diagonal representation (1.5) clearly reveals whether H is positive definite (the fully attractive case, $\alpha_+ = 1, \alpha_- = 0, \mathscr{F} = \mathscr{F}(\mathbf{Q}_+)$), negative definite (fully repulsive, $\alpha_+ = 0, \alpha_- = 1, \mathscr{F} = \mathscr{F}(\mathbf{Q}_-)$), or indefinite (partly repulsive, $\alpha_+ = \alpha_- = 1, \mathscr{F} = \mathscr{F}(\mathbf{Q}_+, \mathbf{Q}_-)$). The nature of the pair-potential H drives the nature of the mean-field free energy $\mathscr{F}_0$. In this section, we show that if H is positive definite, then $\mathscr{F}_0$ is bounded below and attains its global minimum at a critical point, if H is negative definite, then $\mathscr{F}_0$ is bounded above and attains its global maximum at a critical point, and if H is indefinite, then $\mathscr{F}_0$ is neither bounded above nor bounded below but still has a critical point of least free energy, which must be a saddle point (a minimax point). The first case is the most common in practice, and it will be exemplified in Section 1.3. The second case is rather rare, and it is considered here for completeness. The third case occurs in the study of biaxial nematic liquid crystals, as recalled in Section 1.2.6 below. We consider the first two cases in the next subsection and the (more difficult) third case in the subsequent one.

H Positive or Negative Definite

We require some information about the way in which the definiteness/indefiniteness properties of H translate into properties of $\mathscr{F}_0$. The properties that $\mathscr{F}_0$ inherits are types of coercivity.[9] We have the following.

Lemma 1.1. *Let H be a bilinear Hamiltonian in the diagonalized representation* (1.5)*, and let $\mathscr{F}_0$ be the associated mean-field free energy (in its extended interpretation) as constructed in* Section 1.2. *Assume that $\mathbf{q}_-$ and $\mathbf{q}_+$ are bounded on Ω.*

1. *If $\alpha_+ = 1$, then $\mathscr{F}_0(\mathbf{Q}_+, \mathbf{Q}_-)$ is* coercive *with respect to $\mathbf{Q}_+$ for each fixed $\mathbf{Q}_-$.*
2. *If $\alpha_- = 1$, then $\mathscr{F}_0(\mathbf{Q}_+, \mathbf{Q}_-)$ is* negatively coercive[10] *with respect to $\mathbf{Q}_-$ for each fixed $\mathbf{Q}_+$.*

[9] We recall that a real-valued function f on a normed linear space is said to be *coercive* if $f(x) \to \infty$, as $\|x\| \to \infty$.

[10] A real-valued function f is *negatively coercive* if $-f$ is coercive.

Proof. We let $|\cdot|$ denote the induced norm on the finite-dimensional inner-product space $\mathfrak{T}_+$ or $\mathfrak{T}_-$:

$$|\mathbf{Q}_+|^2 = \mathbf{Q}_+ \cdot \mathbf{Q}_+, \quad |\mathbf{Q}_-|^2 = \mathbf{Q}_- \cdot \mathbf{Q}_-.$$

We wish to show that, under the appropriate hypotheses,

$$\mathscr{F}_0(\mathbf{Q}_+, \mathbf{Q}_-) \to \infty, \quad \text{as } |\mathbf{Q}_+| \to \infty,$$

and

$$\mathscr{F}_0(\mathbf{Q}_+, \mathbf{Q}_-) \to -\infty, \quad \text{as } |\mathbf{Q}_-| \to \infty.$$

Since $\mathscr{F}_0 = -\frac{1}{\beta} \ln \mathscr{Z}_0$ and $\mathscr{Z}_0 = Z_0^N$, these are equivalent to

$$Z_0(\mathbf{Q}_+, \mathbf{Q}_-) \to 0, \quad \text{as } |\mathbf{Q}_+| \to \infty,$$

and

$$Z_0(\mathbf{Q}_+, \mathbf{Q}_-) \to \infty, \quad \text{as } |\mathbf{Q}_-| \to \infty.$$

With H in the diagonal representation (1.5), the single-particle Hamiltonian (in its extended interpretation) is given by

$$\begin{aligned} H_0(\omega; \mathbf{Q}_+, \mathbf{Q}_-) = &\frac{U_{\mathfrak{z}}}{2} [\alpha_+ \mathbf{Q}_+ \cdot \mathbf{Q}_+ - \alpha_- \mathbf{Q}_- \cdot \mathbf{Q}_-] \\ &- U_{\mathfrak{z}} \left[\alpha_+ \mathbf{q}_+(\omega) \cdot \mathbf{Q}_+ - \alpha_- \mathbf{q}_-(\omega) \cdot \mathbf{Q}_-\right] \end{aligned}$$

and

$$\begin{aligned} &Z_0(\tilde{\beta}, \mathbf{Q}_+, \mathbf{Q}_-) \\ &\quad = e^{-\frac{\tilde{\beta}}{2}\alpha_+ \mathbf{Q}_+ \cdot \mathbf{Q}_+} e^{\frac{\tilde{\beta}}{2}\alpha_- \mathbf{Q}_- \cdot \mathbf{Q}_-} \int_{\Omega} e^{\tilde{\beta}\alpha_+ \mathbf{q}_+(\omega) \cdot \mathbf{Q}_+} e^{-\tilde{\beta}\alpha_- \mathbf{q}_-(\omega) \cdot \mathbf{Q}_-} d\omega, \end{aligned} \tag{1.51}$$

where

$$\tilde{\beta} := \frac{U_{\mathfrak{z}}}{k_B \theta} = \frac{\mathfrak{z} U_0}{k_B \theta} \tag{1.52}$$

is the reduced (dimensionless) reciprocal temperature associated with the coordination parameter $\mathfrak{z}$ and interaction strength U_0. In this expression, the terms quadratic in $\mathbf{Q}_+$ and $\mathbf{Q}_-$ dominate the behavior for $|\mathbf{Q}_+|$ and $|\mathbf{Q}_-|$ large, which we show now.

The functions $\mathbf{q}_-$ and $\mathbf{q}_+$ are assumed bounded on Ω, and so we can define

$$M_- := \sup_{\omega \in \Omega} |\mathbf{q}_-(\omega)|, \quad M_+ := \sup_{\omega \in \Omega} |\mathbf{q}_+(\omega)|.$$

Then we can bound

$$e^{\tilde{\beta}\alpha_+ \mathbf{q}_+(\omega) \cdot \mathbf{Q}_+} \leqq e^{\tilde{\beta}\alpha_+ M_+ |\mathbf{Q}_+|}, \quad \forall\, \omega \in \Omega,$$

and

$$e^{-\tilde{\beta}\alpha_- \mathbf{q}_-(\omega) \cdot \mathbf{Q}_-} \geqq e^{-\tilde{\beta}\alpha_- M_- |\mathbf{Q}_-|}, \quad \forall\, \omega \in \Omega.$$

With these estimates, it follows from (1.51) that if $\alpha_+ > 0$, then for each fixed $\mathbf{Q}_-$,

$$Z_0 \leqq e^{\frac{\tilde{\beta}}{2}\alpha_-|\mathbf{Q}_-|^2}\left[\int_\Omega e^{-\tilde{\beta}\alpha_-\mathbf{q}_-(\omega)\cdot\mathbf{Q}_-}d\omega\right]e^{-\frac{\tilde{\beta}}{2}\alpha_+\left(|\mathbf{Q}_+|^2-2M_+|\mathbf{Q}_+|\right)} \to 0,$$

$$\text{as } |\mathbf{Q}_+| \to \infty;$$

while if $\alpha_- > 0$, then for each fixed $\mathbf{Q}_+$,

$$Z_0 \geqq e^{-\frac{\tilde{\beta}}{2}\alpha_+|\mathbf{Q}_+|^2}\left[\int_\Omega e^{\tilde{\beta}\alpha_+\mathbf{q}_+(\omega)\cdot\mathbf{Q}_+}d\omega\right]e^{\frac{\tilde{\beta}}{2}\alpha_-\left(|\mathbf{Q}_-|^2-2M_-|\mathbf{Q}_-|\right)} \to \infty,$$

$$\text{as } |\mathbf{Q}_-| \to \infty,$$

which concludes the proof. □

As a consequence, we can conclude that in the cases in which the pair-potential H is either positive definite or negative definite, the mean-field free energy attains its extremal value at a critical point, provided that $\mathbf{q}$ is bounded on Ω. This assumption is normally satisfied, since in most applications the state space Ω is bounded and the tensors collected in $\mathbf{q}$ are continuous functions on Ω.

Theorem 1.3. *Let H be a bilinear two-particle Hamiltonian in the general form* (1.4) *with $\widehat{H}$ as in* (1.5), *and let $\mathscr{F}_0$ be the associated mean-field free energy (in its extended interpretation) constructed as in* Section 1.2. *Assume that the function $\mathbf{q}$ is bounded on Ω.*

1. *If H is positive definite, then $\mathscr{F}_0$ is bounded from below and attains its minimum value at a critical point.*
2. *If H is negative definite, then $\mathscr{F}_0$ is bounded from above and attains its maximum value at a critical point.*

Proof. If H is positive definite, then in its diagonal representation, it takes the form (1.5) with $\alpha_+ = 1$ and $\alpha_- = 0$, and Lemma 1.1 guarantees that $\mathscr{F}_0 = \mathscr{F}_0(\beta, \mathbf{Q})$ is coercive. By construction, $\mathscr{F}_0$ is differentiable, and a differentiable, coercive real-valued function on a normed linear space is necessarily bounded from below and attains its minimum at a critical point (see, for example, Chapter I of [329]). In a similar way, if H is negative definite, then its diagonal representation takes the form (1.5) with $\alpha_+ = 0$ and $\alpha_- = 1$; Lemma 1.1 guarantees that $\mathscr{F}_0$ is negatively coercive, and therefore it must be bounded above and attain its maximum at a critical point. □

As we have noted already, the positive-definite case is the common, familiar one. The case of a negative-definite H is peculiar and corresponds in some sense to an interaction energy that completely discourages any natural ordering. Nevertheless, even in this situation, we are still guaranteed that for every temperature, the mean-field free energy has at least one equilibrium phase $\mathbf{Q}^*$, and all such stationary points of $\mathscr{F}_0$ must still satisfy $\mathscr{F}(\beta) \leqq \mathscr{F}_0(\beta, \mathbf{Q}^*)$, with any one possessing the least value of $\mathscr{F}_0$ still giving the best approximation to the true ensemble free energy at that temperature. The case of an indefinite pair-potential H is more subtle and difficult to analyze; we discuss this next.

H Indefinite

The analysis of our model in the case of an indefinite bilinear Hamiltonian requires some higher-order calculus of the mean-field free energy F_0 and single-particle potential H_0. It also relies upon our ability to construct a "deflated model," which is a function of a reduced set of tensor order parameters that possesses a global minimum (analogous to the case of H positive definite). We include here the study of this case for its relevance to the theory of biaxial liquid crystals; the classical case in which the bilinear Hamiltonian H is positive definite would suffice to treat the MAIER–SAUPE theory, which is the most instructive of all molecular theories of liquid crystals. We shall omit most proofs in our account below, referring the interested reader to the comprehensive study [116].

Calculus

We require some information about the second derivatives of F_0 with respect to the extended order tensors. This will in turn be related to derivatives of H_0. Derivatives of functions on vector spaces can be interpreted in various ways and are in general identified with multilinear forms. For our purposes, for a function such as $H_0 = H_0(\mathbf{Q})$, $\mathbf{Q} \in \mathcal{T}$, it is expedient to identify $\frac{\partial H_0}{\partial \mathbf{Q}}$ with the element of $\mathcal{T}$ that represents the linear form associated with the first derivative and to identify $\frac{\partial^2 H_0}{\partial \mathbf{Q}^2}$ with the linear transformation on $\mathcal{T}$ that represents the quadratic form associated with the second derivative; that is, we set

$$\frac{\partial H_0}{\partial \mathbf{Q}} \in \mathcal{T}, \quad \frac{\partial^2 H_0}{\partial \mathbf{Q}^2} \in L(\mathcal{T}),$$

where $L(\mathcal{T})$ is the space of all linear transformations on $\mathcal{T}$, so that, for all increments $\mathbf{U} \in \mathcal{T}$,

$$H_0(\mathbf{Q} + \varepsilon \mathbf{U}) = H_0(\mathbf{Q}) + \varepsilon \frac{\partial H_0}{\partial \mathbf{Q}}(\mathbf{Q}) \cdot \mathbf{U} + \frac{\varepsilon^2}{2} \frac{\partial^2 H_0}{\partial \mathbf{Q}^2}(\mathbf{Q})\mathbf{U} \cdot \mathbf{U} + o(\varepsilon^2), \quad \text{as } \varepsilon \to 0.$$

For example, consider the case of the bilinear two-particle Hamiltonian in its diagonal representation (1.5),

$$H(\omega, \omega') = -U_0 \left[\alpha_+ \mathbf{q}_+(\omega) \cdot \mathbf{q}_+(\omega') - \alpha_- \mathbf{q}_-(\omega) \cdot \mathbf{q}_-(\omega')\right],$$

for which the mean-field single-particle Hamiltonian is given by (1.41):

$$\begin{aligned} H_0(\omega; \mathbf{Q}_+, \mathbf{Q}_-) = & \frac{U_0}{2} \left[\alpha_+ \mathbf{Q}_+ \cdot \mathbf{Q}_+ - \alpha_- \mathbf{Q}_- \cdot \mathbf{Q}_-\right] \\ & - U_0 \left[\alpha_+ \mathbf{q}_+(\omega) \cdot \mathbf{Q}_+ - \alpha_- \mathbf{q}_-(\omega) \cdot \mathbf{Q}_-\right]. \end{aligned}$$

For this we obtain (see also (1.46))

$$\frac{\partial H_0}{\partial \mathbf{Q}_-}(\omega; \mathbf{Q}_+, \mathbf{Q}_-) = U_0 \alpha_- \left[\mathbf{q}_-(\omega) - \mathbf{Q}_-\right] \in \mathcal{T}_-$$

and

$$\frac{\partial^2 H_0}{\partial \mathbf{Q}_-^2}(\omega; \mathbf{Q}_+, \mathbf{Q}_-) = -U_{\mathfrak{z}}\alpha_- \mathbf{I} \in L(\mathcal{T}_-),$$

where $\mathbf{I}$ is the identity in $L(\mathcal{T}_-)$, such that

$$\frac{\partial^2 H_0}{\partial \mathbf{Q}_-^2}(\omega; \mathbf{Q}_+, \mathbf{Q}_-)\mathbf{U}_- = -U_{\mathfrak{z}}\alpha_- \mathbf{U}_-, \quad \forall \mathbf{U}_- \in \mathcal{T}_-.$$

We say that the linear transformation $\frac{\partial^2 H_0}{\partial \mathbf{Q}^2}$ is *negative definite* if the associated quadratic form is negative definite in the sense

$$\frac{\partial^2 H_0}{\partial \mathbf{Q}^2}\mathbf{U} \cdot \mathbf{U} < 0, \quad \forall \mathbf{U} \in \mathcal{T},\ \mathbf{U} \neq \mathbf{0}.$$

Observe that for $H_0(\omega; \mathbf{Q}_+, \mathbf{Q}_-)$ above, we have (by virtue of the bilinear nature of H_0)

$$\frac{\partial^2 H_0}{\partial \mathbf{Q}_+^2} \text{ and } \frac{\partial^2 H_0}{\partial \mathbf{Q}_-^2} \text{ are both constant,}$$

$$\frac{\partial^2 H_0}{\partial \mathbf{Q}_+^2} \text{ is positive definite (if } \alpha_+ > 0),$$

$$\frac{\partial^2 H_0}{\partial \mathbf{Q}_-^2} \text{ is negative definite (if } \alpha_- > 0).$$

As a final preliminary note, we indicate that we will conform to the following notation for tensor products:

$$\mathbf{V}_1, \mathbf{V}_2 \in \mathcal{T} \;\Rightarrow\; \mathbf{V}_1 \otimes \mathbf{V}_2 \in L(\mathcal{T})$$

via

$$(\mathbf{V}_1 \otimes \mathbf{V}_2)\,\mathbf{W} := (\mathbf{V}_2 \cdot \mathbf{W})\,\mathbf{V}_1, \quad \forall \mathbf{W} \in \mathcal{T}.$$

The first partial derivatives $\frac{\partial F_0}{\partial \mathbf{Q}}$ have already been related to $\frac{\partial H_0}{\partial \mathbf{Q}}$ in (1.45). The second partial derivatives $\frac{\partial^2 F_0}{\partial \mathbf{Q}^2}$ can be handled in a similar way. We obtain the following.

Theorem 1.4. *Let $H_0 = H_0(\omega; \mathbf{Q})$ be a sufficiently regular real-valued function defined on $\Omega \times \mathcal{T}$, Ω a state space, and $\mathcal{T}$ an inner-product space. Let $F_0 = F_0(\beta, \mathbf{Q})$ be the associated (per particle)* HELMHOLTZ *potential defined by*

$$F_0(\beta, \mathbf{Q}) = -\frac{1}{\beta} \ln \int_{\Omega} e^{-\beta H_0(\omega; \mathbf{Q})}\, d\omega. \tag{1.53}$$

The following formula is valid in general:

$$\frac{\partial^2 F_0}{\partial \mathbf{Q}^2} = \left\langle \frac{\partial^2 H_0}{\partial \mathbf{Q}^2} \right\rangle_{\rho_0} - \beta \left\langle \frac{\partial H_0}{\partial \mathbf{Q}} \otimes \frac{\partial H_0}{\partial \mathbf{Q}} \right\rangle_{\rho_0} + \beta \left\langle \frac{\partial H_0}{\partial \mathbf{Q}} \right\rangle_{\rho_0} \otimes \left\langle \frac{\partial H_0}{\partial \mathbf{Q}} \right\rangle_{\rho_0}. \tag{1.54}$$

Proof. The result follows by expansion of

$$F_0(\beta, \mathbf{Q} + \epsilon \mathbf{U}) = F_0(\beta, \mathbf{Q}) + \cdots$$

in (1.53), using the definitions of the distribution functions and averages as in Section 1.2. □

We note that formula (1.54) is equally valid for derivatives with respect to subsets of tensor order parameters, such as $\mathbf{Q}_+$ and $\mathbf{Q}_-$. For our purposes, the important consequence of this formula is that whenever a bilinear pair-potential has at least some repulsive components ($\alpha_- \neq 0$), then $\frac{\partial^2 F_0}{\partial \mathbf{Q}_-^2}$ must be negative definite at all $\mathbf{Q}_-$-critical points.

Corollary 1.2. *Let* H *be a bilinear two-particle Hamiltonian in the diagonalized form* (1.5) *that is either indefinite or negative definite (i.e., partly or fully repulsive), and let* F_0 *be the associated per-particle mean-field* HELMHOLTZ *free energy. Then* $\frac{\partial^2 F_0}{\partial \mathbf{Q}_-^2}$ *is necessarily negative definite at all* $\mathbf{Q}_-$*-critical points:*

$$\frac{\partial F_0}{\partial \mathbf{Q}_-}(\beta, \mathbf{Q}_+^*, \mathbf{Q}_-^*) = \mathbf{0} \Rightarrow \frac{\partial^2 F_0}{\partial \mathbf{Q}_-^2}(\beta, \mathbf{Q}_+^*, \mathbf{Q}_-^*) \text{ negative definite.}$$

Proof. We know that, in general,

$$\frac{\partial F_0}{\partial \mathbf{Q}_-} = \left\langle \frac{\partial H_0}{\partial \mathbf{Q}_-} \right\rangle_{\rho_0};$$

so

$$\frac{\partial F_0}{\partial \mathbf{Q}_-} = \mathbf{0} \Rightarrow \left\langle \frac{\partial H_0}{\partial \mathbf{Q}_-} \right\rangle_{\rho_0} = \mathbf{0},$$

and the formula in Theorem 1.4 simplifies to

$$\frac{\partial^2 F_0}{\partial \mathbf{Q}_-^2} = \left\langle \frac{\partial^2 H_0}{\partial \mathbf{Q}_-^2} \right\rangle_{\rho_0} - \beta \left\langle \frac{\partial H_0}{\partial \mathbf{Q}_-} \otimes \frac{\partial H_0}{\partial \mathbf{Q}_-} \right\rangle_{\rho_0}.$$

Also, we have already observed that for the diagonalized bilinear model, we have (for any ω, $\mathbf{Q}_+$, $\mathbf{Q}_-$)

$$\frac{\partial^2 H_0}{\partial \mathbf{Q}_-^2}(\omega; \mathbf{Q}_+, \mathbf{Q}_-) = -U_{\mathfrak{z}} \alpha_- \mathbf{I},$$

where $U_{\mathfrak{z}}$ and α_- are both positive by assumption. Evaluating the associated quadratic form, we obtain (using also the fact that $\beta > 0$)

$$\begin{aligned} \frac{\partial^2 F_0}{\partial \mathbf{Q}_-^2}(\beta, \mathbf{Q}_+^*, \mathbf{Q}_-^*)\mathbf{U}_- \cdot \mathbf{U}_- &= -U_{\mathfrak{z}} \alpha_- |\mathbf{U}_-|^2 - \beta \left\langle \left[\frac{\partial H_0}{\partial \mathbf{Q}_-}(\cdot; \mathbf{Q}_+^*, \mathbf{Q}_-^*) \cdot \mathbf{U}_- \right]^2 \right\rangle_{\rho_0} \\ &\leqq -U_{\mathfrak{z}} \alpha_- |\mathbf{U}_-|^2 < 0, \quad \text{if } \mathbf{U}_- \neq \mathbf{0}. \end{aligned}$$

We conclude that $\frac{\partial^2 F_0}{\partial \mathbf{Q}_-^2}$ is negative definite at $(\beta, \mathbf{Q}_+^*, \mathbf{Q}_-^*)$. □

We note that a similar conclusion *cannot* be obtained for the Hessian with respect to the attractive variables, $\frac{\partial^2 F_0}{\partial \mathbf{Q}_+^2}$. The calculation analogous to the above (at a $\mathbf{Q}_+$-critical point, $\frac{\partial F_0}{\partial \mathbf{Q}_+} = \mathbf{0}$) gives

$$\frac{\partial^2 F_0}{\partial \mathbf{Q}_+^2}(\beta, \mathbf{Q}_+^*, \mathbf{Q}_-^*)\mathbf{U}_+ \cdot \mathbf{U}_+ = U_3\alpha_+|\mathbf{U}_+|^2 - \beta\left\langle\left[\frac{\partial H_0}{\partial \mathbf{Q}_+}(\cdot; \mathbf{Q}_+^*, \mathbf{Q}_-^*) \cdot \mathbf{U}_+\right]^2\right\rangle_{\rho_0},$$

which is a sum of a nonnegative and a nonpositive term and may be positive, negative, or zero. In this regard, there is an intrinsic asymmetry in the formalism. Finally, it does not matter at this stage whether we consider derivatives of the ensemble mean-field free energy $\mathscr{F}_0$ or the free energy per particle F_0, since they are proportional: $F_0 = \mathscr{F}_0/N$.

Deflated Model

At this point we focus our attention on the indefinite (partly repulsive) case, the case in which $\alpha_+ = \alpha_- = 1$ in the diagonal representation (1.5). We have seen that in this situation the associated mean-field free energy $F_0(\beta, \mathbf{Q}_+, \mathbf{Q}_-)$ inherits three properties that will be essential to our further development: for each fixed $\beta > 0$,

1. $F_0(\beta, \mathbf{Q}_+, \mathbf{Q}_-)$ is coercive with respect to $\mathbf{Q}_+$ for each fixed $\mathbf{Q}_-$,
2. $F_0(\beta, \mathbf{Q}_+, \mathbf{Q}_-)$ is negatively coercive with respect to $\mathbf{Q}_-$ for each fixed $\mathbf{Q}_+$, and
3. $\partial^2 F_0/\partial \mathbf{Q}_-^2$ is negative definite at all $\mathbf{Q}_-$-critical points.

While F_0 possesses neither a global maximum nor a global minimum, the properties above allow us to define a *deflated* free energy that is a function of a reduced set of parameters (β and $\mathbf{Q}_+$ only) and that *does* possess a global minimum.

Theorem 1.5. *Let H be an indefinite* (partly repulsive) *bilinear two-particle Hamiltonian in the diagonal representation* (1.5), *that is, with $\alpha_+ = \alpha_- = 1$, and let F_0 be its associated per-particle mean-field* HELMHOLTZ *function, constructed as in* Section 1.2. *Then for every fixed $\beta > 0$ and $\mathbf{Q}_+ \in \mathfrak{T}_+$, the equation*

$$\frac{\partial F_0}{\partial \mathbf{Q}_-}(\beta, \mathbf{Q}_+, \mathbf{Q}_-) = \mathbf{0}$$

is uniquely solvable for $\mathbf{Q}_-$:

$$\mathbf{Q}_- = G_0(\beta, \mathbf{Q}_+).$$

The reduced function

$$f_0(\beta, \mathbf{Q}_+) := F_0(\beta, \mathbf{Q}_+, G_0(\beta, \mathbf{Q}_+))$$

is well defined for all $\beta > 0$ and $\mathbf{Q}_+ \in \mathfrak{T}_+$, and we term it the deflated mean-field free energy. *For every $\beta > 0$ and $\mathbf{Q}_+ \in \mathfrak{T}_+$, it satisfies*

$$f_0(\beta, \mathbf{Q}_+) = \max_{\mathbf{Q}_- \in \mathfrak{T}_-} F_0(\beta, \mathbf{Q}_+, \mathbf{Q}_-). \tag{1.55}$$

The critical points of f_0 with respect to $\mathbf{Q}_+$ are in one-to-one correspondence with the critical points of F_0 with respect to $(\mathbf{Q}_+, \mathbf{Q}_-)$.

Minimax Characterization

The deflated mean-field free energy f_0 inherits coercivity from the $\mathbf{Q}_+$-coercivity of F_0 by virtue of the characterization (1.55). As a consequence, it attains its global minimum at a critical point, and this point must be a least-free-energy point and must admit a characterization as a minimax point. The following theorem is proved in [116].

Theorem 1.6 (Gartland & Virga [116]). *Let H be an indefinite* (partly repulsive) *bilinear two-particle Hamiltonian in the diagonal representation* (1.5), *and let F_0 and f_0 be the associated mean-field* Helmholtz *free energy (constructed as in* Section 1.2) *and the deflated free energy (constructed as in* Theorem 1.5). *Then for any given $\beta > 0$, f_0 attains its global minimum at a critical point $\mathbf{Q}_+^* \in \mathcal{T}_+$ that admits the characterization*

$$f_0(\beta, \mathbf{Q}_+^*) = \min_{\mathbf{Q}_+ \in \mathcal{T}_+} f_0(\beta, \mathbf{Q}_+) = \min_{\mathbf{Q}_+ \in \mathcal{T}_+} \max_{\mathbf{Q}_- \in \mathcal{T}_-} F_0(\beta, \mathbf{Q}_+, \mathbf{Q}_-).$$

The associated point $(\mathbf{Q}_+^*, \mathbf{Q}_-^*)$, with $\mathbf{Q}_-^* = G_0(\beta, \mathbf{Q}_+^*)$, is a least-free-energy point and characterizes the phase of the mean-field model:

$$\begin{aligned} f_0(\beta, \mathbf{Q}_+^*) &= F_0(\beta, \mathbf{Q}_+^*, \mathbf{Q}_-^*) \\ &= \min\{F_0(\beta, \mathbf{Q}_+, \mathbf{Q}_-) \mid (\mathbf{Q}_+, \mathbf{Q}_-) \text{ a critical point of } F_0\}. \end{aligned}$$

Thus the case of the mean-field model associated with an indefinite bilinear pair-potential (1.5) is completely understood. Existence is guaranteed of at least one stationary point of F_0 that gives the best approximation to the true free energy of the ensemble at a given temperature among all self-consistent tensor order parameters. We conclude our treatment with the introduction of a generalized notion of local stability that stationary points of F_0 must satisfy in order to be viable candidates for global least-free-energy phase points.

1.2.5 Local Stability Criterion

One of the main analyses performed with mean-field models is the construction of associated phase diagrams, which chart out the bulk equilibrium phases of the system in different regions of the parameter space to be explored. Under most circumstances, this must be done numerically. In the computational physics community, this is usually accomplished by generating a variety of initial guesses (for each fixed set of parameters) and relaxing the free energy to local minima from these guesses. The smallest of the values of the local minima defines the phase. If the mean-field free energy possesses a global minimum (the most common case), then this is a viable (and popular) approach.

Numerical analysts generally prefer to follow paths of equilibrium points of the mean-field free energy, using numerical continuation and bifurcation techniques, along each path, classifying points as locally stable or unstable, computing their free energies, and then a posteriori declaring the solution with the minimal free energy (for a given set of parameters) to be the equilibrium phase of the system—completely analogous to the direct free-energy minimization approach in this last regard. In the case of a mean-field free energy with a global minimum, each approach has its advantages and disadvantages, and neither is immune from failing to detect a competing equilibrium solution. In the case of an indefinite mean-field free energy—the case of present interest—only the latter approach is viable.

In the case of a mean-field free energy that possesses a global minimum, the assessment of the local linear stability of an equilibrium point is usually a matter of examining the sign of the minimum eigenvalue of an appropriate Hessian matrix. In the case of an indefinite free energy, this is no longer the situation, and a different approach is required. We develop such an approach now.

Scalar Order Parameters

The practical analysis of specific mean-field models normally necessitates the transformation of $F_0(\beta, \mathbf{Q})$ to a function of a finite number of scalar variables, also called the *scalar order parameters*, that characterize the tensors in $\mathbf{Q}$. The number of scalar order parameters will not be greater than the dimension of $\mathcal{T}$ and is often strictly less, made so by symmetries or degeneracies or additional modeling assumptions.

Consider again the one-particle Hamiltonian H_0 in (1.50) associated with the two-particle Hamiltonian H_u of the uniaxial interactions in (1.6),

$$H_0(\omega, \mathbf{Q}) = U_\mathfrak{z} \left(\frac{1}{2}\mathbf{Q} - \mathbf{q}(\omega) \right) \cdot \mathbf{Q} + c, \tag{1.56}$$

where c is a scaling constant introduced here much in the same spirit as was C in (1.25) and

$$U_\mathfrak{z} = \mathfrak{z} U_0 > 0$$

with U_0 the interaction strength for the pair-potential (1.6) and $\mathfrak{z}$ the coordination parameter for the ensemble, as in (1.35) and (1.39). It then follows from (1.44) that the corresponding one-particle partition function Z_0 is given by

$$Z_0(\beta, \mathbf{Q}) = e^{-\beta U_\mathfrak{z} \frac{1}{2}\mathbf{Q}\cdot\mathbf{Q}} \frac{1}{e^{\beta c}} \int_\Omega e^{\beta U_\mathfrak{z} \mathbf{q}(\omega)\cdot\mathbf{Q}} d\omega, \tag{1.57}$$

where $\mathbf{q}(\omega)$ is as in (1.7). Setting

$$e^{\beta c} = |\Omega| \tag{1.58}$$

makes Z_0 such that $Z_0(\beta, \mathbf{0}) = 1$, and, by (1.44), the one-particle free energy F_0 vanishes in the isotropic phase. For all values of $\mathbf{Q}$, F_0 then reads as

$$F_0(\beta, \mathbf{Q}) = U_\mathfrak{z} \frac{1}{2}\mathbf{Q} \cdot \mathbf{Q} - \frac{1}{\beta} \ln \frac{1}{|\Omega|} \int_\Omega e^{\beta U_\mathfrak{z} \mathbf{q}(\omega)\cdot\mathbf{Q}} d\omega, \tag{1.59}$$

which is more conveniently given the following dimensionless form:

$$\widetilde{F}_0(\tilde{\beta}, \mathbf{Q}) = \frac{1}{2}\mathbf{Q}\cdot\mathbf{Q} - \frac{1}{\tilde{\beta}}\ln\frac{1}{|\Omega|}\int_\Omega e^{\tilde{\beta}\mathbf{q}(\omega)\cdot\mathbf{Q}}d\omega, \tag{1.60}$$

where

$$\widetilde{F}_0 = \frac{F_0}{U_3} \quad \text{and} \quad \tilde{\beta} = \beta U_3 = \frac{3U_0}{k_B\theta}, \tag{1.61}$$

as in (1.52).

The order tensor $\mathbf{Q}$ is real, symmetric, traceless, and of second rank; it has in principle five degrees of freedom. The free energy $\widetilde{F}_0$, however, is frame indifferent and depends only on the two independent eigenvalues of $\mathbf{Q}$. If we fix the frame of $\mathbf{Q}$ to be $(\boldsymbol{e}_x, \boldsymbol{e}_y, \boldsymbol{e}_z)$, then we can express $\mathbf{Q}$ in the form

$$\mathbf{Q} = S\left(\boldsymbol{e}_z \otimes \boldsymbol{e}_z - \frac{1}{3}\mathbf{I}\right) + T\left(\boldsymbol{e}_x \otimes \boldsymbol{e}_x - \boldsymbol{e}_y \otimes \boldsymbol{e}_y\right), \quad S, T \in \mathbb{R}, \tag{1.62}$$

which is the notation used in [315]. By using (1.62) in (1.60), also with the aid of (1.3), we obtain

$$\begin{aligned}\widetilde{F}_0(\tilde{\beta}, \mathbf{Q}) &= \frac{1}{3}S^2 + T^2 - \frac{1}{\tilde{\beta}}\ln\frac{1}{4\pi}\int_0^{2\pi}d\varphi\int_0^{\pi}e^{\tilde{\beta}g_{\mathrm{u}}(\vartheta,\varphi;S,T)}\sin\vartheta\, d\vartheta \\ &=: f_{\mathrm{u}}(\tilde{\beta}, S, T),\end{aligned} \tag{1.63}$$

where

$$g_{\mathrm{u}}(\vartheta, \varphi; S, T) = S\left(\cos^2\vartheta - \frac{1}{3}\right) + T\sin^2\vartheta\cos 2\varphi$$

and the angles ϑ and φ are defined by equations (1.2). The function f_{u} defined by (1.63) will be the basis of our analysis in Section 1.3. Here we see from it that symmetry reduces the five degrees of freedom of the order tensor to two scalar order parameters. The transformation from order tensors to scalar order parameters is not uniquely defined, and to progress in this section, we must make some assumptions concerning the free energy expressed as a function of scalar order parameters such that it retains sufficient structure of the free energy expressed as a function of the order tensor collection $\mathbf{Q}$.

The case of interest is the case of an indefinite bilinear pair-potential as in (1.5), the partly repulsive case, for which the associated mean-field free energy has neither a global minimum nor a global maximum. We assume that, when expressed in terms of scalar order parameters, the free energy takes the form

$$f(\boldsymbol{x}, \boldsymbol{y}), \quad \boldsymbol{x} \in \mathbb{R}^{m_+}, \ \boldsymbol{y} \in \mathbb{R}^{m_-},$$

and that it satisfies the following assumptions.

Assumptions 1.7.

1. f is coercive with respect to $\boldsymbol{x}$ for each fixed $\boldsymbol{y}$,
2. f is negatively coercive with respect to $\boldsymbol{y}$ for each fixed $\boldsymbol{x}$, and
3. the Hessian $\nabla^2_{\boldsymbol{yy}} f$ is negative definite at all $\boldsymbol{y}$-critical points.

Thus we are assuming that the form of $f(\boldsymbol{x}, \boldsymbol{y})$ parallels the form of $F_0(\beta, \mathbf{Q}_+, \mathbf{Q}_-)$ (with dependence on temperature suppressed). The scalar variables in $\boldsymbol{x}$ parameterize the attractive order tensors in the collection $\mathbf{Q}_+$, while those in $\boldsymbol{y}$ parameterize $\mathbf{Q}_-$. It is shown in [116] that the conclusions reached below hold for any parameterization that can be transformed to the above by a sufficiently regular change of variables.

Deflated Free Energy

We adopt the following notational conventions for the gradients and Hessians of $f(\boldsymbol{x}, \boldsymbol{y})$:

$$\nabla f = \begin{bmatrix} \nabla_{\boldsymbol{x}} f \\ \nabla_{\boldsymbol{y}} f \end{bmatrix}, \quad \nabla_{\boldsymbol{x}} f = \begin{bmatrix} \dfrac{\partial f}{\partial x_1} \\ \vdots \\ \dfrac{\partial f}{\partial x_{m_+}} \end{bmatrix}, \quad \nabla_{\boldsymbol{y}} f = \begin{bmatrix} \dfrac{\partial f}{\partial y_1} \\ \vdots \\ \dfrac{\partial f}{\partial y_{m_-}} \end{bmatrix},$$

and

$$\nabla^2 f = \begin{bmatrix} \nabla^2_{\boldsymbol{xx}} f & \nabla^2_{\boldsymbol{xy}} f \\ \nabla^2_{\boldsymbol{yx}} f & \nabla^2_{\boldsymbol{yy}} f \end{bmatrix},$$

where

$$\nabla^2_{\boldsymbol{xy}} f = \nabla_{\boldsymbol{y}} \left(\nabla_{\boldsymbol{x}} f \right) = \begin{bmatrix} \dfrac{\partial^2 f}{\partial y_1 \partial x_1} & \cdots & \dfrac{\partial^2 f}{\partial y_{m_-} \partial x_1} \\ \vdots & & \vdots \\ \dfrac{\partial^2 f}{\partial y_1 \partial x_{m_+}} & \cdots & \dfrac{\partial^2 f}{\partial y_{m_-} \partial x_{m_+}} \end{bmatrix}, \text{ etc.}$$

Under the assumptions we have placed on f above, we can follow exactly the same path as in Section 1.2.4 to deduce that for each $\boldsymbol{x}$,

$$\nabla_{\boldsymbol{y}} f(\boldsymbol{x}, \boldsymbol{y}) = \mathbf{0} \text{ is uniquely solvable for } \boldsymbol{y} = \boldsymbol{g}(\boldsymbol{x}),$$

and the deflated function

$$h(\boldsymbol{x}) := f(\boldsymbol{x}, \boldsymbol{g}(\boldsymbol{x})) = \max_{\boldsymbol{y} \in \mathbb{R}^{m_-}} f(\boldsymbol{x}, \boldsymbol{y}) \tag{1.64}$$

is well defined for all $\boldsymbol{x}$ in $\mathbb{R}^{m_+}$ and is coercive. As a consequence, h is bounded from below and attains its global minimum value at a critical point. As before, the critical points of h and f are in one-to-one correspondence, that is,

$$\nabla h(\boldsymbol{x}^*) = \mathbf{0} \Leftrightarrow \nabla f(\boldsymbol{x}^*, \boldsymbol{y}^*) = \mathbf{0}, \quad \text{with } \boldsymbol{y}^* = \boldsymbol{g}(\boldsymbol{x}^*) \text{ unique.}$$

We need to relate the Hessians of h and f. By definition,

$$\nabla_{\boldsymbol{y}} f(\boldsymbol{x}, \boldsymbol{g}(\boldsymbol{x})) = \mathbf{0}, \quad \forall\, \boldsymbol{x} \in \mathbb{R}^{m_+}.$$

From this follows

$$\mathbf{0} = \nabla_x \left[\nabla_y f(x, g(x))\right] = \nabla^2_{yx} f(x, g(x)) + \nabla^2_{yy} f(x, g(x)) \nabla g(x),$$

which implies

$$\nabla g(x) = -\nabla^2_{yy} f(x, g(x))^{-1} \nabla^2_{yx} f(x, g(x)).$$

In addition,

$$\begin{aligned}\nabla h(x) &= \nabla_x [f(x, g(x))] \\ &= \nabla_x f(x, g(x)) + \nabla g(x)^T \nabla_y f(x, g(x)) = \nabla_x f(x, g(x)),\end{aligned}$$

since $\nabla_y f(x, g(x)) = \mathbf{0}$. Combining these, we obtain

$$\begin{aligned}\nabla^2 h(x) &= \nabla_x \left[\nabla_x f(x, g(x))\right] \\ &= \nabla^2_{xx} f(x, g(x)) + \nabla^2_{xy} f(x, g(x)) \nabla g(x) \\ &= \nabla^2_{xx} (f(x, g(x)) - \nabla^2_{xy} f(x, g(x)) \nabla^2_{yy} f(x, g(x))^{-1} \nabla^2_{yx} f(x, g(x)).\end{aligned} \tag{1.65}$$

Necessary conditions for x^* to be a local minimum point of h are

$$\nabla h(x^*) = \mathbf{0} \text{ and } \nabla^2 h(x^*) \text{ positive semidefinite.}$$

The first-order condition here is equivalent to

$$\nabla_x f(x^*, y^*) = \mathbf{0}, \ \nabla_y f(x^*, y^*) = \mathbf{0}, \ \text{ with } y^* = g(x^*).$$

By virtue of the assumed negative definiteness of $\nabla^2_{yy} f$ at all critical points, from (1.65) it now follows that a sufficient condition to guarantee the second-order condition above is that $\nabla^2_{xx} f(x^*, y^*)$ be positive semidefinite. In fact, more can be said about the relationship between $\nabla^2 h$ and $\nabla^2 f$.

Local Stability Criterion

We recall the notion of the *inertia* of a symmetric matrix M as the triple of integers giving the number of positive, zero, and negative eigenvalues, counting multiplicities:

$$i(M) = (\#\text{ positive eigenvalues}, \#\text{ zero eigenvalues}, \#\text{ negative eigenvalues}).$$

SYLVESTER's law of inertia[11] guarantees that the inertia of a matrix is invariant under congruence transformations:

$$i(M) = i(NMN^{\mathsf{T}}), \quad \text{for any } N \text{ nonsingular.}$$

As a consequence of this, the following relationship concerning the inertia of the Hessian of f has been established in [116].

---

[11] See, for example, Section 4.5 of [148].

Lemma 1.2. *At all critical points* $(\boldsymbol{x}^*, \boldsymbol{y}^*)$ *of the free energy expressed in terms of scalar order parameters,* $f(\boldsymbol{x}, \boldsymbol{y})$*, the inertia of the Hessian of* f *satisfies*

$$i(\nabla^2 f(\boldsymbol{x}^*, \boldsymbol{y}^*)) = i\left(\begin{bmatrix} \nabla^2 h(\boldsymbol{x}^*) & \\ & \nabla^2_{\boldsymbol{yy}} f(\boldsymbol{x}^*, \boldsymbol{y}^*) \end{bmatrix}\right).$$

Here h *is the deflated free energy, as in* (1.64).

At critical points of f, $\nabla^2_{\boldsymbol{yy}} f$ is negative definite by assumption and thus has all negative eigenvalues (m_-, counting multiplicity). It follows that the number of positive and the number of zero eigenvalues of $\nabla^2 f$ must be the same as those of $\nabla^2 h$ at such points. In order for a critical point $(\boldsymbol{x}^*, \boldsymbol{y}^*)$ of f to correspond to a local minimum of h, the Hessian $\nabla^2 h(\boldsymbol{x}^*)$ must be positive semidefinite, and so $\nabla^2 f(\boldsymbol{x}^*, \boldsymbol{y}^*)$ must have m_+ nonnegative eigenvalues (counting multiplicities). If $\nabla^2 f(\boldsymbol{x}^*, \boldsymbol{y}^*)$ were to have fewer than m_+ nonnegative eigenvalues (greater than m_- negative eigenvalues), then $\nabla^2 h(\boldsymbol{x}^*)$ could not be positive semidefinite, and the point could not possibly be a global minimum point of h (a least-free-energy point of f). We say that such a point is *locally unstable*. We have the following.

Theorem 1.8 (Local Stability Criterion [116]). *Let* $(\boldsymbol{x}^*, \boldsymbol{y}^*)$ *be a critical point of the indefinite mean-field free energy* $f(\boldsymbol{x}, \boldsymbol{y})$ *expressed in terms of scalar order parameters* $\boldsymbol{x} \in \mathbb{R}^{m_+}$ (attractive) *and* $\boldsymbol{y} \in \mathbb{R}^{m_-}$ (repulsive) *that is assumed to satisfy the three conditions of* Assumptions 1.7. *If* $\nabla^2 f(\boldsymbol{x}^*, \boldsymbol{y}^*)$ *has fewer than* m_+ *nonnegative eigenvalues (greater than* m_- *negative eigenvalues) counting multiplicities, then the point* $(\boldsymbol{x}^*, \boldsymbol{y}^*)$ *is a* locally linearly unstable *equilibrium point of* f *in the sense that* $(\boldsymbol{x}^*, \boldsymbol{y}^*)$ *cannot possibly correspond to a global least-free-energy point characterizing the phase of the system. Otherwise, the point is* locally linearly stable, *and the associated deflated mean-field free energy* h *satisfies the first-order and second-order necessary conditions for a local minimum at* $\boldsymbol{x}^*$.

1.2.6 Biaxial Nematic Liquid Crystals

As an application of our method, we compute here the mean-field free energy per particle F_0 for the pair-potential of biaxial nematics in equation (1.10). Though, as already remarked in Section 1.1.1, equation (1.10) could be set in the form (1.5), here we deliberately compute F_0 for the pair-potential expressed in the molecular tensors in (1.7) and (1.11). This also serves the purpose of illustrating how the theory presented here requires only that H be reducible to the form (1.5), but it does not necessarily prescribe employing the diagonal variables $\mathbf{q}_\pm$. As pointed out in Section 1.2.5, ultimately only the repulsive dimension m_-, that is, the number of independent scalar order parameters that represent the order tensor collection $\mathbf{Q}_-$, plays a role in our local stability criterion. Note that m_- is necessarily less than or equal to the dimension of $\mathcal{T}_-$, the vector space in which $\mathbf{Q}_-$ resides. Thus, in general, the decomposition in equation (1.5) is crucial to identifying $\mathbf{q}_-$ and $\mathbf{Q}_-$, and so to determine m_-, but this by no means makes $\mathbf{Q}_\pm$ privileged thermodynamic variables.

With the molecular tensors $\mathbf{q}$ and $\mathbf{b}$ as in equations (1.7) and (1.11), we define the order tensors

$$\mathbf{Q} := \langle \mathbf{q}(\cdot) \rangle_{\rho_0}, \quad \mathbf{B} := \langle \mathbf{b}(\cdot) \rangle_{\rho_0}. \tag{1.66}$$

According to equation (1.35), the one-particle Hamiltonian H_0 becomes

$$H_0(\omega; \mathbf{Q}, \mathbf{B}) = \frac{U_{\mathfrak{z}}}{2}(\mathbf{Q}\cdot\mathbf{Q} + 2\gamma\mathbf{Q}\cdot\mathbf{B} + \lambda\mathbf{B}\cdot\mathbf{B}) + U_{\mathfrak{z}}\widetilde{H}_0(\omega; \mathbf{Q}, \mathbf{B}),$$

where $U_{\mathfrak{z}} = \mathfrak{z}U_0$ and

$$\widetilde{H}_0(\omega; \mathbf{Q}, \mathbf{B}) := -\{\mathbf{q}(\omega)\cdot\mathbf{Q} + \gamma[\mathbf{q}(\omega)\cdot\mathbf{B} + \mathbf{b}(\omega)\cdot\mathbf{Q}] + \lambda\mathbf{b}(\omega)\cdot\mathbf{B}\}. \tag{1.67}$$

By (1.30), the partition function Z_0 reads

$$Z_0(\tilde{\beta}, \mathbf{Q}, \mathbf{B}) = e^{-\frac{\tilde{\beta}}{2}(\mathbf{Q}\cdot\mathbf{Q}+2\gamma\mathbf{Q}\cdot\mathbf{B}+\lambda\mathbf{B}\cdot\mathbf{B})}\frac{1}{|\Omega|}\int_\Omega e^{-\tilde{\beta}\widetilde{H}_0(\omega;\mathbf{Q},\mathbf{B})}d\omega, \tag{1.68}$$

where $\tilde{\beta}$ is as in (1.52) and Z_0 has been scaled precisely as in (1.57). By (1.44), we obtain from (1.68) the following dimensionless form of F_0:

$$\widetilde{F}_0(\tilde{\beta}, \mathbf{Q}, \mathbf{B}) := \frac{1}{2}(\mathbf{Q}\cdot\mathbf{Q}+2\gamma\mathbf{Q}\cdot\mathbf{B}+\lambda\mathbf{B}\cdot\mathbf{B}) - \frac{1}{\tilde{\beta}}\ln\frac{1}{|\Omega|}\int_\Omega e^{-\tilde{\beta}\widetilde{H}_0(\omega;\mathbf{Q},\mathbf{B})}d\omega. \tag{1.69}$$

We assume that the order tensors $\mathbf{Q}$ and $\mathbf{B}$, both symmetric and traceless, share the same eigenframe $(\boldsymbol{e}_x, \boldsymbol{e}_y, \boldsymbol{e}_z)$. Under this assumption, which is natural in the absence of any external field acting on the ensemble, we represent $\mathbf{Q}$ as in (1.62) and we give $\mathbf{B}$ a similar form:

$$\mathbf{B} = S'\left(\boldsymbol{e}_z\otimes\boldsymbol{e}_z - \frac{1}{3}\mathbf{I}\right) + T'\left(\boldsymbol{e}_x\otimes\boldsymbol{e}_x - \boldsymbol{e}_y\otimes\boldsymbol{e}_y\right). \tag{1.70}$$

We note that S, T, S', and T' are the scalar order parameters: S and S' describe the uniaxial components of $\mathbf{Q}$ and $\mathbf{B}$, whereas T and T' describe their biaxial components. A biaxial nematic phase is characterized by the growth of either one of the latter from zero. By (1.62), (1.67), and (1.70), the function $\widetilde{F}_0$ in (1.69) can be given the form

$$\begin{aligned}\widetilde{F}_0(\tilde{\beta}, \mathbf{Q}, \mathbf{B}) &= \frac{1}{3}S^2 + T^2 + 2\gamma\left(\frac{1}{3}SS' + TT'\right) + \lambda\left(\frac{1}{3}S'^2 + T'^2\right) \\ &\quad - \frac{1}{\tilde{\beta}}\ln\frac{1}{8\pi^2}\int_0^{2\pi}d\psi\int_0^{2\pi}d\varphi\int_0^{\pi}e^{\tilde{\beta}g_{\mathrm{b}}(\vartheta,\varphi,\psi;S,T,S',T';\gamma,\lambda)}\sin\vartheta\,d\vartheta \\ &=: f_{\mathrm{b}}(\tilde{\beta}, S, T, S', T'; \gamma, \lambda),\end{aligned} \tag{1.71}$$

where

$$\begin{aligned}&g_{\mathrm{b}}(\vartheta, \varphi, \psi; S, T, S', T'; \gamma, \lambda) \\ &:= \left(\cos^2\vartheta - \frac{1}{3}\right)(S + \gamma S') + \sin^2\vartheta\left[(T + \gamma T')\cos 2\varphi + (\gamma S + \lambda S')\cos 2\psi\right] \\ &\quad + \left[(1 + \cos^2\vartheta)\cos 2\varphi\cos 2\psi - 2\cos\vartheta\sin 2\varphi\sin 2\psi\right](\gamma T + \lambda T').\end{aligned}$$

To apply to f_{b} the local stability criterion of the preceding section, we identify the attractive and repulsive dimensions m_+ and m_-. By (1.13), the second-rank tensors $\mathbf{q}_1$ and $\mathbf{q}_2$ constituting $\mathbf{q}_+$ and $\mathbf{q}_-$ in equations (1.14)–(1.16) are linear combinations of $\mathbf{q}$ and $\mathbf{b}$ depending on the model parameters (γ, λ). Similarly, the order tensor collections $\mathbf{Q}_+$ and $\mathbf{Q}_-$ are linear combinations of $\mathbf{Q}$ and $\mathbf{B}$, and so, in view of (1.62) and (1.70), $m = m_+ + m_- \leqq 4$. The attractive and repulsive dimensions for the function f_{b} in (1.71) are easily derived from equations (1.14)–(1.16); they are collected in Table 1.2 for all possible choices of the model parameters.

	$\lambda > \gamma^2$	$\lambda = \gamma^2$	$\lambda < \gamma^2$
m_+	4	2	2
m_-	0	0	2

Table 1.2. Attractive and repulsive dimensions m_+ and m_- for the mean-field theory of biaxial nematic liquid crystals.

The criterion in Section 1.2.5 classifies as locally linearly stable the critical points of f_{b} as a function of (S, T, S', T') whenever the Hessian of f_{b} possesses four nonnegative eigenvalues if $\lambda > \gamma^2$, or two nonnegative and two negative eigenvalues if $\lambda < \gamma^2$. The case $\lambda = \gamma^2$ is singular, but it is historically relevant, since it was the first case studied theoretically [110, 111]. FREISER based his prediction of nematic biaxial phases on the special Hamiltonian in (1.10) with $\lambda = \gamma^2$.

The biaxial phase was first found experimentally in lyotropic liquid crystals [365]. Compelling experimental evidence for biaxial phases in thermotropic liquid crystals is much more recent [2, 205, 219, 232, 298, 299], and though unanimous consensus has not yet been reached [113, 199, 206, 295, 296], the new experiments have considerably revived interest in liquid crystal science [198].

As shown by (1.15), a single molecular tensor, $\widehat{\mathbf{q}} := \mathbf{q} + \gamma\mathbf{b}$, survives in the two-particle Hamiltonian with $\lambda = \gamma^2$; correspondingly, a single order tensor $\widehat{\mathbf{Q}} := \mathbf{Q} + \gamma\mathbf{B}$ represented as in (1.62), with scalar order parameters $\widehat{S} := S + \gamma S'$ and $\widehat{T} := T + \gamma T'$, suffices to express $\widetilde{F}_0$. Accordingly, the scalar order parameters (S, T, S', T') in (1.71) become redundant, and they can be rearranged in the independent order parameters $(\widehat{S}, \widehat{T})$, so that f_{b} reduces to a function $\widehat{f}_{\mathrm{b}}$ of these latter:

$$\widehat{f}_{\mathrm{b}}(\tilde{\beta}, \widehat{S}, \widehat{T}; \gamma) := \frac{1}{3}\widehat{S}^2 + \widehat{T}^2 - \frac{1}{\tilde{\beta}} \ln \frac{1}{8\pi^2} \int_0^{2\pi} d\psi \int_0^{2\pi} d\varphi \int_0^{\pi} e^{\tilde{\beta}\widehat{g}_{\mathrm{b}}(\vartheta,\varphi,\psi;\widehat{S},\widehat{T};\gamma)} \sin\vartheta \, d\vartheta,$$

where

$$\widehat{g}_{\mathrm{b}}(\vartheta, \varphi; \widehat{S}, \widehat{T}; \gamma) := \left(\cos^2\vartheta - \frac{1}{3}\right)\widehat{S} + \sin^2\vartheta\left(\widehat{T}\cos 2\varphi + \gamma\widehat{S}\cos 2\psi\right) + \gamma\left[(1+\cos^2\vartheta)\cos 2\varphi\cos 2\psi - 2\cos\vartheta\sin 2\varphi\sin 2\psi\right]\widehat{T}.$$

Our stability criterion thus classifies as locally linearly stable the critical points of $\widehat{f}_b$ in $(\widehat{S}, \widehat{T})$ where the Hessian of $\widehat{f}_b$ has two nonnegative eigenvalues.

In the general case, the stability criterion in Section 1.2.5 has been implemented numerically and systematically applied in the domain of the model parameter space (γ, λ) defined by the inequalities

$$\lambda > 0, \qquad 1 - |2\gamma| + \lambda > 0, \tag{1.72}$$

where the pair-potential H in equation (1.10) achieves its minimum value (preferred relative orientation) in a completely aligned state $\omega = \omega'$ (i.e., $\mathbf{q} = \mathbf{q}'$ and $\mathbf{b} = \mathbf{b}'$), a domain that goes across the three regimes in Table 1.2 (see Figure 1.3). New types

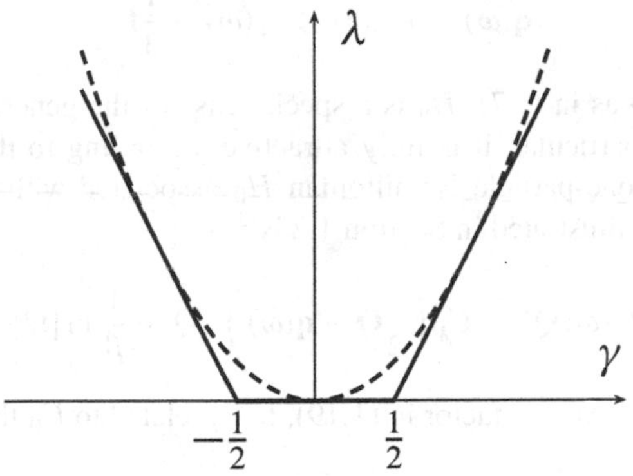

Fig. 1.3. The admissible region in the parameter space (λ, γ) described by the inequalities (1.72). The dashed parabola represents the singular set $\lambda = \gamma^2$.

of uniaxial-to-biaxial transitions were predicted [23, 24, 63, 64, 315], which also unveiled a tricritical point in the phase diagram [66], whose existence was further confirmed by an alternative theoretical approach [370] and detected in two independent experiments [219, 232]. Describing in detail all these consequences would be beyond the scope of this book. The few comments above concerning the mean-field theory of biaxial nematic liquid crystals based on the Hamiltonian in (1.10) should suffice to illustrate the applicability of the method explained here to this still growing field. The following section will be devoted to the derivation within this very setting of the classical MAIER–SAUPE theory for uniaxial nematic liquid crystals, already variously introduced in this section.

Before proceeding further, we indicate how the theory presented here could be further generalized. We believe that the basic ideas of GARTLAND & VIRGA [116] could be used to establish the validity of the minimum principle for indefinite mean-field free energies in functional settings more refined and advanced than that envisaged here. Perhaps the question could also be asked of finding the most general structure of H, and the most general environments $\mathbb{Q}$ and $\mathbb{T}$ compatible with the desired property of $\mathscr{F}_0$ to deliver the best approximation to $\mathscr{F}$ at its least critical value.

1.3 MAIER–SAUPE Theory

The MAIER–SAUPE mean-field theory plays a central role in the molecular description of uniaxial nematic liquid crystals for its formal neatness and its practical effectiveness. Thus, so far in our development it has often served as a test case for our general treatment of the mean-field approximation. The two-particle Hamiltonian H_u of the MAIER–SAUPE theory was already introduced in (1.6) above,

$$H_u(\omega, \omega') = -U_0 \mathbf{q}(\omega) \cdot \mathbf{q}(\omega'),$$

where $U_0 > 0$ is the interaction strength and $\mathbf{q}(\omega)$ is the molecular tensor

$$\mathbf{q}(\omega) = \boldsymbol{e}_1(\omega) \otimes \boldsymbol{e}_1(\omega) - \frac{1}{3}\mathbf{I} \tag{1.73}$$

in the orientation ω as in (1.7). H_u is a special case of the general bilinear Hamiltonian in (1.5); in particular, it is fully attractive, according to the terminology of Section 1.1.1. The one-particle Hamiltonian H_0 associated with H_u by the mean-field approximation illustrated in Section 1.2 is[12]

$$H_0(\omega, \mathbf{Q}) = U_{\mathfrak{z}}\left(\frac{1}{2}\mathbf{Q} - \mathbf{q}(\omega)\right) \cdot \mathbf{Q} + \frac{1}{\beta} \ln |\Omega|, \tag{1.74}$$

where β is the BOLTZMANN factor in (1.19), $U_{\mathfrak{z}}$ is related to U_0 through

$$U_{\mathfrak{z}} = \frac{2n}{N} U_0,$$

n is the total number of interactions in an ensemble of N particles.[13], and use has also been made of (1.58)

The order tensor $\mathbf{Q}$ represents the mean field to which all molecules in the ensemble are subject: by definition, it is a symmetric, traceless tensor in the three-dimensional vector translation space $\mathcal{V}$. The admissible mean fields $\mathbf{Q}$ must satisfy the self-consistency condition[14]

$$\mathbf{Q} = \langle \mathbf{q} \rangle_{\rho_0} = \int_{\Omega} \mathbf{q}(\omega) \rho_0(\omega; \beta, \mathbf{Q}) d\omega, \tag{1.75}$$

where ρ_0 is the probability density function on Ω associated with the one-particle Hamiltonian H_0,

$$\rho_0(\omega; \beta, \mathbf{Q}) = \frac{e^{-\beta H_0(\omega; \mathbf{Q})}}{Z_0(\beta, \mathbf{Q})}. \tag{1.76}$$

In this expression, β is the reciprocal reduced temperature defined as in (1.19) and Z_0 is the mean-field partition function per particle, which in (1.57) was found to be

[12] Cf. equations (1.56) and (1.58).

[13] Cf. equations (1.39) and (1.42) above.

[14] Which is the form of (1.31) appropriate to this case.

$$Z_0(\beta, \mathbf{Q}) = e^{-\beta U_3 \frac{1}{2}\mathbf{Q}\cdot\mathbf{Q}} \frac{1}{|\Omega|} \int_{\Omega} e^{\beta U_3 \mathbf{q}(\omega)\cdot\mathbf{Q}} d\omega, \tag{1.77}$$

where Ω is the orientational state space described in Section 1.1.1. According to (1.44), the mean-field free energy per particle F_0 takes the form in (1.59),

$$F_0(\beta, \mathbf{Q}) = U_3 \frac{1}{2}\mathbf{Q}\cdot\mathbf{Q} - \frac{1}{\beta} \ln \frac{1}{|\Omega|} \int_{\Omega} e^{\beta U_3 \mathbf{q}(\omega)\cdot\mathbf{Q}} d\omega. \tag{1.78}$$

Since H_u is positive definite, if properly rescaled by an additive constant, Theorem 1.3 ensures that F_0 is bounded from below and it attains its minimum value at a critical point. By Theorem 1.2, for any given β, all critical points $\mathbf{Q}$ of $F_0(\beta, \mathbf{Q})$ solve the self-consistency condition (1.75) and those where F_0 attains its minimum realize the best mean-field approximation to the "true" free energy at the given temperature, and so they represent the globally stable phases.

It is worth observing that, for every β, there is at least one solution of (1.75), and so one critical point of F_0: this is $\mathbf{Q} = \mathbf{0}$, which represents the isotropic phase.

Lemma 1.3. $\mathbf{Q} = \mathbf{0}$ *is a solution of* (1.75) *for all values of* β.

Proof. By (1.76) and (1.77),

$$\rho_0(\omega; \beta, \mathbf{0}) \equiv \frac{1}{|\Omega|},$$

and so, also with the aid of (1.73), the right side of (1.75) becomes

$$\mathbf{A} := \frac{1}{|\Omega|} \int_{\Omega} \mathbf{q}(\omega) d\omega = \frac{1}{4\pi} \int_{\mathbb{S}^2} \left(\boldsymbol{e} \otimes \boldsymbol{e} - \frac{1}{3}\mathbf{I} \right) da(\boldsymbol{e}), \tag{1.79}$$

where a is the area measure over the unit sphere $\mathbb{S}^2$. It follows from (1.79) that

$$\mathbf{R}\mathbf{A}\mathbf{R}^{\mathsf{T}} = \frac{1}{4\pi} \int_{\mathbb{S}^2} \left(\mathbf{R}\boldsymbol{e} \otimes \mathbf{R}\boldsymbol{e} - \frac{1}{3}\mathbf{I} \right) da(\boldsymbol{e}) \quad \forall\, \mathbf{R} \in \mathsf{O}(3), \tag{1.80}$$

where the superscript ^{T} denotes transposition. Since $\mathbb{S}^2$ is invariant under the action of the full orthogonal group $\mathsf{O}(3)$, and $da(\boldsymbol{e}) = da(\mathbf{R}\boldsymbol{e})$ for all $\mathbf{R} \in \mathsf{O}(3)$, equation (1.80) becomes

$$\mathbf{R}\mathbf{A}\mathbf{R}^{\mathsf{T}} = \mathbf{A},$$

meaning that $\mathbf{A}$ commutes with all $\mathbf{R} \in \mathsf{O}(3)$,

$$\mathbf{R}\mathbf{A} = \mathbf{A}\mathbf{R} \quad \forall\, \mathbf{R} \in \mathsf{O}(3). \tag{1.81}$$

By (1.81), every eigenspace of $\mathbf{A}$ is invariant under $\mathsf{O}(3)$, since, for an eigenvector $\boldsymbol{a}$ of $\mathbf{A}$ with eigenvalue α,

$$\mathbf{R}\mathbf{A}\boldsymbol{a} = \alpha\mathbf{R}\boldsymbol{a} = \mathbf{A}(\mathbf{R}\boldsymbol{a}), \tag{1.82}$$

and so $\mathbf{R}\boldsymbol{a}$ is an eigenvector of $\mathbf{A}$ with the same eigenvalue as $\boldsymbol{a}$. On the other hand, by (1.79), $\mathbf{A}$ is symmetric, and by the spectral theorem (see Appendix A.1) it possesses a full basis of eigenvectors, all necessarily with the same eigenvalue by (1.81). We thus conclude that $\mathbf{A} = \alpha\mathbf{I}$, whence it follows that $\mathbf{A} = \mathbf{0}$, since, again by (1.79), $\mathbf{A}$ is traceless. □

In a similar way, we prove that F_0 defined as in (1.78) for all symmetric tensors[15] $\mathbf{Q}$ can also be expressed as a function of the invariants of $\mathbf{Q}$,

$$I_1 = \operatorname{tr}\mathbf{Q}, \quad I_2 = \operatorname{tr}\mathbf{Q}^2, \quad I_3 = \operatorname{tr}\mathbf{Q}^3. \tag{1.83}$$

Lemma 1.4. *The function F_0 in* (1.78) *is* isotropic *in $\mathbf{Q}$, that is, it satisfies*

$$F_0(\beta, \mathbf{R}\mathbf{Q}\mathbf{R}^\mathsf{T}) = F_0(\beta, \mathbf{Q}) \qquad \forall\, \mathbf{R} \in \mathrm{O}(3).$$

Proof. By direct substitution in (1.78), we easily obtain that

$$\begin{aligned} F_0(\beta, \mathbf{R}\mathbf{Q}\mathbf{R}^\mathsf{T}) &= U_3 \frac{1}{2}\mathbf{R}\mathbf{Q}\mathbf{R}^\mathsf{T} \cdot \mathbf{R}\mathbf{Q}\mathbf{R}^\mathsf{T} - \frac{1}{\beta}\ln\frac{1}{4\pi}\int_{\mathbb{S}^2} e^{\beta U_3 (\boldsymbol{e}\otimes\boldsymbol{e} - \frac{1}{3}\mathbf{I})\cdot\mathbf{R}\mathbf{Q}\mathbf{R}^\mathsf{T}} da(\boldsymbol{e}) \\ &= U_3 \frac{1}{2}\mathbf{Q}\cdot\mathbf{R}^\mathsf{T}\mathbf{R}\mathbf{Q}\mathbf{R}^\mathsf{T}\mathbf{R} - \frac{1}{\beta}\ln\frac{1}{4\pi}\int_{\mathbb{S}^2} e^{\beta U_3 (\mathbf{R}\boldsymbol{e}\otimes\mathbf{R}\boldsymbol{e} - \frac{1}{3}\mathbf{I})\cdot\mathbf{Q}} da(\boldsymbol{e}) \\ &= F_0(\beta, \mathbf{Q}), \end{aligned}$$

where use has been made of the identieties

$$\mathbf{R}^\mathsf{T}\mathbf{R} = \mathbf{R}\mathbf{R}^\mathsf{T} = \mathbf{I}$$

and of the same change of variables in the integral as in (1.80). □

Corollary 1.3. *There is a function F_0' of both β and the invariants I_i of $\mathbf{Q}$ in* (1.83) *such that*

$$F_0(\beta, \mathbf{Q}) = F_0'(\beta, I_1, I_2, I_3). \tag{1.84}$$

Proof. The existence of F_0' follows from the representation theorem for all scalar isotropic functions of a symmetric tensor [357]. □

Letting $(\lambda_1, \lambda_2, \lambda_3)$ denote the eigenvalues of $\mathbf{Q}$ and $(\boldsymbol{n}_1, \boldsymbol{n}_2, \boldsymbol{n}_3)$ the corresponding eigenvectors, we write $\mathbf{Q}$ in the form

$$\mathbf{Q} = \sum_{i=1}^{3} \lambda_i \boldsymbol{n}_i \otimes \boldsymbol{n}_i, \tag{1.85}$$

so that

$$I_1 = \lambda_1 + \lambda_2 + \lambda_3, \quad I_2 = \lambda_1^2 + \lambda_2^2 + \lambda_3^2, \quad I_3 = \lambda_1^3 + \lambda_2^3 + \lambda_3^3. \tag{1.86}$$

Thus, (1.84) implies that F_0 can also be expressed as a function $\widehat{F}_0$ of the eigenvalues of $\mathbf{Q}$, symmetric under all their exchanges,

$$F_0(\beta, \mathbf{Q}) = \widehat{F}_0(\beta, \lambda_1, \lambda_2, \lambda_3). \tag{1.87}$$

As explained in Section 1.2.5, our strategy now requires determining, for every given β, all critical points of F_0 that are eligible to be the points of least free energy, that is, the points where F_0 is locally stable. Such a strategy is better pursued by scaling F_0 to U_3, as in (1.60) and (1.61), and by setting $\tilde{\beta} := \beta U_3$ in accordance with (1.52).

[15] When $F_0(\beta, \cdot)$ is extended to the whole space of symmetric tensors, so is also Z_0 defined in (1.77). In Theorem 1.9 below, solutions of (1.75) will indeed be sought in such an extended space.

Dimensionless Functions

Henceforth, to avoid clutter, we shall use the same symbol F_0 in (1.78) to denote the dimensionless function $\widetilde{F}_0$ in (1.60) and the same symbol β in (1.19) to denote the dimensionless parameter $\tilde{\beta}$ in (1.52).[16] Thus, with this new interpretation of the old symbols, we rewrite (1.78) and (1.77) as[17]

$$F_0(\beta,\mathbf{Q}) = \frac{1}{2}\mathbf{Q}\cdot\mathbf{Q} - \frac{1}{\beta}\ln\frac{1}{|\Omega|}\int_\Omega e^{\beta\mathbf{q}(\omega)\cdot\mathbf{Q}}d\omega \tag{1.88}$$

and

$$Z_0(\beta,\mathbf{Q}) = e^{-\beta\frac{1}{2}\mathbf{Q}\cdot\mathbf{Q}}\frac{1}{|\Omega|}\int_\Omega e^{\beta\mathbf{q}(\omega)\cdot\mathbf{Q}}d\omega. \tag{1.89}$$

Both equations (1.75) and (1.76) remain formally unaltered, provided there we express H_0 through

$$H_0(\omega,\mathbf{Q}) = \left(\frac{1}{2}\mathbf{Q} - \mathbf{q}(\omega)\right)\cdot\mathbf{Q} + \frac{1}{\beta}\ln|\Omega|, \tag{1.90}$$

instead of (1.74).

We shall represent the order tensor $\mathbf{Q}$ through an appropriate set of scalars. Our choice, already shown in (1.62), will be further illustrated and discussed in the following subsection.

1.3.1 Scalar Order Parameters

The order tensor $\mathbf{Q}$ will be written as

$$\mathbf{Q} = S\left(\boldsymbol{e}_z\otimes\boldsymbol{e}_z - \frac{1}{3}\mathbf{I}\right) + T\left(\boldsymbol{e}_x\otimes\boldsymbol{e}_x - \boldsymbol{e}_y\otimes\boldsymbol{e}_y\right). \tag{1.91}$$

This formula represents the most general symmetric and traceless second-rank tensor $\mathbf{Q}$ in its eigenframe $(\boldsymbol{e}_x,\boldsymbol{e}_y,\boldsymbol{e}_z)$, the corresponding eigenvalues being $T-\frac{1}{3}S$, $-T-\frac{1}{3}S$, and $\frac{2}{3}S$. Here $\mathbf{Q}$ is in general a biaxial tensor, having three unequal eigenvalues; it clearly appears as the superposition of a uniaxial tensor,

$$\boldsymbol{e}_z\otimes\boldsymbol{e}_z - \frac{1}{3}\mathbf{I},$$

with two equal eigenvalues in the plane $(\boldsymbol{e}_x,\boldsymbol{e}_y)$, and a *purely* biaxial tensor

$$\boldsymbol{e}_x\otimes\boldsymbol{e}_x - \boldsymbol{e}_y\otimes\boldsymbol{e}_y,$$

with one zero eigenvalue along the axis $\boldsymbol{e}_z$. Thus, S and T are called the *uniaxial* and *biaxial* scalar order parameters. When both $S \neq 0$ and $T \neq 0$, then $\mathbf{Q}$ is generally

[16] Said prosaically, in $\widetilde{F}_0$ and $\tilde{\beta}$, we drop the tilde.

[17] Formally, equations (1.88) and (1.89) are obtained from (1.78) and (1.77), respectively, by setting $U_3 = 1$.

biaxial, whereas it is purely biaxial when $S = 0$ and $T \neq 0$. When $T = 0$, $\mathbf{Q}$ is uniaxial. It should be noted that the biaxiality of $\mathbf{Q}$ is not at all related to the molecular shape, which is supposed to be uniaxial in the MAIER–SAUPE theory . The possible biaxiality of $\mathbf{Q}$ has its origin in the probability distribution ρ_0, which may fail to be uniaxial, despite each molecule being so. Sometimes, when we wish to distinguish this source of macroscopic biaxiality from that connected with the possible molecular biaxiality, reflected on the macroscopic scale by the order tensor $\mathbf{B}$ in (1.70), we call *phase* biaxiality that embodied by $\mathbf{Q}$ and *intrinsic* biaxiality that embodied by $\mathbf{B}$; correspondingly, T and T' are the phase and intrinsic biaxiality parameters: in principle, they are independent of one another, the former being defined for both uniaxial and biaxial molecules, whereas the latter is defined only for biaxial molecules.

The eigenframe $(\boldsymbol{e}_x, \boldsymbol{e}_y, \boldsymbol{e}_z)$ of $\mathbf{Q}$ has no intrinsic meaning: were the members of $(\boldsymbol{e}_x, \boldsymbol{e}_y, \boldsymbol{e}_z)$ subject to a permutation, with the scalar order parameters S and T left unchanged, the order tensor $\mathbf{Q}$ would represent the same molecular organization, and so the same phase, only relative to a different frame. Conversely, we could envision such an invariance of phase as an equivalence relation involving the scalar order parameters S and T: keeping the eigenframe of $\mathbf{Q}$ fixed, we consider all transformations of the pair (S, T) that leave the spectrum of $\mathbf{Q}$ unchanged. These constitute a six-element group generated by the elementary transformations

$$(S, T) \mapsto (S, -T), \tag{1.92a}$$

$$(S, T) \mapsto \left(\frac{3T - S}{2}, \frac{T + S}{2} \right), \tag{1.92b}$$

$$(S, T) \mapsto \left(\frac{-3T - S}{2}, \frac{T - S}{2} \right), \tag{1.92c}$$

which reflect three elementary exchanges in the eigenvalues of $\mathbf{Q}$: they respectively correspond to exchanging the eigenvalues corresponding to the pairs $(\boldsymbol{e}_x, \boldsymbol{e}_y)$, $(\boldsymbol{e}_x, \boldsymbol{e}_z)$, and $(\boldsymbol{e}_y, \boldsymbol{e}_z)$. The loci of the (S, T) plane invariant under these transformations are the lines correspondingly represented by the equations

$$T = 0, \qquad T = S, \quad \text{and} \quad T = -S. \tag{1.93}$$

The union of these lines is a set that is invariant under the action of all transformations (1.92). Moreover, by its very definition, this set constitutes the whole collection of uniaxial states represented by (1.91). For this reason, we also call *uniaxial lines* the geometric loci represented by (1.93).

When the scalar order parameters S and T describe through (1.91) an order tensor $\mathbf{Q}$ that satisfies the self-consistency condition (1.75), then they obey some bounds that we now proceed to make explicit. It follows from (1.75) that, for any given unit vector $\boldsymbol{e}$,

$$\boldsymbol{e} \cdot \mathbf{Q}\boldsymbol{e} = \int_{\Omega} \left[(\boldsymbol{e}_1(\omega) \cdot \boldsymbol{e})^2 - \frac{1}{3} \right] \rho_0(\omega; \beta, \mathbf{Q}) d\omega,$$

whence, since $0 \leqq (\boldsymbol{e}_1(\omega) \cdot \boldsymbol{e})^2 \leqq 1$ for all $\omega \in \Omega$, one arrives at

$$-\frac{1}{3} \leqq \boldsymbol{e} \cdot \mathbf{Q}\boldsymbol{e} \leqq \frac{2}{3}. \tag{1.94}$$

The lower bound in (1.94) is attained only if $\boldsymbol{e}_1(\omega)\cdot\boldsymbol{e} \equiv 0$, meaning that all molecules lie in the plane orthogonal to $\boldsymbol{e}$, whatever may be their alignment, whereas the upper bound is attained only if $\boldsymbol{e}_1(\omega) \cdot \boldsymbol{e} \equiv 1$, meaning that all molecules are oriented along $\boldsymbol{e}$. We refer to the molecular alignment in these limiting cases as *planar* and *full*, respectively. More generally, a uniaxial state with $S < 0$ will be called *discotic*, while a uniaxial state with $S > 0$ will be called *calamitic*: planar and full alignments are correspondingly the extreme limiting cases of discotic and calamitic states.

It follows immediately from (1.94) and the representation (1.85) that the eigenvalues λ_i of $\mathbf{Q}$ satisfy the same bounds as in (1.94). Moreover, letting $\boldsymbol{e}$ in (1.94) be in turn $\boldsymbol{e}_z$, $\boldsymbol{e}_y$, and $\boldsymbol{e}_x$, we also obtain

$$-\frac{1}{2} \leqq S \leqq 1 \quad \text{and} \quad -\frac{1}{3}(1-S) \leqq T \leqq \frac{1}{3}(1-S). \tag{1.95}$$

These inequalities show that in the limit of planar alignment the biaxial order parameter T ranges in the interval $\left[-\frac{1}{2}, \frac{1}{2}\right]$, reflecting different degrees of anisotropy in the molecular distribution. By contrast, in the limit of full alignment, T necessarily vanishes, since the molecular distribution becomes peaked at a single orientation. Inequalities (1.95) delimit a triangle in the (S, T) plane that comprises all states represented by (1.91) and is compatible with the self-consistency condition (1.75). The uniaxial lines (1.93) divide the admissible triangle into six elementary triangles, one transforming into the other under the action of the transformations in the group generated by (1.92) (see Figure 1.4). One of them suffices to represent all inequivalent states of the order tensor $\mathbf{Q}$ in (1.91). The vertices of the admissible triangle, $\mathsf{U}_+ = (1,0)$, $\mathsf{U}'_+ = \left(-\frac{1}{2}, \frac{1}{2}\right)$, and $\mathsf{U}''_+ = \left(\frac{1}{2}, -\frac{1}{2}\right)$, represent one and the same uniaxial state with full alignment. U_+ is fixed under (1.92a), while U'_+ and U''_+ are its images under (1.92b) and (1.92c), respectively. Similarly, the point $\mathsf{U}_- = \left(\frac{1}{4}, \frac{1}{4}\right)$ is fixed under transformation (1.92b), while $\mathsf{U}'_- = \left(-\frac{1}{2}, 0\right)$ and $\mathsf{U}''_- = \left(\frac{1}{4}, -\frac{1}{4}\right)$ are its images under (1.92c) and (1.92a), respectively. The points U_-, U'_-, and U''_- in the (S, T) plane represent the same planar uniaxial state, though all but one have $T \neq 0$.

As in (1.63), with the aid of (1.91), we now express the dimensionless free energy per particle $F_0(\beta, \mathbf{Q})$ in (1.88) as a function f_{u} of the dimensionless reciprocal temperature β and the scalar order parameters S and T:

$$f_{\mathrm{u}}(\beta, S, T) := \frac{1}{3}S^2 + T^2 - \frac{1}{\beta}\ln\frac{1}{4\pi}\int_0^{2\pi} d\varphi \int_0^{\pi} e^{\beta g_{\mathrm{u}}(\vartheta,\varphi;S,T)} \sin\vartheta \, d\vartheta, \tag{1.96}$$

where

$$g_{\mathrm{u}}(\vartheta, \varphi; S, T) = S\left(\cos^2\vartheta - \frac{1}{3}\right) + T\sin^2\vartheta\cos 2\varphi. \tag{1.97}$$

For given β, the stationary points of f_{u} are the points (S, T) that represent through (1.91) all self-consistent mean-field order tensors $\mathbf{Q}$.

It is instructive to prove directly that f_{u} is positively coercive in (S, T) and so it attains its global minimum at a critical point. To this end, we remark that by (1.97)

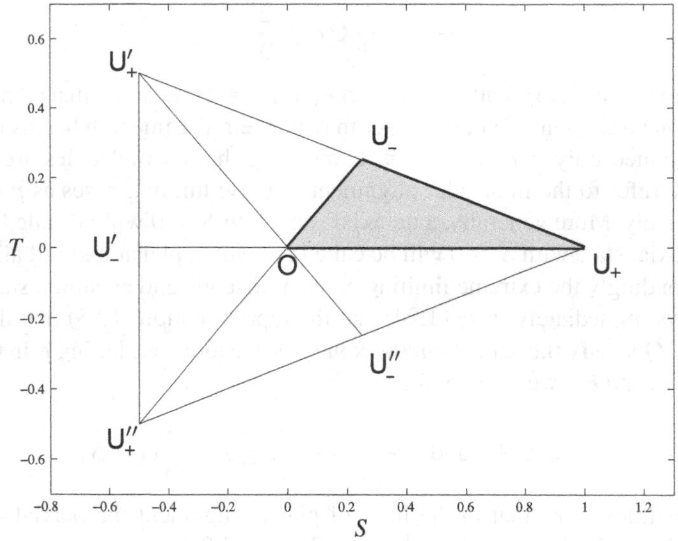

Fig. 1.4. Admissible order parameters in the plane (S,T). The shaded triangle $\mathsf{OU_+U_-}$, where O is the origin of the plane, suffices to describe all possible inequivalent states. $\mathsf{U_+}$, $\mathsf{U'_+}$, and $\mathsf{U''_+}$ represent one and the same state, as do $\mathsf{U_-}$, $\mathsf{U'_-}$, and $\mathsf{U''_-}$. Both these states are uniaxial, though the molecular alignment is full in the former and planar in the latter.

$$|g_{\mathrm{u}}| < \frac{2}{3}|S| + |T|,$$

and so, by (1.96),

$$f_{\mathrm{u}}(\beta, S, T) > \frac{1}{3}S^2 + T^2 - \frac{2}{3}|S| - |T|,$$

whence it follows that $f_{\mathrm{u}} \to \infty$ as $\|(S,T)\|_{\mathbb{R}^2} \to \infty$. We are thus assured that the first of Assumptions 1.7, the only one relevant to a fully attractive interaction Hamiltonian, is indeed satisfied by the function f_{u}, for which $m = m_+ = 2$, in the notation of Section 1.2.5.

There are properties that f_{u} inherits from F_0, which would not be easier to prove directly. Since by Lemma 1.3 F_0 is stationary at $\mathbf{Q} = \mathbf{0}$, f_{u} is stationary for $S = T = 0$, that is, at the origin O of the (S,T) plane. Moreover, since F_0 can be expressed through (1.87) as a symmetric function of the eigenvalues of $\mathbf{Q}$ in the representation (1.85), f_{u} is invariant under the transformations (1.92), which exchange the eigenvalues of $\mathbf{Q}$ in the representation (1.91).

In the following subsection we shall seek the critical points (S,T) of f_{u}. Since the corresponding order tensor $\mathbf{Q}$ in (1.91) necessarily obeys the self-consistency condition (1.75), the critical points of f_{u} are guaranteed to obey the bounds in (1.95), and so they will all fall within the admissible triangle in the (S,T) plane depicted in Figure 1.4. Moreover, it would suffice to find those falling in the closure of the smaller triangle $\mathsf{OU_+U_-}$, since all others will be their images under the transformations (1.92).

1.3.2 Critical Points

A remarkable theorem by FATKULLIN & SLASTIKOV [104] makes our search for the critical points of f_u considerably easier.

Theorem 1.9 (FATKULLIN & SLASTIKOV [104]). *If* $\mathbf{Q}$ *is a symmetric tensor that solves the self-consistency condition* (1.75) *associated with the one-particle Hamiltonian* H_0 *in* (1.90)*, then* $\mathbf{Q}$ *is traceless and either zero or uniaxial.*

Proof. We first rewrite the self-consistency condition (1.75) in an equivalent, more convenient form. We start from rewriting the probability distribution density ρ_0 in (1.76) as follows:

$$\rho_0(\omega;\beta,\mathbf{Q}) = \frac{e^{\beta \boldsymbol{e}_1(\omega)\cdot\mathbf{Q}\boldsymbol{e}_1(\omega)}}{z_0(\beta,\mathbf{Q})}, \tag{1.98}$$

where z_0 is the *reduced* partition function defined by

$$z_0(\beta,\mathbf{Q}) := \int_{\Omega} e^{\beta \boldsymbol{e}_1(\omega)\cdot\mathbf{Q}\boldsymbol{e}_1(\omega)}d\omega = \int_{\mathbb{S}^2} e^{\beta \boldsymbol{e}\cdot\mathbf{Q}\boldsymbol{e}}\,da(\boldsymbol{e}), \tag{1.99}$$

and use has been made of (1.73). For every $\beta > 0$, the function $z_0(\beta,\cdot)$ is defined by (1.99) in the whole space of symmetric tensors $\mathbf{Q}$ (not necessarily traceless). By (1.98), (1.75) then becomes

$$\mathbf{Q} + \frac{1}{3}\mathbf{I} = \frac{1}{\beta}\frac{1}{z_0}\frac{\partial z_0}{\partial \mathbf{Q}}. \tag{1.100}$$

This tensorial equation can be converted into three scalar equations involving the eigenvalues λ_i of $\mathbf{Q}$ in (1.85). Let $\mathbf{R} \in \mathsf{O}(3)$ be any orthogonal tensor. It follows from (1.99) that

$$z_0(\beta,\mathbf{RQR}^{\mathsf{T}}) = \int_{\mathbb{S}^2} e^{\beta \boldsymbol{e}\cdot\mathbf{RQR}^{\mathsf{T}}\boldsymbol{e}}\,da(\boldsymbol{e}) = \int_{\mathbb{S}^2} e^{\beta\mathbf{R}^{\mathsf{T}}\boldsymbol{e}\cdot\mathbf{Q}(\mathbf{R}^{\mathsf{T}}\boldsymbol{e})}da(\boldsymbol{e}). \tag{1.101}$$

Since, as in the proof of Lemma 1.3, $\mathbb{S}^2$ is invariant under the action of the full orthogonal group $\mathsf{O}(3)$, and $da(\boldsymbol{e}) = da(\mathbf{R}\boldsymbol{e})$ for all $\mathbf{R} \in \mathsf{O}(3)$, changing $\boldsymbol{e}$ into $\mathbf{R}\boldsymbol{e}$ does not affect the second integral in (1.101), and so we arrive at

$$z_0(\beta,\mathbf{RQR}^{\mathsf{T}}) = z_0(\beta,\mathbf{Q}) \quad \forall\, \mathbf{R} \in \mathsf{O}(3),$$

which says that z_0 is isotropic in $\mathbf{Q}$. By the representation theorem for the isotropic scalar functions of a symmetric tensor [357], we conclude that z_0 can be expressed as a function of the invariants of $\mathbf{Q}$ in (1.83),

$$z_0(\beta,\mathbf{Q}) = z_0'(\beta,I_1,I_2,I_3), \tag{1.102}$$

and eventually, as a symmetric function of the eigenvalues λ_i of $\mathbf{Q}$,

$$z_0(\beta, \mathbf{Q}) = \hat{z}_0(\beta, \lambda_1, \lambda_2, \lambda_3). \tag{1.103}$$

By differentiating with respect to $\mathbf{Q}$ both sides of equation (1.102), also with the aid of (1.83), we obtain that

$$\frac{\partial z_0'}{\partial \mathbf{Q}} = \frac{\partial z_0'}{\partial I_1}\mathbf{I} + 2\frac{\partial z_0'}{\partial I_2}\mathbf{Q} + 3\frac{\partial z_0'}{\partial I_3}\mathbf{Q}^2. \tag{1.104}$$

By (1.104) and (1.85), since $\mathbf{I}$, $\mathbf{Q}$, and $\mathbf{Q}^2$ all share the same eigenframe $(\boldsymbol{n}_1, \boldsymbol{n}_2, \boldsymbol{n}_3)$, from (1.100) we arrive at

$$\lambda_i + \frac{1}{3} = \frac{1}{\beta U}\frac{1}{z_0'}\left(\frac{\partial z_0'}{\partial I_1} + 2\frac{\partial z_0'}{\partial I_2}\lambda_i + 3\frac{\partial z_0'}{\partial I_3}\lambda_i^2\right), \quad i = 1, 2, 3,$$

which, by (1.83), (1.102), and (1.103), can also be written as

$$\lambda_i + \frac{1}{3} = \frac{1}{\beta}\frac{1}{z_0'}\left(\frac{\partial z_0'}{\partial I_1}\frac{\partial I_1}{\partial \lambda_i} + \frac{\partial z_0'}{\partial I_2}\frac{\partial I_2}{\partial \lambda_i} + \frac{\partial z_0'}{\partial I_3}\frac{\partial I_3}{\partial \lambda_i}\right) = \frac{1}{\beta}\frac{1}{\hat{z}_0}\frac{\partial \hat{z}_0}{\partial \lambda_i}. \tag{1.105}$$

Letting $\boldsymbol{e}$ in the second integral of (1.99) be represented as

$$\boldsymbol{e} = x_1\boldsymbol{n}_1 + x_2\boldsymbol{n}_2 + x_3\boldsymbol{n}_3 \qquad \text{with} \quad x_1^2 + x_2^2 + x_3^2 = 1,$$

by (1.85) we give $\hat{z}_0$ the following form:

$$\hat{z}_0(\beta, \lambda_1, \lambda_2, \lambda_3) = \int_{|\boldsymbol{x}|=1} e^{\beta \sum_{i=1}^{3} \lambda_i x_i^2}\, da(\boldsymbol{x}). \tag{1.106}$$

The proof of the theorem is then completed by the following lemma, which is slightly adapted from [104] (see, in particular, Lemma 1 of Appendix A). □

Lemma 1.5. *Let the real numbers* $(\lambda_1, \lambda_2, \lambda_3)$ *solve equations* (1.105) *with* $\hat{z}_0$ *as in* (1.106). *Then, necessarily,*

(i) $-\frac{1}{3} \leqq \lambda_i \leqq \frac{2}{3}$ *for* $i = 1, 2, 3$,
(ii) $\lambda_1 + \lambda_2 + \lambda_3 = 0$,
(iii) $\lambda_i = \lambda_j$ *for some* $i \neq j$.

The proof of this lemma relies on a semiexplicit representation of the function $\hat{z}_0$ by means of BESSEL functions in SOMMERFELD's representation. Although the proof given in [104] is not very technical, it is indirect and rather intricate. As FATKULLIN & SLASTIKOV [104] suggest, there should exist a proof of Lemma 1.5 based on pure symmetry, translating the intuitive idea that the tensors $\mathbf{Q}$ solving the self-consistency equation (1.75) must retain the uniaxial symmetry of the molecular tensors $\mathbf{q}$ they collectively represent. Such a simpler proof, however, if indeed it exists, has so far remained unknown. Another proof of Lemma 1.5, simpler than FATKULLIN & SLASTIKOV's, though not to the desired degree, was given in [373] as part of an analysis of the steady states of a SMOLUCHOWSKI equation initiated by the studies [51] and [52]. In the same vein, yet another proof of Lemma 1.5 was proposed in [190], which we found as intricate as FATKULLIN & SLASTIKOV's.

We have thus learned from Theorem 1.9 that the search for all critical points of f_u in (1.63) can be restricted to the uniaxial lines (1.93) of the (S, T) plane. In particular, we shall seek the critical points of f_u on the first of these lines, $T = 0$, since the critical points of f_u on the remaining two are their images under transformations (1.92b) and (1.92c). Moreover, since f_u is invariant under (1.92a), for any given β and S,

$$\frac{\partial f_u}{\partial T}(\beta, S, 0) = 0,$$

and so, for given β, every critical point $S = S_0$ of the restricted function

$$\hat{f}_u(\beta, S) := f_u(\beta, S, 0) = \frac{1}{3}S^2 - \frac{1}{\beta}\ln\frac{1}{2}\int_0^{\pi} e^{\beta S\left(\cos^2\vartheta - \frac{1}{3}\right)}\sin\vartheta\, d\vartheta \tag{1.107}$$

corresponds to a critical point $(S_0, 0)$ of f_u. Our search for the critical points of $\hat{f}_u$ will profit from the properties of DAWSON's *integral* daw, a special function related to the error function erf, which arises in many branches of physics. To expedite our analysis of the critical points of $\hat{f}_u$ we now digress slightly to collect the required properties of daw, mainly obtained from Chapter 7 of [1] and Chapter 42 of [316].

DAWSON's Integral

The function daw : $\mathbb{R} \to \mathbb{R}$ defined by

$$\mathrm{daw}(x) := e^{-x^2}\int_0^x e^{t^2}\, dt \tag{1.108}$$

is also called DAWSON's integral. It is related to the error function erf : $\mathbb{C} \to \mathbb{C}$, defined over the complex field $\mathbb{C}$ as

$$\mathrm{erf}(z) := \frac{2}{\sqrt{\pi}}\int_0^z e^{-t^2}\, dt,$$

through the formula

$$\mathrm{daw}(x) = -i\frac{\sqrt{\pi}}{2}e^{-x^2}\,\mathrm{erf}(ix),$$

where i is the imaginary unit. DAWSON's integral can also be extended to the whole of $\mathbb{C}$, and for a purely imaginary argument it evaluates to

$$\mathrm{daw}(iy) = i\frac{\sqrt{\pi}}{2}e^{y^2}\,\mathrm{erf}(y). \tag{1.109}$$

By (1.109), both functions

$$x \mapsto \sqrt{x}\,\mathsf{daw}(\sqrt{x}) \quad \text{and} \quad x \mapsto \frac{\mathsf{daw}(\sqrt{x})}{\sqrt{x}},$$

which will play a role in our development below, can be extended as real-valued functions to the whole of $\mathbb{R}$.

For $x \in \mathbb{R}$, $\mathsf{daw}(x)$ is the unique solution of the differential equation

$$f' = 1 - 2xf$$

that satisfies the condition

$$f(0) = 0.$$

It can be shown that $\mathsf{daw}(x)$ is an odd function with the following asymptotic expansions:

$$\mathsf{daw}(x) = x - \frac{2}{3}x^3 + \frac{4}{15}x^5 + O(x^7) \quad \text{for} \quad x \to 0, \tag{1.110a}$$

$$\mathsf{daw}(x) = \frac{1}{2x} + \frac{1}{4x^3} + \frac{3}{8x^5} + O\left(\frac{1}{x^7}\right) \quad \text{for} \quad x \to \infty. \tag{1.110b}$$

Similarly, by (1.110b),

$$\sqrt{x}\,\mathsf{daw}\left(\sqrt{x}\right) = \frac{1}{2} + \frac{1}{4x} + \frac{3}{8x^2} + O\left(\frac{1}{x^3}\right) \quad \text{for} \quad x \to +\infty. \tag{1.111a}$$

Moreover, since

$$\mathsf{erf}(x) \to 1 \quad \text{for} \quad x \to \infty,$$

it follows from (1.109) that

$$\sqrt{-x}\,\mathsf{daw}\left(\sqrt{-x}\right) \approx -\frac{\sqrt{\pi}}{2}\sqrt{x}e^{x} \quad \text{for} \quad x \to +\infty.$$

A graph of the function daw on the real line is shown in Figure 1.5. The two stationary points of daw, a maximum and a minimum, are at $x = \pm x_0$, respectively, with $x_0 \doteq 0.924$.

We now illustrate the relevance of DAWSON's integral daw to the analysis of the function $\widehat{f}_{\mathrm{u}}$ in (1.107). By setting $u := \sin\vartheta$ in (1.107), we easily arrive at

$$\widehat{f}_{\mathrm{u}}(\beta, S) = \frac{1}{3}S^2 + \frac{1}{3}S - \frac{1}{\beta}\ln\int_0^1 e^{\beta S u^2}\,du. \tag{1.112}$$

Letting $y := \sqrt{\beta S}u$, by (1.108), we also obtain that

$$\int_0^1 e^{\beta S u^2}\,du = \frac{1}{\sqrt{\beta S}}\int_0^{\sqrt{\beta S}} e^{y^2}\,dy = e^{\beta S}\frac{\mathsf{daw}\left(\sqrt{\beta S}\right)}{\sqrt{\beta S}},$$

which gives (1.112) the following concise form:

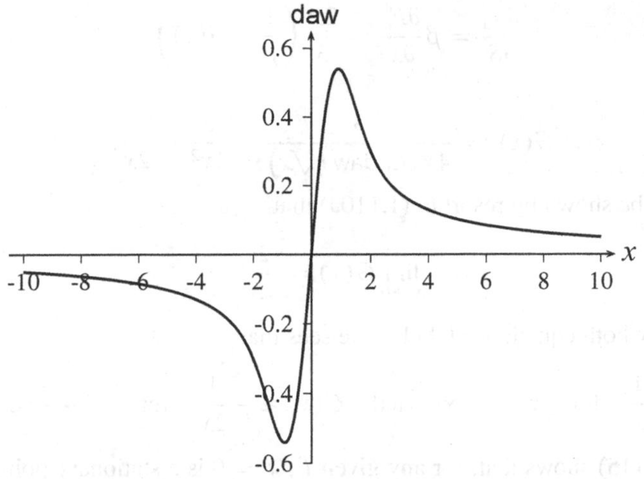

Fig. 1.5. The function $\mathsf{daw}(x)$ is plotted in an interval of $\mathbb{R}$, symmetric with respect to the origin and sufficiently wide to capture the asymptotic behavior in (1.110b).

$$\widehat{f_\mathrm{u}}(\beta, S) = \frac{1}{3}S^2 - \frac{2}{3}S - \frac{1}{\beta}\ln\left(\frac{\mathsf{daw}\left(\sqrt{\beta S}\right)}{\sqrt{\beta S}}\right). \tag{1.113}$$

In the next subsection, this representation of $\widehat{f_\mathrm{u}}$ will be instrumental to the stability analysis of the critical points of the MAIER–SAUPE free energy.

1.3.3 Stability Analysis

We have shown in the preceding subsection that all critical points in $\mathbf{Q}$ of the MAIER–SAUPE dimensionless free energy per particle $F_0(\beta, \mathbf{Q})$ in (1.88) correspond to the critical points of the function $\widehat{f_\mathrm{u}}(\beta, S)$ defined in (1.107) as the restriction to the line $T = 0$ of the function $f_\mathrm{u}(\beta, S, T)$ introduced in (1.96), which expresses F_0 when $\mathbf{Q}$ is represented in terms of the scalar order parameters (S, T) recalled in (1.91). Here we determine all critical points of $\widehat{f_\mathrm{u}}$ for every value of β and probe their stability against perturbations in S; the stability of the corresponding critical points of f_u on the line $T = 0$ will accordingly be probed against perturbations in T.

By letting

$$x := \beta S, \tag{1.114}$$

we easily recast $\widehat{f_\mathrm{u}}$ in (1.113) in the form

$$F_\mathrm{u}(\beta, x) := \widehat{f_\mathrm{u}}\left(\beta, \frac{x}{\beta}\right) = \frac{1}{\beta}\left\{\frac{1}{3\beta}x^2 - \frac{2}{3}x - \ln\left(\frac{\mathsf{daw}\left(\sqrt{x}\right)}{\sqrt{x}}\right)\right\},$$

where, by (1.109), F_u is defined for all $x \in \mathbb{R}$. By the chain rule,

$$\frac{\partial \widehat{f}_{\mathrm{u}}}{\partial S} = \beta \frac{\partial F_{\mathrm{u}}}{\partial x} = \frac{2x}{3}\left(\frac{1}{\beta} - G(x)\right), \tag{1.115}$$

where

$$G(x) := \frac{3}{4x\sqrt{x}\,\mathsf{daw}\left(\sqrt{x}\right)} - \frac{3}{4x^2} - \frac{1}{2x}. \tag{1.116}$$

It can easily be shown by resort to (1.110a) that

$$\lim_{x\to 0} G(x) = \frac{2}{15}. \tag{1.117}$$

Moreover, by both equations (1.111), one sees that

$$G(x) \approx \frac{1}{x} \quad \text{for} \quad x \to +\infty \quad \text{and} \quad G(x) \approx -\frac{1}{2x} \quad \text{for} \quad x \to -\infty. \tag{1.118}$$

Equation (1.115) shows that, for any given β, $x = 0$ is a stationary point of F_{u}; all other stationary points are the roots of the equation

$$G(x) = \frac{1}{\beta}. \tag{1.119}$$

Figure 1.6 illustrates the graph of the function G. By this graph and asymptotic properties of G listed in (1.118), it is apparent that G possesses a single stationary point at $x = x_*$, which is a maximum; a numerical evaluation gives

$$x_* \doteq 2.178 \quad \text{and, correspondingly,} \quad G_* := G(x_*) \doteq 0.149.$$

Equation (1.119) has a rather transparent graphical interpretation: its roots can be identified with the intersections between the graph of G and the straight line $y = \frac{1}{\beta}$ in the Cartesian plane (x, y) of Figure 1.6. For $\frac{1}{\beta} > G_*$, there is no such intersection, and so F_{u} is stationary only at $x = 0$, which through (1.114) corresponds to the stationary point $S = 0$ for $\widehat{f}_{\mathrm{u}}$. For $\frac{1}{\beta} < G_*$, two extra stationary points emerge for F_{u}, since two intersections split on the graph of G from the point (x_*, G_*), $x_1 < x_*$ and $x_2 > x_*$, the former diverging to $-\infty$ and the latter to $+\infty$ as $\frac{1}{\beta}$ is reduced from G_* toward 0. Again by (1.114), two stationary points of $\widehat{f}_{\mathrm{u}}$, S_1 and S_2, correspond to x_1 and x_2.

To probe the local stability of all three stationary points of $\widehat{f}_{\mathrm{u}}$, that is, $S = 0$ and $S = S_{1,2}$, we compute the second derivative of $\widehat{f}_{\mathrm{u}}$ in S. By (1.115) and (1.114), we readily obtain that

$$\frac{\partial^2 \widehat{f}_{\mathrm{u}}}{\partial S^2} = \frac{2\beta}{3}\left\{\left(\frac{1}{\beta} - G(x)\right) - xG'(x)\right\}. \tag{1.120}$$

Evaluating the right-hand side of (1.120) at $x = 0$, by (1.117) we learn that

$$\left.\frac{\partial^2 \widehat{f}_{\mathrm{u}}}{\partial S^2}\right|_{S=0} \geqq 0 \quad \Leftrightarrow \quad \frac{1}{\beta} \geqq \frac{2}{15}.$$

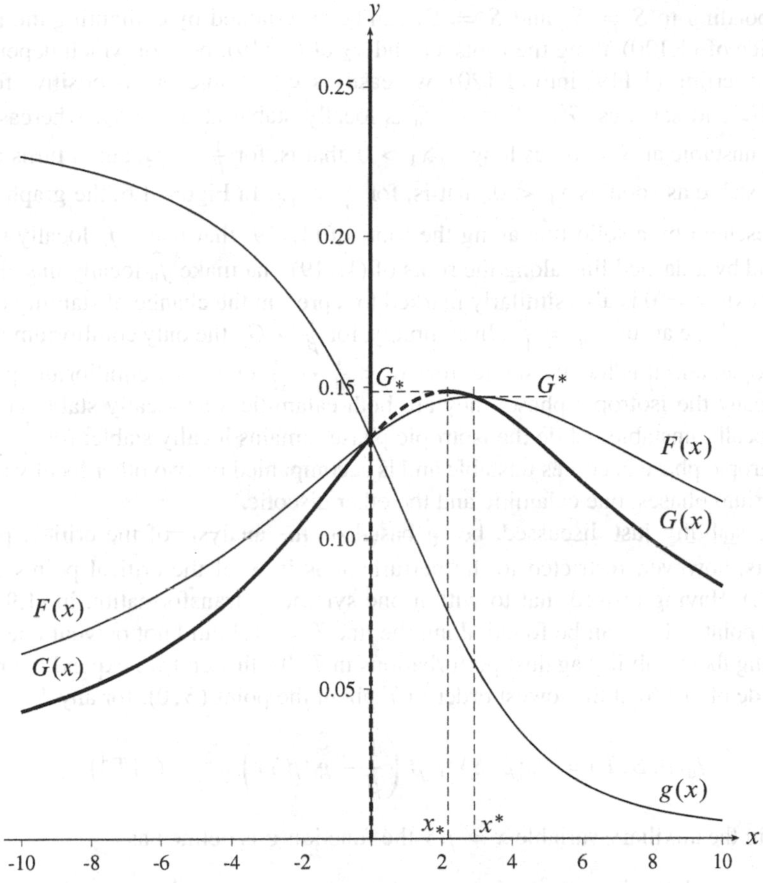

Fig. 1.6. The graphs of the functions G, g, and F defined by (1.116), (1.121), and (1.125), respectively, are plotted against x in the Cartesian plane (x, y). The intersections between the line $y = \frac{1}{\beta}$ and the graph of G represent all possible equilibrium uniaxial phases. The isotropic phase, which is at equilibrium for all values of the reduced temperature $\frac{1}{\beta}$, is represented by the line $x = 0$. Solid lines correspond to equilibria locally stable against uniaxial perturbations; dashed lines represent equilibria unstable against the same perturbations. Where the graph of G lies below the graph of F, the uniaxially stable equilibria bear less energy than the isotropic phase. However, below the graph of g all equilibria are unstable against biaxial perturbations. Thus, only the positive roots of (1.119) can be fully stable. The maximum of G falls at $x = x_* \doteq 2.178$; correspondingly, $G_* := G(x_*) \doteq 0.149$. In the plane (x, y), (x^*, G^*) is the only point besides $\left(0, \frac{2}{15}\right)$ where the graphs of G and F cross; its coordinates are $x^* \doteq 2.923$ and $G^* := G(x^*) \doteq 0.147$.

This shows that the isotropic phase, which according to the MAIER–SAUPE theory is an equilibrium phase for all values of β, is locally stable for $\frac{1}{\beta} > \frac{2}{15}$ and locally unstable for $\frac{1}{\beta} < \frac{2}{15}$. For $\frac{1}{\beta} < G_*$, the local stability of the equilibrium phases corresponding to $S = S_1$ and $S = S_2$ can be ascertained by computing the right-hand side of (1.120) along the roots x_1 and x_2 of (1.119), both of which depend on β. By inserting (1.119) into (1.120), we easily see that since x_2 is positive for all $\frac{1}{\beta} < G_*$ and satisfies $G'(x_2) < 0$, $\widehat{f_u}$ is locally stable at $S = S_2$, whereas it is locally unstable at $S = S_1$ as long as $x_1 > 0$, that is, for $\frac{1}{\beta} > \frac{2}{15}$, and it turns again locally stable as soon as $x_1 < 0$, that is, for $\frac{1}{\beta} < \frac{2}{15}$. In Figure 1.6, the graph of G is represented by a solid line along the roots of (1.119) that make $\widehat{f_u}$ locally stable in S and by a dashed line along the roots of (1.119) that make $\widehat{f_u}$ locally unstable in S. The axis $x = 0$ is also similarly marked to represent the change of stability of the isotropic phase around $\frac{1}{\beta} = \frac{2}{15}$. In summary, for $\frac{1}{\beta} > G_*$ the only equilibrium phase is isotropic, and it is locally stable; for $G_* < \frac{1}{\beta} < \frac{2}{15}$ two other equilibrium phases accompany the isotropic phase; they are both calamitic, one locally stable and the other locally unstable, while the isotropic phase remains locally stable; for $\frac{1}{\beta} < \frac{2}{15}$, the isotropic phase becomes unstable and is accompanied by two other locally stable equilibrium phases, one calamitic and the other discotic.

The stability just discussed, being based on the analysis of the critical points of $\widehat{f_u}$, is, however, restricted to the perturbations in S of the critical points of f_u in (1.96). Having proved that to within one symmetry transformation in (1.92) all critical points of f_u can be found along the line $T = 0$ should not prevent one from exploring their stability against perturbations in T. To this end we expand the right-hand side of (1.96) at the lowest order in T about the point $(S, 0)$, for any S,

$$f_u(\beta, S, T) = \widehat{f_u}(\beta, S) + \beta\left(\frac{1}{\beta} - g(\beta S)\right)T^2 + O\left(T^4\right),$$

where in the auxiliary variable $x = \beta S$ the function g is defined as

$$g(x) := \frac{1}{4} + \frac{1}{4x} + \frac{3}{16x^2} - \frac{1}{8\sqrt{x}\,\mathsf{daw}\left(\sqrt{x}\right)} - \frac{3}{16x\sqrt{x}\,\mathsf{daw}\left(\sqrt{x}\right)}. \tag{1.121}$$

At a critical point S of $\widehat{f_u}$, $x = \beta S$ makes F_u stationary. To assess the stability in T of f_u, we need to see whether

$$\frac{1}{\beta} - g(x) > 0 \tag{1.122}$$

at either $x = 0$ or at the roots of (1.119). It can be checked with the aid of (1.110) and (1.111) that

$$\lim_{x\to 0} g(x) = \frac{2}{15}, \quad \lim_{x\to -\infty} g(x) = \frac{1}{4}, \quad \text{and} \quad g(x) \approx \frac{1}{2x^2} \quad \text{for} \quad x \to +\infty. \tag{1.123}$$

Thus, it follows from (1.122) that the isotropic phase is locally stable and locally unstable against biaxial perturbations precisely in the same ranges of $\frac{1}{\beta}$ where it is locally stable and locally unstable against uniaxial perturbations, that is, for $\frac{1}{\beta} > \frac{2}{15}$

and $\frac{1}{\beta} < \frac{2}{15}$, respectively. For the equilibrium phases represented by the roots of (1.119), by combining (1.119) with (1.122) we can reduce this latter stability inequality to a graphical criterion: the roots of (1.119) that fall where the graph of G lies above the graph of g are locally stable against biaxial perturbations, whereas the roots of (1.119) that fall where the graph of G lies below the graph of g are locally unstable against biaxial perturbations. Figure 1.6 along with the asymptotic behaviors in (1.118) and (1.123) allows us to conclude that the graphs of G and g cross at a single point, that is, at $x = 0$. Moreover, all negative roots of (1.119), corresponding to discotic equilibrium phases locally stable against uniaxial perturbations, are locally unstable against biaxial perturbations; the roots in the interval $]0, x_*[$, which are unstable against uniaxial perturbations, would instead be locally stable against biaxial perturbations, whereas the roots $x_2 > x_*$ of (1.119) are locally stable against both uniaxial and biaxial perturbations. The equilibrium calamitic phases represented by these latter roots are thus competing with the isotropic phase for the absolute minimizer of the MAIER–SAUPE free energy f_{u} for $\frac{1}{\beta}$ in the interval $\left[G_*, \frac{2}{15}\right]$, where they are both locally stable.

To identify the absolute free enery minimizer we need to determine whether, for given β, the root $x_2(\beta)$ of (1.119) is such that

$$F_{\mathrm{u}}(\beta, x_2(\beta)) < \lim_{x \to 0} F_{\mathrm{u}}(\beta, x); \tag{1.124}$$

whenever this inequality is satisfied, the pair $(S, T) = (S_2, 0)$, with $S_2(\beta) := \frac{x_2(\beta)}{\beta}$, prevails over the pair $(S, T) = (0, 0)$ as the absolute minimizer of f_{u}. Since, by (1.110a),

$$\lim_{x \to 0} F_{\mathrm{u}}(\beta, x) = 0,$$

inequality (1.124) becomes

$$\frac{1}{\beta} < \frac{1}{x_2(\beta)} + \frac{3}{x_2(\beta)^2} \ln\left(\frac{\mathrm{daw}\left(\sqrt{x_2(\beta)}\right)}{\sqrt{x_2(\beta)}}\right) =: F(x_2(\beta)). \tag{1.125}$$

Again inserting (1.119) in (1.125), we change the latter into

$$G(x_2) < F(x_2), \tag{1.126}$$

which can be given a simple graphical interpretation: for any prescribed β, the absolute minimizer of f_{u} is $(S_2(\beta), 0)$ whenever $x_2(\beta)$ falls where the graph of G lies below the graph of F, whereas it is $(0, 0)$ whenever $x_2(\beta)$ falls where the graph of G lies above the graph of F. Though it would strictly suffice to explore the crossing of these graphs only within the interval $\left[G_*, \frac{2}{15}\right]$ in their ranges, we find it convenient to widen this study to the whole of their ranges. By use of (1.111a) in the definition (1.125) of F, we obtain the following asymptotic estimate:

$$F(x) \approx \frac{2}{x} \quad \text{for} \quad x \to \infty,$$

while (1.110a) leads us to

$$\lim_{x\to 0} F(x) = \frac{2}{15}.$$

These properties of F, combined with the perusal of the graphs of G and F shown in Figure 1.6, ensure that (1.126) is satisfied for all $x_2 > x^*$, since $(x^*, G(x^*))$ is the only point besides $\left(0, \frac{2}{15}\right)$ that the graphs of G and F have in common. A numerical computation shows that

$$x^* \doteq 2.923 \quad \text{and} \quad G^* := G(x^*) \doteq 0.147.$$

We conclude that upon decreasing the reduced temperature $\frac{1}{\beta}$ through the supercritical value G_*, the minimizer of the MAIER–SAUPE free energy f_u is still the isotropic phase, but the uniaxial calamitic phase described by the pair $(S_2(\beta), 0)$ becomes locally stable; at the critical reduced temperature $\frac{1}{\beta} = G^*$, the minimizer of f_u jumps from the isotropic phase to the uniaxial calamitic phase, while the isotropic phase remains locally stable up to the subcritical temperature $\frac{1}{\beta} = \frac{2}{15}$. An instructive illustration of this *first-order* transition is provided by the graph against the reduced temperature $\frac{1}{\beta}$ of the scalar order parameter S, which describes both the isotropic and the uniaxial equilibrium phases. By (1.114) and (1.119), the function $\frac{1}{\beta} \mapsto S_2(\beta)$ can be given the following parametric form:

$$\frac{1}{\beta} = G(x), \quad S_2(x) = xG(x), \quad \text{for} \quad x > 0. \tag{1.127}$$

The graph of this function is plotted in Figure 1.7: a dashed line represents the uni-

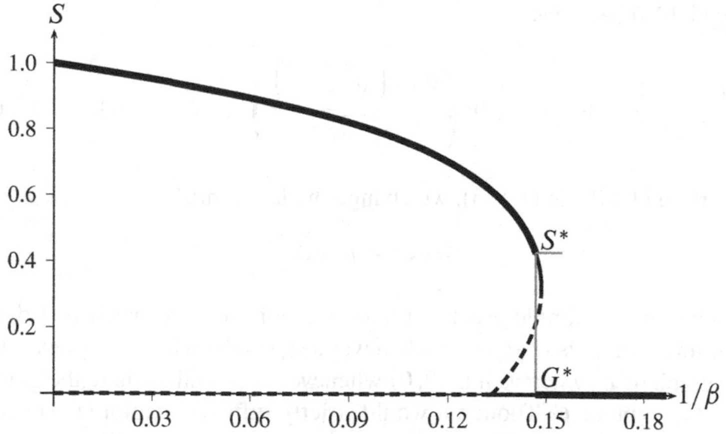

Fig. 1.7. The uniaxial scalar order parameter S as a function of the reduced (dimensionless) temperature $\frac{1}{\beta}$. Solid lines represent locally stable equilibria; dashed lines represent locally unstable equilibria. A heavier line marks the absolute free energy minimizer, which displays a first-order transition at $\frac{1}{\beta} = G^* \doteq 0.147$, where S jumps from 0 to $S^* \doteq 0.429$.

axial phase in the temperature range where it is locally unstable, while a solid line represents it in the temperature range where it is locally stable. The local stability of the isotropic phase is represented likewise. It follows from (1.118) and from taking the limit as $x \to \infty$ in (1.127) that $S_2 \to 1$ as $\frac{1}{\beta} \to 0$, as also shown in Figure 1.7. A heavier line identifies the global free energy minimizer; the first-order transition at $\frac{1}{\beta} = G^*$ is marked by a thin solid line bridging the jump in S from 0 to S^*. By (1.127), one readily computes

$$S^* = x^* G^* \doteq 0.429.$$

The prediction of such a phase transition taking place with a universal increase in the uniaxial scalar order parameter is the most remarkable achievement of the MAIER–SAUPE mean-field theory.

In the following section, we shall briefly review a fundamental criticism raised against the MAIER–SAUPE theory and indicate a way to combine the purely attractive character of the interaction Hamiltonian (1.74) on which the theory is based with some repulsive features. Despite all criticisms, the MAIER–SAUPE theory remains exemplary for both its simplicity and the insight that it provides.

1.4 Steric Effects

Molecular interactions are thought to determine the ability of ordered phases to emerge in certain anisotropic fluids. Perhaps the most telling illustration of this paradigm is the isotropic-to-nematic transition in liquid crystals described in the preceding section, where we substantiated the picture often drawn that describes liquid crystal molecules as rods or ribbons subject to interactions that tend to make them align alike. In general, whenever such a tendency prevails over disorganizing causes, an ordered phase is established from the isotropic, disordered phase. This ordering phase transition, which is usually first-order, induces a local common molecular orientation.

A satisfactory microscopic theory for liquid crystals must be based on the interactions exchanged by the constituent molecules. Different special models for molecular interactions have been proposed in the last decades. We shall be contented in this book with describing in detail only the MAIER–SAUPE theory ; for an account of other theories and the still unceasing debate around them; the interested reader is referred, for example, to the review article [304].

We learned already that in a mean-field approach, a single molecule is envisaged as immersed in a field produced by the averaged action of all other molecules that surround it. The key ingredient to a mean-field theory is the *pair-potential* H, which is the interaction energy U of two molecules averaged over the molecules' relative position. H, also called the two-particle Hamiltonian, has so far been the basis of our development (see Section 1.1.1). In this section, we explore to some extent how H is related to U and the assumptions involved in both positing the latter and deriving the former from it. For molecules described as rigid particles, U in general depends on the vector joining the centers of charge of the interacting molecules and on the relative molecular orientation. For flexible molecules, U is a more complicated function that also depends on the molecular conformations.[18]

The existing interaction models for liquid crystal molecules can be divided into three broad categories: short-range and repulsive, long-range and attractive, and VAN DER WAALS . Correspondingly, these models attribute the collective aligning attitude of molecules to three different mechanisms: to the mutual hindrance of molecules that reflects the anisotropy of their shape, to the dispersion interaction that reflects the anisotropy induced by their oscillating charges [193, 215], and to the coexistence of the former effects in an interaction energy that accounts for both short- and long-range forces.

The ONSAGER hard-core interaction for long rodlike molecules [258] is the most noticeable example in the first category. It is an athermal model, where the ordering transition is driven by increasing density, instead of decreasing temperature. In a way, this model properly describes lyotropic liquid crystals, since the interaction energy mimics the pure steric repulsion of molecules. Nonetheless, it can legitimately be presented on the same footing as the models in the other two categories introduced above, as a limiting case of extremely short-range interactions. Though conceptually appealing, the ONSAGER model fails to represent the isotropic-to-nematic transition faithfully.

Far more successful in this regard is the MAIER–SAUPE model , the simplest and most celebrated example of long-range dispersion models (see Section 1.3). Crucial to the justification of the mean field associated with this model interaction is the assumption that the molecules are isotropically distributed around every probe molecule. As remarked in [70], a relatively small deviation from spherical symmetry already causes the ordering phase transition to disappear.

A recognized limitation of the purely dispersive MAIER–SAUPE theory is its complete neglect of short-range interactions. A remedy to this was provided by the theory of GELBART & BARON [119, 55], where an anisotropic, short-range, repulsive interaction is incorporated in the model alongside a long-range, attractive interaction, which need not be anisotropic. This theory is often referred to as the generalized VAN DER WAALS theory; it is computationally demanding and has been explicitly worked out only for special repulsive potentials. However, it has clearly shown that the anisotropy in the resulting pair-potential is due mostly to the interplay between two components of the parent interaction, namely, the repulsive potential and the isotropic part of the attractive potential [364, 359, 89].

Many other models and generalized theories have been proposed. We refer the interested reader to specialized reviews [118, 117, 304] that also illustrate the intellectual wealth stimulated by the desire to understand in depth the nematic ordering transition.

---

18 Here we consider only molecules described as rigid particles. A mean-field treatment for flexible biaxial molecules can be found in a recent paper by LUCKHURST [201]. In a different vein, molecular shape fluctuations are also considered in [195].

Often a unifying view is gained by a wise blend of symmetry and averaging. If every molecule is isotropically surrounded by all others, the pair-potential H introduced in Section 1.1.1 is obtained by averaging the interaction energy U of a given molecule with respect to all others with the same orientation relative to the selected molecule. Formalizing rigorously this averaging is indeed less trivial than it may appear: the major difficulty resides in handling the divergence of U when the distance between the interacting molecules approaches zero. Such a divergence embodies the ultimate short-range repulsion between molecules: even in the simplest realization of this repulsion, that is, in the hard-core interaction, the average over the intermolecular distance contributes to the dependence of the pair-potential on the relative molecular orientation. This is precisely the avenue taken in this section. We combine the long-range induced dipole–dipole interactions with a short-range, hard-core interaction and we compute the resulting pair-potential H, whose anisotropy stems now from both long- and short-range components of the interaction energy U.

An alternative approach was proposed by LUCKHURST & ZANNONI [203]. They reconciled the antagonism between short-range, repulsive interactions and long-range, attractive interactions, by assuming that the former are responsible for the local organization of molecules in clusters, which in turn are subject to the latter. This syncretic view holds that the molecular clusters bound by short-range interactions are not destroyed at the transition where their long-range organization changes, and thus survive in both the isotropic and nematic phases. According to this view, not molecules but stable clusters would be subject to the pair-potential. In either interpretation, our formal development remains unaffected.

In this section, following [313], we write in a compact form the interaction due to dispersion forces and we describe the excluded region $\mathfrak{R}^*$, the region in space that a molecule cannot access because of the presence of another molecule. A *steric tensor* will then be defined in terms of $\mathfrak{R}^*$ that embodies the anisotropy of the steric interactions. We also show how to construct the excluded region starting from a given molecular shape. For a special class of shapes, this construction is carried out explicitly, and the steric tensor is computed analytically in Section 1.4.4. In Section 1.4.5, the steric effect is finally determined for two classical dispersive interactions: for uniaxial and biaxial molecules, respectively.

1.4.1 Dispersion Forces

Deriving the dispersion energy for the long-range induced dipole–dipole interaction of two molecules from quantum-mechanical perturbation theory requires resorting to a number of approximations if one wishes an explicitly computable formula. In the account given by STONE [320] (see, in particular, Section 4.3.2), one approximation plays a dominant role: this is the UNSÖLD approximation [350] , also called the *average-energy* approximation, as employed by LONDON [192]. In the approach of BUCKINGHAM [38], this approximation amounts to assuming that all states in the molecules that contribute to their dispersion interaction have excitation energies close to the same average, which we correspondingly denote by E and E', for each molecule.

Letting $\mathbf{A}$ and $\mathbf{A}'$ be the symmetric tensors representing the polarizabilities of the interacting molecules, we give the approximate dispersion energy the following compact form:

$$U_{\mathrm{d}} = -\frac{C_0}{r^6}(\mathbf{U} \boxtimes \mathbf{U})[\mathbf{A}] \cdot \mathbf{A}', \tag{1.128}$$

where

$$C_0 = \frac{9EE'}{4(E+E')(4\pi\epsilon_0)^2}$$

with ϵ_0 the dielectric constant in vacuum. In (1.128),

$$r := |p_0' - p_0|$$

is the distance between the charge centers p_0' and p_0 of the two molecules, and the uniaxial, second-rank tensor $\mathbf{U}$ is built from the unit vector

$$\boldsymbol{e}_r := \frac{1}{r}(p_0' - p_0),$$

directed from p_0 to p_0', according to the definition

$$\mathbf{U} = \mathbf{U}(\boldsymbol{e}_r) := \boldsymbol{e}_r \otimes \boldsymbol{e}_r - \frac{1}{3}\mathbf{I}. \tag{1.129}$$

For two given second-rank tensors $\mathbf{A}$ and $\mathbf{B}$, the fourth-rank tensor $\mathbf{A} \boxtimes \mathbf{B}$ is defined by its action on an arbitrary second-rank tensor $\mathbf{C}$: it delivers the second-rank tensor defined by [71, 273]

$$(\mathbf{A} \boxtimes \mathbf{B})[\mathbf{C}] := \mathbf{A}\mathbf{C}\mathbf{B}^{\mathsf{T}} \quad \text{for all } \mathbf{C}\,.$$

Moreover, the inner product denoted in (1.128) by a dot $\cdot$ is defined as in (1.8) (and in Appendix A.1).

Equation (1.128) is valid under the assumption that certain oscillators in one molecule, all with frequencies very close to one another, are coupled with similar oscillators in the other molecule. When the oscillators that contribute to the interaction in each molecule have quite different frequencies, the total dispersion energy U_{d} acquires several terms, all in the form (1.128). In the case of N such distinct oscillators, the dispersion energy is

$$U_{\mathrm{d}} = -\frac{1}{r^6}\sum_{h,k=1}^{N} C_{hk}(\mathbf{U} \boxtimes \mathbf{U})[\mathbf{A}_h] \cdot \mathbf{A}_k', \tag{1.130}$$

where $\mathbf{A}_h$ and $\mathbf{A}_k'$ are the polarizability tensors corresponding in each molecule to the coupled oscillators, and

$$C_{hk} = \frac{9E_hE_k'}{4(E_h+E_k')(4\pi\epsilon_0)^2},$$

E_h and E'_k being the energies of the coupled states.

In the following, we shall build upon (1.128) our explicit representation of the steric effects in dispersion force interactions, assuming, for simplicity, that, in each molecule, essentially a single oscillator is involved in the interaction. The general case would then follow by superimposing all individual dispersion interactions including their steric corrections.

1.4.2 Excluded Region

U_{d} is a potential energy of *soft forces*. For neutral, nonpolar molecules, it is the first term in a multipolar expansion, valid only if p_0 and p'_0 are sufficiently far apart. These long-range forces are complemented by short-range *hard forces*, which represent the steric hindrance to molecular interactions. While dispersion forces are attractive, as are most long-range forces hard, steric forces are repulsive. We imagine a simple picture to describe these latter: we think of the charge centers p_0 and p'_0 as surrounded by three-dimensional regions, $\mathfrak{R}$ and $\mathfrak{R}'$, respectively, which represent the ranges of the repulsive hard forces. These essentially make $\mathfrak{R}$ and $\mathfrak{R}'$ impenetrable to one another, while they are dormant whenever $\mathfrak{R}$ and $\mathfrak{R}'$ are not in contact. $\mathfrak{R}$ and $\mathfrak{R}'$, which we call the VAN DER WAALS regions for the two molecules, reflect the molecular shapes, though they need not coincide with them.

Molecular interactions are ultimately responsible for the mesogenic behavior of some molecules which, unlike others, tend to form ordered phases. Often, a theoretical understanding of these ordering transitions is achieved within the mean-field approximation, as in the MAIER–SAUPE theory illustrated in Section 1.3 above. Replacing the space-dependent dispersion energy (1.128) with a space-independent one [70] is crucial to the success of this theory. This is achieved by assuming that molecules with the same relative orientation are isotropically distributed in space around any given probe molecule and by computing an effective interaction energy between the probe and all other molecules.

The interaction energy U_{d} in (1.128) depends via $\boldsymbol{e}_r$ on the relative position of the two molecules and via $\mathbf{A}$ and $\mathbf{A}'$ on their relative orientation. The relative hindrance of the VAN DER WAALS regions introduces in the effective intermolecular forces a dependence upon the relative molecular orientation subtler than the one explicitly appearing in (1.128). Following [313], we now make this idea more precise. As shown in Figure 1.8, for given $\mathfrak{R}$ and $\mathfrak{R}'$, there is a region $\mathfrak{R}^*$ in space, depending on $\mathfrak{R}$ and $\mathfrak{R}'$, inaccessible to the charge center p'_0 if $\mathfrak{R}$ and $\mathfrak{R}'$ are mutually impenetrable. We call $\mathfrak{R}^*$ the *excluded region*. As suggested by Figure 1.8, the boundary $\partial\mathfrak{R}^*$ of the excluded region is traced by p'_0, while $\partial\mathfrak{R}'$ glides without rolling over $\partial\mathfrak{R}$. Similarly, the region inaccessible to p_0 by the impenetrability of $\mathfrak{R}'$ is traced by all possible trajectories described by p_0, while $\partial\mathfrak{R}$ glides without rolling over $\partial\mathfrak{R}'$. Since in both cases the relative motion between $\mathfrak{R}$ and $\mathfrak{R}'$, regarded as rigid bodies, is purely translational, the excluded regions obtained in these two ways differ simply by a translation.

The molecular distribution in space will be taken to be homogeneous. This allows us to define the effective *dispersion pair-potential* H_{d} as the average dispersion

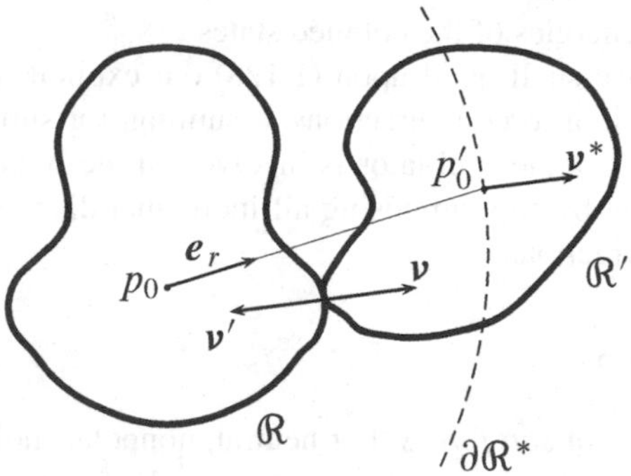

Fig. 1.8. The VAN DER WAALS regions $\mathcal{R}$ and $\mathcal{R}'$ surrounding the charge centers p_0 and p_0' of the interacting molecules. The unit vector $\boldsymbol{e}_r$ is directed from p_0 to p_0'; $\boldsymbol{\nu}$ and $\boldsymbol{\nu}'$ are the unit outer normals to $\partial\mathcal{R}$ and $\partial\mathcal{R}'$, respectively. The boundary $\partial\mathcal{R}^*$ of the excluded region $\mathcal{R}^*$ is enveloped by p_0', while $\partial\mathcal{R}'$ glides without rolling over $\partial\mathcal{R}$. The unit vector $\boldsymbol{\nu}^*$ is the outer normal to $\partial\mathcal{R}^*$. The region enveloped by all possible trajectories described by p_0 while $\partial\mathcal{R}$ glides without rolling over $\partial\mathcal{R}'$ would differ from $\mathcal{R}^*$ only by a translation.

energy U_d exchanged between two molecules with a given relative orientation, while their relative position varies freely in space. Any two such molecules share the same excluded region $\mathcal{R}^*$. To account effectively for the presence of more than a pair of molecules in the system, we imagine that an infinite number of molecules, all equally oriented, are uniformly distributed in space so that the same number of molecules, N_{mac}, will be present in the same macroscopic volume V_{mac}. Let a probe molecule wander about the molecules of this system while keeping its orientation unchanged. For any given molecule in the system, the total energy exchanged with the probe molecule can be computed by imagining this latter exploring an influence ball $\mathbb{B}_i$ with radius R_i around the given molecule and then taking the limit as $R_i \to \infty$. Repeating this argument for each molecule in the system reproduces the same result, given the homogeneity of the molecular distribution, and so the average energy is finally estimated by multiplying the total energy exchanged between a single molecule and its probe companion by the number density $\varrho := N_{mac}/V_{mac}$.

Making precise the above definition for H_d, we obtain from (1.128) that

$$H_d = -C_0\varrho\left(\lim_{R_i\to\infty}\int_{\mathbb{B}_i\setminus\mathcal{R}^*}\frac{1}{r^6}\mathbf{U}\boxtimes\mathbf{U}\,dV\right)[\mathbf{A}]\cdot\mathbf{A}', \tag{1.131}$$

where V is the volume measure in the three-dimensional Euclidean space $\mathcal{E}$. Differently said, (1.131) can be obtained by integrating over the whole admissible space the interaction U_d in (1.128) multiplied by the probability of finding an interacting molecule at any given point in space, that is, the number density ϱ.

The excluded region $\mathcal{R}^*$ defined above through the kinematic construction illustrated in Figure 1.8 is indeed subtler than the definition of H_d in (1.131) may at first glance suggest. This latter could also be valid if $\mathcal{R}^*$ were defined only as the region in space inaccessible to the charge center p_0' when the VAN DER WAALS region $\mathcal{R}'$

is approached in all radial directions e_r until it is in contact with $\mathcal{R}$ while keeping its orientation unaltered. The region thus obtained, which depends on the choice of p_0 and p_0' and on the relative orientation of $\mathcal{R}$ and $\mathcal{R}'$, may be called the *radially* excluded region $\mathcal{R}_\mathrm{r}^*$. It can be seen by example that there are classes of shapes $\mathcal{R}$ and $\mathcal{R}'$ for which $\mathcal{R}_\mathrm{r}^*$ and $\mathcal{R}^*$ differ. A simple example can be obtained when $\mathcal{R}$ is *not* star-shaped[19] relative to p_0 and $\mathcal{R}'$ is a ball of sufficiently small radius. As is easily shown, $\mathcal{R}_\mathrm{r}^*$ is star-shaped by construction, while $\mathcal{R}^*$ may fail to be so.

In principle, $\mathcal{R}^*$ can be defined for two arbitrary regions $\mathcal{R}$ and $\mathcal{R}'$, as in Figure 1.8. However, in our case the two interacting molecules are identical, so that $\mathcal{R}$ and $\mathcal{R}'$ differ only by a rigid rotation $\mathbf{R}$, as do correspondingly $\mathbf{A}$ and $\mathbf{A}' = \mathbf{R}\mathbf{A}\mathbf{R}^\mathsf{T}$. Thus, H_d ultimately depends on $\mathbf{R}$: explicitly through $\mathbf{A}'$, and implicitly through $\mathcal{R}^*$.

When $\mathcal{R}^*$ is a ball of radius R, the integral in (1.131) can be evaluated directly, and one obtains

$$H_\mathrm{d} = -C_0\varrho\left(\int_R^\infty \frac{1}{r^4}\int_{\mathbb{S}^2} \mathbf{U}\boxtimes\mathbf{U}\,dA\right)[\mathbf{A}]\cdot\mathbf{A}' = -\frac{4\pi C_0\varrho}{3R^3}\langle\mathbf{U}\boxtimes\mathbf{U}\rangle_{\mathbb{S}^2}[\mathbf{A}]\cdot\mathbf{A}',$$

where A is the area measure over the unit sphere $\mathbb{S}^2$, and $\langle\cdots\rangle_{\mathbb{S}^2}$ denotes the average over it:

$$\langle\cdots\rangle_{\mathbb{S}^2} := \frac{1}{4\pi}\int_{\mathbb{S}^2}(\cdots)\,dA(e_r).$$

By symmetry, $\langle\mathbf{U}\boxtimes\mathbf{U}\rangle_{\mathbb{S}^2}$ is a linear combination of isotropic fourth-rank tensors. An explicit computation shows that[20]

$$\langle\mathbf{U}\boxtimes\mathbf{U}\rangle_{\mathbb{S}^2}[\mathbf{A}]\cdot\mathbf{A}' = \frac{1}{45}[\mathbf{A}\cdot\mathbf{A}' + 3(\operatorname{tr}\mathbf{A})(\operatorname{tr}\mathbf{A}')],$$

and so, up to a constant that is independent of the relative orientation of the molecules,

$$H_\mathrm{d} = -\frac{4\pi C_0\varrho}{135R^3}\mathbf{A}\cdot\mathbf{A}'. \tag{1.132}$$

This formula can be further simplified by introducing the traceless parts $\mathbf{A}_0$ and $\mathbf{A}_0'$ of $\mathbf{A}$ and $\mathbf{A}'$, respectively, according to

$$\mathbf{A} = \mathbf{A}_0 + \frac{1}{3}(\operatorname{tr}\mathbf{A})\mathbf{I} \quad\text{and}\quad \mathbf{A}' = \mathbf{A}_0' + \frac{1}{3}(\operatorname{tr}\mathbf{A})\mathbf{I}. \tag{1.133}$$

By (1.133),

$$\mathbf{A}\cdot\mathbf{A}' = \mathbf{A}_0\cdot\mathbf{A}_0' + \frac{1}{3}(\operatorname{tr}\mathbf{A})^2,$$

and in (1.132) we can replace $\mathbf{A}$ and $\mathbf{A}'$ by $\mathbf{A}_0$ and $\mathbf{A}_0'$, only altering H_d by an inessential constant. If the polarizability $\mathbf{A}$ is uniaxial about a molecular axis e_1,

[19] A region $\mathcal{R}$ is *star-shaped* relative to a point $p_0 \in \mathcal{R}$ if, for any point $p \in \mathcal{R}$, the whole segment joining p_0 and p is also contained in $\mathcal{R}$. See also equation (1.135) below for an explicit representation of a star-shaped region.

[20] The reader is referred to [313] for the details of this computation.

$$\mathbf{A} = \alpha_\parallel \boldsymbol{e}_1 \otimes \boldsymbol{e}_1 + \alpha_\perp(\mathbf{I} - \boldsymbol{e}_1 \otimes \boldsymbol{e}_1),$$
$$\mathbf{A}_0 = (\alpha_\parallel - \alpha_\perp)\left(\boldsymbol{e}_1 \otimes \boldsymbol{e}_1 - \frac{1}{3}\mathbf{I}\right), \tag{1.134}$$

where $\alpha_\parallel$ and $\alpha_\perp$ denote the polarizabilities along the symmetry axis and perpendicular to it, then (1.132) yields the classical uniaxial MAIER–SAUPE interaction as in (1.9), since

$$H_\mathrm{d} = -\frac{4\pi C_0 \varrho}{135 R^3}(\alpha_\parallel - \alpha_\perp)^2\left[(\boldsymbol{e}_1 \cdot \boldsymbol{e}_1')^2 - \frac{1}{3}\right],$$

where $\boldsymbol{e}_1' = \mathbf{R}\boldsymbol{e}_1$. As we shall see below in Section 1.4.5, a deviation from the spherical shape of the excluded region $\mathcal{R}^*$ will entail a steric correction to the MAIER–SAUPE theory.

Steric Tensor

Henceforth we assume that $\mathcal{R}$ is such that the excluded region $\mathcal{R}^*$ is *star-shaped* relative to p_0, that is, it can be represented as[21]

$$\mathcal{R}^* = \{p_0' \in \mathcal{E} : |p_0' - p_0| < u^*(\boldsymbol{e}_r)\}. \tag{1.135}$$

Here the *shape function* u^* is defined in such a way that the mapping $\boldsymbol{e}_r \mapsto u^*(\boldsymbol{e}_r)\boldsymbol{e}_r$ maps the unit sphere $\mathbb{S}^2$ around p_0 into $\partial\mathcal{R}^*$. In this case, the radial integration in (1.131) can be performed explicitly, and one finds that

$$H_\mathrm{d} = -\frac{4\pi C_0 \varrho}{3}\left\langle \frac{1}{u^{*3}}\mathbf{U} \boxtimes \mathbf{U}\right\rangle_{\mathbb{S}^2}[\mathbf{A}] \cdot \mathbf{A}' = -\frac{4\pi C_0 \varrho}{3}\mathbf{S}_{\mathcal{R}^*}[\mathbf{A}] \cdot \mathbf{A}', \tag{1.136}$$

where we have introduced the fourth-rank tensor

$$\mathbf{S}_{\mathcal{R}^*} := \left\langle \frac{1}{u^{*3}}\mathbf{U} \boxtimes \mathbf{U}\right\rangle_{\mathbb{S}^2}.$$

We call $\mathbf{S}_{\mathcal{R}^*}$ the *steric tensor* because it depends only on the shape of the excluded region and can in principle be computed once u^* is known. The steric tensor also plays a role in expressing the effective pair potential H_d in (1.130), valid when multiple molecular oscillators participate in the interaction. It readily follows from the reasoning that led us to (1.136) that

$$H_\mathrm{d} = -\frac{4\pi\varrho}{3}\sum_{h,k=1}^{N} C_{hk}\mathbf{S}_{\mathcal{R}^*}[\mathbf{A}_h] \cdot \mathbf{A}_k',$$

where $\mathbf{A}_h$ and $\mathbf{A}_k'$ are the polarizability tensors corresponding to each molecular oscillator.

Before we tackle in Section 1.4.4 the problem of computing $\mathbf{S}_{\mathcal{R}^*}$ for a specific class of molecular shapes, we first address the problem of how to determine the shape function u^* for a given molecular shape.

---

[21] In other words, we assume that the excluded region $\mathcal{R}^*$ coincides with the radially excluded region $\mathcal{R}_\mathrm{r}^*$.

1.4.3 Perturbative Method

Obtaining the excluded region $\mathfrak{R}^*$ from the VAN DER WAALS regions $\mathfrak{R}$ and $\mathfrak{R}'$ is not in general an easy task, as also witnessed by some recent work [372, 371], mostly related to liquid crystals. A vast literature has been devoted to computing the *excluded volume* $V(\mathfrak{R}^*)$. We refer the reader to that literature to appreciate the many subtleties involved in the geometric problem of constructing $\mathfrak{R}^*$.[22]

Here, following again [313], we further build upon the kinematic construction of $\mathfrak{R}^*$ and develop an analytic method, which we then apply in a perturbative limit. We consider molecules whose shape can be represented like $\mathfrak{R}^*$ in (1.135), that is,

$$\mathfrak{R} = \{p \in \mathcal{E} \mid p - p_0 = r\boldsymbol{e},\ 0 \leqq r < u(\boldsymbol{e})\}, \tag{1.137}$$

where $\boldsymbol{e}$ is the radial unit vector, and u is the shape function of $\mathfrak{R}$. Like $\mathfrak{R}^*$, the region $\mathfrak{R}$ is star-shaped relative to the charge center p_0. It follows from (1.137) that $\partial\mathfrak{R}$ is the image of $\mathbb{S}^2$ under the mapping $\boldsymbol{u}(\boldsymbol{e}) := u(\boldsymbol{e})\boldsymbol{e}$. Figure 1.9 shows both $\mathfrak{R}$ and the unit sphere $\mathbb{S}^2$ around which $\partial\mathfrak{R}$ is built. If u is continuously differentiable

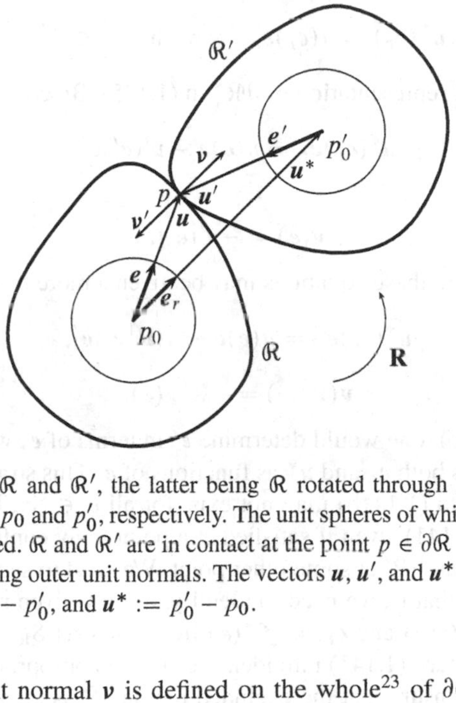

Fig. 1.9. The regions $\mathfrak{R}$ and $\mathfrak{R}'$, the latter being $\mathfrak{R}$ rotated through $\mathbf{R}$. $\mathfrak{R}$ and $\mathfrak{R}'$ are star-shaped with respect to p_0 and p_0', respectively. The unit spheres of which their boundaries are images are also depicted. $\mathfrak{R}$ and $\mathfrak{R}'$ are in contact at the point $p \in \partial\mathfrak{R} \cap \partial\mathfrak{R}'$, where $\boldsymbol{\nu}$ and $\boldsymbol{\nu}'$ denote the corresponding outer unit normals. The vectors $\boldsymbol{u}$, $\boldsymbol{u}'$, and $\boldsymbol{u}^*$ are defined as follows: $\boldsymbol{u} := p - p_0$, $\boldsymbol{u}' := p - p_0'$, and $\boldsymbol{u}^* := p_0' - p_0$.

on $\mathbb{S}^2$, the outer unit normal $\boldsymbol{\nu}$ is defined on the whole[23] of $\partial\mathfrak{R}$, and, as shown in [313], it can be given the following concise form:

[22] There are essentially two methods widely used to determine $\mathfrak{R}^*$ and $V(\mathfrak{R}^*)$ in special classes of shapes; they are based on *convex-body coordinates* and MINKOWSKI *sums*: illustrations of these methods and relevant bibliographic sources can be found in [303, 224].

[23] Thus the topological boundary $\partial\mathfrak{R}$ and the reduced boundary $\partial^*\mathfrak{R}$ coincide.

$$\nu(e) = \frac{ue - \nabla_s u}{\sqrt{u^2 + |\nabla_s u|^2}}, \tag{1.138}$$

where $\nabla_s u$ is the surface gradient of u on $\mathbb{S}^2$.

When $\mathcal{R}$ is subject to the rotation $\mathbf{R} \in \mathsf{SO}(3)$, thus becoming $\mathcal{R}'$, each e on the sphere $\mathbb{S}^2$ around which $\partial\mathcal{R}'$ is built can be seen as the image of $\mathbf{R}^\mathsf{T} e$ under $\mathbf{R}$, so that $\mathcal{R}'$ is represented as in (1.137) with u replaced by

$$u'(e) := u(\mathbf{R}^\mathsf{T} e), \qquad \forall\, e \in \mathbb{S}^2. \tag{1.139}$$

Correspondingly, the outer unit normal ν' to $\partial\mathcal{R}'$ is given by

$$\nu'(e) = \mathbf{R}\nu(\mathbf{R}^\mathsf{T} e), \qquad \forall\, e \in \mathbb{S}^2. \tag{1.140}$$

Figure 1.9 illustrates the situation we envisage. The shapes $\mathcal{R}$ and $\mathcal{R}'$, with their charge centers p_0 and p_0', are in mutual contact at a point p on $\partial\mathcal{R} \cap \partial\mathcal{R}'$, designated correspondingly by e and e' on the unit spheres around which $\partial\mathcal{R}$ and $\partial\mathcal{R}'$ are built. The vector $u^* := p_0' - p_0$ describes the boundary of $\partial\mathcal{R}^*$ of the excluded region, built around the unit sphere centered at p_0; formally, we write

$$u^*(e_r) = u(e_r)e_r \qquad \text{with} \quad e_r \in \mathbb{S}^2,$$

where u^* is the radial representation of $\partial\mathcal{R}^*$ in (1.135). By construction,

$$u^*(e_r)e_r = u(e)e - u'(e')e'$$

and

$$\nu(e) = -\nu'(e').$$

By (1.139) and (1.140), these equations may be given a more transparent form,

$$u^*(e_r)e_r = u(e)e - u(\mathbf{R}^\mathsf{T} e')e', \tag{1.141a}$$

$$\nu(\mathbf{R}^\mathsf{T} e') = -\mathbf{R}^\mathsf{T}\nu(e). \tag{1.141b}$$

In general, by (1.141b), one would determine e' in terms of e, which, once inserted into (1.141a), delivers both e_r and u^* as functions of e. This strategy may, however, fail, since a solution to (1.141b) may not exist for all $e \in \mathbb{S}^2$. Moreover, the local contact conditions (1.141), even if satisfied at a point, may conflict with the mutual impenetrability of $\mathcal{R}$ and $\mathcal{R}'$ at some other point. We need to seek global solvability of equations (1.141), that is, we need to identify, for any given rotation $\mathbf{R} \in \mathsf{SO}(3)$, the mappings $e' = f'(e)$ and $e_r = f^*(e)$ from a subset $\mathbb{S}_\mathbf{R} \subset \mathbb{S}^2$ into and onto $\mathbb{S}^2$, respectively, that turn (1.141) into identities for an appropriate positive u^*. This would, in particular, ensure that the excluded region $\mathcal{R}^*$ is also star-shaped, as desired. Correspondingly, the steric tensor in (1.4.2) could then be converted into an integral over $\mathbb{S}_\mathbf{R}$ through the change of variables induced on $\mathbb{S}^2$ by f^*. Such an analytic program, however, may easily become prohibitive for sufficiently general shapes $\mathcal{R}$. Notable examples are convex shapes, for which (1.141b) is uniquely solvable, with $\mathbb{S}_\mathbf{R} = \mathbb{S}^2$ for all rotations $\mathbf{R}$. Nevertheless, even for this special category of shapes, the functions u^* and f^* may be rather complicated, as illuminated in [313] already for ellipsoids.

For this reason, we resort to a perturbative approach and apply the method outlined above to a special class of molecular shapes. Specifically, we set

$$u(\boldsymbol{e}) = R[1 + \varepsilon v(\boldsymbol{e})], \tag{1.142}$$

where $R > 0$ is now a characteristic molecular length, $\varepsilon > 0$ is a small perturbation parameter, and v is a bounded, smooth mapping defined on $\mathbb{S}^2$. Without loss of generality, we may normalize v by requiring that

$$\langle v \rangle_{\mathbb{S}^2} = 0, \tag{1.143}$$

so that R can be interpreted as the average molecular radius. Equation (1.142) represents a convex spheroidal molecule. It readily follows from (1.142) and (1.138) that for such a molecule the outer unit normal takes the form

$$\boldsymbol{\nu}(\boldsymbol{e}) = \boldsymbol{e} - \varepsilon \nabla_\mathrm{s} v(\boldsymbol{e}) + o(\varepsilon). \tag{1.144}$$

This mapping is clearly one-to-one on $\mathbb{S}^2$ whenever the second surface gradient $\nabla_\mathrm{s}^2 v$ on $\mathbb{S}^2$ is bounded. Under this assumption, which we make henceforth, the shape $\mathfrak{R}$ is convex, and so also will be $\mathfrak{R}^*$. By use of (1.144) in (1.141b), we arrive at an implicit function for $\boldsymbol{e}'$,

$$\boldsymbol{e}' = -\boldsymbol{e} + \varepsilon[\nabla_\mathrm{s} v(\boldsymbol{e}) + \mathbf{R}\nabla_\mathrm{s} v(\mathbf{R}^\mathsf{T}\boldsymbol{e}')] + o(\varepsilon),$$

whence, since $\boldsymbol{e}' = -\boldsymbol{e} + O(\varepsilon)$, it follows that

$$\boldsymbol{e}' = -\boldsymbol{e} + \varepsilon[\nabla_\mathrm{s} v(\boldsymbol{e}) + \mathbf{R}\nabla_\mathrm{s} v(-\mathbf{R}^\mathsf{T}\boldsymbol{e}))] + o(\varepsilon). \tag{1.145}$$

By (1.145), (1.141a) becomes

$$u^*(\boldsymbol{e}_r)\boldsymbol{e}_r = R\{2 + \varepsilon[v(\boldsymbol{e}) + v(\mathbf{R}^\mathsf{T}\boldsymbol{e})]\}\boldsymbol{e} - \varepsilon R[\mathbf{R}\nabla_\mathrm{s} v(-\mathbf{R}^\mathsf{T}\boldsymbol{e}) + \nabla_\mathrm{s} v(\boldsymbol{e})] + o(\varepsilon). \tag{1.146}$$

Since $\mathfrak{R}'$ is described by u' in (1.139), $\mathfrak{R}^*$, which is to be a spheroid like $\mathfrak{R}$, is described by

$$u^*(\boldsymbol{e}_r) = R^*[1 + \varepsilon v^*(\boldsymbol{e}_r)] + o(\varepsilon), \tag{1.147}$$

where both R^* and v^* are unknown. Inserting (1.147) into (1.146) and observing that both $\nabla_\mathrm{s} v(\boldsymbol{e})$ and $\mathbf{R}\nabla_\mathrm{s} v(-\mathbf{R}^\mathsf{T}\boldsymbol{e})$ are orthogonal to $\boldsymbol{e}$, we obtain

$$R^* = 2R\,, \quad v^*(\boldsymbol{e}_r) = \frac{1}{2}[v(\boldsymbol{e}_r) + v(-\mathbf{R}^\mathsf{T}\boldsymbol{e}_r)] \tag{1.148a}$$

and

$$\boldsymbol{e}_r = \boldsymbol{e} - \frac{1}{2}\varepsilon[\nabla_\mathrm{s} v(\boldsymbol{e}) + \mathbf{R}\nabla_\mathrm{s} v(-\mathbf{R}^\mathsf{T}\boldsymbol{e})] + o(\varepsilon). \tag{1.148b}$$

Equations (1.145) and (1.148b) are the perturbative limits of the functions f' and f^*. Thus, for any given R and v representing $\mathfrak{R}$ through (1.142), equations (1.147) and (1.148) determine explicitly the corresponding representation of $\mathfrak{R}^*$ through (1.147). In the following subsection, this representation will lead us to an explicit formula for the steric tensor $\mathbf{S}_{\mathfrak{R}^*}$.

1.4.4 Steric Biaxiality

One assumption in the original derivation of the MAIER–SAUPE interaction [208, 209, 210] is that, for the averaging process, the distribution of the molecules is spherically symmetric. This is a particularly questionable assumption, because the interaction energy decays with the sixth power of the intermolecular distance, and so the most important contributions stem from the nearest molecules. Already MAIER & SAUPE suggest that the steric effect can be taken into account by considering small groups of molecules that would then be roughly spherically symmetric [210]. This leads merely to a renormalization of the constants. Here, without abandoning the assumption on spherical spatial symmetry for the distribution of molecular charge centers, we explore directly the effect of nonspherical molecular shapes on the dispersion interactions.

As we have seen in Section 1.4.3, for spheroidal molecules the excluded region is given by the explicit representation (1.147) with R^* and v^* as in (1.148a). Then, for small ε,

$$\frac{1}{u^{*3}} = \frac{1}{(2R)^3}(1 - 3\varepsilon v^*) + o(\varepsilon),$$

and, by (1.4.2),

$$\mathbf{S}_{\mathbb{R}^*} = \frac{1}{(2R)^3}\left(\langle \mathbf{U} \boxtimes \mathbf{U}\rangle - 3\varepsilon\langle v^*\mathbf{U} \boxtimes \mathbf{U}\rangle\right) + o(\varepsilon), \tag{1.149}$$

where $\mathbf{U}$ is as in (1.129). In the second average in (1.149), we also take advantage of the fact that, at the lowest order in ε, $\boldsymbol{f}^*$ is the identity on $\mathbb{S}^2$; see (1.148b). The first term on the right-hand side of (1.149) is the same that was found in Section 1.4.2 for a spherical excluded region, and the second term gives the steric correction to the dispersion interaction. To make this more explicit, we consider the multipole expansion of v in terms of Cartesian tensors:

$$v(\boldsymbol{e}_r) = \boldsymbol{E}\cdot\boldsymbol{e}_r + \mathbf{E}\cdot\overline{\boldsymbol{e}_r\otimes\boldsymbol{e}_r} + \mathbf{E}^{(3)}\cdot\overline{\boldsymbol{e}_r\otimes\boldsymbol{e}_r\otimes\boldsymbol{e}_r} \tag{1.150}$$

$$+\,\mathbf{E}^{(4)}\cdot\overline{\boldsymbol{e}_r\otimes\boldsymbol{e}_r\otimes\boldsymbol{e}_r\otimes\boldsymbol{e}_r} + \cdots, \tag{1.151}$$

where $\overline{\cdots}$ denotes the (symmetric) irreducible part of a tensor, $\boldsymbol{E}$ is the *shape dipole*, $\mathbf{E}$ is the *shape quadrupole*, and the $\mathbf{E}^{(i)}$ are the higher moments. The gauge (1.143) forbids any constant term in (1.150). Since, by (1.129), $\mathbf{U}\boxtimes\mathbf{U}$ is even in $\boldsymbol{e}_r$, the odd-rank tensors in the expansion (1.150) do not contribute to the steric tensor (1.149). The first relevant term is the shape quadrupole, a symmetric traceless second-rank tensor that can be computed for a given $v(\boldsymbol{e})$ as

$$\mathbf{E} = \frac{15}{2}\langle v(\boldsymbol{e}_r)\,\overline{\boldsymbol{e}_r\otimes\boldsymbol{e}_r}\,\rangle_{\mathbb{S}^2} = \frac{15}{2}\langle v\mathbf{U}\rangle_{\mathbb{S}^2}.$$

From now on, we neglect higher orders and consider

$$v = \mathbf{E}\cdot\overline{\boldsymbol{e}_r\otimes\boldsymbol{e}_r} = \boldsymbol{e}_r\cdot\mathbf{E}\boldsymbol{e}_r, \tag{1.152}$$

bearing in mind that $\operatorname{tr}\mathbf{E} = 0$. By (1.148a), equation (1.152) leads us to

$$v^*(\boldsymbol{e}_r) = \frac{1}{2}\boldsymbol{e}_r \cdot (\mathbf{E} + \mathbf{E}')\boldsymbol{e}_r,$$

where $\mathbf{E}' = \mathbf{R}\mathbf{E}\mathbf{R}^{\mathsf{T}}$. The steric tensor (1.149) can then be found explicitly by noting that

$$\langle (\mathbf{U} \boxtimes \mathbf{U})_{ijkl} e_m e_n \rangle_{\mathbb{S}^2} S_{mn} A_{kl} A'_{ij} = \frac{2}{315}\mathbf{S} \cdot (3 \operatorname{tr}\mathbf{A}(\mathbf{A} + \mathbf{A}') - 2\mathbf{A}\mathbf{A}') \tag{1.153}$$

for symmetric tensors $\mathbf{A}$, $\mathbf{A}'$, and $\mathbf{S}$, with $\operatorname{tr}\mathbf{S} = 0$ and $\operatorname{tr}\mathbf{A} = \operatorname{tr}\mathbf{A}'$.[24] With this, by (1.136) and (1.4.2), the effective dispersion pair-potential H_{d} becomes

$$\begin{aligned} H_{\mathrm{d}} = -\frac{4\pi C_0 \varrho}{945(2R)^3} \{ & 7[\mathbf{A} \cdot \mathbf{A}' + 3(\operatorname{tr}\mathbf{A})^2] \\ & - 3\varepsilon(\mathbf{E} + \mathbf{E}') \cdot [3 \operatorname{tr}\mathbf{A}(\mathbf{A} + \mathbf{A}') - 2\mathbf{A}\mathbf{A}'] \} + o(\varepsilon). \end{aligned} \tag{1.154}$$

It is convenient to introduce in (1.154) the tensors $\mathbf{A}_0$ and $\mathbf{A}'_0$ defined in (1.133); one then obtains

$$\begin{aligned} H_{\mathrm{d}} = -\frac{4\pi C_0 \varrho}{2835(2R)^3} \{ & 7[3\mathbf{A}_0 \cdot \mathbf{A}'_0 + 10(\operatorname{tr}\mathbf{A})^2] \\ & - 3\varepsilon(\mathbf{E} + \mathbf{E}') \cdot [7 \operatorname{tr}\mathbf{A}(\mathbf{A}_0 + \mathbf{A}'_0) - 6\mathbf{A}_0\mathbf{A}'_0] \}, \end{aligned} \tag{1.155}$$

valid up to the first order in ε.

1.4.5 Special Interactions

We now consider the special case in which the shape quadrupole and polarizability tensor share the same eigenframe, $(\boldsymbol{e}_1, \boldsymbol{e}_2, \boldsymbol{e}_3)$. Then both tensors can be represented as linear combinations of the identity $\mathbf{I}$ and the two orthogonal tensors

$$\mathbf{q} := \boldsymbol{e}_1 \otimes \boldsymbol{e}_1 - \frac{1}{3}\mathbf{I} \quad \text{and} \quad \mathbf{b} := \boldsymbol{e}_2 \otimes \boldsymbol{e}_2 - \boldsymbol{e}_3 \otimes \boldsymbol{e}_3,$$

already introduced in (1.7) and (1.11) above. Since the quadrupolar shape tensor $\mathbf{E}$ must be traceless, we write it as

$$\mathbf{E} = \sigma_\parallel \mathbf{q} + \sigma_\perp \mathbf{b},$$

where $\sigma_\parallel$ and $\sigma_\perp$ are scalar parameters, and then also

$$\mathbf{E}' = \sigma_\parallel \mathbf{q}' + \sigma_\perp \mathbf{b}'$$

with $\mathbf{q}' = \mathbf{R}\mathbf{q}\mathbf{R}^{\mathsf{T}}$ and $\mathbf{b}' = \mathbf{R}\mathbf{b}\mathbf{R}^{\mathsf{T}}$.

---

[24] In equation (1.153), e_m denotes the mth Cartesian component of the unit vector $\boldsymbol{e}_r$. Similarly, S_{mn}, A_{kl}, and A'_{ij} are the components of the second-rank tensors $\mathbf{S}$, $\mathbf{A}$, and $\mathbf{A}'$, all in the same basis.

Uniaxial Interaction

For the MAIER–SAUPE interaction the polarizability tensor $\mathbf{A}$ is uniaxial as in (1.134). Equivalently, $\mathbf{A}$ and $\mathbf{A}'$ can also be written as

$$\mathbf{A} = \bar{\alpha}\mathbf{I} + \Delta\alpha\mathbf{q} \quad \text{and} \quad \mathbf{A}' = \bar{\alpha}\mathbf{I} + \Delta\alpha\mathbf{q}',$$

where

$$\bar{\alpha} := \frac{1}{3}\operatorname{tr}\mathbf{A} = \frac{1}{3}(\alpha_\parallel + 2\alpha_\perp) \quad \text{and} \quad \Delta\alpha := \alpha_\parallel - \alpha_\perp. \tag{1.156}$$

In [313], (1.155) was found to be equivalent, up to an additive constant, to

$$H_\mathrm{d} = -\frac{4\pi C_0 \varrho(\Delta\alpha)^2}{135(2R)^3}\left\{\left[1 + 2\varepsilon\sigma_\parallel\left(\frac{2}{7} - \frac{3\bar{\alpha}}{\Delta\alpha}\right)\right]\mathbf{q}\cdot\mathbf{q}' - \varepsilon\sigma_\perp\left(\frac{2}{7} + \frac{3\bar{\alpha}}{\Delta\alpha}\right)(\mathbf{q}\cdot\mathbf{b}' + \mathbf{b}\cdot\mathbf{q}')\right\}. \tag{1.157}$$

This formula embodies the steric correction to the classical MAIER–SAUPE interaction energy. It suggests a few comments.

First, since both $\alpha_\parallel$ and $\alpha_\perp$ are positive and

$$\frac{2}{7} - \frac{3\bar{\alpha}}{\Delta\alpha} = -\frac{5\alpha_\parallel + 16\alpha_\perp}{7\Delta\alpha},$$

the sign of the correction to the coefficient of $\mathbf{q}\cdot\mathbf{q}'$ is opposite to the sign of $\sigma_\parallel\Delta\alpha$, meaning that the molecular long axis interaction is depressed if the molecular shape quadrupole is resonant, as it were, with the anisotropic polarizability tensor, and it is enhanced otherwise. Thus, for $\sigma_\perp = 0$, a uniaxial shape quadrupole prolate along the symmetry axis $\boldsymbol{e}_1$ would depress the bare MAIER–SAUPE interaction when $\Delta\alpha > 0$, whereas it would enhance it when $\Delta\alpha < 0$.

Second, for $\sigma_\perp \neq 0$, that is, for a biaxial shape quadrupole, the dispersion interaction between molecules with uniaxial polarizability tensors becomes effectively biaxial as in (1.10) with

$$\gamma = -\varepsilon\sigma_\perp\left(\frac{2}{7} + \frac{3\bar{\alpha}}{\Delta\alpha}\right) \quad \text{and} \quad \lambda = 0.$$

As first shown by LUCKHURST & ROMANO [202] by simulation and lately confirmed within a general mean-field theory [24], a biaxial interaction potential like (1.157) with $\sigma_\perp \neq 0$ does not promote condensed biaxial phases. However, at variance with the classical MAIER–SAUPE potential, the transition temperature for such a potential would depend on the coefficient of the biaxial correction, which here is a function of the molecular shape.

Biaxial Interaction

We now consider the more general case of an arbitrary, possibly biaxial polarizability with eigenvalues α_{11}, α_{22}, and α_{33}. This can be written as

$$\mathbf{A} = \bar{\alpha}\mathbf{I} + \Delta\alpha\mathbf{q} + \frac{1}{2}\Delta\alpha_\perp\mathbf{b}, \tag{1.158}$$

with the average polarizability

$$\bar{\alpha} := \frac{1}{3}\operatorname{tr}\mathbf{A} = \frac{1}{3}(\alpha_{11} + \alpha_{22} + \alpha_{33})$$

and the polarizability anisotropies

$$\Delta\alpha := \alpha_{33} - \frac{1}{2}(\alpha_{11} + \alpha_{22}) \quad \text{and} \quad \Delta\alpha_\perp := \alpha_{11} - \alpha_{22}.$$

When $\alpha_{11} = \alpha_{22}$, this reduces to the MAIER–SAUPE interaction discussed in the preceding subsection. The effective dispersion pair-potential then takes the form

$$H_\mathrm{d} = -\frac{\pi C_0 \varrho}{945(2R)^3}\{a\,\mathbf{q}\cdot\mathbf{q}' + b\,(\mathbf{q}\cdot\mathbf{b}' + \mathbf{b}\cdot\mathbf{q}') + c\,\mathbf{b}\cdot\mathbf{b}'\} \tag{1.159}$$

with

$$a := \Delta\alpha\{28\Delta\alpha + 8\varepsilon[\sigma_\parallel(2\Delta\alpha - 21\bar{\alpha}) - 3\sigma_\perp\Delta\alpha_\perp]\}, \tag{1.160a}$$

$$b := 14\Delta\alpha\Delta\alpha_\perp - 2\varepsilon[\sigma_\perp(4\Delta\alpha^2 + 3\Delta\alpha_\perp^2) + 21\bar{\alpha}(2\sigma_\perp\Delta\alpha + \sigma_\parallel\Delta\alpha_\perp)], \tag{1.160b}$$

$$c := \Delta\alpha_\perp\{7\Delta\alpha_\perp - 4\varepsilon[\sigma_\perp(2\Delta\alpha + 21\bar{\alpha}) + \sigma_\parallel\Delta\alpha_\perp]\}. \tag{1.160c}$$

Since ε is a small perturbation parameter, it is easily seen that $a > 0$. Thus, by setting

$$U_0 := \frac{\pi C_0 \varrho a}{945(2R)^3}, \quad \gamma := \frac{b}{a}, \quad \text{and} \quad \lambda := \frac{c}{a}, \tag{1.161}$$

equation (1.159) can be given the form (1.10), put forward by STRALEY [321, 315] for general biaxial molecules,

$$H_\mathrm{d} = -U_0\{\mathbf{q}\cdot\mathbf{q}' + \gamma(\mathbf{q}\cdot\mathbf{b}' + \mathbf{b}\cdot\mathbf{q}') + \lambda\mathbf{b}\cdot\mathbf{b}'\}. \tag{1.162}$$

Earlier than STRALEY, FREISER [110, 111] had proposed a model for thermotropic liquid crystals composed of biaxial molecules, which appeared as a natural extension of the MAIER–SAUPE theory. FREISER posited the effective dispersion pair-potential

$$H_\mathrm{d} = -U_0\mathbf{A}\cdot\mathbf{A}', \tag{1.163}$$

where U_0 is a characteristic coupling energy.[25] As shown by (1.132), for spherical molecules, this formula would result from a dispersion interaction involving a single oscillator in each molecule. Clearly, by (1.158), (1.163) is a special case of (1.162), this latter reducing to the former when $\lambda = \gamma^2$. Similarly, again for spherical molecules, (1.162) can be interpreted in the language of dispersion forces if we imagine three independent oscillators at right angles in each molecule [14]. It is

[25] FREISER's pair potential formed the basis of the first mean-field treatment of biaxial nematics [26].

remarkable that for nonspherical molecules a steric quadrupolar correction to a bare FREISER interaction changes it into a STRALEY interaction , which dispersion forces could justify only through multiple oscillators.

The connection between the STRALEY and FREISER interactions is deeper than this illustrates. The effective dispersion pair-potential in (1.162) can be given the diagonal form (1.12), which we recall here:

$$H_{\mathrm{d}} = -U_0 \left(\alpha_1 \mathbf{q}_1 \cdot \mathbf{q}_1' + \alpha_2 \mathbf{q}_2 \cdot \mathbf{q}_2' \right), \tag{1.164}$$

with $\mathbf{q}_1$ and $\mathbf{q}_2$ as in (1.13). Equation (1.164) shows how the STRALEY interaction (1.162) can be viewed as the superposition of two FREISER interactions. We already learned in Section 1.1.1 that for $\lambda > \gamma^2$, both α_1 and α_2 are positive, and so both interactions in the diagonal decomposition (1.164) are attractive. The pair-potential H_{d} is then fully attractive, in the language introduced in Section 1.1.1 above. For $\lambda = \gamma^2$, either α_1 or α_2 vanishes: (1.164) still reduces to a single attractive term. In this specific instance, the potential H_{d} is also called *simply attractive*. For $\lambda < \gamma^2$, either α_1 or α_2 is negative, and (1.164) appears as a superposition of attractive and repulsive interactions. The pair-potential H_{d} is then partly repulsive. The discriminating parabola $\lambda = \gamma^2$ in the (γ, λ) plane has also been referred to as the *dispersion parabola* [24, 63].

For the particular realization (1.159) of the STRALEY interaction, one readily sees from (1.160) and (1.161) that H_{d} is fully attractive, simply attractive, or partly repulsive depending on whether the discriminant $d := b^2 - ac$ is negative, zero, or positive. When $\sigma_\parallel = \sigma_\perp = 0$, that is, when the steric effect is neglected, H_{d} in (1.159) is simply attractive. In general, it is found that d is a perfect square,

$$d = 4 \left[\sigma_\parallel \Delta\alpha_\perp (4\Delta\alpha - 21\bar{\alpha}) + \sigma_\perp \left(42\bar{\alpha}\Delta\alpha + 4\Delta\alpha^2 - 3\Delta\alpha_\perp^2 \right) \right]^2 \geqq 0.$$

This shows that H_{d} in (1.159) can never be represented by a point that lies above the dispersion parabola in the admissible region of the (γ, λ) plane depicted in Figure 1.3. Thus, accounting for the steric effect cannot change a bare FREISER interaction into a fully attractive STRALEY interaction . This outcome supports the intuitive view presented in [24] that partly repulsive interactions reflect somehow steric hindrance. However, d can vanish, thus rendering H_{d} simply attractive, even in the presence of a steric effect. For example, for $\sigma_\perp \neq 0$, $d = 0$ whenever

$$\frac{\sigma_\parallel}{\sigma_\perp} = \frac{3\eta^2 - 42\xi - 4\xi^2}{\eta(4\xi - 21)},$$

where

$$\xi := \frac{\Delta\alpha}{\bar{\alpha}} \quad \text{and} \quad \eta := \frac{\Delta\alpha_\perp}{\bar{\alpha}}$$

are subject to the bounds

$$-\frac{3}{2} < \xi < 3 \quad \text{and} \quad -3 < \eta < 3.$$

On the other hand, if $\sigma_\perp = 0$, so that the shape quadrupole is uniaxial, d vanishes only if $\Delta\alpha_\perp = 0$, that is, only if the polarizability tensor is also uniaxial. Thus, for spheroidal biaxial molecules, the steric hindrance may either map a bare FREISER interaction into another, represented by a new, effective polarizability tensor and possibly a different coupling energy, or transform it into a partly repulsive STRALEY interaction.

1.4.6 Perspective

In this section, we computed the formal contribution of molecular hindrance to the dispersion force interactions of two rigid molecules. Such a steric effect is embodied by the steric tensor $\mathbf{S}_{\mathcal{R}^*}$ defined by (1.4.2) for a star-shaped excluded region $\mathcal{R}^*$, the region that the repulsion between molecular cores makes inaccessible to both. This is an attempt to give a rigorous account of the interplay between attractive, long-range forces and repulsive, short-range forces in molecular interactions.

To explore analytically the steric effect in a specific class of molecular shapes, we considered spheroidal molecules, and, for simplicity, we restrained up to the quadrupolar term the multipolar expansion of their shape representation. We showed how a biaxial quadrupolar shape can turn the classical MAIER–SAUPE interaction potential for uniaxial nematic liquid crystals into a biaxial interaction potential in the family envisaged by STRALEY on the basis of pure symmetry. The specific steric correction to the MAIER–SAUPE interaction is not capable of promoting biaxial phases, but it affects the transition temperature. In a similar way, we explored the consequences of the steric effect on a bare FREISER interaction, a dispersion interaction between single oscillators in molecules with biaxial polarizability tensors . The steric effect transforms this interaction into a partly repulsive STRALEY interaction, thus corroborating the view that the STRALEY interactions represented by the potential (1.162) with $\lambda < \gamma^2$ somehow embody molecular hindrance [24].

It has been known since the seminal paper of ONSAGER [258] that the ordering phase transitions of nematogenic molecules can also be explained by a purely athermic theory based on excluded volume interactions. For biaxial molecules in the family of spherocuboids [224], it was shown in [283] that the quadrupolar component of the excluded volume interaction is partly repulsive for all geometric parameters describing the molecular shape. Such a conclusion reached for purely hard-core repulsive interactions somehow parallels the one reached here on the partly repulsive nature of the steric correction to a bare FREISER interaction. Since hard-core repulsive interactions result in a quadrupolar attraction, albeit partly repulsive, so does the molecular hindrance in a single oscillator dispersion interaction, at least for spheroidal molecular shapes.

A few questions are raised by these conclusions. First, whether the steric tensor $\mathbf{S}_{\mathcal{R}^*}$ can be computed, possibly numerically, for nonspheroidal molecules. Second, as to the nature of the steric correction to a general STRALEY interaction, not necessarily in the spheroidal approximation. Third, whether the pure dispersion model put forward by BATES & LUCKHURST [12] for V-shaped molecules, which in the way it is formulated could apply as well to X-shaped molecules, can be more specifically tailored to V-shaped molecules by accounting for their specific shape in computing the steric correction.

2

Dynamics of Dissipative Fluids

In the first chapter we explored the microscopic origins of orientational order. We now turn to macroscopic continuum theories. These are phenomenological theories that attempt to model real materials. They do not attempt to explain material properties by resorting to the molecular structure of matter, but they can draw inspiration from molecular theories—and the best of modern theories actually do so, in the spirit of a true multiscale approach to materials science. The same continuum theory can describe different materials by means of specific constitutive laws, which being first formulated in accordance with general invariance and symmetry principles, are then corroborated by matching experimental evidence with theoretical predictions, a comparison that eventually determines the phenomenological coefficients of the continuum theory. Often, it is also possible to link microscopic and macroscopic theories by estimating directly on molecular grounds the values of the phenomenological coefficients—for example, through a mean field theory. Whenever this happens, we extract the best from both worlds.

2.1 Continuum Mechanics Fundamentals

This book is not a treatise on continuum mechanics, though it mostly concerns ordered continua. The reader will find, for example, in the textbooks [130], [349], and [131] extensive treatments written in a mathematical language similar to ours. In this section, in an attempt to make our account self-contained, we only recall the basic concepts being used here, with a degree of mathematical rigor that would neither deter the applied scientist nor disappoint the mathematician.

2.1.1 Bodies and Shapes

Here we set forth our language by introducing a number of definitions. Modern continuum mechanics concerns bodies and their motions treated as mathematical abstractions. Since LODGE [191] first distinctively perceived the need of distinguishing between bodies and their placements in space,[1] it has become customary to regard a *body* $\mathcal{B}$ and the whole collection of its parts, or *subbodies* $\mathcal{P} \subseteq \mathcal{B}$, as a measure space, as defined, for example, in [284, p. 217] (see moreover [345, § 15] and, more diffusely, [349, §§ 2–4]). Properly speaking, one should regard $\mathcal{B}$ as a *material universe*[2] of subbodies. Here, given a body $\mathcal{B}$, we shall identify it with an appropriate region in the three-dimensional Euclidean space[3] $\mathcal{E}$, and its subbodies with an appropriate family of subsets, thus becoming guilty of confusing bodies with their *shapes*, in the parlance of TRUESDELL [349].

The issue of selecting the smallest class of shapes in space that a body and its subbodies can occupy while complying with the requirements of constituting a material universe and being at the same time amenable to the analytical transformations typical of continuum mechanics, above all balancing surface and volume integrals through the divergence theorem, has a long history. It starts perhaps with KELLOGG [163] and acquires its clearest mathematical formulation in the work of NOLL [243] (see also [240] and [241] for earlier statements of this problem). Different classes of *fit regions*, as they were called in [249], some unnecessarily wide, have been proposed in the literature.[4] Here we follow [249], to which we also refer the reader for all the technical details that cannot be treated in this short account.

To show how simple regions could be inadequate to serve as fit regions, it was remarked in [249] that open regions with a piecewise smooth boundary would not necessarily have unions or intersections in the same class. Since stability under set-theoretic operations is the first requirement for a material universe, the class of fit regions should be sufficiently large to guarantee this, but hopefully not too large. The second requirement laid down in [249] for the class of fit regions was that it be invariant under smooth diffeomorphisms of $\mathcal{E}$. The third requirement was that a fit region possess a *surface-like* boundary, for which an appropriately extended form of the divergence theorem would be valid. Finally, it was considered "desirable that the class of fit regions include all that can possibly be imagined by an engineer but exclude those that can be dreamt *only* by an ingenious mathematician." The proposal eventually put forward in [249] requires a fit region to be (i) bounded, (ii) regularly open, (iii) with finite perimeter, and (iv) with negligible boundary.

A set $\mathcal{B} \subset \mathcal{E}$ is said *regularly open* if it coincides with the interior of its closure, $\mathcal{B} = \mathring{\overline{\mathcal{B}}}$. The *perimeter* of a set is defined according to DE GIORGI as the total

1 "...it is clear that we have to deal with one continuous geometric manifold (the medium) immersed in and moving through another one (space); we shall refer to these as the 'body manifold' and the 'space manifold,' and we shall call points of the two manifolds 'particles' and 'places' respectively." (reported from [191] in [345, p. 37]).

2 The notion of universe of bodies, which was first introduced by NOLL as a *materially ordered* set in his axiomatics of continuum physics [236, 241] and further developed more recently [247], is presented in great detail in §§ I.2 and I.3 of [349].

3 The reader will find in Appendix A the basic geometric and algebraic notions underlying our language.

4 We record for completeness some other relevant contributions, also witnessing how the interest raised by this issue has propagated to recent times: [68], [301], [72].

variation of its characteristic function.[5] Finally, a set in $\mathcal{E}$ is negligible if its volume measure vanishes.[6] It was already clear in [5] that sets with finite perimeter were potential shapes of continuous bodies. However, the class of sets put forward in [5] did not obey the axioms for a material universe. Also, it was unnecessarily large. Also the class proposed in [133], though it obeys all our desired requirements, is larger than the class of fit regions employed here, as shown in [249].

It worth remarking for later use that the surface-like boundary exhibited by a fit region is smaller than the its topological boundary: it is indeed through the *reduced boundary* $\partial^*\mathcal{P}$ that any subbody $\mathcal{P}$ can be in *contact* with its exterior $\mathcal{P}^{(e)} := \overline{\mathcal{B} \setminus \mathring{\mathcal{P}}}$. This view is justified by the following facts, which are too technical to be proved here (appropriate references are given in [249, § 6]):

(1) The outer unit normal $\boldsymbol{\nu}$ to a fit region $\mathcal{P}$ is defined only on $\partial^*\mathcal{P}$.
(2) The reduced boundary $\partial^*\mathcal{P}$ of a fit region $\mathcal{P}$ has finite area measure.[7]
(3) For every continuous function $f : \overline{\mathcal{P}} \to \mathbb{R}$ differentiable in $\mathcal{P}$ and with integrable gradient ∇f,

$$\int_{\mathcal{P}} \nabla f dV = \int_{\partial^*\mathcal{P}} f \boldsymbol{\nu} dA. \tag{2.1}$$

(4) $\partial^*\mathcal{P}$ differs from the union of a countable collection of compact subsets of C^1-surfaces only by a set of area-measure zero.

Though one can prove that, for a fit region $\mathcal{P}$, the closure $\overline{\partial^*\mathcal{P}}$ of $\partial^*\mathcal{P}$ coincides with the topological boundary $\partial\mathcal{P}$, it may turn out that $A(\partial\mathcal{P} \setminus \partial^*\mathcal{P}) > 0$, as shown in [249] by example. This shows that in our setting the contact between separate subbodies can take place on proper subsets of the boundary they have in common. Formally, one could define the *contact* between two separate subbodies as the intersection of their reduced boundaries. Henceforth, we shall assume that the body $\mathcal{B}$ and any of its subbodies $\mathcal{P}$ are fit regions, according to the definition recalled above.

We have defined a body $\mathcal{B}$ as a measure space identifiable with a fit region in $\mathcal{E}$. Its measure, which is meant to express the bulkiness of its subbodies, is its *mass* M. We further assume that M is *absolutely continuous* (see, for example, [356, p. 75]) with respect to the volume measure, so that

$$M(\mathcal{P}) = \int_{\mathcal{P}} \varrho_0 dV, \tag{2.2}$$

for all subbodies $\mathcal{P}$ of $\mathcal{B}$, where ϱ_0 is the *mass density*. Equation (2.2) shows how the mass density can be regarded as the ultimate ratio between mass and volume (as

5 The theory of sets with finite perimeter, especially the theorems about the reduced boundary, recalled in the text, can be retraced in the original papers of De Giorgi, now also available in English [60], and in textbooks such as [356], Chapter 5, and [61].

6 The volume measure, here denoted by V, is Lebesgue measure on $\mathcal{E}$. The area measure, which shall be denoted by A, is Hausdorff two-dimensional measure. Definitions of these measures can be found, for example, in [356, pp. 78–80] and [374, §§ 1.2, 1.4].

7 which, incidentally, coincides with the perimeter of $\mathcal{P}$.

phrased in [345, p. 38]). More precisely, given a sequence $\mathcal{P}_n$ of fit regions having a point $P \in \mathcal{B}$ in common, the mass density at P results from the limit

$$\varrho_0(P) = \lim_{n\to\infty} \frac{M(\mathcal{P}_n)}{V(\mathcal{P}_n)}.$$

In our development, ϱ_0 will always be a smooth function.

2.1.2 Motion

Formally, given the time interval $I \subseteq \mathbb{R}$, a motion of $\mathcal{B}$ is a mapping

$$\chi : \mathcal{B} \times I \to \mathcal{E} \tag{2.3}$$

such that $\chi(\cdot, t)$ is a diffeomorphism of $\mathcal{B}$ into $\mathcal{E}$ for every $t \in I$ and $\chi(p, \cdot)$ is twice continuously differentiable for every $p \in \mathcal{B}$. For given $t \in I$, we denote by $\mathcal{B}_t$ the image of $\mathcal{B}$ under $\chi(\cdot, t)$, and we call it the *shape* of $\mathcal{B}$ at time t. Similarly, for every subbody $\mathcal{P}$ of $\mathcal{B}$, we denote by $\mathcal{P}_t$ its shape at time t. Often, when t is the time at which the motion is being observed, we call $\mathcal{P}_t$ the *present* shape of the subbody $\mathcal{P}$, and $\mathcal{B}_t$ the present shape of the whole body. Given a point $P \in \mathcal{B}$, its trajectory in space is described by the curve $p_t = \chi(P, t)$, and its *velocity* is then

$$\dot{p}_t := \frac{\partial}{\partial t}\chi(P, t). \tag{2.4}$$

For every $t \in I$, the velocity field $\boldsymbol{v}$ associated with the motion χ delivers the velocities of all trajectories traversing $\mathcal{B}_t$. Formally, $\boldsymbol{v} : \mathcal{C}_\chi \to \mathcal{V}$ is a mapping defined on the collection of shapes $\mathcal{C}_\chi$ induced by χ,

$$\mathcal{C}_\chi := \bigcup_{t\in I}(\mathcal{B}_t, t) \subset \mathcal{E} \times I, \tag{2.5}$$

that takes values in the translation space[8] $\mathcal{V}$ of $\mathcal{E}$, so that $\boldsymbol{v}(\cdot, t) : \mathcal{B}_t \to \mathcal{V}$, for every $t \in I$. This definition reveals that we are adopting the *spatial* description for the flow field,[9] which is most appropriate for fluids. Equivalently, it can phrased as

$$\boldsymbol{v}(\chi(P, t), t) := \frac{\partial}{\partial t}\chi(P, t). \tag{2.6}$$

Given a motion χ of $\mathcal{B}$ as in (2.3) and a time $t \in I$, we call the mapping χ_t : $\mathcal{B}_t \times I \to \mathcal{E}$ defined by (see also § I.8 of [349])

$$\chi_t(p, t') := \chi(\chi^{-1}(p, t), t') \tag{2.7}$$

---

[8] As also recalled in Appendix A.1, the translation space $\mathcal{V} := \mathcal{E} - \mathcal{E}$ is the inner-product space of all vectors acting on the points of $\mathcal{E}$.

[9] Often, such a description is referred to as *Eulerian*, though, as we learn from [349, p. 97], it was indeed introduced by D. BERNOULLI and D'ALEMBERT, showing once more how deceptive traditional attributions can be.

the motion of $\mathcal{B}$ *relative*[10] to $\mathcal{B}_t$. It easily follows from (2.6) and (2.7) that

$$\boldsymbol{v}(p,t) = \left.\frac{\partial}{\partial t'}\chi_t(p,t')\right|_{t'=t}. \tag{2.8}$$

Since this book concerns only fluids, most of our kinematics will be set forth in the generic present part $\mathcal{P}_t$ of $\mathcal{B}_t$. As in (2.8), we shall employ the relative motion χ_t to define all relevant kinematic fields; the acceleration $\boldsymbol{a}$, for example, will thus be

$$\boldsymbol{a}(p,t) := \left.\frac{\partial^2}{\partial t'^2}\chi_t(p,t')\right|_{t'=t}. \tag{2.9}$$

The material points in any given subbody $\mathcal{P}$ of the body $\mathcal{B}$ are the same at all times, and this makes the actual region occupied by $\mathcal{P}$ vary in time. Let $\mathcal{P}_t$ be a part of $\mathcal{B}_t$ for any given $t \in I$. We say that $\mathcal{P}_t$ is *convected* in $\mathcal{P}_{t'}$ by the relative motion χ_t in (2.7) if $\mathcal{P}_{t'} = \chi_t(\mathcal{P}_t, t')$. It follows from (2.7), (2.8), and (2.9) that, for a point p_t in the present part $\mathcal{P}_t$, the point $p_{t+\varepsilon} = \chi_t(p_t, t+\varepsilon)$ in the convected part $\mathcal{P}_{t+\varepsilon}$ can be written as

$$p_{t+\varepsilon} = p_t + \varepsilon \boldsymbol{v}(p_t,t) + \frac{1}{2}\varepsilon^2 \boldsymbol{a}(p_t,t) + o(\varepsilon^2), \tag{2.10}$$

which describes up to second order in the time increment ε the trajectory of the material body point passing through p_t at time t. In (2.10), we assume that the TAYLOR expansion of the mapping $\chi_t(p_t,\cdot)$ exists for all t. Under this assumption, equation (2.10) could also be used to define the fields $\boldsymbol{v}$ and $\boldsymbol{a}$, starting from the trajectories of material body points.[11] In particular, it follows from (2.10) that

$$\boldsymbol{a}(p_t,t) = \frac{d}{dt}\boldsymbol{v}(p_t,t) =: \dot{\boldsymbol{v}}(p_t,t),$$

where a superimposed dot denotes the time derivative taken along trajectories, also called the *material* time derivative. Explicitly, for every $t \in I$, $\boldsymbol{a}(\cdot,t)$ is the vector field on the present shape $\mathcal{B}_t$ of the body represented as

$$\boldsymbol{a} = (\nabla \boldsymbol{v})\,\boldsymbol{v} + \frac{\partial \boldsymbol{v}}{\partial t}, \tag{2.11}$$

in terms of the velocity field $\boldsymbol{v}(p,t)$ on $\mathcal{C}_\chi$, where $\nabla \boldsymbol{v} := \frac{\partial \boldsymbol{v}}{\partial p}$ is the *velocity gradient*.

More generally, for a smooth vector field $\boldsymbol{b}$ defined like $\boldsymbol{v}$ on $\mathcal{C}_\chi$, the material time derivative $\dot{\boldsymbol{b}}$ is defined by (see also Appendix A.4)

---

10 When, as customary for example in solid mechanics, $\mathcal{B}$ is called the *reference* shape (or configuration) for the motion χ, the relative motion χ_t is also said to be referred to the present shape $\mathcal{B}_t$.

11 More generally, in the following we shall define both derivatives and gradients of smooth fields through their TAYLOR expansions, which are assumed to exist, an attitude that we share with [349, pp. 327–328] and [131, p. 43].

$$\dot{\boldsymbol{b}} := (\nabla \boldsymbol{b})\,\boldsymbol{v} + \frac{\partial \boldsymbol{b}}{\partial t}. \tag{2.12}$$

It is immediately seen that

$$\dot{\boldsymbol{b}}(p,t) = \frac{\partial}{\partial t'}\boldsymbol{b}(\chi_t(p,t'),t')\bigg|_{t'=t} \qquad \forall\, p \in \mathcal{B}_t \text{ and } t \in I, \tag{2.13}$$

which could equally be used to define the material time derivative of $\boldsymbol{b}$.

Tangential and Normal Convections of Vectors and Tensors

We now consider different ways to convect vectors and tensors along a motion χ of the body $\mathcal{B}$. To this end, we envisage a curve in the present shape $\mathcal{B}_t$ and study how it evolves in time when its points are considered as material. Imagine, for example, a lace of tiny bubbles or dust particles flowing with a fluid. Formally, for a given time $t \in I$, consider a regular curve[12] $c_t : [0,1] \to \mathcal{B}_t$. Let $\boldsymbol{b}(\cdot,t)$ be the vector field defined along the image $\mathcal{c}_t \in \mathcal{B}_t$ of c_t by

$$\boldsymbol{b}(c_t(\tau),t) := \frac{\partial}{\partial \tau}c_t(\tau) \neq \mathbf{0}, \tag{2.14}$$

where $\tau \in [0,1]$ is the parameter describing the curve c_t. The vector field $\boldsymbol{b}$ is everywhere tangent to c_t by construction. As time elapses, the curve c_t is mapped by the motion χ into the curve

$$c_{t'}(\tau) = \chi_t(c_t(\tau),t'), \quad \tau \in [0,1], \tag{2.15}$$

which clearly coincides with c_t for $t' = t$. The vector field $\boldsymbol{b}(\cdot,t')$ everywhere tangent to $c_{t'}$ is given by

$$\boldsymbol{b}(c_{t'}(\tau),t') = \frac{\partial}{\partial \tau}c_{t'}(\tau) = \nabla\chi_t(c_t(\tau),t')\boldsymbol{b}(c_t(\tau),t), \tag{2.16}$$

where the gradient ∇ operates on the spatial argument only, and use has also been made of both the chain rule and (2.14). Differentiating with respect to t' both sides of equation (2.16), with the aid of (2.8), (2.13), and (2.15) we arrive at

$$\frac{\partial}{\partial t'}\boldsymbol{b}(c_{t'},t')\bigg|_{t'=t} = \dot{\boldsymbol{b}}(c_t(\tau),t) = \nabla\boldsymbol{v}(c_t(\tau),t)\boldsymbol{b}(c_t(\tau),t), \tag{2.17}$$

which is the evolution equation obeyed by a tangent vector field convected by a flow.

Extending (2.17) to a vector field $\boldsymbol{b}$ defined on the whole collection of shapes $\mathcal{C}_\chi$ induced by the motion χ, this evolution equation can easily be written as

$$\overset{\triangledown}{\boldsymbol{b}} := \dot{\boldsymbol{b}} - \mathbf{G}\boldsymbol{b} = \mathbf{0}, \tag{2.18}$$

[12] That is, a curve that possesses one tangent everywhere.

where, as in [349, p. 118], we have set

$$\mathbf{G} := \nabla \boldsymbol{v},$$

and $\overset{\triangledown}{\boldsymbol{b}}$ is the *tangentially* convected derivative[13] of $\boldsymbol{b}$. Equation (2.18) is also called the law of *tangential* convection, to recall the way in which it was derived.

Consider now two curves like $\mathfrak{c}_t$, $\mathfrak{c}_t^{(1)}$ and $\mathfrak{c}_t^{(2)}$, crossing at a point, so that there their tangent vectors $\boldsymbol{b}_1$ and $\boldsymbol{b}_2$ are not parallel. Let both curves be convected by the flow. They will still intersect at a point wandering in space, while their tangent vectors $\boldsymbol{b}_i$ evolve as in (2.16). Since χ_t is a diffeomorphism by assumption, the vectors $\boldsymbol{b}_i$ will never be parallel at any time $t' \neq t$, since they are not so at time t. More generally, consider a system of two fields $(\boldsymbol{b}_1, \boldsymbol{b}_2)$ nowhere parallel on a regular surface[14] $\mathcal{S}_t$ in $\mathcal{B}_t$ that are tangentially convected by the flow.[15] They induce the field $\boldsymbol{f} := \boldsymbol{b}_1 \times \boldsymbol{b}_2$, which locally identifies one of the two normals to $\mathcal{S}_t$, the other being $-\boldsymbol{f}$. The system curves tangent to $\boldsymbol{b}_1$ and $\boldsymbol{b}_2$, which we still denote collectively by $\mathfrak{c}_t^{(1)}$ and $\mathfrak{c}_t^{(2)}$, draw a local web on $\mathcal{S}_t$, which evolves in time convected by the flow; so also does $\boldsymbol{f}$, still designating one normal to the evolved surface $\mathcal{S}_{t'}$.

According to their definition, neither the tangent field $\boldsymbol{b}$ nor the normal field $\boldsymbol{f}$ has unit length: the way their lengths change in time is actually a measure of how length and area are altered in the convecting motion of curves and surfaces. More precisely, let $L(\mathfrak{c}_t)$ be the length of the curve $\mathfrak{c}_t$ in $\mathcal{B}_t$ parameterized by the mapping c_t. It follows from (2.14) that

$$L(\mathfrak{c}_t) = \int_0^1 |\boldsymbol{b}(c_t(\tau), t)| d\tau. \tag{2.19}$$

Similarly, given a regular surface $\mathcal{S}_t$ in $\mathcal{B}_t$ covered by a web of curves $\mathfrak{c}_t^{(1)}$ and $\mathfrak{c}_t^{(2)}$ with tangent fields $\boldsymbol{b}_1$ and $\boldsymbol{b}_2$, the area $A(\mathcal{S}_t)$ of $\mathcal{S}_t$ is given by[16]

$$A(\mathcal{S}_t) = \int_0^1 d\tau_1 \int_0^1 d\tau_2 |\boldsymbol{b}_1(c_t^{(1)}(\tau_1), t) \times \boldsymbol{b}_2(c_t^{(2)}(\tau_2), t)|, \tag{2.20}$$

where τ_1 and τ_2 are the parameters for the mappings $c_t^{(1)}$ and $c_t^{(2)}$ that represent the system of curves $\mathfrak{c}_t^{(1)}$ and $\mathfrak{c}_t^{(2)}$.

We wish now to derive the evolution law for $\boldsymbol{f}$, under the assumption that both $\boldsymbol{b}_i$ evolve according to (2.18). By computing the material time derivative of both sides of the equations $\boldsymbol{f} \cdot \boldsymbol{b}_i = 0$, for $i = 1, 2$, and recalling from (2.18) that $\dot{\boldsymbol{b}}_i = \mathbf{G}\boldsymbol{b}_i$, we easily see that

---

[13] In [131, p. 156], the derivative $\overset{\triangledown}{\boldsymbol{b}}$ defined by (2.18) is called the *contravariant* rate of $\boldsymbol{b}$, and it is given a slightly different geometric interpretation, though closely related to ours.

[14] That is, a surface that can be regarded locally as the graph of a mapping at least twice differentiable from a subset of $\mathbb{R}^2$ into $\mathcal{E}$.

[15] In the parlance of [349, p. 134], curves like $\mathfrak{c}_t$ and surfaces like $\mathcal{S}_t$ are called *substantial*, since they can be thought of as consisting of the same material points at all times t.

[16] It is well known that the length of the vector $\boldsymbol{b}_1 \times \boldsymbol{b}_2$ is the area of the parallelogram delimited by the vectors $\boldsymbol{b}_1$ and $\boldsymbol{b}_2$ (see, for example, [131, p. 4]).

$$\dot{\boldsymbol{f}} \cdot \boldsymbol{b}_i = -\mathbf{G}^{\mathsf{T}} \boldsymbol{f} \cdot \boldsymbol{b}_i, \quad i = 1, 2,$$

whence, since $\boldsymbol{f}$ is orthogonal to both $\boldsymbol{b}_i$, we conclude that

$$\dot{\boldsymbol{f}} = -\mathbf{G}^{\mathsf{T}} \boldsymbol{f} + \alpha \boldsymbol{f}, \tag{2.21}$$

where α is a scalar field that must be determined. Similarly, it follows from the definition of $\boldsymbol{f}$ and (2.18) that

$$\dot{\boldsymbol{f}} = \dot{\boldsymbol{b}}_1 \times \boldsymbol{b}_2 + \boldsymbol{b}_1 \times \dot{\boldsymbol{b}}_2 = \mathbf{G}\boldsymbol{b}_1 \times \boldsymbol{b}_2 + \boldsymbol{b}_1 \times \mathbf{G}\boldsymbol{b}_2. \tag{2.22}$$

By equating the right sides of equations (2.21) and (2.22), we arrive at the following equation for α:

$$\alpha \boldsymbol{f} \cdot \boldsymbol{f} = \boldsymbol{f} \cdot \mathbf{G}\boldsymbol{f} + \mathbf{G}\boldsymbol{b}_1 \cdot \boldsymbol{b}_2 \times \boldsymbol{f} + \mathbf{G}\boldsymbol{b}_2 \cdot \boldsymbol{f} \times \boldsymbol{b}_1, \tag{2.23}$$

where use has also been made of the invariance of the mixed product $\boldsymbol{a} \cdot \boldsymbol{b} \times \boldsymbol{c}$ under cyclic permutations of the vectors $\boldsymbol{a}$, $\boldsymbol{b}$, and $\boldsymbol{c}$ (see Appendix A.1). Since equation (2.23) must be valid for all choices of the vectors $\boldsymbol{b}_1$ and $\boldsymbol{b}_2$, we take them orthogonal and with unit length, so that $(\boldsymbol{b}_1, \boldsymbol{b}_2, \boldsymbol{f})$ is a local orthonormal and positively oriented basis for $\mathcal{V}$. Therefore $\boldsymbol{b}_2 \times \boldsymbol{f} = \boldsymbol{b}_1$ and $\boldsymbol{f} \times \boldsymbol{b}_1 = \boldsymbol{b}_2$, and (2.23) gives $\alpha = \operatorname{tr} \mathbf{G}$. The evolution equation (2.21) for $\boldsymbol{f}$ thus becomes

$$\dot{\boldsymbol{f}} = [(\operatorname{tr} \mathbf{G})\mathbf{I} - \mathbf{G}^{\mathsf{T}}]\boldsymbol{f}, \tag{2.24}$$

which we call the law of *normal* convection. Paralleling (2.18), we also write this equation as

$$\overset{\vartriangle}{\boldsymbol{f}} := \dot{\boldsymbol{f}} + \mathbf{G}^{\mathsf{T}} \boldsymbol{f} - (\operatorname{tr} \mathbf{G}) \boldsymbol{f} = \mathbf{0},$$

where $\overset{\vartriangle}{\boldsymbol{f}}$ is the *normally* convected derivative of the vector field $\boldsymbol{f}$.

From the knowledge of the convection laws for $\boldsymbol{b}$ and $\boldsymbol{f}$, we can now derive the evolution laws for the length $L(\mathcal{C}_t)$ of a convected curve $\mathcal{C}_t$ and the area $A(\mathcal{S}_t)$ of a convected surface $\mathcal{S}_t$. Since

$$|\boldsymbol{b}|^{\cdot} = (\sqrt{\boldsymbol{b} \cdot \boldsymbol{b}})^{\cdot} = \frac{\boldsymbol{b} \cdot \dot{\boldsymbol{b}}}{|\boldsymbol{b}|},$$

by combining (2.18) and (2.19), we arrive at

$$\dot{L}(\mathcal{C}_t) = \int_0^1 \boldsymbol{t} \cdot \mathbf{G}\boldsymbol{t} |\boldsymbol{b}| d\tau = \int_{\mathcal{C}_t} \boldsymbol{t} \cdot \mathbf{G}\boldsymbol{t}\, ds, \tag{2.25}$$

where $\boldsymbol{t} := \frac{\boldsymbol{b}}{|\boldsymbol{b}|}$ is the unit tangent vector to $\mathcal{C}_t$ and s denotes the arc length along $\mathcal{C}_t$. More specifically, since $\mathbf{G} = \nabla \boldsymbol{v}$, equation (2.25) can also be given the form

$$\dot{L}(\mathcal{C}_t) = \int_{\mathcal{C}_t} \boldsymbol{t} \cdot \frac{\partial \boldsymbol{v}}{\partial s} ds, \tag{2.26}$$

where $\frac{\partial \boldsymbol{v}}{\partial s} := (\nabla \boldsymbol{v})\boldsymbol{t}$ is the derivative of $\boldsymbol{v}$ along $\mathcal{C}_t$. Finally, by integrating by parts the integral in (2.26), we also arrive at the following equivalent expression:

$$\dot{L}(\mathcal{C}_t) = \boldsymbol{t}_2 \cdot \boldsymbol{v}_2 - \boldsymbol{t}_1 \cdot \boldsymbol{v}_1 - \int_{\mathcal{C}_t} \sigma \boldsymbol{n} \cdot \boldsymbol{v}\, ds,$$

where $\boldsymbol{t}_2$ and $\boldsymbol{t}_1$ are the unit tangents at the endpoints p_1 and p_2 (with p_2 following p_1 in the ordering induced by increasing s), $\boldsymbol{v}_2$ and $\boldsymbol{v}_1$ are the corresponding velocities, σ is the curvature of $\mathcal{C}_t$, and $\boldsymbol{n}$ its principal unit normal.

Correspondingly, it follows from (2.20) and (2.24) that

$$\begin{aligned} \dot{A}(\mathcal{S}_t) &= \int_0^1 d\tau_1 \int_0^1 d\tau_2 |\boldsymbol{f}|^{\cdot} = \int_0^1 d\tau_1 \int_0^1 d\tau_2 |\boldsymbol{f}| \boldsymbol{\nu} \cdot [(\operatorname{tr} \mathbf{G})\boldsymbol{\nu} - \mathbf{G}^{\mathsf{T}} \boldsymbol{\nu}] \\ &= \int_{\mathcal{S}_t} (\operatorname{tr} \mathbf{G} - \boldsymbol{\nu} \cdot \mathbf{G} \boldsymbol{\nu})\, dA, \end{aligned} \tag{2.27}$$

where $\boldsymbol{\nu} := \frac{\boldsymbol{f}}{|\boldsymbol{f}|}$ is the unit normal to $\mathcal{S}_t$ oriented like $\boldsymbol{f}$. Since $\operatorname{tr} \mathbf{G} = \operatorname{div} \boldsymbol{v}$ and $\boldsymbol{\nu} \cdot \mathbf{G}\boldsymbol{\nu} = \operatorname{div}_\nu \boldsymbol{v}$ is the *normal* divergence[17] of $\boldsymbol{v}$, we derive from (2.27) the following intrinsic equation:

$$\dot{A}(\mathcal{S}_t) = \int_{\mathcal{S}_t} \operatorname{div}_{\mathrm{s}} \boldsymbol{v}\, dA, \tag{2.28}$$

where $\operatorname{div}_{\mathrm{s}}$ denotes the *surface* divergence. By applying the surface-divergence theorem,[18] we also give (2.28) the alternative, more explicit form

$$\dot{A}(\mathcal{S}_t) = \int_{\mathcal{S}_t} (\sigma_1 + \sigma_2)\boldsymbol{v} \cdot \boldsymbol{\nu}\, dA + \int_{\partial \mathcal{S}_t} \boldsymbol{v} \cdot \boldsymbol{\nu}_{\mathcal{S}_t}\, ds,$$

where σ_1 and σ_2 are the principal curvatures of the surface $\mathcal{S}_t$ and $\boldsymbol{\nu}_{\mathcal{S}_t} := \boldsymbol{t} \times \boldsymbol{\nu}$ denotes the *conormal* to its border $\partial\mathcal{S}_t$ with unit tangent $\boldsymbol{t}$.

The notions of tangential and normal convections introduced above for vector fields are easily extended to tensor fields. We say that a tensor field $\mathbf{T}$ is convected *tangentially* on $\mathcal{C}_\chi$ if it transforms tangentially convected vectors into tangentially convected vectors. This means that, also by (2.18),

$$(\mathbf{T}\boldsymbol{b})^{\cdot} = \dot{\mathbf{T}}\boldsymbol{b} + \mathbf{T}\dot{\boldsymbol{b}} = \dot{\mathbf{T}}\boldsymbol{b} + \mathbf{T}\mathbf{G}\boldsymbol{b} = \mathbf{G}\mathbf{T}\boldsymbol{b} \quad \forall\, \boldsymbol{b},$$

whence it follows that a tangentially convected tensor obeys the evolution law

$$\overset{\triangledown}{\mathbf{T}} := \dot{\mathbf{T}} + \mathbf{T}\mathbf{G} - \mathbf{G}\mathbf{T} = \mathbf{0}, \tag{2.29}$$

where $\overset{\triangledown}{\mathbf{T}}$ is the *tangentially* convected derivative of $\mathbf{T}$. Similarly, we say that a tensor field $\mathbf{T}$ is convected *normally* if it transforms normally convected vectors into

[17] The fundamentals of calculus on smooth surfaces can be found in [353, § 2.3.6]; they will also be recalled in Section 5.2.3 below. Here it would suffice to note that on $\mathcal{S}_t$ the divergence $\operatorname{div} \boldsymbol{v}$ splits into the sum of $\operatorname{div}_{\mathrm{s}} \boldsymbol{v}$ and $\operatorname{div}_\nu \boldsymbol{v}$.

[18] This theorem is stated in (5.20) below, where a more extensive use is made of it.

normally convected vectors. The evolution law obeyed by such a tensor field, which follows from the requirement that

$$
\begin{aligned}
(\mathbf{T}f)^{\cdot} &= \dot{\mathbf{T}}f + \mathbf{T}\dot{f} = \dot{\mathbf{T}}f + (\operatorname{tr}\mathbf{G})\mathbf{T}f - \mathbf{T}\mathbf{G}^{\mathsf{T}}f \\
&= (\operatorname{tr}\mathbf{G})\mathbf{T}f - \mathbf{G}^{\mathsf{T}}\mathbf{T}f \quad \forall f,
\end{aligned}
$$

is thus

$$\overset{\vartriangle}{\mathbf{T}} := \dot{\mathbf{T}} - \mathbf{T}\mathbf{G}^{\mathsf{T}} + \mathbf{G}^{\mathsf{T}}\mathbf{T} = \mathbf{0},$$

where $\overset{\vartriangle}{\mathbf{T}}$ is called the *normally* convected derivative of **T**.

Formally, there are two other possible definitions of convected derivatives for a tensor field **T** inspired by our geometric interpretation for tangential and normal convections of vector fields. In one definition, we require **T** to transform tangentially convected vectors into normally convected vectors. The reader will easily show that this requires **T** to obey the evolution law

$$\overset{\blacktriangle}{\mathbf{T}} := \dot{\mathbf{T}} + \mathbf{T}\mathbf{G} + \mathbf{G}^{\mathsf{T}}\mathbf{T} = \mathbf{0}.$$

In the other definition, we require **T** to transform normally convected vectors into tangentially convected vectors. Such a tensor field obeys the following evolution law:

$$\overset{\blacktriangledown}{\mathbf{T}} := \dot{\mathbf{T}} - \mathbf{T}\mathbf{G}^{\mathsf{T}} - \mathbf{G}\mathbf{T} = \mathbf{0}. \tag{2.30}$$

The derivatives $\overset{\blacktriangle}{\mathbf{T}}$ and $\overset{\blacktriangledown}{\mathbf{T}}$ just introduced coincide with the *covariant* and *contravariant* rates of **T**, respectively, as defined, for example, in [131, p. 152].

Transport Theorem

We have seen in the preceding subsection how curves and surfaces are convected by a motion of the body $\mathcal{B}$; in particular, (2.26) and (2.28) describe the convected evolutions of length and area. Here we consider the convection of bulkier parts of $\mathcal{B}$ and are interested in the convection of volume. More generally, the main task of continuum kinematics is to describe how various physical quantities with volume density evolve in time along a motion.

Let a motion χ of $\mathcal{B}$ be given as in (2.3). Consider an arbitrary part $\mathcal{P}_t \subseteq \mathcal{B}_t$ of the shape of $\mathcal{B}$ at time $t \in I$. The volume of this part of the body is given by the integral

$$V(\mathcal{P}_t, \chi) = \int_{\mathcal{P}_t} dV.$$

In general, this volume will change with t, and we can compute the time derivative $\dot{V}(\mathcal{P}_t, \chi)$ of $V(\mathcal{P}_t, \chi)$ explicitly by regarding the relative motion χ_t as a local change of variables. Since $\mathcal{P}_t$ is convected into $\mathcal{P}_{t+\varepsilon}$ by χ_t, we can write

$$V(\mathcal{P}_{t+\varepsilon}, \chi) = \int_{\mathcal{P}_{t+\varepsilon}} dV = \int_{\mathcal{P}_t} \left| \frac{\partial \chi_t}{\partial p_t}(p_t, \varepsilon) \right| dV, \tag{2.31}$$

where $\left|\frac{\partial \chi_t}{\partial p_t}(p_t, \varepsilon)\right|$ denotes the Jacobian determinant of χ_t with respect to p_t. It follows from (2.10) that

$$\left|\frac{\partial \chi_t}{\partial p_t}(p_t, \varepsilon)\right| = |\det \nabla p_{t+\varepsilon}| = |\det(\mathbf{I} + \varepsilon \nabla \boldsymbol{v}(p_t, t))|, \tag{2.32}$$

where ∇ denotes the spatial gradient, operating only on p_t. Since, as easily follows from (A.5) in Appendix A.1, $\det(\mathbf{I} + \varepsilon \mathbf{U}) = 1 + \varepsilon \operatorname{tr} \mathbf{U} + o(\varepsilon)$, for any second-rank tensor $\mathbf{U}$, and since $\operatorname{tr} \nabla \boldsymbol{v} = \operatorname{div} \boldsymbol{v}$, by (2.32), (2.31) becomes

$$V(\mathcal{P}_{t+\varepsilon}, \chi) = \int_{\mathcal{P}_t} (1 + \operatorname{div} \boldsymbol{v}) dV + o(\varepsilon) = V(\mathcal{P}_t, \chi) + \varepsilon \int_{\mathcal{P}_t} \operatorname{div} \boldsymbol{v}\, dV + o(\varepsilon),$$

whence it follows that

$$\dot{V}(\mathcal{P}_t, \chi) = \int_{\mathcal{P}_t} \operatorname{div} \boldsymbol{v}\, dV, \tag{2.33}$$

which by the divergence theorem also reads as

$$\dot{V}(\mathcal{P}_t, \chi) = \int_{\partial^* \mathcal{P}_t} \boldsymbol{v} \cdot \boldsymbol{\nu} dA, \tag{2.34}$$

where $\boldsymbol{\nu}$ denotes the outer unit normal to $\partial^* \mathcal{P}_t$. Equation (2.34) has a clear intuitive interpretation: it shows that the local volume change is the product of the normal velocity $v_\nu := \boldsymbol{v} \cdot \boldsymbol{\nu}$ and the area of the surface element dA.[19]

The motion of a body $\mathcal{B}$ is said to be *isochoric* if the volume of any part $\mathcal{P}_t$ of $\mathcal{B}_t$ is constant at all times. In view of (2.33), this means that

$$\int_{\mathcal{P}_t} \operatorname{div} \boldsymbol{v}\, dV = 0 \quad \text{for all } \mathcal{P}_t.$$

Since the domain of integration $\mathcal{P}_t$ is an arbitrary subset of $\mathcal{B}_t$, this can hold only if $\operatorname{div} \boldsymbol{v} \equiv 0$. Thus, a motion is isochoric whenever

$$\operatorname{div} \boldsymbol{v} = 0 \tag{2.35}$$

throughout the motion. A velocity field $\boldsymbol{v}$ that satisfies (2.35) is also called *solenoidal*. An *incompressible* material comprises bodies that can perform only isochoric motions; for them, (2.35) is a kinematic constraint to be imposed on all admissible motions, since all flows must be solenoidal.

The same argument that led us to (2.33) can be generalized to any integral *shape functional* along the motion χ defined as

---

[19] Properly speaking, dV and dA indicate the measures relative to which integrals are to be evaluated, the former being a three-dimensional volume measure, and the latter a two-dimensional area measure. However, in interpreting equations like (2.33) and (2.34), it might be convenient to think of dV and dA as volumes and areas of the elementary members of a partition approximating the set being measured. The mathematical grounds for such an identification can be found, for example, in [356, § 2.3].

$$\Phi(\mathcal{P}_t, \chi) := \int_{\mathcal{P}_t} \varphi(p_t, t) dV, \tag{2.36}$$

where φ is a scalar field defined on the collection of shapes $\mathcal{C}_\chi$ in (2.5) induced by the motion χ, and the integral is extended over the spatial argument p_t only. Proceeding as in (2.31), we write

$$\Phi(\mathcal{P}_{t+\varepsilon}, \chi) = \int_{\mathcal{P}_{t+\varepsilon}} \varphi(p_{t+\varepsilon}, t+\varepsilon) dV = \int_{\mathcal{P}_t} \varphi(\chi_t(p_t, \varepsilon), t+\varepsilon) \left| \frac{\partial \chi_t}{\partial p_t}(p_t, \varepsilon) \right| dV.$$

Applying the chain rule and letting $p_{t+\varepsilon} = \chi_t(p_t, t+\varepsilon)$, by (2.32) and (2.10), we obtain that

$$\begin{aligned}\Phi(\mathcal{P}_{t+\varepsilon}, \chi) &= \int_{\mathcal{P}_t} \left[\varphi(p_t, t) + \varepsilon\left(\nabla\varphi \cdot \boldsymbol{v} + \frac{\partial \varphi}{\partial t}\right)\right][1 + \varepsilon \operatorname{div} \boldsymbol{v}] dV + o(\varepsilon) \\ &= \Phi(\mathcal{P}_t, \chi) + \varepsilon \int_{\mathcal{P}_t} \left[\varphi \operatorname{div} \boldsymbol{v} + \nabla\varphi \cdot \boldsymbol{v} + \frac{\partial \varphi}{\partial t}\right] dV + o(\varepsilon),\end{aligned}$$

whence it follows that

$$\dot{\Phi}(\mathcal{P}_t, \chi) = \int_{\mathcal{P}_t} (\dot{\varphi} + \varphi \operatorname{div} \boldsymbol{v})\, dV, \tag{2.37}$$

where

$$\dot{\varphi} := \nabla\varphi \cdot \boldsymbol{v} + \frac{\partial \varphi}{\partial t} \tag{2.38}$$

is the material time derivative of $\varphi(p_t, t)$ along a trajectory p_t (see also Appendix A.4 for a direct definition of this derivative). By (2.38) and the divergence theorem, since $\operatorname{div}(\varphi \boldsymbol{v}) = \varphi \operatorname{div} \boldsymbol{v} + \nabla\varphi \cdot \boldsymbol{v}$, (2.37) can also be given the form

$$\dot{\Phi}(\mathcal{P}_t, \chi) = \int_{\mathcal{P}_t} \left(\frac{\partial \varphi}{\partial t} + \operatorname{div}(\varphi \boldsymbol{v})\right) dV \tag{2.39}$$

$$= \int_{\mathcal{P}_t} \frac{\partial \varphi}{\partial t} dV + \int_{\partial^* \mathcal{P}_t} \varphi \boldsymbol{v} \cdot \boldsymbol{\nu}\, dA, \tag{2.40}$$

which shows two distinct contributions to the total rate of change of the functional Φ. The former stems from changes in φ within the region $\mathcal{P}_t$ presently being occupied by the subbody, as if the subbody were not evolving in time; the latter stems from changes in the region occupied by the subbody.

The identities (2.37), (2.39), and (2.40) are equivalent expressions of REYNOLD's *transport theorem.* Equation (2.33) is recovered as a special form of (2.37) for $\varphi = 1$. Similarly, for a vector-valued shape functional $\boldsymbol{\Phi}$ defined by

$$\boldsymbol{\Phi}(\mathcal{P}_t, \chi) := \int_{\mathcal{P}_t} \boldsymbol{\varphi}(p_t, t) dV, \tag{2.41}$$

equation (2.37) is replaced by

$$\dot{\boldsymbol{\Phi}}(\mathcal{P}_t, \chi) = \int_{\mathcal{P}_t} [\dot{\boldsymbol{\varphi}} + (\operatorname{div} \boldsymbol{v})\boldsymbol{\varphi}]\, dV, \tag{2.42}$$

where $\dot{\boldsymbol{\varphi}}$ is the material time derivative of the vector field $\boldsymbol{\varphi}$ defined as in (2.12).

2.1.3 Frame Indifference

Though in its formal definition the motion of a body may well give the illusion of being absolute,[20] it makes sense only relative to other bodies. More precisely, it makes sense only relative to a *frame of reference*, or *observer*. As was suggestively said in [236, p. 278] (also reprised in [345, p. 41]), "physically, a frame of reference is a set of objects whose mutual distances change comparatively little in time, like the walls of a laboratory, the fixed stars, or the wooden horses on a merry-go-round." Moreover, since a motion ultimately describes how material points change position in time, time must concur with space in defining a frame. We may also say that fixing a frame amounts to choosing a representation of the space-time where our description of the mechanical events is set. As in [349, p. 29], we consider an event as a primitive concept whose nature is somewhat clarified by the mathematical structure employed to describe it.

In the same vein, in this book we renounce attributing a precise mathematical meaning to the concept of frame: we shall be contented with giving a rigorous definition for a *change* of frame, which we reckon to be a far more important concept. In *no* way, however, is a frame to be thought of as a coordinate system.

Our theory is based on the assumption that all principles pertaining to physical reality must be independent of the frame of reference or observer that is employed to state them. This axiom, which is essentially an invariance requirement, is called the *principle of frame indifference*, and also the *principle of objectivity*. We next make it explicit.

Change of Frame

Pragmatically, we identify a *frame of reference*, often called simply a frame or an observer, with a rigid body that is endowed with a clock. An *event* in a given frame is identified with where and when it takes place, that is, with a pair (x, t) of a point $x \in \mathcal{E}$ in space and a time $t \in \mathbb{R}$. The set of all events $\{(x, t) | x \in \mathcal{E},\ t \in \mathbb{R}\}$ is then called the *space-time*.

A *change of frame* is an automorphism of space-time in which[21]

1. all distances are preserved,
2. all time intervals are preserved, and
3. the sense of time is preserved.

Indirectly, by defining what a change of frame preserves, we identify what characterizes all admissible observers: they agree on the metric assigned to the Euclidean space $\mathcal{E}$ that hosts mechanical events and they do not mix up future and past.

---

[20] See, in this regard, the discussion in [349, § I.6A] on NEWTON's *absolute* space and time and their *relative* counterparts.

[21] We refer here explicitly and exclusively to classical Euclidean space-time. Similar concepts exist in relativistic mechanics: the pseudo-Euclidean structure appropriate for that space-time is presented in great generality in [238].

The most general form of a change of frame that maps an event (x,t) in one frame to the event (x^*,t^*) as seen in another frame is given by [337][22]

$$x^* = o^*(t) + \mathbf{R}(t)(x - o), \tag{2.43a}$$

$$t^* = t - a. \tag{2.43b}$$

Here, $a \in \mathbb{R}$ is a fixed time difference between the frames, $x, x^* \in \mathcal{E}$ refer to the same point as seen from the two different frames, and $\mathbf{R}(t) \in \mathsf{O}(3)$ is an orthogonal transformation that describes the relative orientation of the frames at time t. The point o is an arbitrary but fixed point in the first frame. The position of its image point o^* as seen from the second frame in general depends on time. The time difference between the frames is immaterial for our further discussions, and so we simply use

$$a = 0 \quad \text{and} \quad t^* = t.$$

We synthetically say that (2.43) changes the frame f into the frame f^*.

The mapping $t \mapsto \mathbf{R}(t)$ is here assumed to be at least differentiable. Since $\det \mathbf{R}$ is then a continuous function of t, it is either identically $+1$ or identically -1, showing that $\mathbf{R}(t)$ either is a rotation, that is, a proper orthogonal tensor (when $\det \mathbf{R} \equiv +1$) or differs from it by a central reflection (when $\det \mathbf{R} \equiv -1$), for all times t.

Indifferent Scalars, Vectors, and Tensors

A quantity is called *frame-indifferent*, or simply indifferent, if two observers in different frames agree on it.

Suppose that one and the same set prescriptions deliver in the frames f and f^* the scalar functions $\phi : \mathcal{E} \times \mathbb{R} \to \mathbb{R}$ and $\phi^* : \mathcal{E} \times \mathbb{R} \to \mathbb{R}$, respectively. They are said to represent an indifferent scalar whenever they satisfy the following identity:

$$\phi^*(x^*,t^*) = \phi(x,t), \tag{2.44}$$

for all pairs (x^*,t^*) related to (x,t) as in (2.43).

The prototype of all indifferent vectors is the translation that connects the same points in both frames. In other words, if $\boldsymbol{u} = q - p$, then $\boldsymbol{u}^* = q^* - p^*$, where starred and unstarred points are related as in (2.43a). An indifferent vector transforms as

$$\boldsymbol{u}^* = \mathbf{R}\boldsymbol{u} \tag{2.45}$$

under the change of frame (2.43). This can be seen as follows:

$$\begin{aligned} \boldsymbol{u}^* &= q^* - p^* = o^* + \mathbf{R}(q - o) - o^* - \mathbf{R}(p - o) \\ &= \mathbf{R}[(q - o) - (p - o)] = \mathbf{R}(p - q) \\ &= \mathbf{R}\boldsymbol{u}. \end{aligned}$$

[22] The proof of (2.43b) is trivial. A proof of (2.43a), which represents all Euclidean isometries, can be found in [349, pp. 344–345]. We refer the interested reader to [238] for the extension of (2.43) to the pseudo-Euclidean spaces relevant to special relativity.

It follows from (2.45) that the inner product of two indifferent vectors is an indifferent scalar, since

$$\boldsymbol{w}^* \cdot \boldsymbol{u}^* = \mathbf{R}\boldsymbol{w} \cdot \mathbf{R}\boldsymbol{u} = \boldsymbol{w} \cdot \mathbf{R}^{\mathsf{T}}\mathbf{R}\boldsymbol{u} = \boldsymbol{w} \cdot \boldsymbol{u}.$$

Indifferent tensors of rank two map indifferent vectors to indifferent vectors. So if $\boldsymbol{u} = \mathbf{T}\boldsymbol{w}$, the transformed vectors $\boldsymbol{u}^*$ and $\boldsymbol{w}^*$ are given by

$$\boldsymbol{w}^* = \mathbf{R}\boldsymbol{w} \quad \text{and} \quad \boldsymbol{u}^* = \mathbf{R}\boldsymbol{u},$$

and then $\boldsymbol{u}^* = \mathbf{T}^*\boldsymbol{w}^*$. To find the way in which $\mathbf{T}$ transforms, consider

$$\boldsymbol{u}^* = \mathbf{R}\boldsymbol{u} = \mathbf{R}\mathbf{T}\boldsymbol{w} = \mathbf{R}\mathbf{T}\mathbf{R}^{\mathsf{T}}\boldsymbol{w}^*$$

and compare this to $\boldsymbol{u}^* = \mathbf{T}^*\boldsymbol{w}^*$. For these expressions of $\boldsymbol{u}^*$ to be equal for all vectors $\boldsymbol{w}^*$, $\mathbf{T}$ must transform according to

$$\mathbf{T}^* = \mathbf{R}\mathbf{T}\mathbf{R}^{\mathsf{T}}. \tag{2.46}$$

Given an orthonormal basis $\mathsf{e} := (\boldsymbol{e}_1, \boldsymbol{e}_2, \boldsymbol{e}_3)$ of $\mathcal{V}$, the tensor $\mathbf{T}$ can be represented as (see Appendix A.1)

$$\mathbf{T} = T_{ij}\boldsymbol{e}_i \otimes \boldsymbol{e}_j,$$

where $T_{ij} := \boldsymbol{e}_i \cdot \mathbf{T}\boldsymbol{e}_j$ are the components of $\mathbf{T}$ in e, and summation over repeated indices is understood. If R_{hk} denote the components in the same basis of the orthogonal tensor $\mathbf{R}$ in (2.46), the components T'_{ij} of $\mathbf{T}^*$ in e are thus

$$T'_{ij} = R_{ih}R_{jk}T_{hk}.$$

On the other hand, by (2.46), the components T^*_{ij} of $\mathbf{T}^*$ in the transformed basis $\mathsf{e}^* := (\boldsymbol{e}^*_1, \boldsymbol{e}^*_2, \boldsymbol{e}^*_3)$ are given by

$$T^*_{ij} = \boldsymbol{e}^*_i \cdot \mathbf{T}^*\boldsymbol{e}^*_j = \mathbf{R}\boldsymbol{e}_i \cdot \mathbf{R}\mathbf{T}\mathbf{R}^{\mathsf{T}}\mathbf{R}\boldsymbol{e}_j = \boldsymbol{e}_i \cdot \mathbf{T}\boldsymbol{e}_j = T_{ij}.$$

Thus the components of $\mathbf{T}^*$ in e^* are the same as the components of $\mathbf{T}$ in e, which shows that they are indifferent scalars. In general, the components of indifferent tensors with arbitrary rank transform as

$$T'_{i'j'k'\ldots} = R_{i'i}R_{j'j}R_{k'k}\cdots T_{ijk\ldots},$$

relative to one and the same basis e of $\mathcal{V}$, while they are indifferent scalars relative to bases e and e^* related by the change of frame (2.43).

Time Derivative of Indifferent Vectors and Tensors

Because the general change of frame (2.43) is time dependent, the time derivative of an indifferent vector is not indifferent. This is to say that $(\boldsymbol{b}^*)^{\cdot} \neq (\dot{\boldsymbol{b}})^*$, where[23]

[23] Since $\boldsymbol{b}$ is a function of t only, its ordinary and material time derivatives are just the same.

$$(\boldsymbol{b}^*)^{\cdot} := \frac{d\boldsymbol{b}^*}{dt} \quad \text{and} \quad (\dot{\boldsymbol{b}})^* := \mathbf{R}\dot{\boldsymbol{b}}.$$

To see this, consider an indifferent vector $\boldsymbol{b}$ depending on t and differentiate $\boldsymbol{b}^* = \mathbf{R}\boldsymbol{b}$:

$$\frac{d\boldsymbol{b}^*}{dt} = \mathbf{R}\dot{\boldsymbol{b}} + \dot{\mathbf{R}}\boldsymbol{b}. \tag{2.47}$$

This shows that

$$(\boldsymbol{b}^*)^{\cdot} - (\dot{\boldsymbol{b}})^* = \dot{\mathbf{R}}\boldsymbol{b}, \tag{2.48}$$

and so $\dot{\boldsymbol{b}}$ is clearly not indifferent for every mapping $t \mapsto \boldsymbol{b}(t)$. In words, (2.48) says that the rate of change in the frame f* of the vector $\boldsymbol{b}^*$, which is the vector $\boldsymbol{b}$ as seen from the frame f*, differs from the rate of change of $\boldsymbol{b}$ in the frame f as seen from the frame f*.

To illustrate the significance of the right side of (2.48), we choose instead of a general vector $\boldsymbol{b}(t)$ a vector $\boldsymbol{d}$ that connects two fixed points in the frame f, so that it is clearly indifferent and constant, $\dot{\boldsymbol{d}} = \mathbf{0}$. With $\boldsymbol{d} = \mathbf{R}^{\mathsf{T}}\boldsymbol{d}^*$, in this case (2.48) simply becomes

$$(\boldsymbol{d}^*)^{\cdot} = \dot{\mathbf{R}}\mathbf{R}^{\mathsf{T}}\boldsymbol{d}^* = \mathbf{\Omega}\boldsymbol{d}^*, \tag{2.49}$$

where we have introduced the *spin tensor*

$$\mathbf{\Omega} := \dot{\mathbf{R}}\mathbf{R}^{\mathsf{T}} \tag{2.50}$$

of the frame f* relative to the frame f. The spin tensor $\mathbf{\Omega}$ is skew-symmetric, which can be seen by differentiating the identity $\mathbf{R}\mathbf{R}^{\mathsf{T}} = \mathbf{I}$ with respect to time to find that $\dot{\mathbf{R}}\mathbf{R}^{\mathsf{T}} + \mathbf{R}\dot{\mathbf{R}}^{\mathsf{T}} = \mathbf{0}$. This shows that $\mathbf{\Omega} = \dot{\mathbf{R}}\mathbf{R}^{\mathsf{T}} = -\mathbf{R}\dot{\mathbf{R}}^{\mathsf{T}} = -\mathbf{\Omega}^{\mathsf{T}}$, and so

$$\mathbf{\Omega} = -\mathbf{\Omega}^{\mathsf{T}}. \tag{2.51}$$

Therefore, as also shown in Appendix A.1, there exits an axial vector $\boldsymbol{\Omega}$ such that[24]

$$\mathbf{\Omega}\boldsymbol{u} = \boldsymbol{\Omega} \times \boldsymbol{u} \quad \forall\, \boldsymbol{u} \in \mathcal{V}. \tag{2.52}$$

This allows us to write (2.49) as

$$(\boldsymbol{d}^*)^{\cdot} = \boldsymbol{\Omega} \times \boldsymbol{d}^*.$$

The spin tensor $\mathbf{\Omega}$ and its associated axial vector $\boldsymbol{\Omega}$ describe the motion of the frame f* as seen from the frame f.

Coming back to the general case (2.48), we have that

$$(\boldsymbol{b}^*)^{\cdot} = (\dot{\boldsymbol{b}})^* + \mathbf{\Omega}\boldsymbol{b}^*. \tag{2.53}$$

The rate of change of the vector $\boldsymbol{b}^*$ has thus two contributions: one comes from the rate of change of $\boldsymbol{b}$ in the frame f as seen from the frame f*, and the other arises from the rotation of the frame f* relative to f.

[24] Often, $\boldsymbol{\Omega}$ is called the *angular velocity* associated with the change of frame (2.43). Here, we refrain from using this name, since we share the concerns of [349, p. 48] as to the false suggestion that it may imply of an angle being varied in time.

Consider now a frame-indifferent tensor $\mathbf{T}$ depending on time. It follows from (2.46) that the time derivative of $\mathbf{T}^*$ in the frame f^* is

$$(\mathbf{T}^*)^{\cdot} = \dot{\mathbf{R}}\mathbf{T}\mathbf{R}^{\mathsf{T}} + \mathbf{R}\dot{\mathbf{T}}\mathbf{R}^{\mathsf{T}} + \mathbf{R}\mathbf{T}\dot{\mathbf{R}}^{\mathsf{T}},$$

whence, by use of (2.50), we obtain

$$(\mathbf{T}^*)^{\cdot} = (\dot{\mathbf{T}})^* + \boldsymbol{\Omega}\mathbf{T}^* - \mathbf{T}^*\boldsymbol{\Omega}, \tag{2.54}$$

where we have set $(\dot{\mathbf{T}})^* := \mathbf{R}\dot{\mathbf{T}}\mathbf{R}^{\mathsf{T}}$, which is easily recognized to represent the rate of change of $\mathbf{T}$ in the frame f as seen from the frame f^*. As for (2.53), the last terms on the right side of (2.54) arise from the rotation of f^* relative to f.

The fact that ordinary time derivatives are not frame indifferent means that they cannot be used in expressing basic principles, which are supposed to be objective. At the same time, in rheology, which is the science of fluid motion, such principles are expected to be phrased in terms of the velocity field, which by either (2.6) or (2.8) is a time derivative. To resolve this dilemma, it is essential to consider time rates that are computed relative to the moving body. Before we can make this explicit, we need to explore how velocity and velocity gradient in a motion transform under a change of frame.

Velocity and Velocity Gradient Transformations

In a motion χ as in (2.3), a material point $P \in \mathcal{B}$ describes the trajectory $p_t = \chi(P,t)$; its velocity in a frame f is given by $\boldsymbol{v} = \dot{p}_t$, as defined by (2.4). Similarly, seen from a different frame f^*, $\boldsymbol{v}^* = (p_t^*)^{\cdot}$. An explicit computation using (2.43) shows that

$$\boldsymbol{v}^* = (p_t^*)^{\cdot} = (o^*)^{\cdot} + \dot{\mathbf{R}}(p_t - o) + \mathbf{R}(\dot{p}_t - \dot{o}) = (o^*)^{\cdot} + \dot{\mathbf{R}}(p_t - o) + \mathbf{R}\boldsymbol{v}, \tag{2.55}$$

where we have used that o is a fixed point in the frame f, and so $\dot{o} = \mathbf{0}$. We recast (2.55) in the form

$$\begin{aligned}
\boldsymbol{v}^* - \mathbf{R}\boldsymbol{v} &= (o^*)^{\cdot} + \dot{\mathbf{R}}(p_t - o) \\
&= (o^*)^{\cdot} + \dot{\mathbf{R}}\mathbf{R}^{\mathsf{T}}(p_t^* - o^*) \\
&= (o^*)^{\cdot} + \boldsymbol{\Omega}(p_t^* - o^*).
\end{aligned}$$

The right-hand side of this latter equation is in general different from zero, which shows that the velocity $\boldsymbol{v}$ does not satisfy (2.45), and so it is not frame indifferent.

We now consider the velocity gradient

$$\mathbf{G} := \nabla \boldsymbol{v} := \frac{\partial \boldsymbol{v}}{\partial p},$$

where $\boldsymbol{v}(\cdot, t) : \mathcal{B}_t \to \mathcal{V}$ is the velocity field at time t on the shape $\mathcal{B}_t$ of the body. It follows from

$$\boldsymbol{v}^* = \mathbf{R}\boldsymbol{v} + (o^*)^{\cdot} + \boldsymbol{\Omega}(p_t^* - o^*) \tag{2.56}$$

and from the chain rule that, in the frame f*,

$$\mathbf{G}^* := \frac{\partial \boldsymbol{v}^*}{\partial p_t^*} = \mathbf{R}\frac{\partial \boldsymbol{v}}{\partial p_t}\frac{\partial p_t}{\partial p_t^*} + \boldsymbol{\Omega}. \tag{2.57}$$

But

$$\frac{\partial p_t}{\partial p_t^*} = \mathbf{R}^{\mathsf{T}}, \tag{2.58}$$

and so

$$\mathbf{G}^* = \mathbf{R}\mathbf{G}\mathbf{R}^{\mathsf{T}} + \boldsymbol{\Omega}. \tag{2.59}$$

This shows that the velocity gradient is not indifferent: the velocity gradient $\mathbf{G}^*$ in the frame f* differs from the velocity gradient $\mathbf{R}\mathbf{G}\mathbf{R}^{\mathsf{T}}$ measured in the frame f as seen by an observer in the frame f*. The difference between the velocity gradients measured by the two observers equals the spin tensor $\boldsymbol{\Omega}$, and so it is entirely due to the relative motion of the two frames.

Because $\boldsymbol{\Omega}$ is skew-symmetric, we find for the symmetric part of the velocity gradient, the *stretching* (or rate of deformation) tensor

$$\mathbf{D} := \frac{1}{2}\left(\mathbf{G} + \mathbf{G}^{\mathsf{T}}\right), \tag{2.60}$$

that

$$\begin{aligned}\mathbf{D}^* &= \frac{1}{2}\left(\mathbf{G}^* + (\mathbf{G}^*)^{\mathsf{T}}\right)\\ &= \frac{1}{2}\left(\mathbf{R}\mathbf{G}\mathbf{R}^{\mathsf{T}} + \boldsymbol{\Omega} + \mathbf{R}\mathbf{G}^{\mathsf{T}}\mathbf{R}^{\mathsf{T}} + \boldsymbol{\Omega}^{\mathsf{T}}\right) = \mathbf{R}\mathbf{D}\mathbf{R}^{\mathsf{T}},\end{aligned} \tag{2.61}$$

and so $\mathbf{D}$ *is* frame-indifferent. The skew-symmetric part of the velocity gradient, the *vorticity* tenso

$$\mathbf{W} := \frac{1}{2}\left(\mathbf{G} - \mathbf{G}^{\mathsf{T}}\right) \tag{2.62}$$

is *not* indifferent, since from (2.59) we find that

$$\mathbf{W}^* = \frac{1}{2}\left(\mathbf{G}^* - (\mathbf{G}^*)^{\mathsf{T}}\right) = \mathbf{R}\mathbf{W}\mathbf{R}^{\mathsf{T}} + \boldsymbol{\Omega}. \tag{2.63}$$

This equation shows how the vorticity tensor $\mathbf{W}^*$ for the flow $\boldsymbol{v}^*$ in the frame f* results from the sum of the vorticity tensor $\mathbf{R}\mathbf{W}\mathbf{R}^{\mathsf{T}}$ for the flow $\boldsymbol{v}$ in the frame f as seen from the frame f* and the spin tensor $\boldsymbol{\Omega}$ of f* relative to f.

Let $\boldsymbol{w}$ be the axial vector associated to the vorticity tensor $\mathbf{W}$, so that, as in (2.52),

$$\mathbf{W}\boldsymbol{u} = \boldsymbol{w} \times \boldsymbol{u} \quad \forall\, \boldsymbol{u} \in \mathcal{V}.$$

We call $\boldsymbol{w}$ the *spin vector*: it is related to the *vorticity vector* $\boldsymbol{\omega} := \operatorname{curl} \boldsymbol{v}$ by

$$w = \frac{1}{2}\omega. \tag{2.64}$$

Let w^* denote the spin vector of $\mathbf{W}^*$. By applying the tensors on both sides of (2.63) to u^* as in (2.45), we easily obtain that

$$w^* \times u^* = \mathbf{R}\mathbf{W}u + \mathbf{\Omega}u^* = \mathbf{R}(w \times u) + \Omega \times u^*, \tag{2.65}$$

where use has also been made of (2.52). Since, as recalled in Appendix A.1,

$$\mathbf{R}(w \times u) = (\det \mathbf{R})\mathbf{R}w \times \mathbf{R}u, \tag{2.66}$$

(2.65) becomes

$$w^* \times u^* = (\det \mathbf{R})\mathbf{R}w \times u^* + \Omega \times u^*,$$

which is valid for all u^* if and only if

$$w^* = (\det \mathbf{R})\mathbf{R}w + \Omega. \tag{2.67}$$

Equation (2.67) is the vectorial counterpart of (2.63), and, by (2.64), it immediately translates for the vorticity vector as

$$\omega^* = (\det \mathbf{R})\mathbf{R}\omega + 2\Omega.$$

Since neither frame f nor frame f* is absolute, their roles can be exchanged with no effect whatsoever on our development. Clearly, were (2.43) to represent the change of f* into f, $\mathbf{R}$ would be replaced by $\mathbf{R}^{\mathsf{T}}$. By also exchanging $\mathbf{R}$ and $\mathbf{R}^{\mathsf{T}}$ in (2.50), we obtain that the spin tensor $\mathbf{\Omega}^*$ of f relative to f* is given by

$$\mathbf{\Omega}^* := \dot{\mathbf{R}}^{\mathsf{T}}\mathbf{R} = -\mathbf{R}^{\mathsf{T}}\dot{\mathbf{R}}. \tag{2.68}$$

By multiplying both sides of (2.63) on the left by $\mathbf{R}^{\mathsf{T}}$ and on the right by $\mathbf{R}$, using both (2.50) and (2.68), we readily arrive at

$$\mathbf{W} = \mathbf{R}^{\mathsf{T}}\mathbf{W}^*\mathbf{R} + \mathbf{\Omega}^*,$$

which mirrors (2.63) and has precisely the same meaning. An easy computation further shows that $\mathbf{\Omega}^*$ in (2.68) can also be expressed in terms of $\mathbf{\Omega}$ as

$$\mathbf{\Omega}^* = -\mathbf{R}^{\mathsf{T}}\mathbf{\Omega}\mathbf{R}. \tag{2.69}$$

Rigid Motion

Perusal of equation (2.56) leads us to represent the velocity field of a *rigid motion*. Following [349, § I.10], we call the motion of a body $\mathcal{B}$ *rigid* if there is a frame f*, also called the *rest frame*, such that the velocity field v_{R}^* vanishes identically for all points $p^* \in \mathcal{B}_t$. In the frame f, the corresponding velocity field v_{R} is then given by (2.56) as

$$v_{\mathrm{R}} = -\mathbf{R}^{\mathsf{T}}(o^*)^{\cdot} - \mathbf{R}^{\mathsf{T}}\mathbf{\Omega}(p^* - o^*), \tag{2.70}$$

which, by (2.50), (2.68), and (2.43a), becomes

$$\boldsymbol{v}_{\mathrm{R}}(p,t) = \boldsymbol{v}_0^* + \boldsymbol{\Omega}^*(p-o), \tag{2.71}$$

where we have set

$$\boldsymbol{v}_0^* := -\mathbf{R}^{\mathsf{T}}(o^*)^{\cdot}. \tag{2.72}$$

By (2.70), equation (2.56) can also be written in the form

$$\boldsymbol{v}^* = \mathbf{R}(\boldsymbol{v} - \boldsymbol{v}_{\mathrm{R}}), \tag{2.73}$$

which suggests that we interpret $\boldsymbol{v}^*$ as the objective transformation into $\mathfrak{f}^*$ of the velocity $\boldsymbol{v}$ relative to the rigid motion in $\mathfrak{f}$ for which $\mathfrak{f}^*$ is the rest frame. Perhaps more transparently, recalling that $\boldsymbol{v}_{\mathrm{R}}^* = \mathbf{0}$, one can rewrite (2.73) as $(\boldsymbol{v}^* - \boldsymbol{v}_{\mathrm{R}}^*) = \mathbf{R}(\boldsymbol{v} - \boldsymbol{v}_{\mathrm{R}})$, which shows that $(\boldsymbol{v} - \boldsymbol{v}_{\mathrm{R}})$ is indeed an objective vector field. An easy consequence of (2.73) is the transformation law for the acceleration field $\boldsymbol{a}$ in a motion of $\mathcal{B}$. Since $\boldsymbol{a}^* := (\boldsymbol{v}^*)^{\cdot}$, computing the material time derivative of both sides of (2.73), by use of (2.50) and (2.73) itself, we arrive at

$$\boldsymbol{a}^* = \mathbf{R}(\boldsymbol{a} - \boldsymbol{a}_{\mathrm{R}}) + \boldsymbol{\Omega}\,\boldsymbol{v}^*, \tag{2.74}$$

where $\boldsymbol{a}_{\mathrm{R}} := \dot{\boldsymbol{v}}_{\mathrm{R}}$ is the acceleration field in $\mathfrak{f}$ of the rigid motion for which $\mathfrak{f}^*$ is the rest frame. Even recalling that $\boldsymbol{a}_{\mathrm{R}}^* = \mathbf{0}$, equation (2.74) shows that $(\boldsymbol{a} - \boldsymbol{a}_{\mathrm{R}})$ fails to be an objective field, though it can still be rewritten in the following somewhat more telling form:

$$\boldsymbol{a}^* - \boldsymbol{a}_{\mathrm{R}}^* = \mathbf{R}(\boldsymbol{a} - \boldsymbol{a}_{\mathrm{R}}) + \boldsymbol{\Omega}\,(\boldsymbol{v}^* - \boldsymbol{v}_{\mathrm{R}}^*).$$

Equation (2.71), where $\boldsymbol{v}_0^*$ and $\boldsymbol{\Omega}^*$ are an arbitrary vector and an arbitrary skew tensor, both depending on time only, represents the most general rigid motion in $\mathfrak{f}$. It is easily seen from (2.71) that

$$\boldsymbol{v}_0^*(t) = \boldsymbol{v}_{\mathrm{R}}(o,t) \quad \text{and} \quad \boldsymbol{\Omega}^*(t) = \nabla \boldsymbol{v}_{\mathrm{R}}(p,t), \tag{2.75}$$

whence, in particular, it follows also by (2.64) that

$$\operatorname{div} \boldsymbol{v}_{\mathrm{R}} = 0, \quad \boldsymbol{\omega}_{\mathrm{R}} = \operatorname{curl} \boldsymbol{v}_{\mathrm{R}} = 2\boldsymbol{\Omega}^*, \tag{2.76}$$

while the spin vector of $\boldsymbol{v}_{\mathrm{R}}$ is simply $\boldsymbol{w}_{\mathrm{R}} := \frac{1}{2}\boldsymbol{\omega}_{\mathrm{R}} = \boldsymbol{\Omega}^*$. By use of (2.75) and (2.76), we give (2.71) the equivalent form

$$\boldsymbol{v}_{\mathrm{R}}(p,t) = \boldsymbol{v}_{\mathrm{R}}(o,t) + \boldsymbol{w}_{\mathrm{R}}(t) \times (p-o), \tag{2.77}$$

from which we easily obtain that

$$\boldsymbol{a}_{\mathrm{R}}(p,t) = \dot{\boldsymbol{v}}_{\mathrm{R}}(o,t) + \dot{\boldsymbol{w}}_{\mathrm{R}} \times (p-o) + \boldsymbol{w}_{\mathrm{R}} \times [\boldsymbol{w}_{\mathrm{R}} \times (p-o)],$$

where $\dot{\boldsymbol{v}}_{\mathrm{R}}(o,t) = \boldsymbol{a}_{\mathrm{R}}(o,t)$ is the acceleration in the rigid motion being considered of the point that at time t coincides with the origin o.

The special structure of a rigid motion, which (2.77) reveals as being determined by the velocity $\boldsymbol{v}_{\mathrm{R}}(o,t)$ at a point o and the spin vector $\boldsymbol{w}_{\mathrm{R}}$ equal at all points, gives a particularly telling expression for the power $\mathscr{W}_{\mathrm{R}}$ expended by a system of forces with density $\boldsymbol{f} : \mathcal{C}_{\chi} \to \mathcal{V}$ defined on the collection of shapes $\mathcal{C}_{\chi}$ induced by the motion. Given any subbody $\mathcal{P}$ of a body $\mathcal{B}$, the power $\mathscr{W}_{\mathrm{R}}(\mathcal{P}_t,t)$ expended by $\boldsymbol{f}$ in the present shape $\mathcal{P}_t$ is

$$\mathscr{W}_{\mathrm{R}}(\mathcal{P}_t,t) := \int_{\mathcal{P}_t} \boldsymbol{f} \cdot \boldsymbol{v}_{\mathrm{R}} dV. \tag{2.78}$$

By inserting (2.77) into (2.78), we readily arrive at

$$\mathscr{W}_{\mathrm{R}}(\mathcal{P}_t,t) = \boldsymbol{v}_{\mathrm{R}}(o,t) \cdot \boldsymbol{F}(\mathcal{P}_t,t) + \boldsymbol{w}_{\mathrm{R}}(t) \cdot \boldsymbol{M}_o(\mathcal{P}_t,t), \tag{2.79}$$

where

$$\boldsymbol{F}(\mathcal{P}_t,t) := \int_{\mathcal{P}_t} \boldsymbol{f}(p,t) dV$$

and

$$\boldsymbol{M}_o(\mathcal{P}_t,t) := \int_{\mathcal{P}_t} (p-o) \times \boldsymbol{f}(p,t) dV$$

are the *resultant* force and torque relative to o exerted on the present shape of $\mathcal{P}$ by $\boldsymbol{f}$.

Rigid motions play a central role in our development, since they ultimately represent all possible observers. It is instructive to see how one such motion also arises in connection with an indifferent scalar ϕ as in (2.44). It follows from (2.44) applied to a trajectory p_t as in (2.55) that

$$[\phi^*(p_t^*,t)]^{\cdot} = \dot{\phi}(p_t,t), \tag{2.80}$$

which can readily be expanded in

$$\frac{\partial \phi^*}{\partial p^*} \cdot \boldsymbol{v}^* + \frac{\partial \phi^*}{\partial t} = \frac{\partial \phi}{\partial p} \cdot \boldsymbol{v} + \frac{\partial \phi}{\partial t}. \tag{2.81}$$

Now, since p_t^* and p_t are related through (2.43a), by the chain rule, (2.44) implies that

$$\frac{\partial \phi^*}{\partial p^*} = \mathbf{R} \frac{\partial \phi}{\partial p},$$

and so, with the aid of (2.56), (2.81) is given the following form:

$$\frac{\partial \phi^*}{\partial t} = \frac{\partial \phi}{\partial t} + \frac{\partial \phi}{\partial p} \cdot \boldsymbol{v}_{\mathrm{R}},$$

where $\boldsymbol{v}_{\mathrm{R}}$ is the rigid velocity field in (2.71). By (2.76), this latter equation becomes

$$\frac{\partial \phi^*}{\partial t} = \frac{\partial \phi}{\partial t} + \operatorname{div}(\phi \boldsymbol{v}_{\mathrm{R}}), \tag{2.82}$$

which shows that the partial time derivative of an indifferent scalar, unlike its material time derivative, fails to be frame-indifferent.

Frame-Indifferent Rates of Vectors and Tensors

We are now in a position to find indifferent time derivatives for indifferent vectors and tensors. The key is equation (2.63): it shows that relative to a frame that rotates with the body at a given point, that is, at a point where $\mathbf{W} = \mathbf{0}$, the vorticity tensor $\mathbf{W}^*$ in any other frame coincides with the spin tensor of the corresponding change of frame, $\mathbf{W}^* = \boldsymbol{\Omega}$. Clearly, normally there is not a single frame in which $\mathbf{W}$ vanishes identically everywhere, but at every individual material point such a frame exists.

This, together with (2.53), suggests that one define an indifferent time derivative for an indifferent vector field $\boldsymbol{b}$ as

$$\overset{\circ}{\boldsymbol{b}} := \dot{\boldsymbol{b}} - \mathbf{W}\boldsymbol{b}. \tag{2.83}$$

The rate $\overset{\circ}{\boldsymbol{b}}$ is the *corotational* derivative, often also called the JAUMANN derivative,[25] of $\boldsymbol{b}$. At any point where $\mathbf{W} = \mathbf{0}$, the corotational derivative (2.83) coincides with the ordinary material time derivative $\dot{\boldsymbol{b}}$. In general, we find with (2.53) and (2.63) that

$$\begin{aligned}(\boldsymbol{b}^*)^\circ &= (\boldsymbol{b}^*)^\cdot - \mathbf{W}^*\boldsymbol{b}^* = (\dot{\boldsymbol{b}})^* + \boldsymbol{\Omega}\boldsymbol{b}^* - \mathbf{R}\mathbf{W}\mathbf{R}^\mathsf{T}\boldsymbol{b}^* - \boldsymbol{\Omega}\boldsymbol{b}^* \\ &= (\dot{\boldsymbol{b}})^* - \mathbf{R}\mathbf{W}\boldsymbol{b} = \mathbf{R}(\dot{\boldsymbol{b}} - \mathbf{W}\boldsymbol{b}) \\ &= \mathbf{R}\overset{\circ}{\boldsymbol{b}} = (\overset{\circ}{\boldsymbol{b}})^*.\end{aligned} \tag{2.84}$$

This shows that the corotational time derivative is indifferent. In any frame, it gives the rate of change of the field $\boldsymbol{b}$ as measured in the particular frame that rotates locally with the body. Essentially, the term $\mathbf{W}^*\boldsymbol{b}^*$ in (2.83) and the term $\boldsymbol{\Omega}\boldsymbol{b}^*$ in (2.53) cancel, because $\boldsymbol{\Omega} = \mathbf{W}^*$ exactly when $\mathbf{W} = \mathbf{0}$.

The corotational time derivative defined in (2.84) is not the only indifferent rate that can be introduced for vectors. Letting $\mathbf{G} = \mathbf{D} + \mathbf{W}$ in (2.18), the tangential convected derivative $\overset{\triangledown}{\boldsymbol{b}}$ becomes

$$\overset{\triangledown}{\boldsymbol{b}} = \dot{\boldsymbol{b}} - (\mathbf{D}\boldsymbol{b} + \mathbf{W}\boldsymbol{b}) = \overset{\circ}{\boldsymbol{b}} - \mathbf{D}\boldsymbol{b},$$

which is indifferent, since, by (2.84) and (2.61),

$$(\boldsymbol{b}^*)^\triangledown = (\boldsymbol{b}^*)^\circ - \mathbf{D}^*\boldsymbol{b}^* = \mathbf{R}\overset{\circ}{\boldsymbol{b}} + \mathbf{R}\mathbf{D}\mathbf{R}^\mathsf{T}\mathbf{R}\boldsymbol{b} = \mathbf{R}\overset{\triangledown}{\boldsymbol{b}} = (\overset{\triangledown}{\boldsymbol{b}})^*.$$

Similarly, the normally convected derivative in (2.24) can be written for $\boldsymbol{b}$ as

$$\overset{\vartriangle}{\boldsymbol{b}} = \overset{\circ}{\boldsymbol{b}} + \mathbf{D}\boldsymbol{b} - (\operatorname{tr}\mathbf{D})\boldsymbol{b},$$

which like $\overset{\triangledown}{\boldsymbol{b}}$ is indifferent.

A further class of invariant rates was introduced by OLDROYD [252]. They can be seen as extensions of the rates $\overset{\triangledown}{\boldsymbol{b}}$ and $\overset{\vartriangle}{\boldsymbol{b}}$ that have above been given a specific geometric interpretation in terms of deformations induced by the flow. A *codeformational* derivative $\overset{\diamond}{\boldsymbol{b}}$ of a vector $\boldsymbol{b}$ is defined by

[25] As pointed out in [345, § 19A], it was indeed first introduced by ZAREMBA [367] (see also [366] and [368]) and later adopted by JAUMANN [152] in their original formulations of what we can now recognize as a precursor of the frame-indifference principle.

$$\overset{\diamond}{\boldsymbol{b}} := \overset{\circ}{\boldsymbol{b}} + \sigma\mathbf{D}\boldsymbol{b} + \tau(\operatorname{tr}\mathbf{D})\boldsymbol{b}, \tag{2.85}$$

where σ and τ are indifferent scalars. The derivative $\overset{\diamond}{\boldsymbol{b}}$ is clearly indifferent since all three terms on the right-hand side of (2.85) are independently indifferent. More generally, it can be proved that the rate $\overset{\star}{\boldsymbol{b}}$ defined by[26]

$$\overset{\star}{\boldsymbol{b}} := \dot{\boldsymbol{b}} + \boldsymbol{\varphi}(\mathbf{G}, \boldsymbol{b})$$

is indifferent if and only if it can be written in the form

$$\overset{\star}{\boldsymbol{b}} = \overset{\circ}{\boldsymbol{b}} + \boldsymbol{\varphi}(\mathbf{D}, \boldsymbol{b}),$$

where $\boldsymbol{\varphi}$ is an *isotropic* vector-valued function, that is, a function such that

$$\boldsymbol{\varphi}(\mathbf{D}^*, \boldsymbol{b}^*) = \boldsymbol{\varphi}(\mathbf{D}, \boldsymbol{b})^* \quad \forall\, \mathbf{D}, \boldsymbol{b}. \tag{2.86}$$

As suggested by the discussion in [131, p. 155], this shows that the corotational derivative $\overset{\circ}{\boldsymbol{b}}$ is the *generic* indifferent rate of a vector $\boldsymbol{b}$, to within an isotropic vector-valued function.

Indifferent rates for tensors can now be found in a similar way. An indifferent tensor $\mathbf{T}$ transforms according to $\mathbf{T}^* = \mathbf{R}\mathbf{T}\mathbf{R}^{\mathsf{T}}$, and its time derivative $\dot{\mathbf{T}}$ transforms as in (2.54). In analogy to (2.47) and (2.83), this suggests defining the corotational time derivative $\overset{\circ}{\mathbf{T}}$ of $\mathbf{T}$ as

$$\overset{\circ}{\mathbf{T}} := \dot{\mathbf{T}} - \mathbf{W}\mathbf{T} + \mathbf{T}\mathbf{W}. \tag{2.87}$$

A computation analogous to (2.84) then shows that $\overset{\circ}{\mathbf{T}}$ is indifferent. Using (2.54) and (2.63), we find that

$$\begin{aligned}
(\mathbf{T}^*)^{\circ} &= (\mathbf{T}^*)^{\cdot} - \mathbf{W}^*\mathbf{T}^* + \mathbf{T}^*\mathbf{W}^* \\
&= (\dot{\mathbf{T}})^* + \boldsymbol{\Omega}\mathbf{T}^* - \mathbf{T}^*\boldsymbol{\Omega} - \mathbf{R}\mathbf{W}\mathbf{R}^{\mathsf{T}}\mathbf{T}^* - \boldsymbol{\Omega}\mathbf{T}^* + \mathbf{T}^*\mathbf{R}\mathbf{W}\mathbf{R}^{\mathsf{T}} + \mathbf{T}^*\boldsymbol{\Omega} \\
&= (\dot{\mathbf{T}})^* - \mathbf{R}\mathbf{W}\mathbf{T}\mathbf{R}^{\mathsf{T}} + \mathbf{R}\mathbf{T}\mathbf{W}\mathbf{R}^{\mathsf{T}} = \mathbf{R}(\dot{\mathbf{T}} - \mathbf{W}\mathbf{T} + \mathbf{T}\mathbf{W})\mathbf{R}^{\mathsf{T}} \\
&= \mathbf{R}\overset{\circ}{\mathbf{T}}\mathbf{R}^{\mathsf{T}} = (\overset{\circ}{\mathbf{T}})^*.
\end{aligned}$$

As we did above for the vector rates $\overset{\triangledown}{\boldsymbol{b}}$ and $\overset{\vartriangle}{\boldsymbol{b}}$, making use of (2.87) in equations (2.29)–(2.30), we can express the rates $\overset{\triangledown}{\mathbf{T}}$, $\overset{\vartriangle}{\mathbf{T}}$, $\overset{\blacktriangle}{\mathbf{T}}$, and $\overset{\blacktriangledown}{\mathbf{T}}$ as

$$\overset{\triangledown}{\mathbf{T}} = \overset{\circ}{\mathbf{T}} + \mathbf{T}\mathbf{D} - \mathbf{D}\mathbf{T}, \tag{2.88a}$$

$$\overset{\vartriangle}{\mathbf{T}} = \overset{\circ}{\mathbf{T}} + \mathbf{D}\mathbf{T} - \mathbf{T}\mathbf{D}, \tag{2.88b}$$

$$\overset{\blacktriangle}{\mathbf{T}} = \overset{\circ}{\mathbf{T}} + \mathbf{T}\mathbf{D} + \mathbf{D}\mathbf{T}, \tag{2.88c}$$

$$\overset{\blacktriangledown}{\mathbf{T}} = \overset{\circ}{\mathbf{T}} - \mathbf{T}\mathbf{D} - \mathbf{D}\mathbf{T}. \tag{2.88d}$$

It is an easy consequence of (2.87) that the corotational derivative $\overset{\circ}{\mathbf{T}}$ is either symmetric or skew-symmetric whenever $\mathbf{T}$ is correspondingly symmetric or skew-symmetric, proving that symmetry is preserved in the corotational evolution of a

[26] See [131, p. 156].

tensor. Equations (2.88) show that such a symmetry-preserving property is enjoyed by the rates $\overset{\blacktriangle}{\mathbf{T}}$ and $\overset{\blacktriangledown}{\mathbf{T}}$, but not by $\overset{\triangledown}{\mathbf{T}}$ and $\overset{\vartriangle}{\mathbf{T}}$. Furthermore, mirroring the pattern of (2.85), a codeformational derivative $\overset{\diamond}{\mathbf{T}}$ of a tensor field $\mathbf{T}$ can be defined as

$$\overset{\diamond}{\mathbf{T}} := \overset{\circ}{\mathbf{T}} + \sigma\,(\mathbf{DT} + \mathbf{TD}) + \tau(\operatorname{tr}\mathbf{D})\mathbf{T}, \tag{2.89}$$

where σ and τ are indifferent scalars as in (2.85) above. The definition of $\overset{\diamond}{\mathbf{T}}$ is such that it preserves the symmetry (or skew symmetry) of $\mathbf{T}$.

As proved in [131, p. 155] by adapting a classical argument of [234, p. 27], the rate $\overset{\bigstar}{\mathbf{T}}$ defined for a tensor field $\mathbf{T}$ by

$$\overset{\bigstar}{\mathbf{T}} := \dot{\mathbf{T}} + \boldsymbol{\Phi}(\mathbf{G}, \mathbf{T})$$

is frame-indifferent if and only if it can be expressed as

$$\overset{\bigstar}{\mathbf{T}} = \overset{\circ}{\mathbf{T}} + \boldsymbol{\Phi}(\mathbf{D}, \mathbf{T}), \tag{2.90}$$

where $\boldsymbol{\Phi}$ is an *isotropic* tensor-valued function, that is, a function such that

$$\boldsymbol{\Phi}(\mathbf{D}^*, \mathbf{T}^*) = \boldsymbol{\Phi}(\mathbf{D}, \mathbf{T})^* \quad \forall\, \mathbf{D}, \mathbf{T}. \tag{2.91}$$

It is easily seen that all the rates in (2.88) and (2.89) are in the form (2.90), and so they are indifferent.[27]

In the same vein, corotational and codeformational derivatives can be defined for higher-order tensors; see [252, 253] for the general case and [311] for a more specific form relevant to dissipative ordered fluids.

Hemi-indifference

Often the transformation laws (2.45) and (2.46) are required to hold only for all $\mathbf{R} \in \mathsf{SO}(3)$. Correspondingly, vector- or tensor-valued functions $\boldsymbol{\varphi}$ or $\boldsymbol{\Phi}$ that behave as in (2.86) and (2.91), but only for $\mathbf{R} \in \mathsf{SO}(3)$, are said to be *hemitropic*, instead of isotropic. An example of a hemitropic vector function was already encountered in (2.66) above, which since $(\det \mathbf{R})^2 = 1$ can also be written more concisely as

$$\boldsymbol{w}^* \times \boldsymbol{u}^* = (\det \mathbf{R})(\boldsymbol{w} \times \boldsymbol{u})^*. \tag{2.92}$$

Inspired by this, more generally we say that a vector $\boldsymbol{v}$ is *hemi-indifferent*[28] if under a change of frame represented by $\mathbf{R} \in \mathsf{O}(3)$ it transforms as

---

[27] Other invariant tensorial rates are found in [29]. We can apply to all of them the words of [345, p. 97]: they are examples "of the infinitely many possible invariant time fluxes that can be used."

[28] Having called a vector frame-indifferent when it is generally called *polar*, we take here the liberty of calling a vector that is usually called *axial* frame-hemi-indifferent. Sure enough, the axial vector $\boldsymbol{w}$ associated with an indifferent skew tensor $\mathbf{W}$ as in Appendix A.1 is hemi-indifferent. A hemi-indifferent vector is elsewhere also called a *pseudovector*.

$$\boldsymbol{v}^* = (\det \mathbf{R})\mathbf{R}\boldsymbol{v}. \tag{2.93}$$

Thus, (2.92) shows how the vector product of two frame-indifferent vectors is only hemi-indifferent. Similarly, we say that a tensor **T** is hemi-indifferent[29] if

$$\mathbf{T}^* = (\det \mathbf{R})\mathbf{R}\mathbf{T}\mathbf{R}^{\mathsf{T}}. \tag{2.94}$$

An example of a hemi-indifferent tensor is given by the spin gradient $\nabla \boldsymbol{w}$. Indeed, by differentiating both sides of equation (2.67) in the frame f*, by (2.58) and since $\boldsymbol{\Omega}$ is uniform in space, we obtain that[30]

$$(\nabla \boldsymbol{w})^* = (\det \mathbf{R})\mathbf{R}(\nabla \boldsymbol{w})\mathbf{R}^{\mathsf{T}}. \tag{2.95}$$

Here, as in (2.57) above and in the rest of the book, $(\nabla \boldsymbol{w})^*$ denotes the gradient of $\boldsymbol{w}^*$ computed in the frame f*.

2.1.4 Axioms of Classical Mechanics

The axiomatics of continuum mechanics is so rich in contributions that a whole monograph could easily be devoted to it, including the many controversies and contrasting views, too often more in their appearance than in their essence. Such a critical exposition would, however, exceed the scope of this chapter. Here, we shall admittedly be partial: following mostly NOLL's views,[31] especially as presented in [349, §§ 12,13], we shall posit a few simple axioms capable of justifying the basic balance laws of classical continuum mechanics and expressed in a form that can easily be extended to the more complex systems we envisage in our development, in the spirit suggested in [271] and [272] to make the essence of classical axioms predictive in nonclassical settings. We shall phrase the basic axioms in terms of indifference requirements.

Axiom 2.1 (Mass indifference). The mass $M(\mathcal{P})$ of any subbody $\mathcal{P}$ of $\mathcal{B}$ is attributed to all shapes $\{\mathcal{P}_t\}_{t\in I}$ that $\mathcal{P}$ takes in any motion χ of $\mathcal{B}$, and it is frame-indifferent.

Formally, Axiom 2.1 can be stated as follows:

$$M(\mathcal{P}_t^*, \chi^*) = M(\mathcal{P}_t, \chi) = M(\mathcal{P}), \quad \forall\, \mathcal{P} \in \mathcal{B} \quad \text{and} \quad \chi, \tag{2.96}$$

where $M(\mathcal{P}_t, \chi)$ is the mass of the shape at time t of the subbody $\mathcal{P}$ of $\mathcal{B}$ along the motion χ in the frame f and $M(\mathcal{P}_t^*, \chi^*)$ is the mass of the same subbody at the same time as seen from the frame f*.

29 A tensor obeying (2.94) is also called a *pseudotensor*.

30 In (2.95).

31 In particular, we refer the reader to the original papers [236] and [237]. NOLL's axiomatics has evolved since these contributions, as witnessed, for example, by [245], [248], and [246], the latter being a short essay on the horizon of mathematics in modern natural philosophy. A frame-free formulation of classical continuum mechanics plays a central role in these new developments.

Axiom 2.2 (Force indifference). All forces assigned to the parts $\mathcal{P}$ of a body $\mathcal{B}$ in whatever shape they take in their motions are frame-indifferent.

In this book we shall consider only forces that can be represented through densities, either relative to the volume measure, for *body forces*, or relative to the area measure, for *contact forces*. Body forces express the action at a distance exerted on the subbody $\mathcal{P}$ in its present shape $\mathcal{P}_t$, while contact forces express the action exerted on $\mathcal{P}_t$ through its reduced boundary $\partial^*\mathcal{P}_t$ by the complement of $\mathcal{P}_t$ in $\mathcal{B}_t$, imagined carved away in an ideal removal of the bonds that constitute the material comprising $\mathcal{B}$. Neither concentrated forces nor forces distributed along edges[32] will ever be treated here.

Letting $\boldsymbol{b}_\mathrm{e}$ denote the total body force density and $\boldsymbol{t}$ the total contact force density, the *resultant external force* $\boldsymbol{f}_\mathrm{e}$ exerted on the present shape $\mathcal{P}_t$ of a subbody $\mathcal{P}$ of $\mathcal{B}$ in a motion $\boldsymbol{\chi}$ is represented by

$$\boldsymbol{f}_\mathrm{e}(\mathcal{P}_t, \boldsymbol{\chi}) := \int_{\mathcal{P}_t} \boldsymbol{b}_\mathrm{e}\, dV + \int_{\partial^*\mathcal{P}_t} \boldsymbol{t}\, dA. \tag{2.97}$$

Axiom 2.2 states that both $\boldsymbol{b}_\mathrm{e}$ and $\boldsymbol{t}$ transform into

$$\boldsymbol{b}_\mathrm{e}^* = \mathbf{R}\boldsymbol{b}_\mathrm{e} \quad \text{and} \quad \boldsymbol{t}^* = \mathbf{R}\boldsymbol{t} \tag{2.98}$$

when the frame $\mathfrak{f}$ where the body performs the motion $\boldsymbol{\chi}$ is changed into the frame $\mathfrak{f}^*$ where the body performs the motion $\boldsymbol{\chi}^*$, so that

$$\boldsymbol{f}_\mathrm{e}^*(\mathcal{P}_t^*, \boldsymbol{\chi}^*) = \mathbf{R}\boldsymbol{f}_\mathrm{e}(\mathcal{P}_t, \boldsymbol{\chi}) \quad \forall\, \mathcal{P} \in \mathcal{B} \quad \text{and} \quad \boldsymbol{\chi}.$$

Axiom 2.2 is indeed more demanding than (2.98) may let us think, since it requires *all* forces to be frame-indifferent: were we able to split either $\boldsymbol{b}_\mathrm{e}$ or $\boldsymbol{t}$ into the sum of forces exerted by different identifiable agencies, each individual force should be subject to the indifference requirement in (2.98). In the light of (2.45), (2.98) says equivalently that all forces, be they body or contact forces, transform as vectors, that is, as translations in the Euclidean space where classical mechanics is hosted. Such an identification between force and space has a long history that goes back at least to NEWTON, who in a manuscript that dates from the late 1710s wrote:

> The forces and speeds of movable bodies do not properly pertain to geometry, but they can be expressed by means of lines, surface-areas, solids and angles, and to that extent reduced to geometry.[33]

Axiom 2.3 (Power indifference). The total power expended by the forces exerted by external agencies in any motion on every subbody $\mathcal{P}$ of $\mathcal{B}$ is frame-indifferent.

---

32 For a theory of edge interactions where surface contact forces are supplemented by line contact forces, we refer the reader to [250] (see also [107] and [73] for related contributions and [69] for an attempt to extend [250] to less regular shapes).

33 See [233, vol. 8, p. 453]. The interested reader is referred to [129, § 4.3] for a description of the role of geometry in NEWTON's *Principia*.

As suggested by the representation in (2.97) for the resultant $f_e(\mathcal{P}_t, \chi)$ of the external forces acting on $\mathcal{P}_t$ along the motion χ, the total power $\mathscr{W}^{(e)}(\mathcal{P}_t, \chi)$ that they expend can be written as

$$\mathscr{W}^{(e)}(\mathcal{P}_t, \chi) := \int_{\mathcal{P}_t} \boldsymbol{b}_e \cdot \boldsymbol{v}\, dV + \int_{\partial^*\mathcal{P}_t} \boldsymbol{t} \cdot \boldsymbol{v}\, dA, \tag{2.99}$$

where $\boldsymbol{v}$ is the velocity field associated with the motion χ. Axiom 2.3 can then be phrased in the following form:

$$\mathscr{W}^{(e)}(\mathcal{P}_t^*, \chi^*) = \mathscr{W}^{(e)}(\mathcal{P}_t, \chi) \quad \forall\, \mathcal{P} \in \mathcal{B} \text{ and } \chi. \tag{2.100}$$

We shall see in the following section how the axioms posited here imply the balance laws of classical continuum mechanics, which are to be regarded as the basic evolution laws of the theory. A preliminary, necessary step to this end is the classical axiom of inertia, which we shall assert in the classical setting, but in a form that may easily inspire its extension to the more general theories to which our study is directed.

We customarily split the density $\boldsymbol{b}_e$ of the external body forces into the sum

$$\boldsymbol{b}_e := \boldsymbol{b} + \boldsymbol{b}_i, \tag{2.101}$$

where traditionally $\boldsymbol{b}$ and $\boldsymbol{b}_i$ are the body force densities of the interactions that can be ascribed to the *near* bodies[34] and to the *far* bodies[35] around $\mathcal{B}_t$, respectively. We shall call $\boldsymbol{b}$ the density of *applied* forces and $\boldsymbol{b}_i$ the density of *inertial* forces. We shall not dwell any further in interpreting the splitting in (2.101); we rather heed that unlike $\boldsymbol{b}$, $\boldsymbol{b}_i$ is affected by the motion of the body $\mathcal{B}$, and so it depends on the frame to which the latter is referred. Whatever may be their representation, both $\boldsymbol{b}$ and $\boldsymbol{b}_i$ must comply with Axiom 2.2, and so they must be frame-indifferent.

The axiom of inertia will identify frames where $\boldsymbol{b}_i$ can be given a constitutive assignment. Once such an assignment is made, we can recover the force of inertia $\boldsymbol{b}_i^*$ in any other frame by requiring, in accordance with Axiom 2.2, that

$$\boldsymbol{b}_i^* := \mathbf{R}\boldsymbol{b}_i, \tag{2.102}$$

where $\mathbf{R}$ is the orthogonal tensor describing the change of frame. Here (2.102) properly appears as a definition.

Axiom 2.4 (Inertia postulate). There is a frame relative to which $\boldsymbol{b}_i$ is independent of $\boldsymbol{v}$ and

$$\int_{\mathcal{P}_t} \boldsymbol{b}_i \cdot \boldsymbol{v}\, dV = -\dot{\mathscr{K}}(\mathcal{P}_t, \chi) \quad \forall\, \mathcal{P} \in \mathcal{B} \text{ and } \chi, \tag{2.103}$$

where $\mathscr{K}(\mathcal{P}_t, \chi)$ is the shape functional traditionally identified with the *kinetic energy* of $\mathcal{P}$ along the motion χ.

34 In TRUESDELL's terminology, these are the bodies in the *great system* [349, § I.13].

35 More properly, these are the bodies outside the great system. This reflects the Machian point of view, which also inspires the modern historical perspective of BARBOUR [8].

In words, equation (2.103) says that the *opposite* to the kinetic energy $\mathscr{K}$ is the potential of inertial forces, since the power expended by $\boldsymbol{b}_\mathrm{i}$ can be expressed as the material time derivative of $\mathscr{K}$. This form of the postulate of inertia requires the constitutive choices for $\boldsymbol{b}_\mathrm{i}$ and $\mathscr{K}$ to be consistent with (2.103), at least in one frame. The requirement that $\boldsymbol{b}_\mathrm{i}$ be independent of $\boldsymbol{v}$ forbids adding to it any powerless contribution[36] that leaves (2.103) unaltered, since such an addition would indeed depend on $\boldsymbol{v}$.

By (2.101) and (2.103), in a frame where Axiom 2.4 is valid, the total external power $\mathscr{W}^{(\mathrm{e})}$ in (2.99) can be expressed as

$$\mathscr{W}^{(\mathrm{e})}(\mathcal{P}_t,\chi) = \mathscr{W}^{(\mathrm{a})}(\mathcal{P}_t,\chi) - \dot{\mathscr{K}}(\mathcal{P}_t,\chi), \tag{2.104}$$

where

$$\mathscr{W}^{(\mathrm{a})}(\mathcal{P}_t,\chi) := \int_{\mathcal{P}_t} \boldsymbol{b}\cdot\boldsymbol{v}\,dV + \int_{\partial^*\mathcal{P}_t} \boldsymbol{t}\cdot\boldsymbol{v}\,dA \tag{2.105}$$

is the power expended by all external agencies *applied* to $\mathcal{P}$, which exclude inertia, for which Axiom 2.3 requires frame-indifference. It is apparent from (2.104) that $\mathscr{W}^{(\mathrm{e})}$ can also be interpreted as *net working*; according to a definition of TRUESDELL [343, p. 9] it represents the external power still available to the subbody after that gone into motion has been taken into account.[37]

The requirement in (2.103) is not frame-indifferent, and its validity is confined to a specific frame, possibly representative of a whole class not large enough to encompass all frames. A frame where Axiom 2.4 is valid is called an *inertial* frame. Inertial frames are characteristic of classical mechanics. EINSTEIN clarified in his general theory of relativity that gravitational and inertial body forces cannot be distinguished in an objective manner [236, p. 279]. A consequence of (2.101) is that $\boldsymbol{b}_\mathrm{e}$ reduces to the inertial force $\boldsymbol{b}_\mathrm{i}$ whenever $\boldsymbol{b} = \boldsymbol{0}$; thus, $\boldsymbol{b}_\mathrm{i}$ appears as the irreducible body force related to the frame. As TRUESDELL [349, p. 70] wrote,

> An essential feature of classical mechanics is the existence of special frames in which the relation between forces and motions they produce is especially *simple*. Since we have these felicitous frames, it would be simply foolish not to use them.

In the following sections, we shall both explore the consequences of the axioms posited here and exploit the formulation we gave them to extend gradually their validity, without betraying their spirit, so as to encompass the scope this book intends to cover.

2.1.5 Classical Balance Equations

The basic laws of continuum mechanics express balances of mechanical quantities evolving in time along a motion: they generally involve both production rates and

---

[36] In a terminology attributed in [271] to SERRIN, such powerless inertial forces, excluded by our formulation of Axiom 2.4, are named after CORIOLIS. In the same terminology, the only inertial forces allowed here are called d'Alembertian.

[37] In general, in this book we employ the words *power* and *working* as synonyms, with a slight preference for the latter when it refers to shape functionals such as in (2.104) and (2.105).

sources, which are separately specified below. In the whole discussion that follows, a motion χ of the body $\mathcal{B}$ is presumed to be assigned in a frame f, as defined in (2.3).

Mass

Since the motion χ maps the measure space $\mathcal{B}$ into another measure space, $\mathcal{B}_t$, at any time $t \in I$, the mass density ϱ_0 defined by (2.2) for any part $\mathcal{P}$ of the body $\mathcal{B}$ can similarly be defined for any part $\mathcal{P}_t$ of $\mathcal{B}_t$, so that the mass of $\mathcal{P}_t$ can be represented as

$$M(\mathcal{P}_t, \chi) = \int_{\mathcal{P}_t} \varrho(p_t, t) dV, \tag{2.106}$$

with $\varrho : \mathcal{C}_\chi \to \mathbb{R}^+$ the mass density along the whole collection of shapes $\mathcal{C}_\chi$ induced by the motion χ. M in (2.106) is a special shape functional in the form (2.36).

Axiom 2.1 states, in particular, that mass is *conserved*, that is, see also (2.96), that $\dot{M}(\mathcal{P}_t, \chi) = 0$ for all $t \in I$ and $\mathcal{P}_t$. Using the transport theorem (2.37), we see that this conservation law implies that

$$\dot{M}(\mathcal{P}_t, \chi) = \int_{\mathcal{P}_t} (\dot{\varrho} + \varrho \operatorname{div} \boldsymbol{v})\, dV = 0. \tag{2.107}$$

Since this has to hold for all $\mathcal{P}_t$, the integrand in (2.107) must vanish identically, which leads to the local form of conservation of mass,

$$\dot{\varrho} + \varrho \operatorname{div} \boldsymbol{v} = 0. \tag{2.108}$$

Since the volume measure V is frame-indifferent, Axiom 2.1 also ensures that the mass density ϱ is an indifferent scalar,

$$\varrho^* = \varrho, \tag{2.109}$$

and so is also its material time derivative, while its partial time derivative transforms as in (2.82). Since, by (2.61), div $\boldsymbol{v}$ is a frame-indifferent field, we see that the whole balance equation (2.108) is frame-indifferent, as it should be.

In the case of an isochoric motion, by (2.35), div $\boldsymbol{v} = 0$ and (2.108) becomes $\dot{\varrho} = 0$, which prescribes the material time derivative of the density to vanish at all times. This means that the density at any material point $P \in \mathcal{B}$ remains the same throughout the motion.

Many significant mechanical quantities have a *specific* density, that is, they are absolutely continuous relative to the mass measure M; this implies that their density relative to the volume measure V is proportional to ϱ, and so they can be written as

$$\Psi(\mathcal{P}_t, \chi) := \int_{\mathcal{P}} \varrho \psi(p_t, t) dV. \tag{2.110}$$

As a consequence of conservation of mass, the transport theorem for such quantities takes a particularly simple form:

$$\dot{\psi}(\mathcal{P}_t,\chi) = \int_{\mathcal{P}_t} [(\varrho\psi)^{\cdot} + \varrho\psi \operatorname{div} \boldsymbol{v}]\, dV = \int_{\mathcal{P}_t} \varrho\dot{\psi}\, dV, \tag{2.111}$$

because, with (2.108),

$$(\varrho\psi)^{\cdot} + \varrho\psi \operatorname{div} \boldsymbol{v} = \varrho\dot{\psi} + \psi\,(\dot{\varrho} + \varrho \operatorname{div} \boldsymbol{v}) = \varrho\dot{\psi}.$$

Similarly, mirroring the definition of the vector-valued shape functional (2.41) in

$$\boldsymbol{\Psi}(\mathcal{P}_t,\chi) := \int_{\mathcal{P}} \varrho\boldsymbol{\psi}(p_t,t)dV,$$

using (2.42) and (2.108), we also arrive at

$$\dot{\boldsymbol{\Psi}}(\mathcal{P}_t,\chi) = \int_{\mathcal{P}_t} \varrho\dot{\boldsymbol{\psi}}\, dV, \tag{2.112}$$

which parallels (2.111).

Forces and Torques

The mathematical statement of Axiom 2.3 in (2.100) is an identity to be satisfied by all subbodies $\mathcal{P}$ in all motions χ. Following [237], we show now that, for (2.100) to hold, the external body force $\boldsymbol{b}_{\mathrm{e}}$ and the traction $\boldsymbol{t}$ cannot be arbitrarily assigned, but they must satisfy appropriate laws. By (2.98) and (2.73), (2.100) becomes

$$\int_{\mathcal{P}_t^*} \mathbf{R}\boldsymbol{b}_{\mathrm{e}} \cdot \mathbf{R}(\boldsymbol{v} - \boldsymbol{v}_{\mathrm{R}})dV + \int_{\partial^*\mathcal{P}_t^*} \mathbf{R}\boldsymbol{t} \cdot \mathbf{R}(\boldsymbol{v} - \boldsymbol{v}_{\mathrm{R}})$$
$$= \int_{\mathcal{P}_t} \boldsymbol{b}_{\mathrm{e}} \cdot \boldsymbol{v} dV + \int_{\partial^*\mathcal{P}_t} \boldsymbol{t} \cdot \boldsymbol{v}\, dA,$$

which, since $\mathbf{R}$ is orthogonal, is equivalent to

$$\boldsymbol{v}_0^* \left(\cdot \int_{\mathcal{P}_t} \boldsymbol{b}_{\mathrm{e}}\, dV + \int_{\partial^*\mathcal{P}_t} \boldsymbol{t}\, dA \right)$$
$$+ \boldsymbol{\Omega}^* \cdot \left(\int_{\mathcal{P}_t} \boldsymbol{b}_{\mathrm{e}} \otimes (p-o)dV + \int_{\partial^*\mathcal{P}_t} \boldsymbol{t} \otimes (p-o)dA \right) = 0, \tag{2.113}$$

where $\boldsymbol{v}_0*$ and $\boldsymbol{\Omega}^*$ are as in (2.72) and (2.69), respectively, and use has been made of (2.71) and of the change of variables $(p^* - o^*) = \mathbf{R}(p-o)$ to reduce $\mathcal{P}_t^*$ to $\mathcal{P}_t$. Since (2.113) must be valid for any vector $\boldsymbol{v}_0$ and any skew tensor $\boldsymbol{\Omega}$, it is satisfied if and only if

$$\int_{\mathcal{P}_t} \boldsymbol{b}_{\mathrm{e}}\, dV + \int_{\partial^*\mathcal{P}_t} \boldsymbol{t}\, dA = \mathbf{0}, \tag{2.114a}$$

$$\int_{\mathcal{P}_t} (p-o) \times \boldsymbol{b}_{\mathrm{e}}\, dV + \int_{\partial^*\mathcal{P}_t} (p-o) \times \boldsymbol{t}\, dA = \mathbf{0}. \tag{2.114b}$$

These equations express the balance of external forces and torques acting on the present shape $\mathcal{P}_t$ of the subbody $\mathcal{P}$. In this form, they are valid in all frames. They will be given the classical form of Cauchy's laws of motion, valid only in inertial frames, once a specific constitutive choice for the inertial force $\boldsymbol{b}_{\mathrm{i}}$ is made.

Inertia

We now make use of Axiom 2.4 to obtain the appropriate representation of inertial forces in classical mechanics. We start by representing $\mathscr{K}$ in an inertial frame (one at least existing by Axiom 2.4) as

$$\mathscr{K}(\mathcal{P}_t, \chi) = \int_{\mathcal{P}_t} \varrho \kappa_0(\boldsymbol{v}) dV, \tag{2.115}$$

where κ_0, which is supposed to depend only on $\boldsymbol{v}$, is the specific density (per unit mass) of kinetic energy. Then, by (2.110), equation (2.103) becomes

$$\int_{\mathcal{P}_t} \boldsymbol{b}_\mathrm{i} \cdot \boldsymbol{v}\, dV = -\int_{\mathcal{P}_t} \varrho \dot{\kappa}_0(\boldsymbol{v}) dV = -\int_{\mathcal{P}_t} \varrho \frac{\partial \kappa_0}{\partial \boldsymbol{v}} \cdot \boldsymbol{a}\, dV,$$

which must be valid for all subbodies $\mathcal{P}$ and all motions χ, and, consequently, since $\boldsymbol{b}_\mathrm{i}$ is independent of $\boldsymbol{v}$, requires $\frac{\partial \kappa_0}{\partial \boldsymbol{v}}$ to be linear in $\boldsymbol{v}$. We set[38]

$$\frac{\partial \kappa_0}{\partial \boldsymbol{v}} = \boldsymbol{v}, \tag{2.116}$$

so that κ_0 acquires the familiar form

$$\kappa_0 = \frac{1}{2} \boldsymbol{v} \cdot \boldsymbol{v}, \tag{2.117}$$

and, consequently,

$$\boldsymbol{b}_\mathrm{i} = -\varrho \boldsymbol{a}. \tag{2.118}$$

As clearly stated by NOLL [244],[39] inertial forces have a constitutive nature; here we derive the constitutive law (2.118) for $\boldsymbol{b}_\mathrm{i}$ from the classical choice for κ_0 in (2.117), which is equally constitutive in nature. All this care might appear as an unnecessary complication to a critical reader, since positing directly (2.118) would suffice to the purposes of classical continuum mechanics. However, in view of our later development, exploring the general consequences of our formulation of the inertia postulate will prove useful in guiding our steps in less familiar territory.

In the same spirit, it is instructive to see how Axiom 2.4 implies in general the existence of a whole class of frames where (2.103) is valid, and which equally deserve to be called inertial. To identify such a class of frames, we require that the kinetic energy $\mathscr{K}$ be indifferent, that is, we define $\mathscr{K}^*$ in such a way that $\mathscr{K}^*(\mathcal{P}_t^*, \chi^*) = \mathscr{K}(\mathcal{P}_t, \chi)$, for all subbodies $\mathcal{P} \subset \mathcal{B}$ and all motions χ. By (2.115) this requirement becomes

$$\int_{\mathcal{P}_t^*} \varrho^* \kappa_0^*(\boldsymbol{v}^*) dV = \int_{\mathcal{P}_t} \varrho \kappa_0(\boldsymbol{v}) dV,$$

---

38 PODIO-GUIDUGLI [271] writes $\frac{\partial \kappa_0}{\partial \boldsymbol{v}} = \mathbf{M}\boldsymbol{v}$ and calls $\mathbf{M}$, a symmetric tensor independent of $\boldsymbol{v}$, the *mass tensor*. Correspondingly, κ_0 is written as $\kappa_0 = \frac{1}{2}\boldsymbol{v} \cdot \mathbf{M}\boldsymbol{v}$. In our choice (2.116), $\mathbf{M}$ is the identity tensor.

39 See also [131, p. 144].

which, since the mass density ϱ is indifferent and $\mathcal{P}_t^*$ is the image of $\mathcal{P}_t$ under an isometry, is equivalent to the local form

$$\kappa_0^*(\boldsymbol{v}^*) = \kappa_0(\boldsymbol{v}), \tag{2.119}$$

where $\boldsymbol{v}^*$ is related to $\boldsymbol{v}$ through (2.73). For $\mathscr{K}^*$ to comply with Axiom 2.4 like $\mathscr{K}$, it must also obey the identity $\dot{\mathscr{K}}^* = \dot{\mathscr{K}}$, which by (2.111) reduces to

$$\dot{\kappa}_0^*(\boldsymbol{v}^*) = \dot{\kappa}_0(\boldsymbol{v}). \tag{2.120}$$

Recalling that (2.119) and (2.73) imply the identity

$$\frac{\partial \kappa_0^*}{\partial \boldsymbol{v}^*} = \mathbf{R}\frac{\partial \kappa_0}{\partial \boldsymbol{v}},$$

using again (2.73), we give (2.120) the following form:

$$\frac{\partial \kappa_0}{\partial \boldsymbol{v}} \cdot \left(\mathbf{R}^{\mathsf{T}}\boldsymbol{a}_{\mathrm{R}} - \boldsymbol{\Omega}^*\boldsymbol{v}\right) = 0,$$

where $\boldsymbol{\Omega}^*$ is as in (2.68), which must be valid for all motions. This requires that $\boldsymbol{\Omega}^* = \mathbf{0}$ and $\boldsymbol{a}_{\mathrm{R}} = \mathbf{0}$, which shows that two inertial frames differ by a constant orthogonal tensor and a translation with constant velocity. Often they are said to differ by a *Galilean transformation*. By (2.74), in such a transformation, the acceleration $\boldsymbol{a}$ is an indifferent vector, and so $\boldsymbol{b}_{\mathrm{i}}$ obeys (2.102).

For (2.102) to be valid for all changes of frame, we *define* the inertial force $\boldsymbol{b}_{\mathrm{i}}^*$ in a noninertial frame $\mathfrak{f}^*$ as

$$\boldsymbol{b}_{\mathrm{i}}^* = \mathbf{R}\boldsymbol{b}_{\mathrm{i}} = -\varrho\mathbf{R}\boldsymbol{a} = -\varrho\boldsymbol{a}^* - \varrho(\mathbf{R}\boldsymbol{a}_{\mathrm{R}} - \boldsymbol{\Omega}\boldsymbol{v}^*), \tag{2.121}$$

where $\boldsymbol{v}^*$ and $\boldsymbol{a}^*$ are velocity and acceleration in $\mathfrak{f}^*$, $\boldsymbol{\Omega}$ is the spin tensor of $\mathfrak{f}^*$ relative to an inertial frame $\mathfrak{f}$, and $\boldsymbol{a}_{\mathrm{R}}$ is the acceleration in $\mathfrak{f}$ of the rigid motion for which $\mathfrak{f}^*$ is the rest frame. Since (2.121) is the form taken by the inertial force in a noninertial frame, it should be no surprise that it depends on the spin tensor relative to an inertial frame. By use of (2.71) and (2.50), we easily arrive at

$$\mathbf{R}\boldsymbol{a}_{\mathrm{R}} - \boldsymbol{\Omega}\boldsymbol{v}^* = -(o^*)^{\cdot\cdot} + (\boldsymbol{\Omega}^2 - \dot{\boldsymbol{\Omega}})(p^* - o^*) - 2\boldsymbol{\Omega}[\boldsymbol{v}^* - (o^*)^{\cdot}],$$

which allows us to rewrite $\boldsymbol{b}_{\mathrm{i}}^*$ in (2.121) as

$$\boldsymbol{b}_{\mathrm{i}}^* = -\varrho\boldsymbol{a}^* + \boldsymbol{b}_{\mathfrak{f}^*} + \boldsymbol{b}_{\mathrm{C}},$$

with

$$\boldsymbol{b}_{\mathfrak{f}^*} := \varrho[(o^*)^{\cdot\cdot} + (\dot{\boldsymbol{\Omega}} - \boldsymbol{\Omega}^2)(p^* - o^*)],$$
$$\boldsymbol{b}_{\mathrm{C}} := 2\varrho\boldsymbol{\Omega}[\boldsymbol{v}^* - (o^*)^{\cdot}],$$

where $\boldsymbol{b}_{\mathfrak{f}^*}$ and $\boldsymbol{b}_{\mathrm{C}}$ are inertial forces different in nature, since only the latter, which is named after CORIOLIS, depends on the velocity $\boldsymbol{v}^*$.

Linear Momentum

It follows from (2.118) and (2.101) that in an inertial frame the balance equation of force (2.114a) can be written as

$$\dot{\boldsymbol{P}}(\mathcal{P}_t, \chi) = \int_{\partial^* \mathcal{P}_t} \boldsymbol{t}\, dA + \int_{\mathcal{P}_t} \boldsymbol{b}\, dV, \tag{2.122}$$

where

$$\boldsymbol{P}(\mathcal{P}_t, \chi) := \int_{\mathcal{P}_t} \varrho \boldsymbol{v}\, dV \tag{2.123}$$

is the total *linear momentum* of the subbody $\mathcal{P}$ in its present shape $\mathcal{P}_t$, and the transport theorem (2.112) has been used to prove that

$$\dot{\boldsymbol{P}}(\mathcal{P}_t, \chi) = \int_{\mathcal{P}_t} \varrho \boldsymbol{a}\, dV. \tag{2.124}$$

By (2.123), (2.122) can also be interpreted as the balance of linear momentum, for which $\boldsymbol{t}$ plays the role of a *flux* and $\boldsymbol{b}$ that of a *supply*.

Stress Tensor

The field $\boldsymbol{t}$, which in (2.114a) represents the external contact action exerted on the present shape $\mathcal{P}_t$ of the subbody $\mathcal{P}$ through its reduced boundary $\partial^* \mathcal{P}_t$, is properly to be regarded as a functional of the set $\partial^* \mathcal{P}_t$. Following the pioneering work of CAUCHY (see, for example, [345, p. 40] and [349, Chapter III]), we assume that at a point $p \in \partial^* \mathcal{P}_t$ the traction $\boldsymbol{t}$ is the same for all shapes that share with $\partial^* \mathcal{P}_t$ the same tangent plane at p. In other words, we assume that all like-oriented contacts with the same contact plane share the same traction. This assumption, which is often called the CAUCHY *postulate* or *stress principle*, is expressed formally by writing $\boldsymbol{t} = \boldsymbol{t}(p, \boldsymbol{\nu})$, where $\boldsymbol{\nu}$ is the outer unit normal to $\partial^* \mathcal{P}_t$ at p. As shown in [349, p. 172], such a statement could also be derived from some weaker assumptions,[40] which forms the content of the HAMEL–NOLL theorem . Here we are interested only in its consequences, the most fundamental of which is the existence of the stress tensor.

Theorem 2.5 (CAUCHY). *If the balance equation* (2.114a) *is satisfied and* $\boldsymbol{t}(\cdot, \boldsymbol{\nu})$ *is a continuous mapping, then there exists a tensor field* $\mathbf{T}$ *such that*

$$\boldsymbol{t}(p, \boldsymbol{\nu}) = \mathbf{T}(p)\boldsymbol{\nu}. \tag{2.125}$$

Proofs of this theorem can be found in most continuum mechanics textbooks; see, for example, [349, p. 174] and [131, p. 137]. The same conclusion (2.125) can also be reached under weaker hypotheses than that advanced in Theorem 2.5: it would exceed the scope of this book to review even the most relevant contributions to this field

[40] It is enough to assume that the traction field $\boldsymbol{t}$ in (2.97) is given by a *functional* of the set $\partial^* \mathcal{P}_t$.

of continuum mechanics that borders on analysis.[41] Like any vector in the translation space $\mathcal{V}$, the normal $\boldsymbol{\nu}$ transforms in an indifferent manner under a change of frame, that is, like $\boldsymbol{u}$ in (2.45). Therefore, CAUCHY's stress tensor is frame-indifferent and transforms as in (2.46), because the traction $\boldsymbol{t}$, being a force, is frame-indifferent by Axiom 2.2.[42]

The tensor $\mathbf{T}$ is called the CAUCHY *stress tensor*: if $\mathbf{T}$ is positive definite, then $\boldsymbol{\nu} \cdot \boldsymbol{t} = \boldsymbol{\nu} \cdot \mathbf{T}\boldsymbol{\nu} > 0$, and so the traction exerts a *tension*; if $\mathbf{T}$ is negative definite, then $\boldsymbol{t}$ exerts a *pressure*.

The existence of the stress tensor allows us to transform the surface integral in (2.122) into a volume integral:

$$\int_{\partial^* \mathcal{P}_t} \boldsymbol{t}\, dA = \int_{\partial^* \mathcal{P}_t} \mathbf{T}\boldsymbol{\nu}\, dA = \int_{\mathcal{P}_t} \operatorname{div} \mathbf{T}\, dV. \tag{2.126}$$

With (2.126) and (2.124), the balance of forces in (2.122) can now be written as a single integral,

$$\int_{\mathcal{P}_t} (\varrho \boldsymbol{a} - \operatorname{div} \mathbf{T} - \boldsymbol{b})\, dV = \mathbf{0}.$$

This can hold for an arbitrary subbody $\mathcal{P}$ only if the integrand vanishes identically, which yields the local form of the linear momentum balance,

$$\varrho \boldsymbol{a} = \operatorname{div} \mathbf{T} + \boldsymbol{b}. \tag{2.127}$$

Equation (2.127) is CAUCHY's *first law of motion*, valid in any inertial frame. This equation is frame-indifferent, provided, with the aid of (2.121), in a noninertial frame we write

$$\mathbf{R}\boldsymbol{a} = \boldsymbol{a}^* + \mathbf{R}\boldsymbol{a}_{\mathrm{R}} - \boldsymbol{\Omega}\boldsymbol{v}^*.$$

Rotational Momentum

The balance equation derived in (2.114b) from our assumption on the frame-indifference of the external power is now interpreted as the balance equation for the *rotational momentum* (also called the *moment of momentum*), thus paralleling the interpretation of (2.114a) as the balance equation of linear momentum.

The total rotational momentum of a part $\mathcal{P}$ in the motion χ relative to an arbitrarily chosen but fixed point of origin o is

$$\boldsymbol{K}_o(\mathcal{P}_t, \chi) := \int_{\mathcal{P}_t} \varrho \boldsymbol{x} \times \boldsymbol{v}\, dV, \tag{2.128}$$

where $\boldsymbol{x} = p - o$ is the position vector of the point p relative to o. It easily follows from (2.128) that the rate of change $\dot{\boldsymbol{K}}_o$ of $\boldsymbol{K}_o$ is found using (2.112) to be

41 For the interested reader, we record here without comment a list of contributors referred to in the short, effective review that opens thc introduction of [282]: GURTIN & MARTINS [132], ŠILHAVÝ [300, 301], DEGIOVANNI, MARZOCCHI, & MUSESTI [68], FOSDICK & VIRGA [108], SEGEV [291, 292], SEGEV & RODNAY [293].

42 The reader is advised to recall the reasoning leading to (2.46) on page 85.

$$\dot{\boldsymbol{K}}_o = \int_{\mathcal{P}_t} \varrho(\dot{\boldsymbol{x}} \times \boldsymbol{v} + \boldsymbol{x} \times \dot{\boldsymbol{v}}) dV = \int_{\mathcal{P}_t} \varrho \boldsymbol{x} \times \dot{\boldsymbol{v}}\, dV = \int_{\mathcal{P}_t} \varrho \boldsymbol{x} \times \boldsymbol{a}\, dV, \tag{2.129}$$

because, for a fixed origin o, $\dot{\boldsymbol{x}} = \boldsymbol{v}$. Thus, (2.114b) can be written as

$$\dot{\boldsymbol{K}}_o = \int_{\partial^* \mathcal{P}_t} \boldsymbol{x} \times \boldsymbol{t}\, dA + \int_{\mathcal{P}_t} \boldsymbol{x} \times \boldsymbol{b}\, dV \tag{2.130}$$

and readily interpreted as the balance of rotational momentum, with the torque of the contact force $\boldsymbol{t}$ as flux and the torque of the body force $\boldsymbol{b}$ as supply. To transform the surface integral in (2.130) into a volume integral, we use the stress tensor $\mathbf{T}$ from (2.125) to obtain

$$\int_{\partial^* \mathcal{P}_t} \boldsymbol{x} \times \boldsymbol{t}\, dA = \int_{\partial^* \mathcal{P}_t} \boldsymbol{x} \times \mathbf{T}\boldsymbol{\nu}\, dA = \int_{\partial^* \mathcal{P}_t} \mathbf{A}\boldsymbol{\nu}\, dA = \int_{\mathcal{P}_t} \operatorname{div} \mathbf{A}\, dV,$$

where we have defined the tensor $\mathbf{A}$ such that

$$\mathbf{A}\boldsymbol{u} = \boldsymbol{x} \times \mathbf{T}\boldsymbol{u}, \quad \forall\, \boldsymbol{u} \in \mathcal{V}.$$

The components A_{il} of $\mathbf{A}$ in any basis $(\boldsymbol{e}_1, \boldsymbol{e}_2, \boldsymbol{e}_3)$ of $\mathcal{V}$ are $A_{il} = \epsilon_{ijk} x_j T_{kl}$, which means that in the same basis the components of $\operatorname{div} \mathbf{A}$ are

$$(\operatorname{div} \mathbf{A})_i = A_{il,l} = \epsilon_{ijk}(x_{j,l} T_{kl} + x_j T_{kl,l}) = \epsilon_{ijk}(T_{kj} + x_j T_{kl,l}),$$

where we have used that $x_{j,l} = \delta_{jl}$, and so, in intrinsic notation,

$$\operatorname{div} \mathbf{A} = 2\boldsymbol{\tau} + \boldsymbol{x} \times \operatorname{div} \mathbf{T}, \tag{2.131}$$

where $\boldsymbol{\tau}$ is the axial vector associated with the skew-symmetric part $\operatorname{skw}(\mathbf{T})$ of $\mathbf{T}$ via[43]

$$\operatorname{skw}(\mathbf{T})\boldsymbol{u} = \boldsymbol{\tau} \times \boldsymbol{u} \quad \forall\, \boldsymbol{u} \in \mathcal{V}. \tag{2.132}$$

Hence the total torque transmitted by the contact forces takes the form

$$\int_{\partial^* \mathcal{P}_t} \boldsymbol{x} \times \boldsymbol{t}\, dA = \int_{\mathcal{P}_t} (2\boldsymbol{\tau} + \boldsymbol{x} \times \operatorname{div} \mathbf{T}) dV$$

and the balance (2.130) becomes

$$\int_{\mathcal{P}_t} (\varrho \boldsymbol{x} \times \boldsymbol{a} - 2\boldsymbol{\tau} - \boldsymbol{x} \times \operatorname{div} \mathbf{T} - \boldsymbol{x} \times \boldsymbol{b})\, dV = \mathbf{0},$$

where use has also been made of (2.129). The usual localization argument shows that the integrand has to vanish identically, which leads to

$$\boldsymbol{x} \times (\varrho \boldsymbol{a} - \operatorname{div} \mathbf{T} - \boldsymbol{b}) = 2\boldsymbol{\tau}.$$

---

[43] See Appendix A.1.

Since the term in the parentheses on the left-hand side equals zero by CAUCHY's first law of motion (2.127), it follows that

$$\boldsymbol{\tau} = \mathbf{0}. \tag{2.133}$$

By the definition (2.132) of $\boldsymbol{\tau}$, equation (2.133) is true if and only if $\mathbf{T}$ is symmetric, that is,

$$\mathbf{T} = \mathbf{T}^{\mathsf{T}}. \tag{2.134}$$

This is CAUCHY's *second law of motion*, which says that the balance of rotational momentum simply implies that the stress tensor is symmetric.

CAUCHY's two laws of motion (2.127) and (2.134) govern all *simple* continua. We shall see in Section 2.1.6 below how they need to be modified to encompass the more general continua we envisage in this book.

Power

Equations (2.127) and (2.134) guarantee that the total external power $\mathscr{W}^{(e)}$ is frame-indifferent. We now make use of these equations to find the explicit frame-indifferent form that $\mathscr{W}^{(e)}$ acquires when they are valid.

By use of (2.125) and applying the divergence theorem, we give $\mathscr{W}^{(a)}$ in (2.105) the following form:

$$\mathscr{W}^{(a)}(\mathcal{P}_t, \chi) = \int_{\mathcal{P}_t} [\boldsymbol{v} \cdot \boldsymbol{b} + \operatorname{div}(\mathbf{T}\boldsymbol{v})]\, dV,$$

where also (2.134) has been employed. Since $\operatorname{div}(\mathbf{T}\boldsymbol{v}) = \operatorname{div}\mathbf{T} \cdot \boldsymbol{v} + \mathbf{T} \cdot \nabla\boldsymbol{v}$, by (2.127), we see that

$$\mathscr{W}^{(a)}(\mathcal{P}_t, \chi) = \int_{\mathcal{P}_t} (\mathbf{T} \cdot \mathbf{D} + \varrho\boldsymbol{a} \cdot \boldsymbol{v})\, dV, \tag{2.135}$$

where (2.134) has been used again along with (2.60). By (2.116) and (2.104), we immediately derive from (2.135) that

$$\mathscr{W}^{(e)}(\mathcal{P}_t, \chi) = \int_{\mathcal{P}_t} \mathbf{T} \cdot \mathbf{D}\, dV =: \mathscr{W}^{(i)}(\mathcal{P}_t, \chi), \tag{2.136}$$

where $\mathscr{W}^{(i)}$ is interpreted as the power expended by the *internal forces*.[44] By (2.61) and (2.46), $\mathscr{W}^{(i)}$ is frame-indifferent, as it should be, since (2.136) shows that it equals $\mathscr{W}^{(e)}$ along all motions that obey CAUCHY's laws. Moreover, it readily follows from (2.75) that $\mathscr{W}^{(i)}$ vanishes identically for all subbodies in any rigid motion.

We shall see below how (2.136) needs to be modified when other internal agencies are able to expend power. At this stage, it suffices to remark that, by (2.136), equation (2.104) becomes

[44] An alternative name for $\mathscr{W}^{(i)}$ is the *power stress*, which is often used in the literature (see, for example, [343] and [131]).

$$\mathscr{W}^{(\mathrm{i})}(\mathcal{P}_t,\chi)+\dot{\mathscr{K}}(\mathcal{P}_t,\chi)=\mathscr{W}^{(\mathrm{a})}(\mathcal{P}_t,\chi), \tag{2.137}$$

which says that the power expended in an inertial frame by the external forces applied to any subbody $\mathcal{P}$ in any motion χ is balanced by the power expended by the internal forces and the rate of change in the kinetic energy.

Constitutive Laws

In deriving the balance equations we have made only fairly general assumptions about the nature of the underlying continuum. Consequently, they hold for a wide range of different materials. To describe the behavior of a particular body, additional assumptions are needed about the nature of the specific material that constitutes the body. These assumptions are usually stated in the form of *constitutive equations*. These equations describe the relationship between the motion of the body and the stress tensor, and where applicable the couple stress tensor.

The nature of any material is independent of the frame of reference from which it is observed. A constitutive equation specifies the stress tensor in terms of the motion of the material. But the stress tensor relates exclusively to forces *within* a body, and so constitutive equations must be written in such a way that all observers agree on the stress tensor for any given motion of the body.

As we pointed out above, for the stress tensor to be objective, it has to depend on indifferent quantities only. For example, in a simple fluid the only indifferent quantity is $\mathbf{D}$, and so the stress $\mathbf{T}$ has to be a function that depends only on $\mathbf{D}$. We will see simple examples of this in Section 2.3 below.

We also insisted in this section on the constitutive nature of inertia. For general continua, this means that the intrinsic rotational moment $\boldsymbol{k}_{\mathrm{i}}$ must be given a constitutive definition in terms of the motion. We shall show in the following section that the principle we posit there will also serve the purpose of deriving the constitutive choices for both stresses and inertia from those for simpler scalar functions with a direct physical interpretation.

2.1.6 General Balance Equations

The axiomatic scheme recalled above, from which we derived CAUCHY's classical laws of motion, was extended by BEATTY [13] to the more general situation in which interactions may transfer to subbodies torques that are not moments of forces. Perhaps the first coherent mathematical theory for bodies capable of such interactions was proposed by the COSSERATs [53, 54]: it was based on an appropriate Hamiltonian principle, and was not concerned with the effects of dissipation.[45]

Here we resume the presentation of the general balance equations valid in this case from the integral forms of the balances of linear and rotational momenta: it is shown in [13] how these balance equations can indeed be derived from an extension

[45] The interested reader is referred to [345, § 98] for a modern presentation of this theory and its extension due to TOUPIN [338, 339]. Attention to the COSSERATs' theory had already been drawn by TRUESDELL [341].

of Axiom 2.3 on the indifference of the power of external actions. Actually, only the balance equation of torques is affected by the consequences of assuming that a distribution of contact couples supplements the moments of contact forces and a distribution of body couples supplements the moments of body forces, including their inertial components. No assumption needs in principle to be made on the origin of these couples, though here we ascribe them to the nature of the order structure that inhabits what classical continuum mechanics simply regards as a body-point.

The balance of rotational momentum in a continuous body takes the particularly simple form in (2.134) if the material points are treated exactly as that: points without any structure attached to them. However, for continua with internal microstructure, such as all types of liquid crystals, the local spin of the "points" needs to be taken into account. We show now how the balance of rotational momentum must be augmented in the case of materials with a microstructure.

Such an internal structure contributes a local spin, and so the total rotational momentum in (2.128) has to be augmented by the rotational momentum of the material element. Though we could derive such an intrinsic rotational momentum from a proper extension of Axiom 2.4, for brevity, we assume that it possesses the specific density $\boldsymbol{k}_{\mathrm{i}}$, so that $\boldsymbol{K}_o$ in (2.128) is eventually replaced by

$$\boldsymbol{K}_o = \int_{\mathcal{P}} \varrho(\boldsymbol{x} \times \boldsymbol{v} + \boldsymbol{k}_{\mathrm{i}})dV. \tag{2.138}$$

In addition to moments of forces, there also are couples. With a body couple per unit volume $\boldsymbol{k}$ and a surface contact couple per unit area $\boldsymbol{c}$, the rotational momentum balance (2.130) becomes

$$\dot{\boldsymbol{K}}_o = \int_{\partial\mathcal{P}} (\boldsymbol{x} \times \boldsymbol{t} + \boldsymbol{c})dA + \int_{\mathcal{P}} (\boldsymbol{x} \times \boldsymbol{b} + \boldsymbol{k})dV. \tag{2.139}$$

The surface couple has a representation in terms of a second-rank tensor similar to that of the the traction in (2.125). If the surface couple $\boldsymbol{c}$ is, at any point p, a function of the normal $\boldsymbol{\nu}$ to $\partial^*\mathcal{P}$, then it can be shown that it is a linear function of the normal. It can hence be expressed in terms of a *couple stress* tensor $\mathbf{L}$ according to[46]

$$\boldsymbol{c} = \mathbf{L}\boldsymbol{\nu}. \tag{2.140}$$

Together with div $\mathbf{A}$ as in (2.131) the balance (2.139) then becomes

$$\dot{\boldsymbol{K}}_o = \int_{\mathcal{P}} [\mathrm{div}(\mathbf{A} + \mathbf{L}) + \boldsymbol{x} \times \boldsymbol{b} + \boldsymbol{k}]dV.$$

With $\boldsymbol{K}_o$ as in (2.138), by (2.112), we have

$$\dot{\boldsymbol{K}}_o = \int_{\mathcal{P}} \varrho(\boldsymbol{x} \times \dot{\boldsymbol{v}} + \dot{\boldsymbol{k}}_{\mathrm{i}})dV,$$

and comparing the last two equations then yields

[46] See also TRUESDELL & TOUPIN [348, § 200].

$$\boldsymbol{x} \times (\varrho \boldsymbol{a} - \operatorname{div} \mathbf{T} - \boldsymbol{b}) + \varrho \dot{\boldsymbol{k}}_{\mathrm{i}} = 2\boldsymbol{\tau} + \operatorname{div} \mathbf{L} + \boldsymbol{k}.$$

Using the local balance of linear momentum (2.127) then results in

$$\varrho \dot{\boldsymbol{k}}_{\mathrm{i}} = 2\boldsymbol{\tau} + \operatorname{div} \mathbf{L} + \boldsymbol{k}. \tag{2.141}$$

This is the local form of the balance equation of rotational momentum that replaces (2.134) for bodies subject to general interactions, which convey couples along with moments of forces. In this book we shall make the internal order structure liable for such extended interactions: equation (2.141) must be satisfied along every admissible motion of the body. The vector $\boldsymbol{\tau}$, still as in (2.132), is no longer necessarily zero, and so the stress tensor will in general have a skew-symmetric part.

We derive now from the balance laws for general continua (2.127) and (2.141) the balance of power that is to replace (2.137) in this setting. We shall, in particular, learn how to write for these bodies the kinetic energy $\mathscr{K}$ in an inertial frame. Inspired by (2.79) and interpreting the local body point as formally endowed with an extended structure, we write the power $\mathscr{W}^{(\mathrm{a})}$ expended by the actions applied to the present shape of a subbody $\mathcal{P}$ in a motion $\boldsymbol{\chi}$ as

$$\mathscr{W}^{(\mathrm{a})}(\mathcal{P}_t, t) = \int_{\partial^* \mathcal{P}_t} (\boldsymbol{t} \cdot \boldsymbol{v} + \boldsymbol{c} \cdot \boldsymbol{w}) dA + \int_{\mathcal{P}_t} (\boldsymbol{b} \cdot \boldsymbol{v} + \boldsymbol{k} \cdot \boldsymbol{w}) dV, \tag{2.142}$$

where $\boldsymbol{w} := \frac{1}{2} \operatorname{curl} \boldsymbol{v}$ is the spin vector associated with the flow $\boldsymbol{v}$, that is, the axial vector of the vorticity tensor $\mathbf{W} = \operatorname{skw}(\nabla \boldsymbol{v})$ in (2.62). Since the body $\mathcal{B}$ envisaged in the theory of general interactions is still classical, that is, deprived of the internal degrees of order that confer further structure to it, here the couples $\boldsymbol{c}$ and $\boldsymbol{k}$ expend power against the fluid vorticity $\boldsymbol{w}$. We shall see in the following chapters how these internal degrees will expend power against other independent rates. In the present case, ignoring such an order structure, it is as if it were frozen and rigidly conveyed by the local vorticity of the fluid, like pebbles in a creek.

Equation (2.142) can also be written in Cartesian components relative to a basis $(\boldsymbol{e}_1, \boldsymbol{e}_2, \boldsymbol{e}_3)$ of $\mathcal{V}$ as

$$\mathscr{W}^{(\mathrm{a})}(\mathcal{P}_t, t) = \int_{\partial^* \mathcal{P}_t} \left(T_{ij} \nu_j v_i + \frac{1}{2} L_{ih} \nu_h \epsilon_{ijk} v_{k,j} \right) dA + \int_{\mathcal{P}_t} (b_i v_i + k_i w_i) dV, \tag{2.143}$$

where use has also been made of (2.125) and (2.140). By the divergence theorem, we convert the surface integral in (2.143) into

$$\int_{\mathcal{P}_t} \left(T_{ij,j} v_i + T_{ij} v_{i,j} + \frac{1}{2} L_{ih,h} \epsilon_{ijk} v_{k,j} + \frac{1}{2} L_{ih} \epsilon_{ijk} v_{k,jh} \right) dV. \tag{2.144}$$

Since in Cartesian components (2.64) reads as

$$w_i = \frac{1}{2} \epsilon_{ijk} v_{k,j},$$

we easily realize that the last integrand in (2.144) can also be written as $L_{ih} w_{i,h}$. Thus, reverting to absolute notation and recalling (2.127) and (2.141), which are valid only in an inertial frame, we obtain that

$$\mathscr{W}^{(a)}(\mathcal{P}_t, t) = \int_{\mathcal{P}_t} \left[\varrho(\boldsymbol{a}\cdot\boldsymbol{v} + \dot{\boldsymbol{k}}_\mathrm{i}\cdot\boldsymbol{w}) + \mathbf{T}\cdot\mathbf{D} + \mathbf{L}\cdot\nabla\boldsymbol{w}\right] dV, \tag{2.145}$$

having used the identities (see also Appendix A.1)

$$\mathbf{T}\cdot\nabla\boldsymbol{v} = \mathrm{sym}(\mathbf{T})\cdot\mathbf{D} + \mathrm{skw}(\mathbf{T})\cdot\mathbf{W} = \mathbf{T}\cdot\mathbf{D} + 2\boldsymbol{\tau}\cdot\boldsymbol{w},$$

where sym(**T**) denotes the symmetric part of **T**. To reduce (2.145) to a form similar to (2.137), we readily identify the power $\mathscr{W}^{(\mathrm{i})}$ of the internal actions as

$$\mathscr{W}^{(\mathrm{i})}(\mathcal{P}_t, t) := \int_{\mathcal{P}_t} (\mathbf{T}\cdot\mathbf{D} + \mathbf{L}\cdot\nabla\boldsymbol{w}) dV, \tag{2.146}$$

which shows how the couple stress **L**, which expends power against the gradient of the spin vector $\boldsymbol{w}$, contributes to the internal power. We further define the kinetic energy $\mathscr{K}$ in such a way that

$$\dot{\mathscr{K}}(\mathcal{P}_t, t) = \int_{\mathcal{P}_t} \varrho(\boldsymbol{a}\cdot\boldsymbol{v} + \dot{\boldsymbol{k}}_\mathrm{i}\cdot\boldsymbol{w}) dV.$$

It is easily seen, also with the aid of (2.110), that this equation follows directly from the linear constitutive law

$$\boldsymbol{k}_\mathrm{i} = \mathbf{M}\boldsymbol{w},$$

with **M** a symmetric constant tensor, and the definition

$$\mathscr{K}(\mathcal{P}_t, t) := \int_{\mathcal{P}_t} \frac{1}{2}\varrho(\boldsymbol{v}\cdot\boldsymbol{v} + \mathbf{M}\boldsymbol{w}\cdot\boldsymbol{w}) dV. \tag{2.147}$$

With $\mathscr{K}$ as in (2.147) and $\mathscr{W}^{(\mathrm{i})}$ as in (2.146), the balance of powers in (2.137) is also formally valid for the bodies described by the general balance laws (2.127) and (2.141). In turn, as shown in [13], these balance laws could be obtained from the requirement that the total external power $\mathscr{W}^{(\mathrm{e})} = \mathscr{W}^{(\mathrm{a})} - \dot{\mathscr{K}}$ be frame-indifferent. Thus, for (2.137) to hold, the internal power $\mathscr{W}^{(\mathrm{i})}$ in (2.146) should be frame-indifferent as well. Since both **T** and **D** are frame-indifferent tensors, whereas, as shown by (2.95), $\nabla\boldsymbol{w}$ is only hemi-indifferent, $\mathscr{W}^{(\mathrm{i})}$ is frame-indifferent only if the couple stress tensor **L** is also hemi-indifferent, and so under a change of frame represented by $\mathbf{R} \in \mathsf{O}(3)$ it transforms as

$$\mathbf{L}^* = (\det\mathbf{R})\mathbf{R}\mathbf{L}\mathbf{R}^\mathsf{T};$$

see (2.94). This, together with (2.140), shows that the contact couple $\boldsymbol{c}$ is hemi-indifferent, and so it transforms like the vector product of indifferent vectors, see (2.92), which is the transformation law that holds for the moment of forces.

Often, the bodies for which the balance equation (2.141) is valid are called *polar continua* (see, for example, [345, § 98]). Equation (2.141) has been further generalized to *multipolar* continua, by imagining that subbodies may exchange interactions described by generalized forces that expend power against higher gradients of the velocity field. Apart from a limited digression in the closing Chapter 5, our development will not concern such bodies; here we are contented with citing the work of GREEN & RIVLIN [127] that spawned a literature too vast to be mentioned.

Even when the underlying order microstructure is only moderately complex, equations (2.127) and (2.141) do not suffice to describe the time evolution of the relevant order tensors introduced in Chapter 1. A more compelling principle is then required to deliver the general evolution equations, which must still be compatible with the classical balances expressed by equations (2.127) and (2.141). Several proposals have been made to this end in the literature; in Section 2.2, we shall present the one we prefer.

2.2 Dissipation Principle

Our aim is to posit a variational principle that would allow us to derive equations of motion for a *dissipative system*. To this end, we shall make two central assumptions: first, that the total mechanical power, excluding dissipation, can be written as a product of generalized forces by generalized velocities, and second that these forces are balanced by frictional forces of the simplest conceivable type, namely those that possess a quadratic velocity potential.[47] A quadratic velocity potential was originally introduced by RAYLEIGH in analytical mechanics, but he envisaged from the start its application to continuous bodies [323, footnote on page 364].[48]

Conceptually, dissipation is closely related to irreversibility. A question that has long been debated is how to reconcile the microscopic dynamics, which are governed by time-reversible laws of motion, such as those advanced in the classical dynamics of mass-points, and the macroscopic dynamics, which, established here over NOLL's axiomatic construction, describe resistive phenomena through appropriate constitutive laws for either stress and couple stress tensors. The conceptual separation between microscopic reversibility and macroscopic dissipation was clearly perceived at the close of the nineteenth century, as witnessed, for example, by the following excerpt from J. J. THOMPSON's book [331, p. 281], where an explanation is also attempted:

> But if every physical phenomenon can be explained by means of frictionless dynamical systems each of which is reversible, then it follows that if we could only control the phenomenon in all its details, it would be reversible, so that as was pointed out by MAXWELL, the irreversibility of any system is due to the limitation of our powers of manipulation. The reason we can not reverse every process is because we only possess the power of dealing with the molecules *en masse* and not individually, while the reversal of some processes would require the reversal of the motion of each individual molecule.

---

47 In the following chapters this scheme will grow to include the power both expended and dissipated against the time rate of the collective order tensor **Q** introduced in Chapter 1.

48 In general, for RAYLEIGH's work we also refer the reader to his collected papers [324].

This question is interwoven with the concept of *entropy* in the statistical mechanics formulation given by BOLTZMANN, who identified the obscure thermodynamic notion introduced by CLAUSIUS with a relative measure for the trend toward equilibrium, where entropy is maximized. In loose terms, a system evolves in time toward macroscopic states with more possible microscopic realizations, and finally to the state with most possible realizations, where it remains, having thus reached its equilibrium.[49]

W. THOMSON (LORD KELVIN) [334] also attempted an explanation for the irreversibility riddle: this was particularly disturbing his general view of dissipation as an irremediable source of energy degradation [332]. KELVIN argued that a complete reversal of the velocities of all molecules cannot be achieved and that a necessarily incomplete reversal could at most induce a temporary decrease in entropy:

> The number of molecules being finite, it is clear that small finite deviations from absolute precision in the reversal we have supposed would not obviate the resulting disequalisation of the distribution of energy. But the greater the number of molecules, the shorter will be the time during which the disequalising will continue; and it is only when we regard the number of molecules as practically infinite that we can regard spontaneous disequalisation as practically impossible.[50]

BOLTZMANN had to endure two major objections against his statistical interpretation of entropy: *reversibility* and *recurrence*. The latter, perhaps more serious than the former, arose essentially from a theorem of POINCARÉ [274] stating that a dynamical system of pointwise atoms subject to interaction forces depending only on their mutual distances returns infinitely close to its initial configuration in phase space after a sufficiently long time.[51] Here we take TRUESDELL's attitude toward such disputes [343, p. 121]:

> In fact, it requires no great mathematician to see that the reversibility theorem and POINCARÉ's recurrence theorem make irreversible behavior impossible for dynamical systems in the classical sense. Something must be added to the dynamics of conservative systems, something not consistent with it, in order to get irreversibility at all.[52]

Different choices can be made that embody irreversibility for macroscopic systems, each founding a different thermodynamic theory. In Section 2.2.2 below we shall

---

[49] As MÜLLER [227, p. 101] puts it, BOLTZMANN discovered the *strategy of nature*, though "it is not much of a strategy, because it consists of letting things happen and of permitting blind chance to take its course." When applied to the whole universe, such a fatal increase in entropy has stimulated many teleological extrapolations of this concept from which here we abstain.

[50] See [333] and [227, p. 104]. We also refer the reader to [266, Chapter 27] for a stimulating cosmological interpretation of this concept.

[51] A terse and witty account on the controversy that on these themes opposed BOLTZMANN to ZERMELO, an assistant to PLANCK who was to make fundamental contributions to axiomatic set theory, can be found again in MÜLLER's book [227, pp. 103–107].

[52] TRUESDELL mainly addressed this criticism against the use made in [256, 257] of a vague *principle of microscopic reversibility* to justify a symmetry request for certain bilinear constitutive laws for the entropy production.

briefly describe the conceptual bases of continuum thermodynamics. This theory will aid us in putting in a wider perspective the mechanical variational principle that in this book constitutes the conceptual basis for the dynamics of macroscopic dissipative systems. We go first to the roots of classical mechanics, interpreting in a variational fashion the first general dynamical equations encompassing systematically resistive forces, albeit of a special nature.

We recall in Section 2.2.1 the standard LAGRANGE–RAYLEIGH equations of analytical mechanics and then show how they can be reinterpreted as a balance of generalized forces and frictional forces, and how the actual evolution prescribed by these equations minimizes the total dissipation with respect to a class of constrained variations of the velocities. We then formulate an appropriate variational principle that is equivalent to the LAGRANGE–RAYLEIGH equations and we generalize it in Section 2.2.3 to make it applicable also to continua, after having revived in Section 2.2.2 the basic concepts of continuum thermodynamics needed to phrase our principle in that language. We conclude in Section 2.2.5 by putting the dissipation principle into the context of other related principles.

2.2.1 LAGRANGE–RAYLEIGH Equations

For the reader's ease, we now retrace the main steps needed to complete the program so neatly announced by WHITTAKER [362, § 93]:

> When a system is subject to external resisting forces which are directly proportional to the velocities of their points of application, it is possible to express the equations of motion of the system in general coordinates in terms of the kinetic and potential energies and of a single new function.

Our treatment of the finite-dimensional case, which is rooted in classical analytical mechanics, will inspire our extension to the continuum case, which is the main objective of this book; some peculiar technical aspects may there appear more intricate, but the conceptual structure will remain the same.

HAMILTON's Principle

We consider a holonomic dynamical system described by m generalized coordinates $q_1, \dots, q_m$. We denote by q and $\dot{\mathsf{q}}$ the vectors in $\mathbb{R}^m$ of the generalized coordinates and the generalized velocities. While q lives in the subspace $\mathfrak{Q} \in \mathbb{R}^m$ of admissible Lagrangian configurations, $\dot{\mathsf{q}} \in \mathbb{R}^m$ is not subject to further restrictions.[53] When all active forces are conservative, the system possesses a potential energy $V = V(q)$. Its kinetic energy $T = T(\mathsf{q}, \dot{\mathsf{q}})$ is assumed to be a positive-definite quadratic form in the

[53] Generalized velocities are further restrained only in *anholonomic* systems.

velocities[54] $\dot{\mathsf{q}}$. Both V and T are supposed to be smooth functions[55] of q. The evolution of the system is then determined by a single scalar function, the LAGRANGE *function*

$$L(\mathsf{q}, \dot{\mathsf{q}}) := T - V. \tag{2.148}$$

The equations of motion can be derived from HAMILTON*'s principle,* which requires that the *action integral*

$$A[\mathsf{q}] = \int_{t_1}^{t_2} L(\mathsf{q}, \dot{\mathsf{q}}) dt \tag{2.149}$$

between any two prescribed states of the system at times t_1 and t_2 be stationary. This means that any variation of the true evolution will imply a variation in the action A.

To make this notion precise, we consider variations of the functions $q_i(t)$; such variations are arbitrary functions $\delta q_i(t)$ with $\delta q_i(t_1) = \delta q_i(t_2) = 0$, so that the prescribed states at t_1 and t_2 remain unchanged. The variations are added to the q_i, so that these latter are replaced by

$$\hat{q}_i(t) = q_i(t) + \varepsilon \delta q_i(t), \tag{2.150}$$

where ε is a small parameter. Any variation of the q_i consequently entails also a variation of the $\dot{q}_i$, since

$$\frac{d}{dt} \hat{q}_i(t) = \dot{q}_i(t) + \varepsilon \frac{d}{dt} \delta q_i(t). \tag{2.151}$$

For simplicity, the notation

$$\delta \dot{q}_i = \frac{d}{dt} \delta q_i(t) \tag{2.152}$$

is commonly used. The variation δA of A is then defined by

$$\delta A := \left. \frac{dA[\hat{\mathsf{q}}]}{d\varepsilon} \right|_{\varepsilon=0}. \tag{2.153}$$

Principle (HAMILTON). The true evolution of a Lagrangian system is characterized by having $\delta A = 0$.

Computing the derivative in (2.153) explicitly shows that

$$\delta A = \int_{t_1}^{t_2} \sum_{i=1}^{m} \left[\frac{\partial L}{\partial q_i} \delta q_i + \frac{\partial L}{\partial \dot{q}_i} \delta \dot{q}_i \right] dt = \int_{t_1}^{t_2} \sum_{i=1}^{m} \left[\frac{\partial L}{\partial q_i} - \frac{d}{dt} \frac{\partial L}{\partial \dot{q}_i} \right] \delta q_i dt, \tag{2.154}$$

---

54 Such an assumption is compatible only with holonomic systems subject to time-independent constraints; any such system is also called *scleronomic*, while a system subject to time-dependent constraints is called *rheonomic*.

55 Here we are guilty of some abuse of language: we use the same symbol for both a function and the value it delivers. Whenever confusion is unlikely to arise, we indulge in this attitude to avoid an unduly hypertrophic notation.

where the second form is obtained after integrating by parts and using the fact that the δq_i vanish at t_1 and t_2. Since the δq_i are arbitrary, the integral in (2.154) can vanish only if the EULER–LAGRANGE equations

$$\frac{\partial L}{\partial q_i} - \frac{d}{dt}\frac{\partial L}{\partial \dot{q}_i}, = 0, \qquad i \in \{1, \ldots, m\}, \tag{2.155}$$

hold.

Total Mechanical Power

The EULER–LAGRANGE equations (2.155) can be interpreted as balances of generalized forces. If, using the definition of the LAGRANGE function (2.148), we write them as

$$-\frac{\partial V}{\partial q_i} + \frac{\partial T}{\partial q_i} - \frac{d}{dt}\frac{\partial T}{\partial \dot{q}_i} = 0,$$

we can identify the active forces

$$F_i = -\frac{\partial V}{\partial q_i}$$

as the negative "gradient" of the potential energy V and the intertial forces

$$I_i = \frac{\partial T}{\partial q_i} - \frac{d}{dt}\frac{\partial T}{\partial \dot{q}_i}.$$

The balances

$$F_i + I_i = 0 \tag{2.156}$$

are the form that NEWTON's second law takes in the Lagrangian formulation. These equations can also be derived from D'ALEMBERT's *principle*, which is the classical extension to dynamics of the principle of virtual work of statics [173, Chapter IV]. A formulation of D'ALEMBERT's principle in line with the general spirit of this book involves virtual velocities instead of virtual displacements. In the Lagrangian formalism it can be stated as follows.

Principle (D'ALEMBERT). The true motion of a Lagrangian system is characterized by having the total virtual power

$$\hat{W} = \sum_{i=1}^{m} (F_i + I_i)\hat{\dot{q}}_i \tag{2.157}$$

vanishing for all virtual velocities $\hat{\dot{q}}_i$.

The virtual velocities $\hat{\dot{q}}_i$ in (2.157) must not be confused with the velocities $\frac{d}{dt}\hat{q}_i$ in (2.151) associated with the varied trajectories $\hat{q}_i$ in (2.150); they are characterized by being compatible with only the present state of all constraints, as if these were time-independent.[56] Virtual velocities, in general, are to be associated with imaginary, explorative motions taking place, as it were, at time scales much shorter than that characterizing the actual evolution of the system. Most variational principles of classical mechanics are stipulations about the power expended in these motions by an appropriate system of forces. For holonomic systems, the virtual velocities $\hat{\dot{q}}_i$ are all completely arbitrary if the configuration presently traversed by the motion lies in the interior of the configuration space $\mathfrak{Q}$, and so D'ALEMBERT's principle is immediately recognized to be equivalent to (2.156).

D'ALEMBERT's principle can be given different equivalent formulations. As shown, for example, in [173, Chapter IV], (2.157) is the ultimate form taken by the total virtual power defined by

$$\hat{W} := \sum_{h=1}^{N} (\boldsymbol{f}_h - m_h \boldsymbol{a}_h) \cdot \hat{\boldsymbol{v}}_h, \tag{2.158}$$

where now the sum is extended over all N mass-points that constitute the mechanical system, $\boldsymbol{f}_h$ being the resultant active force applied to the hth mass m_h, $\boldsymbol{a}_h$ its acceleration, and $\hat{\boldsymbol{v}}_h$ its virtual velocity. This latter, just as for the components of $\hat{\dot{\mathsf{q}}}$, is any vector kinematically compatible with the state of the constraints at the present time, under the assumption that these are ideally *frozen* in such a state.[57] Writing $\hat{W}$ in the form (2.158) is not a pedantic digression: it shows that for both the active forces $\boldsymbol{f}_h$ independent of the velocity $\boldsymbol{v}_h$ and the inertial forces $-m_h\boldsymbol{a}_h$, no ambiguity may arise in computing the virtual power: they can indeed be considered as given by the true motion, while their powers are computed against fictitious velocities. Care must instead be used with forces that, like the most elementary resistive forces, depend on velocities: considering them as given by the true motion while the velocities are varied in a virtual motion is not justified. For this reason, the validity of D'ALEMBERT's principle is confined to nondissipative systems. Correspondingly, the LAGRANGE function accounts only for active and inertial forces, not for dissipative ones.

Irrespective of its formulation, D'ALEMBERT's principle is often mistaken as a triviality. Unfortunately, this obscures the real value of the principle, which lies in deriving the evolution equations for a mechanical system from the supposed property of an invariant scalar, the total virtual power.

In the true motion, the one that obeys the evolution equations (2.156), the *total power* W of the system is given by

$$W := \sum_{i=1}^{m} X_i \dot{q}_i = \mathsf{X} \cdot \dot{\mathsf{q}}, \tag{2.159}$$

---

[56] Virtual velocities are defined in general for both holonomic and anholonomic systems, subject to both time-dependent and time-independent constraints.

[57] To relate (2.158) to NEWTON's equations of motion, it suffices to recall the postulate on which Lagrangian mechanics is based, that is, that the reactive forces expend zero total virtual power. Constraints enjoying this property are often said to be *perfect*.

where the generalized forces X_i are the total resultant forces,

$$X_i = F_i + I_i = \frac{\partial L}{\partial q_i} - \frac{d}{dt}\frac{\partial L}{\partial \dot{q}_i}, \tag{2.160}$$

including both inertial and active forces. X is the m-vector of those forces, and the dot denotes the usual scalar product in $\mathbb{R}^m$. In this conservative system, the energy balance takes the form

$$W + \dot{H} = 0, \tag{2.161}$$

where $\dot{H}$ is the rate of change of the total mechanical energy

$$H = T + V.$$

That is,

$$\dot{H} = \dot{T} + \dot{V} = -\mathsf{X} \cdot \dot{\mathsf{q}}. \tag{2.162}$$

To prove (2.162), we start by rewriting the time derivative of the kinetic energy T. Since T is quadratic in $\dot{\mathsf{q}}$, it is a homogeneous function of degree two, for which

$$\sum_{i=1}^{m} \frac{\partial T}{\partial \dot{q}_i} \dot{q}_i = 2T. \tag{2.163}$$

By direct computation, one finds that

$$\dot{T} = \sum_{i=1}^{m} \left(\frac{\partial T}{\partial q_i} \dot{q}_i + \frac{\partial T}{\partial \dot{q}_i} \ddot{q}_i \right) = \sum_{i=1}^{m} \frac{\partial T}{\partial q_i} \dot{q}_i + \frac{d}{dt} \left(\sum_{i=1}^{m} \frac{\partial T}{\partial \dot{q}_i} \dot{q}_i \right) - \sum_{i=1}^{m} \frac{d}{dt} \left(\frac{\partial T}{\partial \dot{q}_i} \right) \dot{q}_i,$$

whence, by (2.163), it readily follows that

$$\dot{T} = \sum_{i=1}^{m} \left[\frac{d}{dt} \frac{\partial T}{\partial \dot{q}_i} - \frac{\partial T}{\partial q_i} \right] \dot{q}_i. \tag{2.164}$$

The potential energy V is independent of the velocities, and so we simply have

$$\dot{V} = \sum_{i=1}^{m} \frac{\partial V}{\partial q_i} \dot{q}_i. \tag{2.165}$$

Equations (2.164) and (2.165) together show that

$$\dot{H} = \dot{T} + \dot{V} = \sum_{i=1}^{m} \left[\frac{d}{dt} \frac{\partial L}{\partial \dot{q}_i} - \frac{\partial L}{\partial q_i} \right] \dot{q}_i, \tag{2.166}$$

which is (2.162) with the generalized forces in the form (2.160),

$$X_i = \frac{\partial L}{\partial q_i} - \frac{d}{dt} \frac{\partial L}{\partial \dot{q}_i}. \tag{2.167}$$

In summary, the equations of motion (2.155) simply require all generalized forces to be zero. It follows from (2.159) and (2.166) that, along a motion, both

$$W = 0 \quad \text{and} \quad \dot{H} = 0, \tag{2.168}$$

so the total mechanical power is zero and the total mechanical energy is conserved. For a holonomic system with a finite number of degrees of freedom, (2.168) is the analogue to the balance of power already encountered in (2.137) for continuum systems: W is easily identified with the external power $\mathscr{W}^{(\mathrm{e})}$, see (2.104), while the internal power $\mathscr{W}^{(\mathrm{i})}$ vanishes identically.

Frictional Forces

An important class of nonconservative frictional forces is given by forces that are proportional and directed opposite to the velocities. Such forces can be derived from a potential that is a positive semidefinite quadratic form in the velocities. This potential is called the RAYLEIGH *dissipation function* $R(q, \dot{q})$, and it yields the frictional forces Y_i via

$$Y_i = -\frac{\partial R}{\partial \dot{q}_i}. \tag{2.169}$$

From this it is evident that the RAYLEIGH function has the dimension of a generalized power, and since it is a homogeneous function of degree two in the velocities, it satisfies

$$\sum_{i=1}^{m} \frac{\partial R}{\partial \dot{q}_i} \dot{q}_i = 2R. \tag{2.170}$$

If the frictional forces are added to the generalized forces (2.167), one obtains the balance of forces

$$X_i + Y_i = 0, \tag{2.171}$$

which becomes, upon inserting the expressions (2.167) and (2.169),

$$\frac{\partial L}{\partial q_i} - \frac{d}{dt}\frac{\partial L}{\partial \dot{q}_i} - \frac{\partial R}{\partial \dot{q}_i} = 0. \tag{2.172}$$

This extended version of the EULER–LAGRANGE equations (2.155) is the standard textbook form of the RAYLEIGH equations (see, for example, [362, p. 231]).

Multiplying both sides of (2.172) by $\dot{q}_i$ and summing up yields the energy balance; by (2.159) and (2.170), we find that

$$2R = W. \tag{2.173}$$

Since (2.162) is a purely kinematical identity, (2.173) shows that the change in the mechanical energy is

$$\dot{H} = -W = -2R. \tag{2.174}$$

What qualifies the frictional forces Y_i as being *dissipative* is the requirement that R be positive semidefinite. The *total dissipation*

$$D := 2R \tag{2.175}$$

is thus nonnegative. We refer to (2.173) as the *dissipation identity*; since (2.168) is the analogue to (2.137) in the finite-dimensional setting, (2.174) heralds a new balance of power to be set forward in the continuum setting with the aid of the thermodynamic arguments discussed in Section 2.2.2 below. For the time being, we keep in mind the simple meaning borne by (2.173) and (2.174): the power of all external forces goes into dissipation. One may easily anticipate that, in the presence of an internal power, this direct, simple balance gets richer in options.

Variational Formulation

We want to posit a variational principle that allows one to derive the equations (2.172) from two scalar functions: the total mechanical power W and the dissipation function R. This principle would then be generalized to continua in a straightforward fashion, while a direct generalization of (2.172) to continua would pose various problems with, for example, boundary conditions or material frame-indifference.

The common physical interpretation of the action functional A in (2.149) suggests that HAMILTON's principle of least action could not in general account for frictional forces. However, the problem of phrasing the evolution equations for a dissipative system within a Hamiltonian principle formalism is indeed subtler than usually believed on purely physical grounds: it is a special instance of the general *inverse* problem of the calculus of variations, which asks whether a given differential equation can be regarded as the EULER–LAGRANGE equation of an appropriate functional. Reading a paper by BATEMAN [11], we learn that this question was already asked by TOLMAN[58] and that WHITTAKER had proposed a simple model for a dissipative system with two degrees of freedom that he conjectured could not be derived from a Hamiltonian principle. BATEMAN states in [11] that CASSEN showed that WHITTAKER's equations do not strictly derive from a Lagrangian function, though a Lagrangian function can be found whose associated equations of motion are compatible with WHITTAKER's equations, since the set of solutions of the former includes the solutions of the latter.[59]

Similarly, as already remarked above, D'ALEMBERT's principle of virtual power is not suitable for extension to forces depending on the velocities. We posit instead the following.

Principle (of Minimum Constrained Dissipation). For a system with total mechanical power W and RAYLEIGH dissipation function R that obey the energy balance (2.173), the true velocity $\dot{\mathsf{q}}$ traversing a given configuration q is such that the dissipation function attains its minimum with respect to all virtual velocities $\hat{\dot{\mathsf{q}}} = \dot{\mathsf{q}} + \delta\dot{\mathsf{q}}$ once both the generalized forces X and the total mechanical power $W = \mathsf{X} \cdot \dot{\mathsf{q}}$ are held fixed.

---

58 See also Sections 10, 11 of [335] for a general Lagrangian formulation of chemical systems.

59 We shall comment again on this issue in Section 2.3.4 below, in connection with the NAVIER–STOKES equations of motion for a linearly viscous fluid conducting heat.

The identity (2.173) is valid in every true motion: Since W is prescribed by the true motion and held fixed while the motion is varied, the minimum value of R is prescribed as well, but R is virtually freed, and its value prescribed by the true motion must be the constrained minimum for the varied motion.

It should be recalled that variations here are different from those employed in HAMILTON's principle in Section 2.2.1 in an important respect. There, initial and final states are frozen, and the configuration in between those states is varied by a time-dependent $\delta \mathsf{q}$. Here, we freeze the configuration q at a given instant in time and leave it unaltered. The main players are variations $\delta \dot{\mathsf{q}}$ of the velocities $\dot{\mathsf{q}}$, which are not required to be "small" in any regard. These are instantaneous variations in their own right and should not to be confused with the time derivatives of variations of the configuration as in (2.152). Additionally, as we shall soon see, we will also need instantaneous variations $\delta \ddot{\mathsf{q}}$ of the accelerations. Again, these are not time-dependent and in particular not time derivatives of the $\delta \dot{\mathsf{q}}$, but they can and have to be chosen completely independent of the $\delta \dot{\mathsf{q}}$.

To see how the dissipation principle delivers the evolution equations (2.172), we require R to be stationary with respect to this special class of variations. The constraint on the mechanical power W can be treated in the standard way through a LAGRANGE multiplier, so that $\dot{\mathsf{q}}$ may be arbitrarily perturbed. More specifically, with no loss of generality, we may represent R as

$$R(\mathsf{q}, \dot{\mathsf{q}}) = \frac{1}{2} \sum_{i,j=1}^{m} R_{ij} \dot{q}_i \dot{q}_j = \frac{1}{2} \dot{\mathsf{q}} \cdot \mathsf{R} \dot{\mathsf{q}}, \tag{2.176}$$

where $\mathsf{R} = \mathsf{R}(\mathsf{q})$ is a matrix of $\mathbb{R}^{m \times m}$ with entries R_{ij}, which we assume to depend continuously on q. It follows from (2.176) that

$$R_{ij} = \frac{\partial^2 R}{\partial \dot{q}_i \partial \dot{q}_j},$$

and so R can be taken to be symmetric. The variation δR of R, which is defined as $\delta R(\mathsf{q}, \dot{\mathsf{q}}) := R(\mathsf{q}, \hat{\dot{\mathsf{q}}}) - R(\mathsf{q}, \dot{\mathsf{q}})$, is simply

$$\delta R = \frac{1}{2} \delta \dot{\mathsf{q}} \cdot \mathsf{R} \delta \dot{\mathsf{q}} + \dot{\mathsf{q}} \cdot \mathsf{R} \delta \dot{\mathsf{q}},$$

where use has also been made of the symmetry of R. The condition of constrained minimality for R prescribed by the principle then reads

$$\delta R + \lambda \delta W = \delta R + \lambda \delta (\mathsf{X} \cdot \dot{\mathsf{q}}) = \frac{1}{2} \delta \dot{\mathsf{q}} \cdot \mathsf{R} \delta \dot{\mathsf{q}} + (\mathsf{R} \dot{\mathsf{q}} + \lambda \mathsf{X}) \cdot \delta \dot{\mathsf{q}} \geqq 0, \tag{2.177}$$

where λ is a LAGRANGE multiplier and $\delta \dot{\mathsf{q}}$ is now an arbitrary vector in $\mathbb{R}^m$. To obtain (2.177) it is crucial that X be kept fixed. However, because X in general does depend on $\dot{\mathsf{q}}$, a system of variations $\delta \dot{\mathsf{q}}$ of the actual velocity vector $\dot{\mathsf{q}}$ that leaves also the generalized forces X unchanged implies variations $\delta \ddot{\mathsf{q}}$ of $\ddot{\mathsf{q}}$ to be chosen accordingly. That this can indeed be done follows from the fact that the kinetic energy

T is a positive-definite quadratic form in $\dot{\mathsf{q}}$ and V is independent of $\dot{\mathsf{q}}$. By (2.160), X is thus linear in $\ddot{\mathsf{q}}$ and

$$\det \frac{\partial \mathsf{X}}{\partial \ddot{\mathsf{q}}} \neq 0;$$

this means that $\ddot{\mathsf{q}}$ can also be expressed in terms of X, q, and $\dot{\mathsf{q}}$ as

$$\ddot{\mathsf{q}} = \mathsf{B}(\mathsf{q}, \dot{\mathsf{q}})\mathsf{X} + \mathsf{a}(\mathsf{q}, \dot{\mathsf{q}})$$

(cf. [362, p. 40]), where B is an invertible matrix in $\mathbb{R}^{m \times m}$ and a is a vector in $\mathbb{R}^m$. It follows that, for any variations $\delta\dot{\mathsf{q}}$, appropriate variations $\delta\ddot{\mathsf{q}}$ that ensure that X remains fixed are given by

$$\delta\ddot{\mathsf{q}} = \mathsf{B}(\mathsf{q}, \dot{\mathsf{q}} + \delta\dot{\mathsf{q}})\mathsf{X} + \mathsf{a}(\mathsf{q}, \dot{\mathsf{q}} + \delta\dot{\mathsf{q}}) - \ddot{\mathsf{q}}.$$

Since δR vanishes for $\delta\dot{\mathsf{q}} = 0$ and R is positive semidefinite, a necessary and sufficient condition for (2.177) to be satisfied is that the linear form $(R\dot{\mathsf{q}} + \lambda\mathsf{X}) \cdot \delta\dot{\mathsf{q}}$ vanish identically; this is the case if and only if

$$R\dot{\mathsf{q}} + \lambda\mathsf{X} = 0,$$

which by (2.176) can also be written as

$$\frac{\partial R}{\partial \dot{\mathsf{q}}} + \lambda\mathsf{X} = 0. \tag{2.178}$$

The value of λ can be determined by taking the inner product of both sides of (2.178) with $\dot{\mathsf{q}}$ and requiring that the resulting equation agree with the energy balance (2.173). This shows that $\lambda = -1$, and so we find that

$$\mathsf{X} - \frac{\partial R}{\partial \dot{\mathsf{q}}} = 0, \tag{2.179}$$

which by (2.169) is equivalent to (2.171). Thus, the minimality condition for R in (2.178) just becomes the system of equations of motion (2.172).

In light of the formal reasoning that just led us to show that the principle of minimum constrained dissipation implies RAYLEIGH's equations of motion, one can easily reformulate that principle in an equivalent way. Let the function $\widetilde{R}$ be defined by

$$\widetilde{R} := R - W. \tag{2.180}$$

It easily follows from (2.180) that (2.179) is precisely the minimality condition for $\widetilde{R}$ subject to variations that leave only X unchanged but may affect W. While by (2.173) the minimum of R is $\frac{1}{2}W$ and it is attained on the true motion, the minimum of $\widetilde{R}$, similarly attained on the true motion, is $-\frac{1}{2}W$. We shall call $\widetilde{R}$ the *reduced* RAYLEIGH function and we shall rephrase the principle of minimum constrained dissipation in the following equivalent form.

Principle (of Minimum Reduced Dissipation). For a system with total mechanical power W and RAYLEIGH dissipation function R that obey the energy balance (2.173), the true velocity $\dot{\mathsf{q}}$ traversing a given configuration q is such that the reduced dissipation function $\widetilde{R} = R - W$ attains its minimum with respect to all virtual velocities $\hat{\dot{\mathsf{q}}} = \dot{\mathsf{q}} + \delta\dot{\mathsf{q}}$ once the generalized forces X are held fixed.

This formulation of the dissipation principle has also the merit of showing directly how it reduces to D'ALEMBERT's principle for nondissipative systems: When R vanishes identically, $\widetilde{R}$ reduces to $-W$, which is simply required to be stationary on the true motion.

The principle of minimum constrained dissipation for Lagrangian systems was introduced in essentially the same form adopted here[60] by BIOT [19], who also invoked it in his Lagrangian formulation of thermodynamics presented in the book [20].[61] He also proposed in [19], among others, the alternative formulation of this variational principle in terms of the reduced RAYLEIGH function $\widetilde{R}$, though he did not give it this name. By analogy with (2.175), we also call

$$\widetilde{D} := 2\widetilde{R} = D - 2W \tag{2.181}$$

the *total reduced dissipation*.

Nonquadratic Potentials

If the objective of our variational formulation of the evolution equations for dissipative systems with finite degrees of freedom were only justifying (2.179) as a minimality (or just a stationarity) condition, we could also easily consider nonquadratic potentials R. Let indeed $R = R(\mathsf{q}, \dot{\mathsf{q}})$ be any smooth functions in the pair $(\mathsf{q}, \dot{\mathsf{q}})$ and let $\widetilde{R}$ be defined as in (2.180). By subjecting $\widetilde{R}$ to free variations in $\dot{\mathsf{q}}$ only, while keeping the generalized force X fixed, we easily obtain (2.179) as a stationarity condition for $\widetilde{R}$. By taking the inner product of both sides of (2.179) with $\dot{\mathsf{q}}$, we readily arrive at

$$W = \frac{\partial R}{\partial \dot{\mathsf{q}}} \cdot \dot{\mathsf{q}} = D, \tag{2.182}$$

where D is again to be interpreted as the dissipated energy. Though we might easily imagine the dissipative mechanisms at work in the system that would suggest one form or another for D, arriving at the corresponding potential R would in general require solving the partial differential equation displayed in (2.182). In the case that D is homogeneous of degree n in $\dot{\mathsf{q}}$, by the EULER formula for the derivative of homogeneous functions, the trivial solution of this equation is

$$R = \frac{1}{n} D,$$

[60] A justification of this principle was also provided in [311].

[61] Further thoughts, indeed more allusive than conclusive, were presented almost at the same time in [21].

but in general the relationship between R and D is not expected to be so simple.

Strictly speaking, much of what we have said (and shall say) could apply to all homogeneous dissipation functions[62] D, though in this book we shall consider only the classical quadratic case.

A Simple Example

We illustrate the principle with a simple example in which the generalized forces are indeed real forces. We consider a particle of mass m moving in two dimensions under the influence of gravity and linear drag. If the position of the particle is given by

$$\boldsymbol{r} = x\boldsymbol{e}_x + y\boldsymbol{e}_y,$$

then the kinetic energy is

$$T = \frac{1}{2} m \dot{\boldsymbol{r}}^2$$

and the potential energy

$$V = mgy = mg\,\boldsymbol{r} \cdot \boldsymbol{e}_y.$$

The RAYLEIGH function is

$$R = \frac{1}{2} \mu \dot{\boldsymbol{r}}^2$$

with the positive drag coefficient μ. With these functions, we find the inertial force $\boldsymbol{I}$, the active force $\boldsymbol{F}$, and the friction force $\boldsymbol{Y}$ to be

$$\boldsymbol{I} = -m\ddot{\boldsymbol{r}}, \qquad \boldsymbol{F} = -mg\boldsymbol{e}_y, \text{ and } \qquad \boldsymbol{Y} = -\mu\dot{\boldsymbol{r}}.$$

With

$$\boldsymbol{X} = \boldsymbol{I} + \boldsymbol{F} = -m(\ddot{\boldsymbol{r}} + g\boldsymbol{e}_y), \tag{2.183}$$

the total mechanical power is

$$W = \boldsymbol{X} \cdot \dot{\boldsymbol{r}} = -m(\ddot{\boldsymbol{r}} + g\boldsymbol{e}_y) \cdot \dot{\boldsymbol{r}} = -\dot{T} - \dot{V}.$$

The dissipation principle requires that

$$0 = \delta(R + \lambda W) = \frac{\partial R}{\partial \dot{\boldsymbol{r}}} \cdot \delta\dot{\boldsymbol{r}} + \lambda\delta(\boldsymbol{X} \cdot \dot{\boldsymbol{r}}),$$

where λ is a LAGRANGE multiplier and $\boldsymbol{X}$ is to be kept constant. In this case, where $\boldsymbol{X}$ does not depend on $\dot{\boldsymbol{r}}$, this simply means that the variation $\delta\ddot{\boldsymbol{r}}$ of the acceleration in (2.183) is zero. We hence find that

$$0 = \left(\frac{\partial R}{\partial \dot{\boldsymbol{r}}} + \lambda \boldsymbol{X}\right) \cdot \delta\dot{\boldsymbol{r}},$$

62 With even degree n if we want to salvage the minimality of D, supported by the thermodynamic interpretation that shall be proposed in Section 2.2.2.

and so

$$0 = \frac{\partial R}{\partial \dot{\boldsymbol{r}}} + \lambda \boldsymbol{X} = \mu \dot{\boldsymbol{r}} - \lambda m(\ddot{\boldsymbol{r}} + g\boldsymbol{e}_y). \tag{2.184}$$

Multiplying this by $\dot{\boldsymbol{r}}$ shows that

$$0 = 2R + \lambda W = 2R - \lambda(\dot{T} + \dot{V}),$$

and so we need $\lambda = -1$ to recover the energy balance as

$$2R = -(\dot{T} + \dot{V}).$$

With this value of λ, (2.184) becomes

$$m\ddot{\boldsymbol{r}} = -mg\boldsymbol{e}_y - \mu \dot{\boldsymbol{r}},$$

which is simply NEWTON's equation of motion.

The total dissipation on the true motion is given by

$$D = 2R = \mu \dot{\boldsymbol{r}}^2.$$

On any virtual motion with virtual velocity

$$\hat{\boldsymbol{v}} = \dot{\boldsymbol{r}} + \delta \dot{\boldsymbol{r}},$$

the total dissipation is

$$\hat{D} = \mu \hat{\boldsymbol{v}} \cdot \hat{\boldsymbol{v}} = \mu(\dot{\boldsymbol{r}}^2 + 2\dot{\boldsymbol{r}} \cdot \delta \dot{\boldsymbol{r}} + \delta \dot{\boldsymbol{r}}^2). \tag{2.185}$$

From (2.184), on the true motion, $\boldsymbol{X} = \mu \dot{\boldsymbol{r}}$. But because both $\boldsymbol{X}$ and its power $W = \boldsymbol{X} \cdot (\dot{\boldsymbol{r}} + \delta \dot{\boldsymbol{r}})$ are fixed under the constrained variation, this implies that $\boldsymbol{X} \cdot \delta \dot{\boldsymbol{r}} = 0$ for all admissible variations. It follows that also $\dot{\boldsymbol{r}} \cdot \delta \dot{\boldsymbol{r}} = 0$, and so

$$\hat{D} = \mu(\dot{\boldsymbol{r}}^2 + \delta \dot{\boldsymbol{r}}^2),$$

which shows that the total dissipation D of the true motion is indeed a minimum. Likewise, by (2.181) and (2.185), the total reduced dissipation on the varied motion is

$$\widetilde{D} = \mu(-\dot{\boldsymbol{r}}^2 + \delta \dot{\boldsymbol{r}}^2),$$

which is again minimized by the true motion.

2.2.2 Glimpses of Continuum Thermodynamics

Though this book is concerned mostly with purely mechanical theories, the principles we employ to derive the evolution equations are better phrased and understood within the broader context of continuum thermodynamics. The main difficulty here is that this is not a single established science; rather, it still appears as a collection of doctrines, with precepts and prohibitions, praised by different schools with opposed adherents. In the little space we can devote to this subject we shall neither attempt to draw a historical sketch of how different theories have unraveled nor venture a comparison between the merits and drawbacks they all have.[63] We shall rather broaden our perspective by the amount strictly needed to relate dissipation to entropy production, and in doing so we shall stay close to the spirit of the continuum mechanics infrastructure we have so far adhered to.

Homogeneous Processes

We start by stating the axioms of thermodynamics in the simplest possible context that grants a sufficient level of generality for them to be further extended by analogy. We assume that a body $\mathfrak{B}$ undergoes a *homogeneous process*, where all quantities of thermodynamic interest are uniform in space, while they may change in time.

Formally, a homogeneous thermodynamic process is defined by a pair of functions $(\theta(t), \mathsf{q}(t))$ depending on the time t, where θ is the *absolute temperature*,[64] assumed to be *positive*, and $\mathsf{q} = (q_1, \ldots, q_m)$ is a vector in an admissible subset $\mathfrak{Q}$ of $\mathbb{R}^m$ representing here a collection of scalar parameters accounting for the present *state* of the body, such as its volume V. Along a thermodynamic process, the time derivatives $\dot{\theta}$ and $\dot{\mathsf{q}}$ are the temperature and state *rates*, respectively; together they define the *process rate* $(\dot{\theta}, \dot{\mathsf{q}})$.

Thermodynamics essentially stems from realizing that mechanical power is *not* the only means able to increase the *internal energy* $\mathscr{U}$ of $\mathfrak{B}$. This energy $\mathscr{U}$, whose existence is here assumed, is a function of time given by a functional $\mathfrak{U}$ of the process:

$$\mathscr{U}(t) = \mathfrak{U}[(\theta, \mathsf{q})].$$

The *net working* $\mathscr{W}^{(e)}$, which was given a precise meaning in (2.99) along the motion of a generic subbody $\mathfrak{P}$ of $\mathfrak{B}$, denotes here the power expended by all external agencies in a homogeneous process of the whole body $\mathfrak{B}$. Inspired by (2.99), we assume that $\mathscr{W}^{(e)}$ is a function of t given by a linear form of the state rate $\dot{\mathsf{q}}$, so that it vanishes at equilibrium.

$\mathscr{W}^{(e)}$ is not the only source of changes for $\mathscr{U}$. TRUESDELL [343, p. 9] has well described the classical equivalence of work and heat, regarded as independent, concurrent causes of energy changes:

> If we recognize that doing work may change the energy of a body, and that heating may effect work, at this level of generality the best we can say is that to describe this *equivalence of heat and work*, we assume the existence of a second kind of working, $\mathscr{Q}$, called the *heating*, which is not identified with anything from mechanics.

---

[63] Only in Section 2.2.5 below shall we not resist making an exception, which we deem necessary.

[64] Temperature is here a *primitive* quantity that assigns to a body the measure of how *hot* it is. In the scale adopted here, θ is a positive measure: the greater lower bound of all measurable temperatures is $\theta = 0$, by definition.

Unlike $\mathscr{W}^{(\mathrm{e})}$, $\mathscr{Q}$ is given by a linear functional of the whole process rate $(\dot{\theta}, \dot{\mathsf{q}})$, and like $\mathscr{W}^{(\mathrm{e})}$ it vanishes at equilibrium.[65]

We state now for homogeneous processes an axiom that will be further extended below to more general processes undergone by deformable bodies.

Axiom 2.6 (First Law of Thermodynamics). In a thermodynamic process (θ, q), the internal energy $\mathscr{U}$, the net working $\mathscr{W}^{(\mathrm{e})}$, and the heating $\mathscr{Q}$ are related through the equation

$$\dot{\mathscr{U}} = \mathscr{W}^{(\mathrm{e})} + \mathscr{Q}. \tag{2.186}$$

Resort to (2.104) allows us to write (2.186) in the form

$$(\mathscr{U} + \mathscr{K})^{\cdot} = \mathscr{W}^{(\mathrm{a})} + \mathscr{Q},$$

which shows that the internal energy $\mathscr{U}$ and the kinetic energy $\mathscr{K}$ are both affected within the body $\mathfrak{B}$ by the power $\mathscr{W}^{(\mathrm{a})}$ expended by the external agencies applied to $\mathfrak{B}$ and by the heating $\mathscr{Q}$ transferred to $\mathfrak{B}$ from without.

A second axiom describes natural processes and provides the little conceptual *quantum* that makes irreversibility arise in systems with infinitely many particles.

Axiom 2.7 (Second Law of Thermodynamics). There is an upper bound to the heating $\mathscr{Q}$ that can be expressed as $\theta\dot{\mathscr{S}}$, where $\mathscr{S}$ is the *entropy* of the body, that is,

$$\mathscr{Q} \leqq \theta\dot{\mathscr{S}}, \tag{2.187}$$

in all homogeneous processes.

The entropy $\mathscr{S}$ is a function of t given by a functional $\mathfrak{S}$ of the process (θ, q). The inequality (2.187) is meant to be satisfied by all admissible homogeneous processes. It should perhaps be noted that both $\mathscr{U}$ and $\mathscr{S}$, which enter (2.186) and (2.187) only through their time derivatives, are determined to within an arbitrary additive constant.

In the light of Axiom 2.7, a homogeneous process is said to be *reversible* whenever the equality sign holds in (2.187); otherwise, it is irreversible, as indeed are most processes. To justify (2.187) as a statement about irreversibility, TRUESDELL [343, pp. 9–10] proposes that

> The irreversibility of natural processes is represented by the existence of an *a priori* least upper bound $[\theta\dot{\mathscr{S}}]$ for the heating $\mathscr{Q}$. The term "irreversibility" is justified because $\mathscr{Q}$, the rate of increase of energy not accompanied by mechanical working, is bounded above but not necessarily below. There is no limit to the magnitude of negative values of $\mathscr{Q}$, which represents conversion of energy into heating without performance of work, but there certainly is to positive ones. Work and energy may always be converted into heat, but there is a limit to the rate at which heat may be converted into energy without doing mechanical work.

---

[65] Explicit expressions for $\mathscr{W}^{(\mathrm{e})}$ and $\mathscr{Q}$ along homogeneous processes will be given in Section 2.2.3 below, where we shall also derive the equations that govern their thermodynamic evolution.

We renounce justifying Axioms 2.6 and 2.7 any further; rather, we accept them as the basis of our understanding of thermodynamic processes. Though formulated for homogeneous processes, slight formal alterations will suffice to extend their validity to more general processes of continuum mechanics.

According to this theory, thermodynamic equilibrium is reached when neither of the thermodynamic functions introduced so far changes in time any longer. Thus, by the linearity of $\mathscr{W}^{(e)}$ in $\dot{\mathrm{q}}$ and $\mathscr{Q}$ in $(\dot{\theta}, \dot{\mathrm{q}})$, (2.186) and (2.187) are identically satisfied at equilibrium, the latter as an equality.

We say that a process is *adiabatic* whenever $\mathscr{Q} \equiv 0$ and *isoentropic* whenever $\dot{\mathscr{S}} \equiv 0$. It is clear from (2.187) that such processes are equivalent only if they are reversible.

The quantity that fills the gap created by the second law of thermodynamics in inequality (2.187) is the *total dissipation* $\mathscr{D}$,

$$\mathscr{D} := \theta\dot{\mathscr{S}} - \mathscr{Q} \geqq 0. \tag{2.188}$$

By (2.188), a process is reversible if and only if $\mathscr{D}$ vanishes along it. A deeper interpretation of $\mathscr{D}$ can be gained by introducing the *free energy* $\mathscr{F}$ as

$$\mathscr{F} := \mathscr{U} - \theta\mathscr{S}.$$

By differentiating both sides of this equation with respect to time and making use of (2.186), we readily arrive at

$$\mathscr{D} = \mathscr{W}^{(e)} - \dot{\mathscr{F}} - \dot{\theta}\mathscr{S}, \tag{2.189}$$

which clearly appears as an extension of the dissipation identity (2.173) already encountered for mass-point systems, provided we introduce the RAYLEIGH *dissipation function* $\mathscr{R}$ through

$$\mathscr{R} := \frac{1}{2}\mathscr{D}, \tag{2.190}$$

which mimics (2.175). Here the role of the total power W is played by

$$\mathscr{W} := \mathscr{W}^{(e)} - \dot{\mathscr{F}} - \dot{\theta}\mathscr{S}, \tag{2.191}$$

which is the total power of the external agencies (including inertia) diminished by the rate of change of the free energy, $\dot{\mathscr{F}}$, and by what we call the *thermal production* of energy,[66]

$$\mathscr{T} := \dot{\theta}\mathscr{S}, \tag{2.192}$$

which vanishes identically in any isothermal process. We call $\mathscr{W}$ in (2.191) the *total working*. In words, equation (2.189) says that the input of external power that increases neither the free energy nor the thermal production of energy is lost in dissipation.

66 We borrow this name from [131, p. 189].

Requiring that $\mathscr{D}$ in (2.189) be positive semidefinite embodies the *reduced* dissipation inequality.[67] Such an inequality is classically[68] required to be valid for all possible processes and thus employed to place restrictions upon the constitutive laws that may express $\mathscr{W}^{(\mathrm{e})}$, $\mathscr{F}$, and $\mathscr{S}$ in terms of thc process (θ, q): these laws must make $\mathscr{D}$ as *defined* by (2.189) positive definite on *all* processes.

Here our attitude will be different. Inspired by the perspective offered by the classical mass-point mechanics to regard R (and D) in Section 2.2.1 as a potential for RAYLEIGH dissipative forces in Lagrangian dynamics, in Section 2.2.3 we shall consider $\mathscr{R}$ in (2.190) to be constitutively assigned as a positive semidefinite quadratic potential of the state rate $\dot{\mathsf{q}}$. We shall thus reformulate the principle of minimum constrained dissipation as a means to derive the equations that govern the thermodynamic evolution.

In the rest of this section, we restate the axioms of thermodynamics formulated here for homogeneous processes so as to encompass the more general processes that a deforming continuous body may suffer.

Deformable Bodies

For a body $\mathscr{B}$ undergoing a motion χ defined as in (2.3), a thermodynamic process is characterized by fields defined on the collection of shapes $\mathscr{C}_\chi$ in (2.5). For classical continuous bodies, a process is defined by the pair of fields (θ, χ), where θ is the absolute temperature. Equilibrium is reached wherever neither θ nor χ depends on time. For any subbody $\mathscr{P}$ of $\mathscr{B}$ in the motion χ, we call $\mathscr{U}(\mathscr{P}_t, \chi)$ the internal energy stored in the present shape $\mathscr{P}_t$ and we assume that it is absolutely continuous with respect to the mass measure, so that

$$\mathscr{U}(\mathscr{P}_t, \chi) := \int_{\mathscr{P}_t} \varrho \upsilon \, dV, \tag{2.193}$$

where ϱ is the mass density as above and υ is the density per unit mass of internal energy stored in the present shape $\mathscr{B}_t$ of the body. The energy density υ is here assumed to be a smooth function of position and time: no specific constitutive assumption is made at this stage that relates it to the thermodynamic process (θ, χ). The internal energy in (2.193) is thus a shape functional that is posited as a primitive quantity of the theory.

Similarly, the heating $\mathscr{Q}$ is a shape functional defined as

$$\mathscr{Q}(\mathscr{P}_t, \chi) := -\int_{\partial^* \mathscr{P}_t} \boldsymbol{q} \cdot \boldsymbol{\nu} \, dA + \int_{\mathscr{P}_t} \sigma \, dV, \tag{2.194}$$

where $\boldsymbol{q}$ is the *heat flux* and σ is the *heat supply*, the former expressing the rate at which heat enters $\mathscr{P}_t$ through its boundary[69] and the latter the rate at which heat is

---

67 Here "reduced" simply means that the heating $\mathscr{Q}$ has been disposed of.

68 See TRUESDELL [343, pp. 13–14].

69 The sign of the first integral in (2.194) is chosen so as to interpret $\boldsymbol{q}$ as the *influx* of heat when it makes an obtuse angle with the outer unit normal $\boldsymbol{\nu}$ and as the *efflux* of heat when

supplied to its interior.[70] Definition (2.194) has the general appearance of an integral balance law for a continuous body: both a surface flux and a volume supply contribute to it, though $\mathscr{Q}$ is *not* itself the time derivative of a function. In the thermodynamics of homogeneous processes, $\mathscr{Q}$ is assumed to be a linear form of the process rate $(\dot{\theta}, \dot{\mathsf{q}})$, though it is not a rate itself, so as to make it vanish in equilibrium. Likewise, we assume that in equilibrium the shape functional $\mathscr{Q}$ in (2.194) vanishes identically, that is, for all shapes $\mathcal{P}_t$ and at all times t. Such an assumption requires that both $\boldsymbol{q}$ and σ vanish in $\mathcal{B}_t$, for all t.

Axiom 2.6 is here replaced by the requirement that the time rate of the internal energy $\mathscr{U}$ in (2.193) be balanced by the net working $\mathscr{W}^{(\mathrm{e})}$ in (2.99) and the heating $\mathscr{Q}$ in (2.194):

$$\dot{\mathscr{U}}(\mathcal{P}_t, \boldsymbol{\chi}) = \mathscr{W}^{(\mathrm{e})}(\mathcal{P}_t, \boldsymbol{\chi}) + \mathscr{Q}(\mathcal{P}_t, \boldsymbol{\chi}), \tag{2.195}$$

for all shapes $\mathcal{P}_t$ and all motions $\boldsymbol{\chi}$. By (2.105), (2.195) can also be given the form

$$[\mathscr{U}(\mathcal{P}_t, \boldsymbol{\chi}) + \mathscr{K}(\mathcal{P}_t, \boldsymbol{\chi})]^{\cdot} = \mathscr{W}^{(\mathrm{a})}(\mathcal{P}_t, \boldsymbol{\chi}) + \mathscr{Q}(\mathcal{P}_t, \boldsymbol{\chi}), \tag{2.196}$$

which expresses how both the power expended by all external agencies applied to the subbody $\mathcal{P}$ and the heating provided to it determine the rate at which both the internal and kinetic energies stored in $\mathcal{P}$ increase. Equations (2.195) and (2.196) are equivalent formulations of the *first law of thermodynamics* for deformable bodies: they introduce the balance of energy alongside the other classical balances of linear and rotational momenta posited in Section 2.1.5.

To formulate the *second law of thermodynamics* for deformable bodies, we assume that the *entropy* content of a subbody $\mathcal{P}$ in its present shape $\mathcal{P}_t$ produced by the motion $\boldsymbol{\chi}$ is expressed by the shape functional

$$\mathscr{S}(\mathcal{P}_t, \boldsymbol{\chi}) := \int_{\mathcal{P}_t} \varrho\eta \, dV,$$

where η, which is a function of position and time, is the entropy density per unit mass. $\mathscr{S}$ is a phenomenological quantity whose time rate is to be related to an upper bound for the heating, much in the spirit of what (2.186) requires for homogeneous processes. To grant such an interpretation, we further assume that $\mathscr{S}$ obeys the growth law

$$\dot{\mathscr{S}}(\mathcal{P}_t, \boldsymbol{\chi}) \geqq -\int_{\partial^*\mathcal{P}_t} \frac{1}{\theta}\boldsymbol{q}\cdot\boldsymbol{\nu}\, dA + \int_{\mathcal{P}_t} \frac{\sigma}{\theta} dV \tag{2.197}$$

for all shapes $\mathcal{P}_t$ and all motions $\boldsymbol{\chi}$. For a temperature field θ uniform in space, (2.197) reduces to (2.187), which we postulated for homogeneous processes. In general, (2.197), which is also called the CLAUSIUS–DUHEM inequality, shows how the sources of $\mathscr{Q}$, both flux and supply, provide corresponding sources for a lower bound

it makes an acute angle with $\boldsymbol{\nu}$. Convenient as these names may appear, they will never be used in this book: the contribution that the heat flux $\boldsymbol{q}$ makes to the total heating $\mathscr{Q}$ would be properly defined mathematically whatever sign precedes the first integral in (2.194).

[70] Provided, for example, by the absorption of external radiation.

on the growth of $\mathscr{S}$. As above, a process will be said to be reversible whenever the equality sign holds in (2.197) throughout it; otherwise, it is irreversible.

An equivalent way to write (2.197) is to set

$$\dot{\mathscr{S}}(\mathcal{P}_t, \chi) + \int_{\partial^* \mathcal{P}_t} \frac{1}{\theta} \boldsymbol{q} \cdot \boldsymbol{\nu} \, dA - \int_{\mathcal{P}_t} \frac{\sigma}{\theta} dV =: \int_{\mathcal{P}_t} \sigma_\mathrm{i} \, dV$$

and to require σ_i to be a positive semidefinite function of position and time. Then (2.197) can be rewritten as

$$\dot{\mathscr{S}}(\mathcal{P}_t, \chi) = - \int_{\partial^* \mathcal{P}_t} \frac{1}{\theta} \boldsymbol{q} \cdot \boldsymbol{\nu} \, dA - \int_{\mathcal{P}_t} \left(\frac{\sigma}{\theta} + \sigma_\mathrm{i} \right) dV, \tag{2.198}$$

which appears in the classical form of an integral balance law, provided we interpret $\frac{1}{\theta}\boldsymbol{q}$ as the *entropy flux* and $\left(\frac{\sigma}{\theta} + \sigma_\mathrm{i}\right)$ as the *entropy supply*. While the entropy flux derives from the heat flux, the entropy supply has a heat component $\frac{\sigma}{\theta}$ proportional to the heat supply, and an intrinsically irreversible component σ_i that is called the *entropy production*. The second law of thermodynamics for deformable bodies requires the entropy production to be positive semidefinite and identifies it as a source of irreversibility. A thermodynamic process is thus said to be reversible whenever the entropy production σ_i vanishes identically along it. In general, if (2.198) represents the entropy balance, we may say, as also suggested by [131, p. 187], that (2.197) represents the entropy *imbalance*.

In principle, there is no reason why in the entropy imbalance (2.197) the entropy flux should be determined by the heat flux. MÜLLER [226], perhaps inspired by the fact that entropy and heat supplies already differ, proposed a theory in which the entropy flux $\boldsymbol{p}$ and heat flux $\boldsymbol{q}$ are independent constitutive quantities. The limited room we can devote to thermodynamics in this book does not permit exploring further the consequences of such an assumption, which are indeed more telling for materials richer in structure than for "classical materials."[71] Here the entropy flux will always be $\boldsymbol{p} = \frac{1}{\theta}\boldsymbol{q}$.

The statements of both first and second laws of thermodynamics for deformable bodies just obtained by analogy from Axioms 2.6 and 2.7 posited above for homogeneous processes need to be invariant under a change of frame. To ensure this, we shall assume that θ, υ, η, σ, and σ_i are all indifferent scalar functions, that is, in the notation of (2.44),

$$\theta^*(p^*, t^*) = \theta(p, t), \tag{2.199a}$$

$$\upsilon^*(p^*, t^*) = \upsilon(p, t), \quad \eta^*(p^*, t^*) = \eta(p, t), \tag{2.199b}$$

$$\sigma^*(p^*, t^*) = \sigma(p, t), \quad \sigma_\mathrm{i}^*(p^*, t^*) = \sigma_\mathrm{i}(p, t), \tag{2.199c}$$

and that $\boldsymbol{q}$ is an indifferent vector field,

$$\boldsymbol{q}^*(p^*, t^*) = \mathbf{R}\boldsymbol{q}(p, t), \tag{2.200}$$

[71] See, for example, [181].

where $\mathbf{R}$ is the orthogonal time-dependent tensor representing the change of frame. We have not yet specified the constitutive laws that make υ, η, σ_{i}, and $\boldsymbol{q}$ depend on the thermodynamic process for a given category of bodies. All (2.199) and (2.200) tell us is that these laws must be frame-indifferent. As a consequence, the laws of thermodynamics are frame-indifferent provided that $\mathscr{W}^{(\mathrm{e})}$ is as well, as indeed required by Axiom 2.3. As shown in Section 2.1.5, this is the case whenever CAUCHY's laws of motion (2.127) and (2.134) are satisfied. These latter in turn imply (2.136), by which both balances (2.195) and (2.198) can easily be reduced to local form.

By (2.193), the definition of heating in (2.194), the transport theorem in (2.111), and the divergence theorem, under standard smoothness assumptions, (2.195) becomes

$$\int_{\mathcal{P}_t} \varrho \dot{\upsilon}\, dV = \int_{\mathcal{P}_t} (\mathbf{T}\cdot\mathbf{D} - \operatorname{div}\boldsymbol{q} + \sigma)\, dV,$$

which is valid for all shapes $\mathcal{P}_t$ if and only if

$$\varrho\dot{\upsilon} = \mathbf{T}\cdot\mathbf{D} - \operatorname{div}\boldsymbol{q} + \sigma. \tag{2.201}$$

This is the local form of the balance of energy. Similarly, equation (2.198) is reduced to

$$\varrho\dot{\eta} = -\operatorname{div}\left(\frac{\boldsymbol{q}}{\theta}\right) + \frac{\sigma}{\theta} + \sigma_{\mathrm{i}},$$

which easily becomes

$$\varrho\dot{\eta} = \frac{1}{\theta^2}\boldsymbol{q}\cdot\nabla\theta - \frac{1}{\theta}\operatorname{div}\boldsymbol{q} + \frac{\sigma}{\theta} + \sigma_{\mathrm{i}}, \tag{2.202}$$

whence, by eliminating $\operatorname{div}\boldsymbol{q}$ with the aid of (2.201), we easily arrive at

$$\varrho\theta\dot{\eta} = \varrho\dot{\upsilon} - \mathbf{T}\cdot\mathbf{D} + \frac{1}{\theta}\boldsymbol{q}\cdot\nabla\theta + \theta\sigma_{\mathrm{i}}. \tag{2.203}$$

Let the function ψ, defined by

$$\psi := \upsilon - \theta\eta, \tag{2.204}$$

be the *free energy* per unit mass. A simple computation shows that (2.203) can be given the following equivalent form:

$$\theta\sigma_{\mathrm{i}} = \mathbf{T}\cdot\mathbf{D} - \varrho\dot{\psi} - \left(\varrho\dot{\theta}\eta + \frac{1}{\theta}\boldsymbol{q}\cdot\nabla\theta\right). \tag{2.205}$$

Integrating both sides of this equation over the shape $\mathcal{P}_t$ of the subbody $\mathcal{P}$ conveyed by the motion $\boldsymbol{\chi}$ and making use of (2.136), we obtain that

$$\mathscr{D}(\mathcal{P}_t,\boldsymbol{\chi}) = \mathscr{W}^{(\mathrm{e})}(\mathcal{P}_t,\boldsymbol{\chi}) - \dot{\mathscr{F}}(\mathcal{P}_t,\boldsymbol{\chi}) - \mathscr{T}(\mathcal{P}_t,\boldsymbol{\chi}) \geqq 0, \tag{2.206}$$

where

$$\mathscr{D}(\mathcal{P}_t,\boldsymbol{\chi}) := \int_{\mathcal{P}_t} \theta\sigma_{\mathrm{i}}\, dV \tag{2.207}$$

is the *total dissipation* functional, $\mathscr{W}^{(e)}$ is the power expended by all external agencies (including inertia), and $\mathscr{F}$ and $\mathscr{T}$ are the *free energy* and *thermal production* functional defined by

$$\mathscr{F}(\mathcal{P}_t, \chi) := \int_{\mathcal{P}_t} \varrho\psi \, dV \tag{2.208}$$

and

$$\mathscr{T}(\mathcal{P}_t, \chi) := \int_{\mathcal{P}_t} \left(\varrho\dot{\theta}\eta + \frac{1}{\theta}\boldsymbol{q} \cdot \nabla\theta \right) dV, \tag{2.209}$$

respectively.

As for $\mathscr{D}$ in (2.188), requiring $\mathscr{D}$ in (2.207) to be positive semidefinite embodies the *reduced* dissipation inequality. Paralleling closely (2.188), (2.206) says that the total external power that contributes neither to the free energy $\mathscr{F}$ nor to the thermal production of energy $\mathscr{T}$ is lost in dissipation. It is worth remarking here for later use that $\mathscr{D}$ vanishes identically in equilibrium.

Equation (2.206) clearly extends (2.188) to the thermodynamics of deformable bodies, and so it further generalizes the dissipation identity (2.173), provided we replace there W with the total working

$$\mathscr{W} := \mathscr{W}^{(e)} - \dot{\mathscr{F}} - \mathscr{T}. \tag{2.210}$$

The RAYLEIGH dissipation functional $\mathscr{R}$, formally defined by (2.190), is here a shape functional, which by (2.207) vanishes identically on all reversible processes. It is easily checked that for a homogeneous process for which all fields are uniform throughout the shape $\mathcal{B}_t$ of the whole body $\mathcal{B}$, (2.206) and (2.209) reduce to (2.188) and (2.192), respectively.

In the following section, we shall formulate a minimum principle for the reduced dissipation

$$\widetilde{\mathscr{R}} := \mathscr{R} - \mathscr{W} = \frac{1}{2}\mathscr{D} - \mathscr{W}, \tag{2.211}$$

where $\mathscr{W}$ is as in (2.210) and $\mathscr{D}$ is a quadratic, positive semidefinite functional in the appropriate dissipation measures.

2.2.3 Principle of Minimum Reduced Dissipation

Here we state the general principle that we set as the basis of our development in the rest of the book. Our formulation will be first phrased in the language of the thermodynamics of homogeneous processes introduced in Section 2.2.2. The form of the principle appropriate for the more general processes undergone by deformable bodies will then present itself as a rather natural extension, also in the light of the foregoing discussion. Section 2.3 will present the application of this principle to the linearly viscous fluids of the NAVIER–STOKES theory, also in the presence of heat conduction. The following chapters will be based on suitable reformulations of the general principle in different settings of ordered fluids, particularly apt for describing liquid crystals in their diverse dynamical manifestations.

Homogeneous Processes

For a body $\mathfrak{B}$ performing a homogeneous thermodynamic process (θ, q), with temperature θ and state variable q in the admissible set $\mathfrak{Q} \in \mathbb{R}^m$ thought of as functions of time t, we assume that the net working $\mathscr{W}^{(\mathrm{e})}$ is given by the following linear form in the state rate:

$$\mathscr{W}^{(\mathrm{e})} = \sum_{i=1}^{m} p_i(\theta, \mathsf{q})\dot{q}_i = \mathsf{p} \cdot \dot{\mathsf{q}}, \tag{2.212}$$

where $\mathsf{p} = (p_1, \ldots, p_m)$ is a vector of functions of the process (θ, q) representing generalized external forces. Similarly, we assume that the heating $\mathscr{Q}$ is given the form

$$\mathscr{Q} = \kappa(\theta, \mathsf{q})\dot{\theta} + \mathsf{k}(\theta, \mathsf{q}) \cdot \dot{\mathsf{q}}, \tag{2.213}$$

where κ is a scalar constitutive function of the process representing the specific heat in a constant state and $\mathsf{k} = (k_1, \ldots, k_m)$ is a constitutive vector of the process, expressing generalized latent heats.

Furthermore, we assume that the free energy $\mathscr{F}$ and the entropy $\mathscr{S}$ are given by the following smooth[72] functions of the process (θ, q):

$$\mathscr{F}(t) = \Psi(\theta(t), \mathsf{q}(t)), \tag{2.214a}$$

$$\mathscr{S}(t) = \mathrm{H}(\theta(t), \mathsf{q}(t)), \tag{2.214b}$$

so that the internal energy $\mathscr{U}$ is given by the function

$$\mathscr{U}(t) = \Upsilon(\theta(t), \mathsf{q}(t)) \tag{2.214c}$$

with

$$\Upsilon(\theta, \mathsf{q}) := \Psi(\theta, \mathsf{q}) + \theta \mathrm{H}(\theta, \mathsf{q}). \tag{2.214d}$$

We also assume that the RAYLEIGH function in (2.190) is expressed by

$$\mathscr{R} = \frac{1}{2}\dot{\mathsf{q}} \cdot R(\theta, \mathsf{q})\dot{\mathsf{q}}, \tag{2.215}$$

where $R(\theta, \mathsf{q})$ is a continuous mapping of the pair (θ, q) into the space of symmetric $m \times m$ real matrices. We take $R(\theta, \mathsf{q})$ as positive semidefinite on all admissible processes.

By (2.212), (2.213), and (2.214c), the first law of thermodynamics in (2.186) requires that

$$\frac{\partial \Upsilon}{\partial \theta}\dot{\theta} + \frac{\partial \Upsilon}{\partial \mathsf{q}} \cdot \dot{\mathsf{q}} = \kappa\dot{\theta} + (\mathsf{p} + \mathsf{k}) \cdot \dot{\mathsf{q}}. \tag{2.216}$$

Likewise, the total working $\mathscr{W}$ defined by (2.191) becomes

$$\mathscr{W} = \Xi\dot{\theta} + \mathsf{X} \cdot \dot{\mathsf{q}}, \tag{2.217}$$

[72] At least twice continuously differentiable.

where Ξ and X are functions of the process (θ, q) only, defined by

$$\Xi(\theta, \mathsf{q}) := -\left(\frac{\partial \Psi}{\partial \theta} + \mathrm{H}\right), \tag{2.218a}$$

$$\mathsf{X}(\theta, \mathsf{q}) := \mathsf{p} - \frac{\partial \Psi}{\partial \mathsf{q}}. \tag{2.218b}$$

The pair (Ξ, X) defines the generalized *thermodynamic forces* that expend power against the process rate $(\dot{\theta}, \dot{\mathsf{q}})$.

It is apparent from (2.217) that $\mathscr{W}$ is a linear form of the process rate $(\dot{\theta}, \dot{\mathsf{q}})$ with coefficients Ξ and X depending on the process. Precisely as (2.217) parallels (2.159), the definition of $\mathscr{R}$ in (2.215) parallels that of R in (2.176). Letting $\widetilde{\mathscr{R}}$ be defined by analogy to (2.180) as $\widetilde{\mathscr{R}} := \mathscr{R} - \mathscr{W}$, with $\mathscr{R}$ and $\mathscr{W}$ as in (2.215) and (2.217), respectively, we are in a position to posit for homogeneous processes the following principle, which extends the one formulated on page 122 for a dynamical system governed by the LAGRANGE–RAYLEIGH equations of motion.

Principle of Minimum Reduced Dissipation (for Homogeneous Processes). For a body $\mathfrak{B}$ performing the homogeneous process (θ, q) with total working $\mathscr{W}$ as in (2.217) and RAYLEIGH dissipation function $\mathscr{R}$ as in (2.215) that obey the second law of thermodynamics in the form (2.189) with $\mathscr{D} = 2\mathscr{R}$, the true thermodynamic evolution $(\dot{\theta}, \dot{\mathsf{q}})$ through the instantaneous value $(\theta(t), \mathsf{q}(t))$ at time t is such that the reduced dissipation function $\widetilde{\mathscr{R}} = \mathscr{R} - \mathscr{W}$ attains its minimum with respect to all virtual process rates $(\hat{\dot{\theta}}, \hat{\dot{\mathsf{q}}})$ with $\hat{\dot{\theta}} = \dot{\theta} + \delta\dot{\theta}$ and $\hat{\dot{\mathsf{q}}} = \dot{\mathsf{q}} + \delta\dot{\mathsf{q}}$ once the generalized thermodynamic forces (Ξ, X) are held fixed.

We now apply this principle to derive the equations that govern the time evolution of a homogeneous process. By repeating verbatim the reasoning that in Section 2.2.1 led us to (2.179), here we readily arrive at the following condition of minimality:

$$\left(\frac{\partial \mathscr{R}}{\partial \dot{\theta}}, \frac{\partial \mathscr{R}}{\partial \dot{\mathsf{q}}}\right) = (\Xi, \mathsf{X}). \tag{2.219}$$

Since we assumed in (2.215) that $\mathscr{R}$ is independent of $\dot{\theta}$, thus *not* regarding the temperature rate as a measure of dissipation, equation (2.219) is equivalent to the pair

$$\Xi = 0, \qquad \frac{\partial \mathscr{R}}{\partial \dot{\mathsf{q}}} = \mathsf{X}.$$

By (2.218a), the former of these equations requires that[73]

$$\mathrm{H} = -\frac{\partial \Psi}{\partial \theta}, \tag{2.220}$$

[73] Equation (2.220) is essentially an equilibrium condition, since it is independent of the thermodynamic evolution: it follows from requiring that $\dot{\theta}$ *not* be a measure of dissipation, and so it has no right to affect R.

while, by (2.218b) and (2.215), the latter becomes

$$R(\theta, \mathsf{q})\dot{\mathsf{q}} = \mathsf{p} - \frac{\partial \Psi}{\partial \mathsf{q}}. \tag{2.221}$$

Equations (2.220) and (2.221) must be valid in all homogeneous processes of $\mathfrak{B}$ that comply with the principle of minimum reduced dissipation stated above. In particular, (2.220), which is *not* an evolution equation but rather appears in the form of an equilibrium condition, must be valid for all initial conditions $(\theta(0), \mathsf{q}(0))$ admissible for a process. Assuming that the admissible set $\mathfrak{Q}$ is open and that a process can start from any point in $\mathbb{R}^+ \times \mathfrak{Q}$, we regard (2.220) as an equilibrium identity valid throughout the domain $\mathbb{R}^+ \times \mathfrak{Q}$, which assigns the entropy function H from the free energy Ψ. By (2.220), (2.214b), and (2.214d), Ψ deserves the name of *thermodynamic potential*, since both the entropy $\mathscr{S}$ and the internal energy $\mathscr{U}$ of a homogeneous process derive from it.

If the validity of the second law of thermodynamics is guaranteed by (2.221), making use of (2.220) and (2.214d) in (2.216), we easily obtain that the first law of thermodynamics requires that

$$\left(\theta \frac{\partial^2 \Psi}{\partial \theta^2} + \kappa\right) \dot{\theta} + \left(\mathsf{p} - \frac{\partial \Psi}{\partial \mathsf{q}} + \mathsf{k} + \theta \frac{\partial^2 \Psi}{\partial \theta \partial \mathsf{q}}\right) \cdot \dot{\mathsf{q}} = 0. \tag{2.222}$$

Equations (2.221) and (2.222) govern the evolution of a homogeneous process in this formal thermodynamic scheme. They suggest a few remarks about equilibrium and reversibility, the former being reached whenever the process rate $(\dot{\theta}, \dot{\mathsf{q}})$ vanishes identically and the latter characterizing the processes for which $\mathscr{R} \equiv 0$. First, since the matrix R is only positive semidefinite, it is not in general invertible in the whole domain $\mathbb{R}^+ \times \mathfrak{Q}$, and so

$$\mathsf{p} = \frac{\partial \Psi}{\partial \mathsf{q}} \tag{2.223}$$

is only a *necessary* equilibrium condition. For it to become sufficient at a given pair (θ_0, q_0), the following conditions must also be satisfied:

$$\det R(\theta_0, \mathsf{q}_0) > 0, \qquad \left(\theta \frac{\partial^2 \Psi}{\partial \theta^2} + \kappa\right)_{(\theta_0, \mathsf{q}_0)} \neq 0, \tag{2.224}$$

which imply that both $\dot{\theta}$ and $\dot{\mathsf{q}}$ vanish at (θ_0, q_0). In general, wherever $\det R = 0$ in $\mathbb{R}^+ \times \mathfrak{Q}$, a new process may branch off and a detailed bifurcation analysis is needed to characterize all possible processes that meet there. If $\det R = 0$ at a putative equilibrium point that satisfies (2.223), this same bifurcation analysis characterizes its stability. Dwelling any further on this matter would, however, exceed the scope of this book.

Equation (2.223) must also be valid along all reversible processes, for which $\dot{\mathsf{q}} \neq 0$ but $R\dot{\mathsf{q}} \equiv 0$. Clearly, such a process may exist only if $\det R \equiv 0$ along it. By contrast, along an irreversible process, (2.221) can be solved for $\dot{\mathsf{q}}$, and (2.222) then determines $\dot{\theta}$, provided that the second inequality in (2.224) is satisfied. For an irreversible process, (2.223) determines the equilibrium states, and, by (2.222), also $\dot{\theta}$ vanishes as soon as an equilibrium state is reached. If both inequalities in (2.224) are satisfied in the whole domain $\mathbb{R}^+ \times \mathcal{Q}$, then all admissible processes are irreversible and equations (2.221) and (2.222) describe how the body $\mathcal{B}$ tends toward local equilibria. Bifurcations in the solutions of the evolution equations, which make the thermodynamics of homogeneous processes undergone by $\mathcal{B}$ richer in options, are related to the local failure of inequalities (2.224). In words, one may also say that local reversibility makes the homogeneous processes of $\mathcal{B}$ more intriguing.

Deformable Bodies

The thermodynamics of homogeneous processes of a body served the purpose of illustrating the general use of the principle of minimum reduced dissipation that we shall make in this book. Clearly, for more complex processes, such as those undergone by a deformable body, either classical or ordered, the mathematical appearance of the principle will be different, though its formal structure will remain just the one already seen above. The reduced dissipation $\widetilde{\mathscr{R}}$ will now be the shape functional defined by (2.211), and similarly the total working $\mathscr{W}$ will be given by (2.210).

As in Section 2.2.2 above, a process is now identified by the pair $(\theta, \boldsymbol{\chi})$, to which there corresponds the process rate $(\dot{\theta}, \boldsymbol{v})$. We assume that the entropy production σ_i is a quadratic positive semidefinite function of certain fields collected in a list d, defined over the collection of shapes $\mathcal{C}_{\boldsymbol{\chi}}$, which are frame-indifferent measures of local dissipation. The fields in d may be scalars, vectors, or tensors: they translate into mathematical terms our perception of the mechanisms of internal dissipation at work for a particular material. We assume that $\boldsymbol{g} := \nabla\theta$ is included[74] in d and that all dissipation measures other than $\boldsymbol{g}$ are linearly related to the velocity field $\boldsymbol{v}$, so as to vanish at equilibrium. We also require that $\dot{\theta}$ cannot be a local measure of dissipation, and so it is excluded from d. Thus, we represent $\mathscr{R}$ as

$$\mathscr{R}(\mathcal{P}_t, \boldsymbol{\chi}) = \int_{\mathcal{P}_t} R \, dV, \tag{2.225}$$

where the density R is given as

$$R = \frac{1}{2}\mathsf{d} \cdot \mathsf{R}[\mathsf{d}] \tag{2.226}$$

and $\mathsf{R} = \mathsf{R}(\theta, \boldsymbol{\chi})$ represents a symmetric bilinear form in the space of all lists like d. By (2.207), equation (2.226) amounts to writing

$$\theta\sigma_\mathrm{i} = \mathsf{d} \cdot \mathsf{R}[\mathsf{d}] = 2R. \tag{2.227}$$

By varying instantaneously the velocity field $\boldsymbol{v}$ by $\delta\boldsymbol{v}$ and the temperature gradient[75] field $\boldsymbol{g}$ by $\delta\boldsymbol{g}$, subject to $\operatorname{curl}\delta\boldsymbol{g} = \mathbf{0}$, without affecting the shape $\mathcal{P}_t$, as in

[74] That for a nonisothermal process $\nabla\theta$ is an intrinsic source of dissipation is suggested by the form (2.209) for the thermal production of energy $\mathscr{T}$ balancing $\mathscr{D}$ in (2.206).

[75] Since by constitutive choice $\dot{\theta}$ is not a dissipation measure, an instantaneous variation $\delta\dot{\theta}$ of $\dot{\theta}$, which will soon play a role, has no effect here.

the spirit of all variations considered here, we produce a variation $\delta\mathbf{d}$, which in turn varies $\mathscr{R}$ into

$$\hat{\mathscr{R}} = \mathscr{R} + \frac{1}{2}\int_{\mathcal{P}_t} \delta\mathbf{d}\cdot\mathbf{R}[\delta\mathbf{d}]dV + \delta\mathscr{R}, \tag{2.228}$$

where

$$\delta\mathscr{R} := \int_{\mathcal{P}_t} \mathbf{d}\cdot\mathbf{R}[\delta\mathbf{d}]dV. \tag{2.229}$$

We further assume that the total working $\mathscr{W}$ in (2.210) can in general be given the form

$$\mathscr{W}(\mathcal{P}_t,\chi) = \int_{\mathcal{P}_t}\{\Xi\dot{\theta} + \boldsymbol{G}\cdot\nabla\theta + \boldsymbol{B}\cdot\boldsymbol{v}\}dV + \int_{\partial^*\mathcal{P}_t}\boldsymbol{T}\cdot\boldsymbol{v}\,dA, \tag{2.230}$$

where Ξ, $\boldsymbol{G}$, $\boldsymbol{B}$, and $\boldsymbol{T}$ are appropriate fields, scalar the first and vectorial the others, defined over the present shape $\mathcal{P}_t$ of the generic subbody $\mathcal{P}$. In particular, we can recognize in $\boldsymbol{B}$ a generalized body force and in $\boldsymbol{T}$ a generalized surface traction, which is in general a functional of $\partial^*\mathcal{P}_t$. Letting $\delta\dot{\theta}$ denote an instantaneous variation of the temperature rate $\dot{\theta}$, independent of the gradient variation $\delta\boldsymbol{g}$, we *define* the variation induced in $\mathscr{W}$ by $(\delta\dot{\theta},\delta\boldsymbol{g},\delta\boldsymbol{v})$ as

$$\delta\mathscr{W} := \int_{\mathcal{P}_t}\{\Xi\delta\dot{\theta} + \boldsymbol{G}\cdot\delta\boldsymbol{g} + \boldsymbol{B}\cdot\delta\boldsymbol{v}\}dV + \int_{\partial^*\mathcal{P}_t}\boldsymbol{T}\cdot\delta\boldsymbol{v}\,dA. \tag{2.231}$$

Now, since $\mathbf{R}$ is positive semidefinite, by (2.228), the reduced dissipation functional $\widetilde{\mathscr{R}} = \mathscr{R} - \mathscr{W}$ attains its minimum under all variations that keep the generalized forces Ξ, $\boldsymbol{G}$, $\boldsymbol{B}$, and $\boldsymbol{T}$ fixed if and only if the variations $\delta\mathscr{R}$ and $\delta\mathscr{W}$ defined in (2.229) and (2.231) are identically equal.

We can finally formulate the principle of minimum reduced dissipation appropriate for the thermodynamics of classical deformable bodies.

Principle of Minimum Reduced Dissipation (for Deformable Bodies). For a classical deformable body $\mathcal{B}$ undergoing a thermodynamic process (θ,χ) with total working $\mathscr{W}$ as in (2.230) and RAYLEIGH dissipation functional $\mathscr{R}$ as in (2.225) that obey the second law of thermodynamics in the form (2.206) with $\mathscr{D} = 2\mathscr{R}$, the true thermodynamic evolution through the pair of instantaneous fields (θ,χ) at time t is such that the reduced dissipation functional $\widetilde{\mathscr{R}} = \mathscr{R} - \mathscr{W}$ attains its minimum with respect to all virtual process rates $(\hat{\dot{\theta}},\hat{\boldsymbol{v}})$, with $\hat{\dot{\theta}} = \dot{\theta} + \delta\dot{\theta}$ and $\hat{\boldsymbol{v}} = \boldsymbol{v} + \delta\boldsymbol{v}$, and with respect to all instantaneous virtual temperature gradients $\hat{\boldsymbol{g}}$, with $\hat{\boldsymbol{g}} = \boldsymbol{g} + \delta\boldsymbol{g}$ and $\operatorname{curl}\delta\boldsymbol{g} = \mathbf{0}$, once the fields Ξ, $\boldsymbol{G}$, $\boldsymbol{B}$, and $\boldsymbol{T}$ in (2.230) are held fixed. That is, the true evolution is characterized by the requirement that

$$\delta\mathscr{R} = \delta\mathscr{W} \quad \text{for all } \mathcal{P}_t, \tag{2.232}$$

where $\delta\mathscr{R}$ and $\delta\mathscr{W}$ are as in (2.229) and (2.231), respectively.

In Section 2.3, as an illustration of this principle, we shall describe the equations valid in the classical theory of the NAVIER–STOKES fluid with heat conduction, while in Section 2.2.5 below this very fluid will appear as a special case in the context of other competing theories. We now comment on constitutive relations for fluids in general.

2.2.4 Simple and Nonsimple Fluids

In Sections 2.1.5 and 2.1.6 we have illustrated the balance equations that govern classical continuum mechanics. They are not sufficient to describe the evolution of a body as long as $\mathbf{T}$ and $\mathbf{L}$, the stress and couple stress tensors, remain unrelated to the motion χ.

According to the general ideas put forward by NOLL [235], where only classical media were considered, unable to carry couples,[76] this is achieved by specifying the stress tensor at time t as a functional of the whole history up to time t of the motion: such a functional dependence would distinguish one material from another, thus being *constitutive* for it. The general conceptual framework for material behavior formulated by NOLL is effectively summarized in TRUESDELL & NOLL [345, § C.III] and TRUESDELL [349, § IV.2].

The materials thus formally defined are called *simple*. Since here we are concerned only with fluids, it will suffice to recall that a simple fluid is defined by formalizing the requirements that

(i) the present stress be determined by the history of the gradient of the deformation, the tensor of $\mathsf{L}(\mathcal{V})$ defined as $\mathbf{F} := \nabla_P \chi(P,t)$ in terms of the motion (2.3), and

(ii) the symmetry group G, under which the material response is invariant, be the largest possible.[77]

In NOLL's theory of simple fluids, requirement (ii) is fulfilled by choosing G coincident with the unimodular group[78] $\mathsf{U}(3)$. Since a simple fluid, like any simple material, cannot sustain couples, its couple stress tensor $\mathbf{L}$ vanishes identically, as do the intrinsic rotational momentum $\boldsymbol{k}_{\mathrm{i}}$ and the body couple $\boldsymbol{k}$, so that the balance equation for torques (2.141) simply reduces to requiring that the CAUCHY stress tensor $\mathbf{T}$ be symmetric. Such a requirement is fulfilled by the theory of simple fluids by assuming that the range of the functional delivering $\mathbf{T}$ is the space of symmetric tensors in $\mathsf{L}(\mathcal{V})$.

There are several ways to go beyond the theory of simple materials. TRUESDELL [342, p. 33] in one of his lectures had already pointed out two possible ways, both originally indicated by TOUPIN [338, 339]:[79]

> In the theory of simple materials, presented in Lecture 1, the stress is assumed to be determined by the history of the deformation gradient. In oriented materials, the body manifold itself is generalized by the addition of directors at the points, but the point deformation still affects the stress through

[76] Often these are also called CAUCHY media.

[77] See [345, p. 427].

[78] The unimodular group comprises all tensors $\mathbf{U} \in \mathsf{L}(\mathcal{V})$ such that $|\det \mathbf{U}| = 1$ (see also Appendix A.3).

[79] We cannot resist transcribing TRUESDELL's comment to the effect that two different nonsimple theories, similar to some extent in their results, were proposed by the same author: "Since both theories were developed by the same man, we cannot even attribute their diverse motivations to socio-economic injustice, wrangling among colleagues, or an unhappy love affair" [342, p. 34].

> its gradient. A different generalization is equally plausible. Leaving the body manifold as it is, a set of points only, we may allow the stress to depend in a more delicate way upon the deformation. NOLL's general theory of materials allows all properties of the deformation in a neighborhood of a point to influence the stress there, but from the outset the stress tensor is assumed symmetric. Mr. TOUPIN has constructed a theory of elastic materials that may have unsymmetric stress tensors—polar-elastic materials of second grade.[80]

The ordered fluids we consider in this book are not simple: they fall within the first category outlined in TRUESDELL's excerpt above, though being dissipative, they are described by neither of TOUPIN's theories. For them we retain the prescription on symmetry to be the largest possible, and this explains why we are seldom concerned explicitly with material symmetry.

Nonsimplicity, at least in our theory, also bears on objectivity, since the notion of observer is also affected by the extra descriptors of molecular order adopted by the theory. A rather general attempt to extend the notion of observer to this nonsimple context was made by CAPRIZ [41, § 3], who introduced a differential manifold $\mathfrak{M}$ to describe the local order and devised actions on it that represent changes of observers in the ambient Euclidean space $\mathcal{E}$. Here, since the local molecular order is described by a list $\mathbf{Q}$ of tensors, possibly of different ranks, we need not keep that level of generality. We only emphasize the role that improper orthogonal transformations $\mathbf{R}$, which are legitimate in establishing a change of frame as in (2.43a), may play on the molecular assembly comprising a body-point in the Euclidean space.

Take $\mathbf{R} = -\mathbf{I}$ and imagine that the molecular interactions are sensitive to the molecules' chirality, that is, that two molecules and their inverted images have different interaction energies.[81] If the macroscopic free-energy functional $\mathscr{F}$ is to reflect the molecular origin of the interactions responsible for the ordering of the fluid, it cannot be invariant under changes of frames involving improper orthogonal transformations. This will be the basis for the different requirements imposed in Section 3.1.1 below on the elastic free-energy density W for nematic (nonchiral) and cholesteric (chiral) liquid crystals. As suggested by NOLL [244, p. 14], this assumption can alternatively be given a more traditional interpretation:

> Rather, the assumption expresses a certain kind of material symmetry. Roughly, it states that the director-field interacts isotropically (or hemitropically in the cholesteric case) with the underlying body. In other words, the body has no implicit preferred directions in addition to the explicit one given by the director-field itself.

It remains, however, true that the classical notion of observer needs to be somehow altered when nonsimple fluids are considered with the meaning used in this book.

---

[80] These are materials whose free energy density is allowed to depend upon both the deformation gradient $\mathbf{F}$ and its gradient $\nabla\mathbf{F}$.

[81] It is sufficient to this end that $-\mathbf{I}$ *not* be a member of the molecular symmetry group $\mathcal{G}$ in (1.1). See [137] for a review on molecular chirality and its role in interactions.

This was already recognized by TRUESDELL [342, p. 117], where he defined in general an observer as "*a non-sentient invariant*." While geometrically all orthogonal tensors, proper and not, are allowed to define a change of frame, as soon as body-points are richer in structure than ordinary points, they become sensitive to chirality.

As will be repeatedly emphasized, our theory requires assigning only two material functionals, the free energy $\mathscr{F}$ and the RAYLEIGH functional $\mathscr{R}$, defined as integrals on the current shape $\mathcal{P}_t$ of a subbody $\mathcal{P}$ of the densities ψ and R. These are the ultimate elements of our theory. Prescribing them in a frame-indifferent way that reflects the symmetries of the molecular interactions will be our primary endeavor in the rest of the book: all our constitutive choices will concern only ψ and R.

We renounce formalizing here the principle of minimum reduced dissipation for general ordered fluids, because, as has already transpired, progressing toward higher generality dims the clarity of our formal statements. The reader can easily imagine how both (2.229) and (2.231) would be inflated if a general collection $\mathbf{Q}$ of order tensors—like the ones encountered in Chapter 1—were to contribute to the state of the body and its time derivative $\dot{\mathbf{Q}}$ to the process rate. Accordingly, in such a case the list $\mathbf{d}$ of invariant dissipation measures would also depend linearly on $\dot{\mathbf{Q}}$, so as to vanish in equilibrium.[82]

The essence of our method is clear: it converts the second law of thermodynamics into a means to determine the relaxation toward equilibrium of a continuous body. Though its consequences may be different in different contexts, a universal feature is anticipated by (2.232): the principle invoked here reduces to D'ALEMBERT's principle of virtual power in the absence of dissipation.

2.2.5 Related and Unrelated Variational Principles

Many variational principles have been posited in various attempts to describe the behavior of dissipative systems and to provide a rational setting to explain irreversibility in natural processes. Some have survived, others have been found illusory, all have been questioned. Here we cannot present a complete account on them, but we feel the need to mention some other perspectives that would contrast with the one we have privileged in this book.

As should already be clear to the reader, dissipation cannot be separated from thermodynamics, even in a purely mechanical theory. Thus, changing our perspective on dissipation will imply changing, possibly implicitly, our thermodynamic setting. This we do below by following first the ideas put forward by TRUESDELL (see, for example, [343] and [342])[83] and then contrasting them with those advanced in a series of papers by ECKART [83, 84, 85, 86]. While the former theory is often called rational thermodynamics, the latter is traditionally referred to as the thermodynamics

[82] An attempt to embrace a broader generality has been made in both [311] and [312]. Here we prefer a more inductive development for our narrative.

[83] These accounts build on earlier work of TRUESDELL [340] and TRUESDELL & TOUPIN [348, pp. 703–704].

of irreversible processes.[84] In the little room we can afford here, we can recapitulate in detail neither of these theories; we shall rather show how they can be formulated in agreement with the basic thermodynamic concepts we have so far established. In essence, rational thermodynamics is based on the CLAUSIUS–DUHEM inequality, while the thermodynamics of irreversible processes is based on the GIBBS equation, both of which we recall below.

CLAUSIUS–DUHEM Inequality

We have already encountered in (2.197) the integral version of this inequality, whose local counterpart is obtained by requiring that σ_{i} as expressed by (2.205) be positive semidefinite. In their seminal paper, COLEMAN & NOLL [49] first exploited such a local formulation of the second law of thermodynamics to enforce restrictions on the admissible constitutive relations of the theory. Here, mainly following [48], we apply their general ideas to a larger class of fluids than the one originally considered in [49]. We shall also see that in the alternative formulation of TRUESDELL [342, p. 43], the CLAUSIUS–DUHEM inequality becomes a *variational* inequality, similar in spirit to the variational approach adopted throughout this book.

Suppose that, being the material that constitutes the body a fluid, the stress tensor $\mathbf{T}$ is a function of the mass density ϱ, the absolute temperature θ, its spatial gradient $\nabla\theta$, and the stretching tensor $\mathbf{D}$, defined by (2.60), which we write in the form

$$\mathbf{T} = \mathfrak{T}(\varrho, \theta, \boldsymbol{g}, \mathbf{D}), \tag{2.233}$$

where we have set $\boldsymbol{g} := \nabla\theta$ as above to abbreviate our notation, and $\mathfrak{T}$ is a smooth function subject to the requirement of frame-indifference:

$$\mathfrak{T}(\varrho, \theta, \mathbf{R}\boldsymbol{g}, \mathbf{R}\mathbf{D}\mathbf{R}^{\mathsf{T}}) = \mathbf{R}\mathfrak{T}(\varrho, \theta, \boldsymbol{g}, \mathbf{D})\mathbf{R}^{\mathsf{T}}, \tag{2.234}$$

for all orthogonal tensors $\mathbf{R}$, where use has also been made of the first equation in (2.199) and (2.109). $\mathfrak{T}$ is assumed to deliver symmetric tensors, so that (2.233) defines a special class of simple fluids, which we call *perfect*.

It is expressive to rewrite (2.233) as follows:

$$\mathbf{T} = \mathbf{T}_0 + \mathbf{T}^{(\mathrm{e})}, \tag{2.235}$$

where $\mathbf{T}_0 := \mathfrak{T}(\varrho, \theta, \mathbf{0}, \mathbf{0})$ is the *equilibrium* stress and

$$\mathbf{T}^{(\mathrm{e})} = \mathfrak{T}^{(\mathrm{e})}(\varrho, \theta, \boldsymbol{g}, \mathbf{D}) := \mathfrak{T}(\varrho, \theta, \boldsymbol{g}, \mathbf{D}) - \mathfrak{T}(\varrho, \theta, \mathbf{0}, \mathbf{0}) \tag{2.236}$$

is the *extra* stress.

---

[84] For acronym addicts, we note that the former theory is often abbreviated as RT, while the latter is referred to as TIP (see, for example, [227]). Sometimes, to distinguish TIP from some of its more recent formulations, it is referred to as CIT, meaning "classical irreversible thermodynamics," while the recent extensions are called EIT, where E stands for "extended" (see, for example, [154]).

It follows from (2.234) that

$$\mathfrak{T}(\varrho,\theta,\mathbf{0},\mathbf{0}) = \mathbf{R}\mathfrak{T}(\varrho,\theta,\mathbf{0},\mathbf{0})\mathbf{R}^{\mathsf{T}} \quad \forall\, \mathbf{R} \in \mathsf{O}(3),$$

which in turn implies that the equilibrium stress $\mathbf{T}_0$ is a spherical tensor, which we write as

$$\mathbf{T}_0(\varrho,\theta) = -p_0(\varrho,\theta)\mathbf{I}, \tag{2.237}$$

where p_0 is the *equilibrium* pressure.

We also assume that both the free-energy density ψ and the entropy density η are given by scalar functions in the same variables entering (2.233):

$$\psi = \Psi(\varrho,\theta,\boldsymbol{g},\mathbf{D}) \quad \text{and} \quad \eta = \mathrm{H}(\varrho,\theta,\boldsymbol{g},\mathbf{D}). \tag{2.238}$$

Since both Ψ and H are required to be frame-indifferent, by the representation theorem in [357] (see, in particular, the table on p. 196), they must both be functions of the following list of scalar invariants:

$$(\varrho,\theta,\boldsymbol{g}\cdot\boldsymbol{g},\boldsymbol{g}\cdot\mathbf{D}\boldsymbol{g},\boldsymbol{g}\cdot\mathbf{D}^2\boldsymbol{g},\operatorname{tr}\mathbf{D},\operatorname{tr}\mathbf{D}^2,\operatorname{tr}\mathbf{D}^3). \tag{2.239}$$

Similarly, the heat flux $\boldsymbol{q}$ is assumed to given in terms of a smooth constitutive function $\mathfrak{q}$ as

$$\boldsymbol{q} = \mathfrak{q}(\varrho,\theta,\boldsymbol{g},\mathbf{D}). \tag{2.240}$$

It follows from the requirement that $\boldsymbol{q}$ be a frame-indifferent vector that the function $\mathfrak{q}$ must obey the property

$$\mathfrak{q}(\varrho,\theta,\mathbf{R}\boldsymbol{g},\mathbf{R}\mathbf{D}\mathbf{R}^{\mathsf{T}}) = \mathbf{R}\mathfrak{q}(\varrho,\theta,\boldsymbol{g},\mathbf{D}) \qquad \forall\, \mathbf{R} \in \mathsf{O}(3). \tag{2.241}$$

By the representation theorem in [358] for isotropic vector-valued functions, (2.241) translates into the explicit formula

$$\mathfrak{q}(\varrho,\theta,\boldsymbol{g},\mathbf{D}) = (q_0\mathbf{I} + q_1\mathbf{D} + q_2\mathbf{D}^2)\boldsymbol{g}, \tag{2.242}$$

where q_0, q_1, and q_2 are scalar isotropic functions of $(\varrho,\theta,\boldsymbol{g},\mathbf{D})$, reducible to functions of the invariants in (2.239).

In requiring $\mathfrak{T}$, Ψ, H, and $\mathfrak{q}$ to depend on the same list of variables $(\varrho,\theta,\boldsymbol{g},\mathbf{D})$, we have adhered to TRUESDELL's *principle of equipresence*:

> A quantity present as an independent variable in one constitutive equation is so present in all, to the extent that its appearance is not forbidden by the general laws of physics or rules of invariance[85] (TRUESDELL [342, p. 42]).

[85] In addition, TRUESDELL [342, p. 43] remarks that "[The principle of equipresence] may be regarded as a natural extension of OCKHAM's razor as restated by NEWTON: 'We are to admit no more causes of natural things than such as are both true and sufficient to explain their appearances, for nature is simple and affects not the pomp of superfluous causes.' "

By inserting (2.235), (2.237), and (2.238) into (2.205), we readily give $\theta\sigma_{\rm i}$ the following form:

$$\theta\sigma_{\rm i} = -p_0\,\mathrm{tr}\,\mathbf{D}+\mathbf{T}^{(\mathrm{e})}\cdot\mathbf{D}-\varrho\left[\frac{\partial\Psi}{\partial\varrho}\dot{\varrho}+\frac{\partial\Psi}{\partial\theta}\dot{\theta}+\frac{\partial\Psi}{\partial\boldsymbol{g}}\cdot\dot{\boldsymbol{g}}+\frac{\partial\Psi}{\partial\mathbf{D}}\cdot\dot{\mathbf{D}}\right]-\varrho\mathrm{H}\dot{\theta}-\frac{1}{\theta}\boldsymbol{q}\cdot\boldsymbol{g},$$

which, since $\operatorname{div}\boldsymbol{v} = \mathrm{tr}\,\mathbf{D}$, the continuity equation (2.108) transforms into

$$\theta\sigma_{\rm i} = \left(\frac{p_0}{\varrho}-\varrho\frac{\partial\Psi}{\partial\varrho}\right)\dot{\varrho}-\varrho\left(\frac{\partial\Psi}{\partial\theta}+\mathrm{H}\right)\dot{\theta}-\varrho\left[\frac{\partial\Psi}{\partial\boldsymbol{g}}\cdot\dot{\boldsymbol{g}}+\frac{\partial\Psi}{\partial\mathbf{D}}\cdot\dot{\mathbf{D}}\right] \tag{2.243}$$

$$+\mathbf{T}^{(\mathrm{e})}\cdot\mathbf{D}-\frac{1}{\theta}\boldsymbol{q}\cdot\boldsymbol{g}. \tag{2.244}$$

It is in the spirit of rational thermodynamics that we obtain restrictions for the constitutive functions $\mathfrak{T}$, Ψ, H, and $\mathfrak{q}$, in addition to those already imposed by frame-indifference, by requiring that the entropy production $\sigma_{\rm i}$ in (2.243) be non-negative for all admissible processes. These latter include all processes that can be merely conceived, though they would require assigning arbitrary body forces $\boldsymbol{b}$ and heat supplies σ in (2.127) and (2.195) to be compatible with the balance equations of the theory. Here, as in the rest of this book, instead of invoking processes whose existence could easily be questioned, we prefer interpreting the requirement that

$$\theta\sigma_{\rm i} \geqq 0 \tag{2.245}$$

as a variational inequality valid for all virtual rates. This is not unprecedented, as we learn from the following excerpt from TRUESDELL's booklet [342, p. 38]:

> Since the body force $\boldsymbol{b}$ and the heat absorption q [our σ] here are regarded as assignable arbitrarily in principle, while in any particular application they will be specified uniquely as a part of the definition of the problem, we may express COLEMAN & NOLL's basic assumption in the following alternative form, using the classical concept of "virtual" changes: *Every constitutive equation must be such as to satisfy both the principle of material frame-indifference and the Clausius–Duhem inequality for all virtual histories of deformation and temperature.*

In other words, this interpretation makes the second law of thermodynamics valid for both real and virtual processes. In our setting, we thus regard the rates $\dot{\varrho}$, $\dot{\theta}$, $\dot{\boldsymbol{g}}$, and $\dot{\mathbf{D}}$ in (2.243) as virtual, and so arbitrary, irrespective of the restrictions dictated by the balance equations, including the mass continuity equation (2.108), which prescribes the *real* $\dot{\varrho}$ in term of ϱ and $\mathbf{D}$. While the rates $\dot{\varrho}$, $\dot{\theta}$, $\dot{\boldsymbol{g}}$, and $\dot{\mathbf{D}}$ are regarded as virtual, and so freely alterable, the fields $(\varrho, \theta, \boldsymbol{g}, \mathbf{D})$ are *frozen* in their real value.

With this remark in mind, we easily see that all terms in (2.243) that are linear in the virtual rates could jeopardize the sign of $\theta\sigma_{\rm i}$ unless they vanish identically. Since this must be the case for all possible values of the frozen fields $(\varrho, \theta, \boldsymbol{g}, \mathbf{D})$, which may represent arbitrary initial states, we conclude that Ψ can depend only on ϱ and θ. Moreover, the equilibrium pressure is given by

$$p_0 = \varrho^2 \frac{\partial \Psi}{\partial \varrho} \tag{2.246}$$

and the entropy function H by

$$\mathrm{H} = -\frac{\partial \Psi}{\partial \theta}. \tag{2.247}$$

While the former equation is often called the *thermal* equation of state, the latter is also called the *caloric* equation of state.[86]

We learn from these equations that the free energy Ψ and the entropy H are not affected by the irreversibility of the processes that the fluid may undergo: they are independent of both $\boldsymbol{g}$ and $\mathbf{D}$ and deliver equations of state formally identical to those valid at equilibrium. All signs of irreversibility are confined to the residual form of (2.243) and (2.245) after use has been made in it of (2.246) and (2.247):

$$\mathbf{T}^{(\mathrm{e})} \cdot \mathbf{D} - \frac{1}{\theta} \boldsymbol{q} \cdot \boldsymbol{g} \geqq 0. \tag{2.248}$$

Correspondingly, the entropy production σ_{i} ultimately reads as

$$\sigma_{\mathrm{i}} = \frac{1}{\theta}\left(\mathbf{T}^{(\mathrm{e})} \cdot \mathbf{D} - \frac{1}{\theta} \boldsymbol{q} \cdot \boldsymbol{g}\right). \tag{2.249}$$

Since $\mathbf{T}^{(\mathrm{e})}$ and $\boldsymbol{q}$ depend through (2.236) and (2.240) on $(\varrho, \theta, \boldsymbol{g}, \mathbf{D})$, (2.248) is the dissipation inequality that must hold for all choices of these fields, since they define initial states that can be selected arbitrarily.

In general, since both $\mathbf{T}^{(\mathrm{e})}$ and $\boldsymbol{q}$ depend on both $\boldsymbol{g}$ and $\mathbf{D}$, (2.248) does not split into two independent inequalities, one thermal and the other mechanical in nature. However, it follows from (2.248) that for $\boldsymbol{g} = \mathbf{0}$,

$$\mathfrak{T}^{(\mathrm{e})}(\varrho, \theta, \mathbf{0}, \mathbf{D}) \cdot \mathbf{D} \geqq 0, \tag{2.250}$$

which, as in [48], we call the *mechanical* dissipation inequality. Likewise, for $\mathbf{D} = \mathbf{0}$, (2.248) becomes

$$\mathfrak{q}(\varrho, \theta, \boldsymbol{g}, \mathbf{0}) \cdot \boldsymbol{g} \leqq 0, \tag{2.251}$$

which was also called in [48] the *heat conduction* inequality. By (2.235) and (2.237) we can define the *mean* pressure[87] as

$$\bar{p} := -\frac{1}{3} \operatorname{tr} \mathbf{T} = p_0 - \frac{1}{3} \operatorname{tr} \mathbf{T}^{(\mathrm{e})}, \tag{2.252}$$

which shows how dissipation affects the pressure out of equilibrium.

---

86 Strictly speaking, the caloric equation of state is the equation that delivers the internal energy density υ in terms of (ϱ, θ). Clearly, once it is proved that Ψ is a function of (ϱ, θ) only, by (2.204), (2.247) is equivalent to such an equation.

87 This is the name given to $\bar{p}$ in (2.252) by SERRIN [297, p. 234] (see also [348, § 204] and [345, p. 72]). Another name often used in the literature for it is the *dynamic* pressure (see, for example, [229, p. 181]).

The general dissipation inequality (2.248) restricts the choices for the constitutive functions $\mathfrak{T}^{(e)}$ and $\mathfrak{q}$ that make them compatible with the CLAUSIUS–DUHEM inequality in the variational interpretation given here. To illustrate these restrictions in a classical case, we follow [48] in requiring that both $\mathfrak{T}^{(e)}$ and $\mathfrak{q}$ be linear in the pair $(\boldsymbol{g}, \mathbf{D})$. It easily follows from (2.236), (2.239), and (2.242) that

$$\mathfrak{T}^{(e)} = \lambda(\operatorname{tr}\mathbf{D})\mathbf{I} + 2\mu\mathbf{D}, \tag{2.253a}$$

$$\mathfrak{q} = -\kappa\boldsymbol{g}, \tag{2.253b}$$

where λ, μ, and κ are functions of (ϱ, θ), the former two representing the viscosity coefficients of the fluid and the latter its heat conductivity.[88] While (2.253a) is the constitutive law characteristic of NAVIER–STOKES fluids, (2.253b) is FOURIER's law of heat conduction for all isotropic media. By (2.253), regarding $\boldsymbol{g}$ and $\mathbf{D}$ as independent variables, we transform (2.248) into the pair of inequalities (2.250) and (2.251), whence we conclude that

$$3\lambda(\operatorname{tr}\mathbf{D})^2 + 2\mu\operatorname{tr}\mathbf{D}^2 \geqq 0 \quad \text{and} \quad \kappa\boldsymbol{g}\cdot\boldsymbol{g} \geqq 0,$$

for all symmetric tensors $\mathbf{D}$ and all vectors $\boldsymbol{g}$. While the latter inequality readily reduces to

$$\kappa \geqq 0, \tag{2.254}$$

a classical argument[89] shows that the former inequality is equivalent to

$$\mu \geqq 0, \quad 3\lambda + 2\mu \geqq 0. \tag{2.255}$$

These very inequalities will guarantee in Section 2.3.2 that the RAYLEIGH functional for a linearly viscous fluid is positive semidefinite, whereas (2.254) will coincide with the extra condition that in Section 2.3.3 guarantees that the dissipation remains positive semidefinite also in the presence of heat conduction.

In Section 2.3 we shall start afresh, considering the station of both inviscid and viscous fluids in the general theory developed in this book. It will there become clearer what marks the difference between our method and that based on the CLAUSIUS–DUHEM inequality recalled above. In essence, the RAYLEIGH functional $\mathscr{R}$ is our major constitutive ingredient: requiring it to be a quadratic, positive semidefinite form in the dissipative measures is an assumption that embodies the second law of thermodynamics. The principle of minimum reduced dissipation, which also requires that the free-energy functional $\mathscr{F}$ be constitutively assigned, will then determine both the evolution equations and the subsidiary constitutive relations compatible with that law. By necessity, the evolution equations determined from the principle of minimum reduced dissipation obey the CLAUSIUS–DUHEM inequality of rational thermodynamics. It will be of comfort to realize in Section 2.3.2 that we easily recover in our setting the conclusions reached here for a linearly viscous fluid conducting heat.

---

[88] Interesting constitutive equations for $\mathfrak{T}^{(e)}$ and $\mathfrak{q}$ different from (2.253) are derived in [48] by requiring that $\mathfrak{T}$ be linear in $\mathbf{D}$ for fixed $\boldsymbol{g}$ and that $\mathfrak{q}$ be linear in $\boldsymbol{g}$ for fixed $\mathbf{D}$.

[89] Which can be found, for example, in [347, p. 237].

Being bound to linearly viscous effects is the only real limitation of our method: we can use it to explore new territories of ordered, nonsimple fluids, but nonlinear irreversibility remains excluded from our theory, as it was from ECKART's, which we now outline.

GIBBS Equation

We showed above that, according to rational thermodynamics, Ψ complies with the CLAUSIUS–DUHEM variational inequality, provided it is a function of (ϱ, θ) only. Thus, it follows from (2.204) that, in a process, the time rate of the internal energy density υ is given by

$$\dot{\upsilon} = \dot{\psi} + \dot{\theta}\eta + \theta\dot{\eta} = \frac{\partial \Psi}{\partial \varrho}\dot{\varrho} + \left(\frac{\partial \Psi}{\partial \theta} + \mathrm{H}\right)\dot{\theta} + \theta\dot{\eta},$$

and hence, by (2.246) and (2.247), since $\theta > 0$,

$$\dot{\eta} = \frac{1}{\theta}\left(\dot{\upsilon} - \frac{p_0}{\varrho^2}\dot{\varrho}\right), \tag{2.256}$$

where p_0 is the equilibrium pressure. Equation (2.256) is called the GIBBS equation.[90]

It is remarkable that the entropy growth predicted by (2.256) is just the same as that valid at equilibrium, since it requires only the equilibrium thermal and caloric equations of state to be assigned. This amply justifies calling (2.256) the *local equilibrium* hypothesis, as has become customary in the thermodynamics of irreversible Processes,[91], where (2.256) is indeed assumed at the start and not derived as a consequence of the CLAUSIUS–DUHEM inequality, as we have done here. ECKART [83], who is rightly regarded as the founder of TIP,[92] actually derived (2.256) from FOURIER's law of conduction (2.253b), which he posited for a class of fluids and employed as a KELVIN hypothesis to show that there is a unique definition (to within sign and unit) for the absolute temperature. In contemporary TIP, also called EIP in [154], the GIBBS equation is simply assumed to hold, with no detriment to the original derivation of ECKART, which applies to a special case.

Here, mainly following the work of ECKART [83], we show how for a viscous fluid conducting heat the GIBBS equation implies the validity of the reduced CLAUSIUS–DUHEM inequality (2.248). For definiteness, we assume again that **T** is

---

[90] According to MÜLLER [227, p. 69], this equation was actually written down and exploited by CLAUSIUS, though GIBBS extended its validity to mixtures with the changes that were in order (see also [154, p. 15]).

[91] See, for example, [154, p. 14] or [229, p. 180] where it is also called the principle of local equilibrium.

[92] As MÜLLER [227, p. 246] remarks, "Eckart never received much credit for his work, because shortly after his publication Josef Meixner (1908–1994) published a very similar theory [218], and so did Ilya Prigogine (1917–[2003]) [276]. In contrast to Eckart the latter authors stayed in the field and monopolized the subject, as it were."

as in (2.233) and (2.235) and that $\boldsymbol{q}$ is represented as in (2.240). We also assume that the first law of thermodynamics is valid in the local form (2.201), where the balance of both linear and rotational momenta expressed by equations (2.127) and (2.134) are already implicit. By making use in (2.201) of (2.256), (2.235), and (2.237), we easily obtain that

$$\varrho\theta\dot{\eta} + \frac{p_0}{\varrho}\dot{\varrho} = -p_0 \operatorname{tr} \mathbf{D} + \mathbf{T}^{(\mathrm{e})} \cdot \mathbf{D} - \operatorname{div} \boldsymbol{q} + \sigma,$$

which, by the continuity equation (2.108), reduces to (2.202), provided we write there σ_{i} as in (2.249). Assuming, as did ECKART in [83], that $\mathbf{T}^{(\mathrm{e})}$ is linearly related to $\mathbf{D}$ and that $\boldsymbol{q}$ is linearly related to $\boldsymbol{g}$, one then recovers (2.253a) and (2.253b), and with them also the classical inequalities (2.254) and (2.255).

Since upon integrating over the present shape $\mathcal{P}_t$ of an arbitrary subbody $\mathcal{P}$ of $\mathcal{B}$, equation (2.202) becomes (2.198), and thus it is in turn equivalent to (2.197) whenever σ_{i} is required to be positive semidefinite as in (2.245). In particular, this shows that for simple viscous fluids the hypothesis of local thermodynamical equilibrium, embodied by the GIBBS equation (2.256), implies that the entropy flux $\boldsymbol{p}$ is related to the heat flux $\boldsymbol{q}$ through $\boldsymbol{p} = \frac{\boldsymbol{q}}{\theta}$, as is customarily assumed in rational thermodynamics.

Many disputes have opposed in past years the adherents of the theories of rational thermodynamics and the thermodynamics of irreversible processes, though, apart from a different care for rigor,[93] they were in essence less dissimilar than they might have appeared. MÜLLER's [227, p. 250] judgment is perhaps sufficiently objective to be shared:

> If the truth were known and admitted, rational thermodynamics is not at all different from TIP. Both theories employ the Clausius–Duhem inequality and the Gibbs equation. It is true that arguments are shuffled around some: The Curie principle of TIP is replaced by the principle of material frame indifference, and the Gibbs equation of rational thermodynamics is a *result*, whereas in TIP it is the basic *hypothesis*. With the Clausius–Duhem inequality it is the other way round. When applied to linear viscous, heat conducting fluids, both theories lead to the same results. This is a good thing for both, because the field equations for such fluids were perfectly well known before either theory was formulated, and they were known to be reliable.

---

[93] Exemplary is in this respect TRUESDELL's fustigation of the infamous CURIE principle, which according to PRIGOGINE & MAZUR [278], who at first did not even attribute it to CURIE, amounts to "all coupling between quantities of different tensorial character being forbidden." This requirement is indirectly referred to in [342] as "a non-existent theorem of algebra" (see also the first footnote on p. 138 of [343]). It is completely deprived of the rank of principle by representation formulas such as (2.239) and (2.242), which can actually be proved.

Unrelated Principles

As will be shown in the following section, the principle of minimum reduced dissipation that we adopt in this book is also capable of reproducing the correct field equations for linearly viscous fluids conducting heat. Its major virtue is, however, that of deriving from two potentials, the free energy $\mathscr{F}$ and the RAYLEIGH functional $\mathscr{R}$, both field equations and auxiliary constitutive relations for nonsimple, ordered fluids.

Many principles have appeared in the literature that might easily be confused with this. Here we mention a few that may appear to be the closest to it, though they are indeed not. First in this list comes the principle of *minimum entropy production* generally attributed to PRIGOGINE [277];[94] it requires the total entropy production, which in our language would be the integral of σ_i over the present shape $\mathscr{B}_t$ of the body $\mathscr{B}$, to be minimum, albeit for special processes. This principle has been subject to rather severe criticism, and it was found that it would contradict the first law of thermodynamics, even for stationary processes. In particular, it was shown in [6] and reprised in MÜLLER [229, p. 182] that for a fluid obeying FOURIER's law (2.253b), at rest between two parallel walls kept at different temperatures, the requirement that the total entropy production be minimum would dictate an equation for the equilibrium distribution of temperature θ that agrees with the classical one obtained from the local form of the balance equation of energy, only for a thermal conductivity κ proportional to the reciprocal of θ^2, which has no physical motivation.[95]

Second is the case of the *minimax* principle for the entropy production density σ_i. This was proposed by STRUCHTRUP & WEISS [322] to enforce appropriate boundary conditions for higher moments in the theory of extended thermodynamics, which to describe a fluid, employs many other fields besides the density ϱ, the temperature θ, the equilibrium pressure p_0, and the velocity $\boldsymbol{v}$. Among the additional fields are the extra stress $\mathbf{T}^{(\mathrm{e})}$, the heat flux $\boldsymbol{q}$, and other quantities with no direct physical meaning, the higher moments. In this theory, all quantities, be they directly interpretable on physical grounds or not, are not specified by constitutive relations, but are independent fields obeying a system of balance laws and an entropy inequality.[96] The minimax principle for the entropy production density prescribes that the boundary conditions for the unphysical moments be determined so as to minimize the maximum of σ_i over the body's shape. It was shown in [322] how such a strategy could be successful in making the field equations determinate in a simple stationary one-dimensional case. It was, however, questioned in [45] by showing how in a dynamical, two-dimensional case the minimax principle would indeed select an unstable solution of the equations for a linearly viscous fluid conducting heat.[97] A detailed

---

94 See also the paper [120] and the book [121].

95 Similar results were also proved in [262] and [147].

96 The reader is referred to the monograph by MÜLLER & RUGGERI [228] as well as to [154, Part I] and [227, Chapter 8] for accounts different in length and depth on this class of thermodynamic theories, which we cannot describe here.

97 In such a case, strictly speaking, the minimax principle should not apply, since higher moments are not needed. However, its inapplicability to such a classical case casts serious doubts on its validity.

account on the attempts made to overcome this and similar difficulties related to the minimax entropy production principle can be found in [154, pp. 213–215]. Whatever may be the reader's view on the success of these attempts, in one basic respect the minimax entropy production principle differs from the principle of minimum reduced dissipation, in that it is intended to find boundary conditions and not balance equations.

The last principle we quote that might be thought of as related to the principle of minimum reduced dissipation, whereas it is not, is not a purely macroscopic principle. It is ONSAGER's principle of *microscopic reversibility*. In what were essentially papers on statistical mechanics, ONSAGER [256, 257] explored the consequences of a statistical formulation of this microscopic principle[98] on the macroscopic laws of irreversible approach to equilibrium of a thermodynamic system. ONSAGER's ideas were partly reprised and partly criticized by [44], whose treatment was more rigorous than ONSAGER's. Though no objection could be raised against the reversibility of microscopic motions, its reverberations on a macroscopic scale are invariably the object of an assumption, in one fashion or another. In ONSAGER's case, such an assumption concerns the *mean regression* of fluctuations, which is assumed to obey the same macroscopic law of decay obtained by averaging out at the outset all stochastic agencies. Though this can be justified[99] for the fluctuations of Brownian particles, it remains a true assumption in all other contexts, as clearly recognized by CASIMIR [44].[100]

This, however, is not the only assumption: the thermodynamic variables $\mathsf{q} = (q_1, \dots, q_m)$ fluctuating about zero in $\mathbb{R}^m$ are uniform in space, and they obey macroscopic decay laws that are assumed to be linear. Letting these latter be represented as

$$\dot{\mathsf{q}} = M\mathsf{q}, \tag{2.257}$$

where M is a matrix in $\mathbb{R}^{m\times m}$, constant in time, and letting the entropy $\mathscr{S}$ of the system be defined (to within an additive constant) by

$$\mathscr{S} = \frac{1}{2}\mathsf{q}\cdot G\mathsf{q}, \tag{2.258}$$

where G is a symmetric, negative definite matrix in $\mathbb{R}^{m\times m}$, constant in time, ONSAGER [256, 257] and CASIMIR [44] concluded that the matrix $L := MG^{-1}$ must be symmetric. Correspondingly, the equations

$$L_{ij} = L_{ji}, \quad \text{for} \quad i, j = 1, \dots, m,$$

are universally known as ONSAGER–CASIMIR *reciprocal relations*.

---

98 Simply stating that paths described by microscopic particles are inverted under time reversal, nearly a truism.

99 With the reservations pointed out in [227, p. 281].

100 Who writes that "The acceptance of Eq. (15) [the formal definition of the regression hypothesis, too technical to be reproduced here] is really a new hypothesis, and although the same hypothesis is made in the theory of Brownian motion, we do not think that it can rigorously be proved without referring in some way or another to kinetic theory."

A plausibility argument, though not a proof, for the symmetry of L can also be found at the phenomenological level. By differentiating with respect to time both sides of equation (2.258), we easily see that the entropy time rate $\dot{\mathscr{S}}$ is given by

$$\dot{\mathscr{S}} = G\mathsf{q} \cdot \dot{\mathsf{q}} = G\mathsf{q} \cdot M\mathsf{q} = \mathsf{q} \cdot GM\mathsf{q}, \tag{2.259}$$

where use has been made of (2.257) and the symmetry of G. Requiring the matrix $A := GM$ to be symmetric and positive semidefinite, we guarantee that $\dot{\mathscr{S}} \geqq 0$ along all solutions of (2.257). Since $L = G^{-1}AG^{-1}$, then also L turns out to be symmetric and positive semidefinite.

The connection between the symmetry of L and a principle of minimal dissipation can be made by defining with COLEMAN & TRUESDELL [50] a RAYLEIGH function

$$\mathscr{R} = \frac{1}{2}\dot{\mathsf{q}} \cdot R\dot{\mathsf{q}},$$

where R is a symmetric matrix in $\mathbb{R}^{m\times m}$, and requiring that the function $\dot{\mathscr{S}} - \mathscr{R}$ be a maximum under all variations $\delta\dot{\mathsf{q}}$, while q is kept fixed. By (2.259), we easily compute the variation

$$\delta(\dot{\mathscr{S}} - \mathscr{R}) = (G\mathsf{q} - R\dot{\mathsf{q}}) \cdot \delta\dot{\mathsf{q}},$$

which vanishes identically for all $\delta\dot{\mathsf{q}}$ whenever

$$R\dot{\mathsf{q}} = G\mathsf{q}. \tag{2.260}$$

This is a macroscopic evolution equation for q, which can be given the form (2.257) only if R is invertible. In such a case, we can set $M = R^{-1}G$, whence it follows that $L = R^{-1}$, which is clearly symmetric. For R just symmetric and positive semidefinite, inserting (2.260) into (2.259), we arrive at

$$\dot{\mathscr{S}} = 2\mathscr{R},$$

whence we see that requiring $\mathscr{R}$ to be positive semidefinite amounts to ensuring that $\mathscr{S}$ grows along all solutions of (2.260).

As remarked in [50], all the above phenomenological motivations for the symmetry of the matrix L, as plausible as they may be, are not conclusive. Its proof remains rooted in statistical mechanics and rests upon the major hypothesis on the mean regression of fluctuation, which is by no means a theorem.[101]

---

[101] The situation is iconically described by MÜLLER [227, p. 282]: "Physicists have a way to quickly become very defensive on Onsager when challenged, probably because of the precariousness of the proof of the *theorem*, or because they do not understand it, or because Onsager has been canonized with the Nobel prize in 1968 ... There is some uneasiness, however." This is apparently also felt by DE GROOT & MAZUR [62, p. 102], who qualify ONSAGER's hypothesis as "not altogether unreasonable." Just as mere fact, we reproduce here the motivation for ONSAGER's Nobel prize: "For the discovery of the reciprocal relations bearing his name, which are fundamental for the thermodynamics of irreversible processes."

2.3 Isotropic Perfect Fluids

This book is concerned only with fluids. We find it instructive to start our illustration of the principle of minimum reduced dissipation from the simplest dissipative fluid, that is, the linearly viscous fluid of the NAVIER–STOKES theory. Actually, in progressing toward increasing degrees of complexity, we start from the inviscid limit, where the principle of minimum reduced dissipation is nothing but D'ALEMBERT's principle. We also do so to show the ability of the principle employed here to identify the balance equations of the theory, even in the classical case. To this end, we shall not assume as in Axiom 2.3 that the net working $\mathscr{W}^{(e)}$ is frame-indifferent: this will rather follow as a consequence of the balance equations.

Throughout this section an inertial frame is assumed to be given, to which the motion of the body $\mathcal{B}$ is referred. Moreover, all constitutive laws we shall consider are of a local nature, both in space and time:[102] in particular, no hereditary phenomena will be considered either in this section or in the subsequent chapters, where the inner degrees of freedom that make fluids *ordered* will also be considered. While, as special simple fluids,[103] the fluids considered in this section are *isotropic*,[104] the ordered fluids studied below will not deserve this name. We shall, however, still call them perfect, since only local measures of either distortion or dissipation will be allowed in ψ or R.

2.3.1 Inviscid Fluids

An inviscid fluid by its very nature does not involve any dissipation. Its motion is determined entirely by the balance of internal and external powers.

Compressible Fluids

The free energy stored in an arbitrary part $\mathcal{P}$ of a body $\mathcal{B}$ in the motion χ is given by the shape functional

$$\mathscr{F}(\mathcal{P}_t, \chi) = \int_{\mathcal{P}_t} \varrho\psi \, dV,$$

where the free energy density per unit mass ψ is a function of only the mass density ϱ: $\psi = \Psi(\varrho)$. Likewise, the kinetic energy $\mathscr{K}$ is defined in the classical form

$$\mathscr{K}(\mathcal{P}_t, \chi) = \frac{1}{2} \int_{\mathcal{P}_t} \varrho \boldsymbol{v} \cdot \boldsymbol{v} \, dV.$$

The power expended by the actions applied by the external world on the present shape $\mathcal{P}_t$ of the subbody $\mathcal{P}$ has two contributions. One stems from external body

---

102 As on page 141, we call *perfect* the fluids that comply with such a restrictive assumption.

103 Classical simple fluids have been defined in Section 2.2.4 above.

104 Since their symmetry group G, being the full unimodular group U(3), contains O(3) (see [349, p. 243].

forces with volume density $\boldsymbol{b}$ and the other from contact forces, exerted on $\mathcal{P}_t$ by the material in $\mathcal{B}_t \setminus \mathcal{P}_t$ via a surface traction $\boldsymbol{t}$, which is itself regarded as a *functional* of the reduced boundary $\partial^* \mathcal{P}_t$ of $\mathcal{P}_t$. It is thus

$$\mathscr{W}^{(a)}(\mathcal{P}_t, \chi) = \int_{\mathcal{P}_t} \boldsymbol{b} \cdot \boldsymbol{v} \, dV + \int_{\partial^* \mathcal{P}_t} \boldsymbol{t} \cdot \boldsymbol{v} \, dA. \tag{2.261}$$

Since no thermal effect is considered at first, the total working $\mathscr{W}$ in (2.210) can be written as[105]

$$\mathscr{W} = \mathscr{W}^{(a)} - (\mathscr{F} + \mathscr{K})^{\cdot}. \tag{2.262}$$

Clearly, Ξ and $\boldsymbol{G}$ in (2.230) can be set equal to zero, while $\boldsymbol{b}$ and $\boldsymbol{t}$ shall contribute to $\boldsymbol{B}$ and $\boldsymbol{T}$, respectively. To determine these latter in full, we need to compute $(\mathscr{F} + \mathscr{K})^{\cdot}$. By the transport theorem in (2.111), we obtain that

$$\begin{aligned} \dot{\mathscr{F}}(\mathcal{P}_t, \chi) + \dot{\mathscr{K}}(\mathcal{P}_t, \chi) &= \int_{\mathcal{P}_t} \varrho(\dot{\psi} + \dot{\boldsymbol{v}} \cdot \boldsymbol{v}) dV = \int_{\mathcal{P}_t} \varrho(\Psi' \dot{\varrho} + \dot{\boldsymbol{v}} \cdot \boldsymbol{v}) dV \\ &= \int_{\mathcal{P}_t} \varrho(\dot{\boldsymbol{v}} \cdot \boldsymbol{v} - \varrho \Psi' \operatorname{div} \boldsymbol{v}) dV. \end{aligned} \tag{2.263}$$

Here, we have denoted by $\dot{\psi}$ the total time derivative of ψ and by Ψ' the derivative of Ψ with respect to its argument. The second equality follows from conservation of mass (2.108).

The first term in the integrand of (2.263) is already in the form of a generalized force $\varrho \dot{\boldsymbol{v}}$, here an inertial force, times the velocity $\boldsymbol{v}$. To compute the variation of the second term, we first have to cast it in the proper form by applying the divergence theorem. With $\operatorname{div}(\varrho^2 \Psi' \boldsymbol{v}) = \varrho^2 \Psi' \operatorname{div} \boldsymbol{v} + \boldsymbol{v} \cdot \nabla(\varrho^2 \Psi')$, we find that

$$\begin{aligned} (\mathscr{F} + \mathscr{K})^{\cdot} &= \int_{\mathcal{P}_t} [\varrho \dot{\boldsymbol{v}} \cdot \boldsymbol{v} + \boldsymbol{v} \cdot \nabla(\varrho^2 \Psi') - \operatorname{div}(\varrho^2 \Psi' \boldsymbol{v})] dV \\ &= \int_{\mathcal{P}_t} [\varrho \dot{\boldsymbol{v}} + \nabla(\varrho^2 \Psi')] \cdot \boldsymbol{v} \, dV - \int_{\partial^* \mathcal{P}_t} \varrho^2 \Psi' \boldsymbol{\nu} \cdot \boldsymbol{v} \, dA, \end{aligned} \tag{2.264}$$

where $\boldsymbol{\nu}$ is the outer unit normal to $\partial^* \mathcal{P}_t$ Thus, combining (2.261) and (2.264), we identify $\boldsymbol{B}$ and $\boldsymbol{T}$ in (2.230) as

$$\boldsymbol{B} = \boldsymbol{b} - \varrho \dot{\boldsymbol{v}} - \nabla p, \tag{2.265a}$$

$$\boldsymbol{T} = \boldsymbol{t} + p \boldsymbol{\nu}, \tag{2.265b}$$

where p is defined by

$$p := \varrho^2 \Psi', \tag{2.266}$$

which is the isothermal counterpart of (2.246). By the arbitrariness of the shape $\mathcal{P}_t$ and the continuity of the integrands $\boldsymbol{B}$ and $\boldsymbol{T}$ in (2.265), for $\delta \mathscr{W}$ in (2.231) to vanish both $\boldsymbol{B}$ and $\boldsymbol{T}$ have to vanish identically. From $\boldsymbol{B} \equiv \boldsymbol{0}$, we obtain the balance equation

[105] Since both $\dot{\theta} \equiv 0$ and $\nabla \theta \equiv \boldsymbol{0}$, the thermal production of energy $\mathscr{T}$ vanishes identically.

$$\varrho\dot{\boldsymbol{v}} = -\nabla p + \boldsymbol{b}, \tag{2.267}$$

and from $\boldsymbol{T} \equiv \boldsymbol{0}$ we obtain the traction condition

$$\boldsymbol{t} = -p\boldsymbol{\nu}. \tag{2.268}$$

Because in this derivation the shape $\mathcal{P}_t$ is an arbitrary part of the present shape $\mathcal{B}_t$ of the body $\mathcal{B}$, both equations hold everywhere in the interior of $\mathcal{B}_t$, and the traction condition is valid also on $\partial^*\mathcal{B}_t$. If no kinematic constraint is required on a part of $\partial^*\mathcal{B}_t$, there the traction $\boldsymbol{t}$, still given by (2.268), must be provided by the external world. Thus, (2.268) becomes the *natural* boundary condition for the dynamic equation (2.267).

Equation (2.268) shows that the traction $\boldsymbol{t}$ on $\partial^*\mathcal{P}_t$ is a function of the outer unit normal $\boldsymbol{\nu}$ to $\partial^*\mathcal{P}_t$, and is indeed a linear function. We can write it as $\boldsymbol{t} = \mathbf{T}\boldsymbol{\nu}$, see equation (2.125), with the stress tensor $\mathbf{T}$ simply a multiple of the identity, $\mathbf{T} = -p\mathbf{I}$. Thus the scalar in (2.266) can be interpreted as the *pressure* of the fluid. By letting ψ be given by a function $\hat{\Psi}$ of the specific volume $1/\varrho$, so that $\Psi(\varrho) =: \hat{\Psi}(1/\varrho)$, we easily give p in (2.266) the more familiar form

$$p = -\hat{\Psi}'.$$

In (2.266), the pressure is a function of the density alone.

The balance equation (2.267) can thus be written in the form of CAUCHY's first law of motion as

$$\varrho\dot{\boldsymbol{v}} = \operatorname{div}\mathbf{T} + \boldsymbol{b} = -\nabla p + \boldsymbol{b}. \tag{2.269}$$

Equation (2.269) together with the continuity equation (2.108) forms the standard EULER equations for an inviscid isotropic elastic fluid.[106]

Incompressible Fluids

In the case of an incompressible fluid there is no elastic energy associated with the density, so that $\mathscr{F} \equiv 0$ and only the kinetic energy $\mathscr{K}$ survives in (2.262). However, in an incompressible fluid $\operatorname{div}\boldsymbol{v} = 0$, as required by the continuity equation (2.108), and to enforce this constraint we introduce a space-dependent LAGRANGE multiplier p and add the power $\mathscr{W}^{(c)}$ of the constraint,

$$\mathscr{W}^{(c)}(\mathcal{P}_t, \chi) = \int_{\mathcal{P}_t} p \operatorname{div}\boldsymbol{v}\, dV, \tag{2.270}$$

to the total working $\mathscr{W}$, precisely as if it were expended by appropriate agencies reacting against the internal constraint. That is, $\mathscr{W}$ in (2.262) becomes

$$\mathscr{W} = \mathscr{W}^{(a)} + \mathscr{W}^{(c)} - \dot{\mathscr{K}},$$

106 See, for example, [297] and [347, Chapter 9].

which we now need to vary according to the prescription in (2.231). To compute the variation of the reactive working $\mathscr{W}^{(c)}$, we make use of the divergence theorem to write it in the form

$$\mathscr{W}^{(c)}(\mathcal{P}_t, \chi) = \int_{\mathcal{P}_t} p \operatorname{div} \boldsymbol{v} \, dV = -\int_{\mathcal{P}_t} \nabla p \cdot \boldsymbol{v} \, dV + \int_{\partial^* \mathcal{P}_t} p \boldsymbol{\nu} \cdot \boldsymbol{v} \, dA. \quad (2.271)$$

Interpreting the terms ∇p and $p\boldsymbol{\nu}$ as reactive forces, we arrive at the same generalized forces $\boldsymbol{B}$ and $\boldsymbol{T}$ as in (2.265) and at the same balance equation (2.269) and boundary condition (2.268), which again allow us to write the stress tensor as $\mathbf{T} = -p\mathbf{I}$. While in the case of a compressible fluid, the pressure p was given as an explicit function of the density, here it is a LAGRANGE multiplier that is determined by the equation $\operatorname{div} \boldsymbol{v} = 0$, which has to be solved along with (2.269).

Synopsis

In both cases we have ultimately written the total working $\mathscr{W}$, possibly including the reactive power $\mathscr{W}^{(c)}$ expended by internal constraints, in the form (2.230),

$$\mathscr{W}(\mathcal{P}_t, \chi) = \int_{\mathcal{P}_t} \boldsymbol{B} \cdot \boldsymbol{v} \, dV + \int_{\partial^* \mathcal{P}_t} \boldsymbol{T} \cdot \boldsymbol{v} \, dA = 0, \quad (2.272)$$

with generalized volume and surface forces $\boldsymbol{B}$ and $\boldsymbol{T}$. Computing the variation of $\mathscr{W}$ in (2.272) as prescribed by (2.231), we readily arrived at the balance equation

$$\boldsymbol{B} = \mathbf{0} \quad \text{in } \mathcal{P}_t$$

and the traction condition

$$\boldsymbol{T} = \mathbf{0} \quad \text{on } \partial^* \mathcal{P}_t.$$

In both cases, the generalized surface force $\boldsymbol{T}$ took the specific form

$$\boldsymbol{T} = \boldsymbol{t} - \mathbf{T}\boldsymbol{\nu},$$

which revealed the character of the shape functional hidden in $\boldsymbol{T}$ and allowed us to identify the stress tensor $\mathbf{T}$.

While the generalized forces $\boldsymbol{B}$ and $\boldsymbol{T}$ here are simply zero, we will see in the following subsection how for a viscous fluid they are balanced by corresponding frictional volume and surface forces.

2.3.2 Viscous Fluids

In a viscous fluid undergoing an isothermal process, a part of the net working $\mathscr{W}^{(e)}$ is dissipated, and in general $\mathscr{W}^{(e)} - \dot{\mathscr{F}} \geqq 0$, as required by (2.206) with $\mathscr{T} \equiv 0$, which here reads

$$2\mathscr{R} = \mathscr{W}^{(e)} - \dot{\mathscr{F}} \geqq 0.$$

We look for the most general RAYLEIGH functional $\mathscr{R}$ that is frame indifferent and quadratic in the velocity $\boldsymbol{v}$. Since $\boldsymbol{v}$ itself is not frame indifferent, we take $\mathscr{R}$ in the form (2.225) with density R as a function of the stretching tensor $\mathbf{D}$ defined in (2.60) as

$$\mathbf{D} = \frac{1}{2}[\nabla\boldsymbol{v} + (\nabla\boldsymbol{v})^{\mathsf{T}}],$$

which is the simplest indifferent measure of dissipation in this context. Formally, the dissipation list $\mathbf{d}$ of Section 2.2.3 here consists only of $\mathbf{D}$. The RAYLEIGH functional is then given by

$$\mathscr{R}(\mathcal{P}_t, \boldsymbol{\chi}) = \int_{\mathcal{P}_t} R(\varrho, \mathbf{D}) dV,$$

where the mass density ϱ enters the dissipation density R as a parameter, not as a dissipation measure.

Before considering specific forms of $R(\varrho, \mathbf{D})$, we derive some general consequences. By (2.232), the principle of minimum reduced dissipation requires the validity of the identity

$$\delta\mathscr{R} = \delta\mathscr{W}.$$

Finding the variation of $\delta\mathscr{R}$ is rather straightforward, because it is an unconstrained variation. We write formally

$$\begin{aligned}\delta\mathscr{R} &= \int_{\mathcal{P}_t} \frac{\partial R}{\partial \nabla\boldsymbol{v}} \cdot \delta\nabla\boldsymbol{v}\, dV \\ &= -\int_{\mathcal{P}_t} \operatorname{div}\left(\frac{\partial R}{\partial \nabla\boldsymbol{v}}\right) \cdot \delta\boldsymbol{v}\, dV + \int_{\partial^*\mathcal{P}_t} \left(\frac{\partial R}{\partial \nabla\boldsymbol{v}}\boldsymbol{\nu}\right) \cdot \delta\boldsymbol{v}\, dA,\end{aligned} \tag{2.273}$$

where we have used the divergence theorem together with the identity

$$\delta\nabla\boldsymbol{v} = \nabla\delta\boldsymbol{v}.$$

The derivative of R with respect to $\nabla\boldsymbol{v}$ can be found by the chain rule; it is simply

$$\frac{\partial R}{\partial \nabla\boldsymbol{v}} = \frac{\partial R}{\partial \mathbf{D}} \circ \frac{\partial \mathbf{D}}{\partial \nabla\boldsymbol{v}} = \frac{\partial R}{\partial \mathbf{D}}.$$

Here, we have used the symbol $\circ$ to denote a specific composition: in Cartesian components,

$$\left(\frac{\partial R}{\partial \mathbf{D}} \circ \frac{\partial \mathbf{D}}{\partial \nabla\boldsymbol{v}}\right)_{ij} = \frac{\partial R}{\partial D_{kl}} \frac{\partial D_{kl}}{\partial v_{i,j}}.$$

From the definition of $\mathbf{D}$ it follows that

$$\frac{\partial D_{kl}}{\partial v_{i,j}} = \frac{1}{2}(\delta_{ki}\delta_{lj} + \delta_{kj}\delta_{li}).$$

This fourth-rank tensor, when operating on a second-rank tensor, simply projects out the symmetric part of that tensor, but the derivative of R with respect to the symmetric tensor $\mathbf{D}$ is already intrinsically symmetric.

Using (2.272) and (2.273) in (2.232) now leads to the two equations

$$B = -\operatorname{div}\left(\frac{\partial R}{\partial \mathbf{D}}\right) \quad \text{in } \mathcal{P}_t, \tag{2.274a}$$

$$T = \frac{\partial R}{\partial \mathbf{D}}\nu \quad \text{on } \partial^*\mathcal{P}_t, \tag{2.274b}$$

with B and T as in (2.265).

Compressible Fluids

Like any second-rank tensor, $\mathbf{D}$ has three basic invariants,[107] $\operatorname{tr}\mathbf{D}$, $\operatorname{tr}\mathbf{D}^2$, and $\operatorname{tr}\mathbf{D}^3$. Since the last invariant is already cubic in the velocity, the most general quadratic RAYLEIGH density R is

$$R(\varrho, \mathbf{D}) = \mu \operatorname{tr}\mathbf{D}^2 + \frac{1}{2}\lambda(\operatorname{tr}\mathbf{D})^2, \tag{2.275}$$

with the two viscosity coefficients μ and λ depending only on ϱ in the isothermal case. It is a classical algebraic result[108] that R in (2.275) is positive semidefinite for all symmetric tensors $\mathbf{D}$ whenever the viscosity coefficients obey the inequalities already encountered in (2.255):

$$\mu \geqq 0, \quad 3\lambda + 2\mu \geqq 0.$$

For R as in (2.275),

$$\frac{\partial R}{\partial \nabla \boldsymbol{v}} = \frac{\partial R}{\partial \mathbf{D}} = 2\mu\mathbf{D} + \lambda(\operatorname{tr}\mathbf{D})\mathbf{I}, \tag{2.276}$$

so that

$$\operatorname{div}\frac{\partial R}{\partial \nabla \boldsymbol{v}} = \mu(\Delta\boldsymbol{v} + \nabla \operatorname{div}\boldsymbol{v}) + \lambda\nabla\operatorname{div}\boldsymbol{v} = \mu\Delta\boldsymbol{v} + (\lambda+\mu)\nabla\operatorname{div}\boldsymbol{v}, \tag{2.277}$$

where $\Delta\boldsymbol{v}$ is the Laplacian of the velocity field, a vector field whose Cartesian components are $(\Delta\boldsymbol{v})_i = v_{i,jj}$. Here, we have assumed that μ and λ are simply constants. Allowing them to be space-dependent through the mass density ϱ would result in two additional terms, $\mathbf{D}\nabla\mu$ and $\operatorname{div}\boldsymbol{v}\nabla\lambda$, where $\nabla\mu = \mu'\nabla\varrho$ and $\lambda\mu = \lambda'\nabla\varrho$, μ' and λ' denoting the derivatives with respect to ϱ.

Using (2.275) in (2.274) and (2.277) in (2.274b) with B and T read off from (2.265) yields

$$\varrho\dot{\boldsymbol{v}} = -\nabla p + \boldsymbol{b} + \mu\Delta\boldsymbol{v} + (\lambda+\mu)\nabla\operatorname{div}\boldsymbol{v} \qquad \text{in } \mathcal{P}_t \tag{2.278}$$

[107] An invariant is a frame-indifferent scalar. The polynomial invariants of a tensor $\mathbf{D}$ are the basic constituents of any analytic scalar function of $\mathbf{D}$.

[108] See, for example, [297, p. 237].

and

$$t = [-p + 2\mu\mathbf{D} + \lambda(\operatorname{div} \boldsymbol{v})\mathbf{I}]\boldsymbol{\nu} \qquad \text{on } \partial^*\mathcal{P}_t,$$

where the pressure p is the same as in (2.266). The traction condition once again yields the stress tensor $\mathbf{T}$ such that $\mathbf{T}\boldsymbol{\nu} = \boldsymbol{t}$, and we can write the equations of motion in the standard form (2.127),

$$\varrho\dot{\boldsymbol{v}} = \operatorname{div} \mathbf{T} + \boldsymbol{b} \tag{2.279}$$

with

$$\mathbf{T} = [-p + \lambda(\operatorname{div} \boldsymbol{v})]\mathbf{I} + 2\mu\mathbf{D}. \tag{2.280}$$

Often the stress tensor in (2.280) is written in the form

$$\mathbf{T} = -p\mathbf{I} + \mathbf{T}^{(e)},$$

where p is as in (2.266) and

$$\mathbf{T}^{(e)} := \lambda(\operatorname{div} \boldsymbol{v})\mathbf{I} + 2\mu\mathbf{D} \tag{2.281}$$

is the extra stress defined in complete agreement with (2.235) and (2.253a). Clearly, $\mathbf{T}^{(e)}$ possesses an isotropic component $-\varpi\mathbf{I}$, where

$$\varpi := -\left(\lambda + \frac{2}{3}\mu\right)\operatorname{div} \boldsymbol{v}$$

is a dynamic pressure of viscous origin. If, as in (2.252), we call $\bar{p} := p + \varpi$ the mean pressure, p can then be designated as the equilibrium pressure, in analogy with the pressure p_0 defined in (2.237) and given by (2.246).

In conclusion, (2.278) is the classical NAVIER–STOKES equation for the motion of compressible viscous fluids, and (2.281) is the corresponding constitutive law for viscous stresses already encountered in (2.253a).

Incompressible Fluids

Here $\operatorname{tr} \mathbf{D} = \operatorname{div} \boldsymbol{v} = 0$, and so the dissipation density is simply

$$R(\mathbf{D}) = \mu \operatorname{tr} \mathbf{D}^2.$$

Consequently,

$$\frac{\partial R}{\partial \nabla \boldsymbol{v}} = \frac{\partial R}{\partial \mathbf{D}} = 2\mu\mathbf{D}$$

and

$$\operatorname{div} \frac{\partial R}{\partial \nabla \boldsymbol{v}} = \mu(\Delta\boldsymbol{v} + \nabla \operatorname{div} \boldsymbol{v}) = \mu\Delta\boldsymbol{v}.$$

With $\boldsymbol{B}$ and $\boldsymbol{T}$ taken from (2.265) we obtain

$$\varrho\dot{\boldsymbol{v}} = -\nabla p + \boldsymbol{b} + \mu\Delta\boldsymbol{v} \qquad \text{in } \mathcal{P}_t$$

and

$$t = (2\mu\mathbf{D} - p\mathbf{I})\boldsymbol{\nu} \quad \text{on } \partial^*\mathcal{P}_t.$$

As before, the equation of motion is the standard linear momentum balance (2.279), but with the stress tensor now given by

$$\mathbf{T} = -p\mathbf{I} + 2\mu\mathbf{D},$$

where the pressure field p needs to be determined by enforcing the incompressibility constraint $\operatorname{div} \boldsymbol{v} \equiv 0$ on the solutions of the equation of motion.

2.3.3 Heat Conduction

We now suppose that the temperature θ of a viscous fluid is neither uniform in space nor constant in time and thus contributes to the dissipation of the net working. For simplicity, we shall consider only the case of a compressible fluid, since the constraint of incompressibility would be treated precisely in the way already illustrated for isothermal processes.

The RAYLEIGH density R must be a quadratic positive semidefinite scalar function of both $\mathbf{D}$ and $\nabla\theta$. Though mixed terms involving both $\mathbf{D}$ and θ, such as $\nabla\theta \cdot \mathbf{D}\nabla\theta$, would be allowed in R by mere frame-indifference, they are ruled out by the assumption that R is a quadratic form in the collection d of all dissipation measures, which here comprises $\mathbf{D}$ and $\nabla\theta$. Thus, to the dissipation density in (2.275) we may add only a single term, quadratic in $\nabla\theta$:

$$R(\varrho, \mathbf{D}, \theta, \nabla\theta) = \mu \operatorname{tr} \mathbf{D}^2 + \frac{1}{2}\lambda(\operatorname{tr} \mathbf{D})^2 + \frac{1}{2}\bar{\kappa}|\nabla\theta|^2, \tag{2.282}$$

where now μ, λ, and $\bar{\kappa}$ are material coefficients depending on both the temperature θ and the mass density ϱ, which enter (2.282) as parameters, not as dissipation measures. Since $\mathbf{D}$ and $\nabla\theta$ are independent fields, R in (2.282) is positive semidefinite whenever both inequalities in (2.255) are satisfied and

$$\bar{\kappa} \geqq 0. \tag{2.283}$$

Since now neither $\dot{\theta}$ nor $\nabla\theta$ vanishes identically, the total working $\mathscr{W}$ must be written as in (2.210) with a nonvanishing thermal production $\mathscr{T}$ as in (2.209), where the entropy density η is a field on the collection of shapes $\mathcal{C}_\chi$ induced by the motion χ by the constitutive law

$$\eta = \mathrm{H}(\theta, \varrho). \tag{2.284}$$

Likewise, the free-energy density per unit mass ψ is now the field

$$\psi = \Psi(\theta, \varrho).$$

Thus, by (2.208), $\dot{\mathscr{F}}$ acquires the form

$$\dot{\mathscr{F}} = \int_{\mathcal{P}_t} \varrho \left(\frac{\partial \Psi}{\partial \theta}\dot{\theta} - \varrho \frac{\partial \Psi}{\partial \varrho} \operatorname{div} \boldsymbol{v} \right) dV, \tag{2.285}$$

which differs in part from (2.263) because Ψ now also depends on θ. With the aid of (2.284), use of (2.285) and (2.209) in (2.230) reduces $\mathscr{W}$ to the form (2.230) with

$$\Xi = -\varrho\left(\mathrm{H} + \frac{\partial\Psi}{\partial\theta}\right), \tag{2.286a}$$

$$\boldsymbol{G} = -\frac{1}{\theta}\boldsymbol{q}, \tag{2.286b}$$

$$\boldsymbol{B} = \boldsymbol{b} - \varrho\dot{\boldsymbol{v}} - \nabla p, \tag{2.286c}$$

$$\boldsymbol{T} = \boldsymbol{t} + p\boldsymbol{\nu}, \tag{2.286d}$$

where p, which has still the meaning of a thermodynamic equilibrium pressure, is now given by

$$p = \varrho^2\frac{\partial\Psi}{\partial\varrho}, \tag{2.287}$$

as was p_0 in (2.246). Comparing (2.286c) and (2.286d) with (2.265a) and (2.265b), respectively, we see that they are formally identical, differing only in the expression for the pressure p, which is given by (2.266) for the latter and by (2.287) for the former.

With the aid of equations (2.286), $\delta\mathscr{W}$ is readily obtained from (2.231) for all shapes $\mathscr{P}_t$. To enforce the principle of minimum reduced dissipation in the form (2.232), we need only compute $\delta\mathscr{R}$. By (2.282), the form of $\delta\mathscr{R}$ appropriate here differs from that in (2.273) with (2.276) by the addition of the integral

$$\int_{\mathscr{P}_t} \bar{\kappa}\nabla\theta\cdot\delta\boldsymbol{g}\,dV. \tag{2.288}$$

Putting together (2.286), (2.288), and (2.273) in (2.232), and invoking both the arbitrariness and the independence of the variation fields $\delta\dot{\theta}$ and $\delta\boldsymbol{v}$, we easily conclude that $\Xi = 0$, which by (2.286a) implies again (2.220), and that the balance equation (2.279) still holds with the stress tensor as in (2.280), where now p is as in (2.287). Furthermore, for (2.232) to be valid, the following condition must also hold:

$$\int_{\mathscr{P}_t}\left(\bar{\kappa}\nabla\theta + \frac{1}{\theta}\boldsymbol{q}\right)\cdot\delta\boldsymbol{g}\,dV = 0 \quad \forall\,\mathscr{P}_t,$$

where the variation $\delta\boldsymbol{g}$ is subject to $\operatorname{curl}\delta\boldsymbol{g} = \boldsymbol{0}$. To account for this constraint, we set $\delta\boldsymbol{g} = \nabla(\delta\gamma)$, for a smooth scalar field γ, and by applying the divergence theorem, we write (2.288) in the equivalent form

$$\int_{\partial^*\mathscr{P}_t}\left(\bar{\kappa}\nabla\theta + \frac{1}{\theta}\boldsymbol{q}\right)\cdot\boldsymbol{\nu}\delta\gamma\,dA - \int_{\mathscr{P}_t}\operatorname{div}\left(\bar{\kappa}\nabla\theta + \frac{1}{\theta}\boldsymbol{q}\right)\delta\gamma\,dV = 0,$$

which is valid for all shapes $\mathscr{P}_t$, provided that

$$\operatorname{div}\left(\bar{\kappa}\nabla\theta + \frac{1}{\theta}\boldsymbol{q}\right) = 0 \quad \text{in } \mathscr{P}_t \quad \text{and} \quad \left(\bar{\kappa}\nabla\theta + \frac{1}{\theta}\boldsymbol{q}\right)\cdot\boldsymbol{\nu} = 0 \quad \text{on } \partial^*\mathscr{P}_t.$$

By the arbitrariness of $\partial^* \mathcal{P}_t$ (that is, the arbitrariness of its normal $\boldsymbol{\nu}$ and any point in $\mathcal{B}_t$), we conclude that

$$\boldsymbol{q} = -\theta \bar{\kappa} \nabla \theta, \tag{2.289}$$

which is the same as FOURIER's law of heat conduction in (2.253b), provided that $\theta\bar{\kappa}$, which is not negative by (2.283), is interpreted as the heat conductivity of the fluid:

$$\kappa = \theta \bar{\kappa}.$$

The time evolution of the temperature θ is also subject to the balance of energy stated in (2.195). There, by (2.204) and (2.220), which is also valid here, we write

$$\varrho \dot{v} = -\varrho \theta \frac{\partial^2 \Psi}{\partial \theta^2} \dot{\theta} + \left(-p + \varrho^2 \theta \frac{\partial^2 \Psi}{\partial \theta \partial \varrho} \right) \operatorname{div} \boldsymbol{v}, \tag{2.290}$$

where use has also been made of (2.287) and the mass continuity equation (2.108). By (2.290) and (2.280), (2.201) becomes

$$\varrho \theta \frac{\partial^2 \Psi}{\partial \theta^2} \dot{\theta} - \varrho^2 \theta \frac{\partial^2 \Psi}{\partial \theta \partial \varrho} \operatorname{div} \boldsymbol{v} + \lambda (\operatorname{div} \boldsymbol{v})^2 + 2\mu \operatorname{tr} \mathbf{D}^2 + \operatorname{div}(\kappa \nabla \theta) + \sigma = 0, \tag{2.291}$$

where the source σ is the heat supply, which needs to be assigned either as a prescribed field on $\mathcal{B}_t$ or through a constitutive law of θ and ϱ. Equation (2.291) together with the balance equation (2.279) and the continuity equation (2.108) constitute the set of evolution equations for the fields θ, $\boldsymbol{v}$, and ϱ on the collection of shapes $\mathcal{C}_\chi$ that describe a perfect fluid with linear viscosity and heat conduction.

For these fluids, it is instructive to express the entropy production σ_i in terms of the viscous extra stress and the heat flux that we have determined through the requirement of minimum reduced dissipation (2.232). With the aid of (2.281) and (2.289), it follows from (2.227) and (2.282) that

$$\sigma_i = \frac{1}{\theta} \left(\mathbf{T}^{(e)} \cdot \mathbf{D} - \frac{1}{\theta} \boldsymbol{q} \cdot \nabla \theta \right) \geqq 0,$$

which coincides with the classical expression of the reduced CLAUSIUS–DUHEM inequality for linearly viscous fluids with heat conduction in (2.243).

2.3.4 Variational Formulations

The classical NAVIER–STOKES equation for compressible viscous fluids arrived at in (2.278) from the variational principle of minimum reduced dissipation has been the object of several studies attempting to derive it from a Hamiltonian variational principle. Here, for completeness, we review these attempts mostly following the paper by MOBBS [222] that extends earlier work of SERRIN [297].

We shall assume that the viscosity coefficients μ and λ and the heat conductivity κ are independent of both ϱ and θ, so as to be constitutive constants of the fluid. By comparing (2.290) and the GIBBS equation (2.256), which is valid here with the equilibrium pressure p_0 replaced by p in (2.287), also by the continuity equation (2.108), we easily see that (2.291) becomes[109]

$$\varrho\theta\dot{\eta} = \lambda(\operatorname{div} \boldsymbol{v})^2 + 2\mu \operatorname{tr} \mathbf{D}^2 + \kappa\Delta\theta, \tag{2.292a}$$

once the heat source σ has been set equal to zero. If similarly, we assume that in (2.278) the body force $\boldsymbol{b}$ has a potential U, so as to be written as $\boldsymbol{b} = \nabla U$, the equation of motion reduces to

$$\varrho\dot{\boldsymbol{v}} = -\nabla(p - U) + \mu\Delta\boldsymbol{v} + (\lambda + \mu)\nabla \operatorname{div} \boldsymbol{v}. \tag{2.292b}$$

Equations (2.292a) and (2.292b) must be supplemented with the continuity equation (2.108), which here we rewrite in the form

$$\dot{\varrho} = -\varrho \operatorname{div} \boldsymbol{v}. \tag{2.292c}$$

We further assume that the function H defined by (2.220) is globally invertible in θ, so that there is a function $T(\varrho, \eta)$ for which $\theta \equiv T(\varrho, H(\varrho, \theta))$ and $\theta = T(\varrho, \eta)$. By expressing θ thus in (2.292a) and transforming likewise p in (2.292b) into a function of (ϱ, θ), all three equations (2.292) can be regarded as evolution equations for the triple $(\eta, \boldsymbol{v}, \varrho)$. To derive these equations from a Hamiltonian principle, we need to find a Lagrangian function $\mathscr{L}$ depending on the fields $(\eta, \boldsymbol{v}, \varrho)$ such that (2.292) are the EULER–LAGRANGE equations in the present shape $\mathcal{P}_t$ of the subbody $\mathcal{P}$ for the action functional $\mathscr{A}$ defined by

$$\begin{aligned} \mathscr{A}(\mathcal{P})[\eta, \boldsymbol{v}, \varrho] &:= \int_{t_0}^{t_1} \mathscr{L}(\mathcal{P}_t)[\eta, \boldsymbol{v}, \varrho]dt, \\ \text{with} \quad \mathscr{L}(\mathcal{P}_t)[\eta, \boldsymbol{v}, \varrho] &:= \int_{\mathcal{P}_t} \ell(\eta, \boldsymbol{v}, \varrho)dV, \end{aligned} \tag{2.293}$$

subject to arbitrary variations that vanish at the endpoints of the time interval $[t_0, t_1]$.

In the inviscid limit, where $\mu = \lambda = 0$, and in the absence of heat conduction, so that we can formally set $\kappa = 0$, equations (2.292) describe the isoentropic evolution of an EULER fluid. HERIVEL [141][110] showed that for these equations to derive from $\mathscr{A}$ in (2.293), ℓ should be chosen as

$$\ell(\eta, \boldsymbol{v}, \varrho) = \frac{1}{2}\varrho\boldsymbol{v} \cdot \boldsymbol{v} - \varrho\upsilon + U,$$

where υ is the internal energy per unit mass, expressed in terms of (ϱ, θ) through (2.204) and by use of the function $\theta = T(\varrho, \eta)$. The HERIVEL–LIN variational

[109] As above, in the following equations Δ denotes the Laplacian.

[110] See also SERRIN [297, pp. 147–149], who quotes unpublished work of C.C. LIN.This principle, which is often named after both HERIVEL and LIN, was anticipated by ECKART [82], though he proved it valid only for irrotational flows and when the internal energy of the fluid depends only on its density.

principle prescribes $\mathscr{A}$ to be stationary for all $\mathcal{P}$ subject to the differential constraints (2.292c) and

$$\dot{\eta} = 0 \quad \text{and} \quad (\chi^{-1})^{\cdot} = 0, \tag{2.294}$$

where χ^{-1} is the inverse of the mapping χ defining a motion in (2.3). While the first constraint in (2.294) prescribes the entropy to remain constant along the trajectories described by each fluid particle, the second phrases in the Eulerian formalism the requirement that the variations of $\mathscr{A}$ maintain unaltered the identity of all fluid particles.[111] MOBBS [222] reviewed the proof of this variational principle and interpreted in physical terms the LAGRANGE multipliers that need to be introduced to free the variations of all fields $(\eta, \boldsymbol{v}, \varrho)$ in computing $\delta\mathscr{A}$. As suggested by SERRIN [297, p. 149], MOBBS [222] extended this method so as to apply it to equations (2.292), and he concluded that no analogue of the HERIVEL–LIN principle can be valid in the viscous case.

This is not the only negative result as to the existence of a variational formulation of equations (2.292). To illustrate it better we start from a positive result, which will soon be shown to be nearly the only one possible. We show how in the purely mechanical case ($\kappa \equiv 0$) a variational principle in the form

$$\delta\mathscr{L}(\mathcal{P}_t)[\boldsymbol{v}] = 0 \quad \text{with} \quad \mathscr{L}(\mathcal{P}_t)[\boldsymbol{v}] = \int_{\mathcal{P}_t} \ell(\boldsymbol{v}, \nabla\boldsymbol{v}) dV, \tag{2.295}$$

where ℓ is now a frame-indifferent function of $\boldsymbol{v}$ and $\nabla\boldsymbol{v}$, implies the equation

$$\mu\Delta\boldsymbol{v} = \nabla(p - U), \tag{2.296}$$

which follows from (2.292b) when $\operatorname{div}\boldsymbol{v} = 0$ and the inertia $\varrho\dot{\boldsymbol{v}}$ vanishes identically. We say that (2.296) describes the *quasistatic* motion of an incompressible NAVIER–STOKES fluid.

It is easy to check that for a solenoidal flow $\boldsymbol{v}$, the vorticity $\boldsymbol{\omega} = \operatorname{curl}\boldsymbol{v}$ is such that

$$\operatorname{curl}\boldsymbol{\omega} = -\Delta\boldsymbol{v},$$

and so if $\boldsymbol{v}$ solves (2.296), $\boldsymbol{\omega}$ satisfies

$$\operatorname{curl}\boldsymbol{\omega} = \nabla\Sigma,$$

where, to within an inessential additive constant,

$$\Sigma = \frac{1}{\mu}(U - p). \tag{2.297}$$

A classical theorem of HELMHOLTZ says the following.

[111] A Hamiltonian principle for the equation of the isoentropic motion of an EULER fluid was also obtained by ECKART [88] in the Lagrangian formalism, where the particles' trajectories need to be varied. The reader is referred to [140] for an energy principle of HERIVEL for the dynamics of inviscid fluids.

Theorem 2.8 (HELMHOLTZ). *A quasistatic motion of an incompressible viscous fluid governed by (2.296) is characterized by the property that the dissipation*

$$\mathscr{R}(\mathcal{P}_t, \chi) = \mu \int_{\mathcal{P}_t} \operatorname{tr} \mathbf{D}^2 dV \tag{2.298}$$

in the present shape $\mathcal{P}_t$ of any subbody $\mathcal{P}$ is less than or equal to any other flow consistent with the same values of $\boldsymbol{v}$ on $\partial^* \mathcal{P}_t$.

Proof. Let $\hat{\boldsymbol{v}}$ be the velocity field associated with the varied motion $\hat{\chi}$. It can be represented in the form

$$\hat{\boldsymbol{v}} = \boldsymbol{v} + \delta \boldsymbol{v},$$

where the variation $\delta\boldsymbol{v}$ vanishes on $\partial^* \mathcal{P}_t$ at any arbitrarily selected time t. Letting $\hat{\mathbf{D}}$ and $\mathbf{D}$ denote the stretching tensors associated with the flows $\hat{\boldsymbol{v}}$ and $\boldsymbol{v}$, respectively, we readily see that

$$\mathscr{R}(\mathcal{P}_t, \hat{\chi}) = \mathscr{R}(\mathcal{P}_t, \chi) + 2\mu \int_{\mathcal{P}_t} \mathbf{D} \cdot \nabla \delta \boldsymbol{v}\, dV + \mu \int_{\mathcal{P}_t} \operatorname{tr}(\hat{\mathbf{D}} - \mathbf{D})^2 dV,$$

whence, since $\mu \geqq 0$, it follows that

$$\delta \mathscr{R} := \mathscr{R}(\mathcal{P}_t, \hat{\chi}) - \mathscr{R}(\mathcal{P}_t, \chi) \geqq 2\mu \int_{\mathcal{P}_t} \mathbf{D} \cdot \nabla \delta \boldsymbol{v}\, dV, \tag{2.299}$$

the equality sign holding if and only if $\delta\boldsymbol{v}$, which must vanish on $\partial^* \mathcal{P}_t$, vanishes identically in $\mathcal{P}_t$. Denoting by D_{ij} and δv_i the Cartesian components of $\mathbf{D}$ and $\delta\boldsymbol{v}$ in a given frame $(\boldsymbol{e}_1, \boldsymbol{e}_2, \boldsymbol{e}_3)$, we can write

$$\int_{\mathcal{P}_t} \mathbf{D} \cdot \nabla \delta \boldsymbol{v}\, dV = \int_{\mathcal{P}_t} D_{ij} \delta v_{i,j}\, dV = \int_{\mathcal{P}_t} \left[(D_{ij} \delta v_i)_{,j} - D_{ij,j} \delta v_i \right] dV, \tag{2.300}$$

where, as usual, a comma denotes differentiation with respect to spatial variables. Since $D_{ij} = \frac{1}{2}(v_{i,j} + v_{j,i})$, it follows from $\operatorname{div} \boldsymbol{v} = v_{j,j} = 0$ and the symmetry of second derivatives that $D_{ij,j} = \frac{1}{2} v_{i,jj}$. Thus, by (2.296) and (2.297), applying the divergence theorem, from (2.300) we conclude that[112]

$$\int_{\mathcal{P}_t} \mathbf{D} \cdot \nabla \delta \boldsymbol{v}\, dV = \int_{\partial^* \mathcal{P}_t} \left(\mathbf{D}\boldsymbol{\nu} + \tfrac{1}{2} \Sigma \boldsymbol{\nu} \right) \cdot \delta \boldsymbol{v}\, dA, \tag{2.301}$$

where use has also been made of the requirement that $\operatorname{div} \delta \boldsymbol{v} = 0$, which is needed for $\hat{\boldsymbol{v}}$ to be solenoidal like $\boldsymbol{v}$. Since $\delta\boldsymbol{v}$ vanishes identically on $\partial^* \mathcal{P}_t$, it follows from (2.301) and (2.299) that $\delta \mathscr{R} \geqq 0$, which is the desired result. □

By comparing (2.295) and (2.298), we readily see that in this case the functional $\mathscr{R}$ plays the role of the Lagrangian $\mathscr{L}$, but as the proof of Theorem 2.8 clearly suggests, such a conclusion may not hold in general.

---

112 Equation (2.301) is often called the HELMHOLTZ–RAYLEIGH formula. See also [327].

To explore to what extent Theorem 2.8 may be extended, by possibly choosing a Lagrangian $\mathscr{L}$ different from $\mathscr{R}$, we remark that by (2.11) and the identity

$$(\nabla \boldsymbol{v})\boldsymbol{v} = \frac{1}{2}\nabla(\boldsymbol{v} \cdot \boldsymbol{v}) - \boldsymbol{v} \times \operatorname{curl} \boldsymbol{v},$$

the acceleration field for a steady flow (for which $\frac{\partial \boldsymbol{v}}{\partial t} = \mathbf{0}$) can be written as

$$\dot{\boldsymbol{v}} = \frac{1}{2}\nabla(\boldsymbol{v} \cdot \boldsymbol{v}) - \boldsymbol{v} \times \operatorname{curl} \boldsymbol{v}.$$

This, in particular, shows that the proof of Theorem 2.8 could be repeated verbatim, with only a different scalar field Σ, for all steady motions such that the vorticity $\boldsymbol{\omega}$ is parallel to the flow. Thus, the NAVIER–STOKES equation for steady solenoidal flows admits a variational formulation with Lagrangian $\mathscr{R}$ when either

$$(\nabla \boldsymbol{v})\boldsymbol{v} = \mathbf{0} \quad \text{or} \quad \boldsymbol{v} \times \operatorname{curl} \boldsymbol{v} = \mathbf{0}. \tag{2.302}$$

MILLIKAN [220] first proved that the steady motions of an incompressible NAVIER–STOKES fluid obeys a variational principle in the form (2.295) if and *only if* one of the conditions in (2.302) holds.[113] In words, one may summarize as in [105] the variational status of classical hydrodynamics by saying that variational principles exist when inertial forces are important and viscous forces are not and when, conversely, viscous forces are important but inertial forces are not. When both types of forces need to be taken into account, no variational formulation is indeed possible.

The limited validity of the classical Theorem 2.8 already in a purely mechanical context casts further doubts on PRIGOGINE's principle of minimum entropy production already criticized in Section 2.2.5.

[113] FINLAYSON [105] simplified considerably MILLIKAN's proof and gave it a rather elegant form. He credits J. BRILL for having first realized that Theorem 2.8 also applies to steady flows that satisfy $(2.302)_2$. This was later also remarked by RAYLEIGH [327].

3

Director Theories

We have seen in Chapter 1 that the nematic phase is most naturally described by two order parameter tensors $\mathbf{Q}$ and $\mathbf{B}$ that can be obtained as macroscopic averages of molecular tensors. However, different phenomenological theories were first developed motivated by the unique optical properties of the nematic phase. We postpone to Chapter 4 the investigation of continuum theories based on the order tensors and begin here by looking at *director* theories that are motivated by the observation that a nematic, although liquid, behaves like a crystal in that it exhibits optically distinguished local directions. These directions are the main protagonists of director theories.

The optical properties of a crystal are primarily determined by the nature of the relationship between an electric field $\boldsymbol{E}$ and the displacement $\boldsymbol{D}$ it induces. In general, this relationship can be written as [175]

$$\boldsymbol{D} = \boldsymbol{\epsilon}(\omega)\boldsymbol{E}, \tag{3.1}$$

where the dielectric tensor $\boldsymbol{\epsilon}$ normally depends on the frequency ω of the electric field. In the limit as ω goes to zero, the dielectric tensor describing the behavior of the crystal in a static electric field is obtained. In the absence of an external magnetic field, $\boldsymbol{\epsilon}$ is symmetric and hence has three real eigenvalues.

In a crystal with cubic symmetry, all three eigenvalues are equal, and so $\boldsymbol{\epsilon}$ is simply a multiple of the identity. Such a crystal behaves optically like an isotropic material.

When exactly two eigenvalues of its dielectric tensor are equal, a crystal is optically uniaxial. There exists then a single unique direction, determined by the eigenvector to the third, distinct eigenvalue, along which the crystal behaves like an isotropic material. In a *uniaxial liquid crystal*, this direction can vary in time and space and is described by a unit vector field called the *nematic director* $\boldsymbol{n}$. Because the director can be identified with an eigenvector of $\boldsymbol{\epsilon}$, it merely describes an axis and as such is a "headless" vector. This fact, that $\boldsymbol{n}$ is to be identified with $-\boldsymbol{n}$, is termed the *nematic symmetry*. Mathematically, it means that $\boldsymbol{n}$ is to be regarded not as an element of the unit sphere $\mathbb{S}^2$ but rather of the real projective plane $\mathbb{RP}^2$. Ordinary nematic liquid crystals are nonpolar materials. In a polar material, there is usually a polarization $\boldsymbol{P}$ that contributes to the displacement $\boldsymbol{D}$ and that is independent of $\boldsymbol{E}$, so that, in that case, (3.1) is to be replaced by $\boldsymbol{D} = \boldsymbol{P} + \epsilon(\omega)\boldsymbol{E}$. If light falls onto a uniaxial crystal along any direction oblique to the optical axis, birefringence (double refraction) occurs: the incoming beam splits into an *ordinary* and an *extraordinary* beam. Two indices of refraction can be defined connected with the velocities of light parallel and perpendicular to the optic axis.

When all three eigenvalues of its dielectric tensor are different, a crystal is optically *biaxial*. The optical behavior in this case is rather complex [175, 33]. There are then three different mutually perpendicular directions associated with the three distinct eigenvalues. While in a solid single crystal these directions are determined by the crystal symmetry, in *biaxial liquid crystals* they vary in space and ultimately depend on the local orientational distribution of the molecules.

In the case of constant scalar order parameter, the dynamical equations for uniaxial nematic liquid crystals have long been established: they were obtained as balance equations for linear and rotational momenta, the latter also including the microstructural contributions [91, 180, 265]. The dynamics of uniaxial nematics with variable order was treated much later by ERICKSEN [99], who posited an additional balance equation for the scalar order parameter. The first theories for biaxial nematics were formulated in terms of a triad of three mutually perpendicular directors [287, 165, 123]. It is sufficient to use just two directors, a route that we follow below.

We start by treating in Section 3.1 the classical uniaxial case with constant scalar order. We then extend our results in two different directions. In Section 3.2 we retain the focus on a uniaxial phase but allow a variable degree of scalar order; in Section 3.3, we introduce a secondary director needed for the description of a biaxial phase while reinstating the assumption of constant scalar order parameters.

3.1 The ERICKSEN–LESLIE Theory

The most common nematic liquid crystal is formed by effectively uniaxial molecules. Even if the molecules do not posses perfect cylindrical symmetry, any deviation therefrom does not manifest itself in the macroscopic properties of the phase they form. It is therefore most natural to idealize the molecules and describe their orientation by the single direction in which their main axis points. If this axis is further assumed to be nonpolar, then, as we have seen in Chapter 1, the average orientation of such molecules can be represented by the order tensor $\mathbf{Q}$, which can be written in the form (1.91) in terms of two scalar order parameters and three orthonormal vectors. We write it here as

$$\mathbf{Q} = S\left(\boldsymbol{n}\otimes\boldsymbol{n} - \frac{1}{3}\mathbf{I}\right) + T\left(\boldsymbol{m}\otimes\boldsymbol{m} - \boldsymbol{l}\otimes\boldsymbol{l}\right) \tag{3.2}$$

with the scalar order parameters S and T and a set of three orthornormal directors $\boldsymbol{n}$, $\boldsymbol{m}$, and $\boldsymbol{l}$. The uniaxial nematic phase is characterized by $T = 0$ and hence can be described by just $\boldsymbol{n}$ and the scalar order parameter S. Since the order parameter S is primarily determined by temperature (or by concentration in lyotropic nematics), it is often assumed to be constant, and so the state of the liquid crystal is determined by $\boldsymbol{n}$ alone. This is the starting point of the ERICKSEN–LESLIE theory. In accordance with the interpretation of the director $\boldsymbol{n}$ as identifying the unique optical axis of symmetry of the nematic, it is defined to be a unit vector and as such needs to satisfy the constraint

$$\boldsymbol{n} \cdot \boldsymbol{n} \equiv 1. \tag{3.3}$$

Furthermore, because it represents an axis, the further identification

$$\boldsymbol{n} \sim -\boldsymbol{n}, \tag{3.4}$$

known as the nematic symmetry, needs to be made. The nematic symmetry (3.4) will in particular entail invariance of all relevant scalar constitutive functions of the theory under reversal of the orientation of $\boldsymbol{n}$.

The mathematical theory of nematic liquid crystals was first phrased as a variational theory in the seminal works of OSEEN [259] and FRANK [109]: the nematic texture represented by the director field $\boldsymbol{n}$ was meant to minimize an elastic distortion energy with density W per unit volume depending on both $\boldsymbol{n}$ and $\nabla \boldsymbol{n}$. Both boundary conditions representing various anchoring mechanisms for $\boldsymbol{n}$ and applied electric or magnetic fields are the external agents that would antagonize the natural tendency of a nematic texture to be uniform in space, oriented in whatever direction. FRANK [109] found the most general function W at most quadratic in $\nabla \boldsymbol{n}$ that obeys (3.4). His explicit formula is reproduced below in (3.32); here, as in most of this book, we are more interested in a general infrastructure in which FRANK's theory, like any other, can be phrased.

The classical dynamical theory of nematic liquid crystals resulted from the separate efforts of J.L. ERICKSEN and F.M. LESLIE.[1] ERICKSEN [90] started by proposing the balance equations of a simpler theory for anisotropic fluids with neither couple stress nor intrinsic torque acting on the director. In [92] he also reformulated the variational theory of OSEEN and FRANK in the language of the then revived continuum mechanics and found that his earlier theory for anisotropic fluids does not reduce to OSEEN and FRANK's in the static limit. OSEEN [259] had also advanced a dynamical theory that simply rephrased ANZELIUS's incomplete attempt.[2] Building upon his earlier hydrostatic theory [92], ERICKSEN proposed

[1] Other accounts on this theory, phrased in a mathematical language different from ours, can be found, for example, in [351], [352], and [319].

[2] OSEEN cites the work of A. ANZELIUS as published in 1931 in the Annual of the University of Uppsala with the title *Über die Bewegung der anisotropen Flüssigkeiten*. As explained in [42], this was indeed ANZELIUS's dissertation, written under OSEEN's supervision. A short account on its contents is presented in [42], whence we draw the following appreciation of ANZELIUS's work: "The thesis, which consists of an eighty-four page booklet, represents the only research which Anzelius published on liquid crystals. Nevertheless, this work was the first serious and consistent attempt to derive a dynamical theory for nematics" [42, p. 1271].

in [91] a general system of balance laws that generalized OSEEN's and much inspired LESLIE's theory.[3] LESLIE [179] reexamined the earlier theory of ERICKSEN for anisotropic fluids applying the CLAUSIUS–DUHEM inequality to those constitutive equations. Later, broadening the constitutive assumptions on viscous dissipative actions, LESLIE [180, 181] obtained a general dynamical theory that in the static limit reduces to ERICKSEN's hydrostatics.

LESLIE's viscous torque and stress were postulated independently of any RAYLEIGH dissipation potential; it was PARODI [265] who first showed how requiring LESLIE's dissipative actions to derive from a dissipation potential embodies a relationship[4] among the six phenomenological viscosity coefficients of the theory.

In the following section we shall start afresh to illustrate how the classical dynamical theory of nematic liquid crystals can be derived from the general dissipation principle posited in Chapter 2. In the closing Sections 3.1.4 and 3.1.5, we shall examine the compatibility of this theory with two extreme neighboring domains: on one side, the original variational theory of OSEEN and FRANK, and on the other, the thermodynamic setting in which we have already placed the NAVIER–STOKES fluid in Section 2.3.3.

3.1.1 Nondissipative Dynamics

We first consider a uniaxial nematic in the absence of viscous dissipation. As in the case of the inviscid isotropic fluid of Section 2.3.1, the evolution equations can be derived from D'ALEMBERT's principle that requires that the variation $\delta\mathscr{W}$ of the total working $\mathscr{W}$ vanish. The relevant velocities are $\boldsymbol{v}$ and the material time derivative of the director $\dot{\boldsymbol{n}}$. The total working $\mathscr{W}$ in (2.210) takes the form

$$\mathscr{W} = \mathscr{W}^{(a)} + \mathscr{W}^{(c)} - \dot{\mathscr{K}} - \dot{\mathscr{F}}, \tag{3.5}$$

where $\mathscr{W}^{(a)}$ is the power of the external agents, $\mathscr{W}^{(c)}$ is the power of the constraints, $\mathscr{K}$ is the kinetic energy, and $\mathscr{F}$ is the free energy. We will specify each of these below. Once all the contributions of the power are written in their appropriate form as a product of generalized forces and velocities, the evolution equations can be obtained by requiring that

$$\delta\mathscr{W} = 0, \tag{3.6}$$

where the variation is defined as in (2.231) so that the generalized forces are fixed while the velocities are arbitrarily varied.

Since most processes connected with the reorientation of the director are slow compared with the frequency of sound waves, we consider here the nematic fluid as incompressible so that the mass density ϱ is constant and $\operatorname{div}\boldsymbol{v} = 0$. The compressible case will be treated within the wider scope of Chapter 5.

---

[3] See also [97] for an effective summary of these early contributions.

[4] Being such a relation phrased in the then popular language of ONSAGER's recoprocity, it is often called the ONSAGER–PARODI relation, a usage that we do not follow here.

External Agents

The external power expended on a uniaxial nematic has two contributions. One is the same power as that expended by external forces acting on the material element[5] of an isotropic fluid; see (2.261). In addition to this, power can also be expended on the director, so that the total external power takes the form

$$\mathscr{W}^{(a)}(\mathcal{P}_t,\chi) = \int_{\mathcal{P}_t} (\boldsymbol{b}\cdot\boldsymbol{v} + \boldsymbol{k}_{\mathsf{n}}\cdot\dot{\boldsymbol{n}})\,dV + \int_{\partial^*\mathcal{P}_t} (\boldsymbol{t}\cdot\boldsymbol{v} + \boldsymbol{c}_{\mathsf{n}}\cdot\dot{\boldsymbol{n}})\,dA, \tag{3.7}$$

where $\boldsymbol{k}_{\mathsf{n}}$ and $\boldsymbol{c}_{\mathsf{n}}$ are generalized body and contact force densities acting on the director $\boldsymbol{n}$. While $\boldsymbol{b}$ and $\boldsymbol{k}_{\mathsf{n}}$ are assigned sources, generally depending on $\boldsymbol{n}$, $\boldsymbol{t}$ and $\boldsymbol{c}_{\mathsf{n}}$ are to be regarded as shape functionals of $\partial^*\mathcal{P}_t$, also depending on $\boldsymbol{n}$.

Power of the Constraints

In the present setting there are two constraints. One is incompressibility of the material, which leads to the requirement that the velocity fields $\boldsymbol{v}$ remain solenoidal. The other is the requirement that the director retain unit length throughout its evolution. Incompressibility is treated as in (2.270) with the pressure p as a LAGRANGE multiplier. To ensure that $\boldsymbol{n}$ remains normalized, we introduce a further LAGRANGE multiplier γ. The power of the constraint (3.3) is obtained by differentiating it with respect to time, which leads to the requirement

$$\boldsymbol{n}\cdot\dot{\boldsymbol{n}} \equiv 0, \tag{3.8}$$

which simply implies that $\dot{\boldsymbol{n}}$ needs to be orthogonal to $\boldsymbol{n}$. Upon multiplying (3.8) by γ and adding the powers of the constraints, we find their total power $\mathscr{W}^{(c)}$ to be

$$\begin{aligned}\mathscr{W}^{(c)}(\mathcal{P}_t,\chi) &= \int_{\mathcal{P}_t} (\gamma\,\boldsymbol{n}\cdot\dot{\boldsymbol{n}} + p\,\mathrm{div}\,\boldsymbol{v})\,dV \\ &= \int_{\mathcal{P}_t} (\gamma\,\boldsymbol{n}\cdot\dot{\boldsymbol{n}} - \nabla p\cdot\boldsymbol{v})\,dV + \int_{\partial^*\mathcal{P}_t} p\boldsymbol{\nu}\cdot\boldsymbol{v}\,dA,\end{aligned} \tag{3.9}$$

where the second form follows after the same integration by parts used to arrive at (2.271).

Kinetic Energy

The kinetic energy density has the usual contribution $\frac{1}{2}\varrho v^2$, but because the material now has internal structure, there is also *microinertia.* This part of the kinetic energy is related to director rotation, but since the director stems from the average orientation of the constituent molecules, it is not normally possible to recover the complete

[5] Here we call generically a body-point that possesses an internal order structure a *material element.*

microinertia from knowledge of the director rotation alone. However, because of the small moment of inertia connected with molecular rotations, the overall microinertia is necessarily small and usually negligible. We take this point of view and neglect microinertia, so that the kinetic energy takes its usual form

$$\mathscr{K}(\mathcal{P}_t, \chi) = \int_{\mathcal{P}_t} \frac{1}{2}\varrho v^2 dV.$$

Accordingly, the rate of change of the kinetic energy is

$$\dot{\mathscr{K}}(\mathcal{P}_t, \chi) = \int_{\mathcal{P}_t} \varrho \dot{\boldsymbol{v}} \cdot \boldsymbol{v}\, dV. \tag{3.10}$$

Elastic Free Energy

In the absence of compressibility, the free energy is independent of the density, and its only contribution stems from the tendency of the director field to oppose local variations. This *curvature elasticity* is usually assumed to have a density W per unit volume that is a function of the director and its first gradient,[6] $W = W(\boldsymbol{n}, \nabla\boldsymbol{n})$. For W to be compatible with the nematic symmetry (3.4), it has to satisfy

$$W(\boldsymbol{n}, \nabla\boldsymbol{n}) = W(-\boldsymbol{n}, -\nabla\boldsymbol{n}). \tag{3.11}$$

Furthermore, because there is no distinguished direction other than $\boldsymbol{n}$ itself, the energy density needs to satisfy

$$W(\boldsymbol{n}, \nabla\boldsymbol{n}) = W(\mathbf{R}\boldsymbol{n}, \mathbf{R}\nabla\boldsymbol{n}\mathbf{R}^{\mathsf{T}}), \tag{3.12}$$

where $\mathbf{R}$ is an arbitrary proper orthogonal transformation.[7] For nonchiral nematic liquid crystals, the requirement (3.12) needs to hold for arbitrary orthogonal transformations $\mathbf{R}$. We postpone giving W a specific form and look at the general case first.

With the total free energy

$$\mathscr{F}(\mathcal{P}_t, \chi) = \int_{\mathcal{P}_t} W(\boldsymbol{n}, \nabla\boldsymbol{n}) dV, \tag{3.13}$$

we first observe that, by the transport theorem (2.37) and $\operatorname{div} \boldsymbol{v} = 0$, we simply have

$$\dot{\mathscr{F}}(\mathcal{P}_t, \chi) = \int_{\mathcal{P}_t} \dot{W}\, dV, \tag{3.14}$$

where as usual the dot denotes the material time derivative. By the chain rule,

---

[6] Here we are guilty of some abuse of notation, since we denote by W both the elastic free-energy density and the function delivering it. Where there is no risk of confusion, we prefer this venial sin to then unwarranted pedantry.

[7] See also [353, § 3.1.1] and the quotation on p. 139.

$$\dot{W} = \frac{\partial W}{\partial \boldsymbol{n}} \cdot \dot{\boldsymbol{n}} + \frac{\partial W}{\partial \nabla \boldsymbol{n}} \cdot (\nabla \boldsymbol{n})^{\cdot}. \tag{3.15}$$

To compute $(\nabla \boldsymbol{n})^{\cdot}$, we observe that the material time derivative $\dot{\boldsymbol{n}}$ is written in index notation as

$$\frac{d}{dt} n_i = \frac{\partial}{\partial t} n_i + n_{i,k} v_k$$

and its gradient $\nabla \dot{\boldsymbol{n}}$ as

$$\left(\frac{d}{dt} n_i \right)_{,j} = \frac{\partial}{\partial t} n_{i,j} + n_{i,kj} v_k + n_{i,k} v_{k,j}.$$

At the same time, the material time derivative of the director gradient $(\nabla \boldsymbol{n})^{\cdot}$ reads as

$$\frac{d}{dt} \left(n_{i,j} \right) = \frac{\partial}{\partial t} n_{i,j} + n_{i,jk} v_k.$$

Comparing these last two expressions yields, for a twice continuously differentiable director field $\boldsymbol{n}$, the identity

$$(\nabla \boldsymbol{n})^{\cdot} = \nabla \dot{\boldsymbol{n}} - (\nabla \boldsymbol{n}) \nabla \boldsymbol{v}. \tag{3.16}$$

Using this in equation (3.15), the change in free energy (3.14) becomes

$$\dot{\mathscr{F}}(\mathcal{P}_t, \chi) = \int_{\mathcal{P}_t} \left\{ \left(\frac{\partial W}{\partial \boldsymbol{n}} - \operatorname{div} \frac{\partial W}{\partial \nabla \boldsymbol{n}} \right) \cdot \dot{\boldsymbol{n}} - \left[(\nabla \boldsymbol{n})^{\mathsf{T}} \frac{\partial W}{\partial \nabla \boldsymbol{n}} \right] \cdot \nabla \boldsymbol{v} \right\} dV$$
$$+ \int_{\partial^* \mathcal{P}_t} \left(\frac{\partial W}{\partial \nabla \boldsymbol{n}} \boldsymbol{\nu} \right) \cdot \dot{\boldsymbol{n}} \, dA,$$

where we have performed an integration by parts and used that

$$\frac{\partial W}{\partial \nabla \boldsymbol{n}} \cdot [(\nabla \boldsymbol{n}) \nabla \boldsymbol{v}] = \left[(\nabla \boldsymbol{n})^{\mathsf{T}} \frac{\partial W}{\partial \nabla \boldsymbol{n}} \right] \cdot \nabla \boldsymbol{v}.$$

After a further integration by parts, $\dot{\mathscr{F}}$ takes the required form of a product of the generalized velocities $\boldsymbol{v}$ and $\dot{\boldsymbol{n}}$ and the corresponding generalized forces:

$$\dot{\mathscr{F}}(\mathcal{P}_t, \chi) = \int_{\mathcal{P}_t} \left\{ \left(\frac{\partial W}{\partial \boldsymbol{n}} - \operatorname{div} \frac{\partial W}{\partial \nabla \boldsymbol{n}} \right) \cdot \dot{\boldsymbol{n}} + \operatorname{div} \left[(\nabla \boldsymbol{n})^{\mathsf{T}} \frac{\partial W}{\partial \nabla \boldsymbol{n}} \right] \cdot \boldsymbol{v} \right\} dV$$
$$+ \int_{\partial^* \mathcal{P}_t} \left\{ \left(\frac{\partial W}{\partial \nabla \boldsymbol{n}} \boldsymbol{\nu} \right) \cdot \dot{\boldsymbol{n}} - \left[(\nabla \boldsymbol{n})^{\mathsf{T}} \frac{\partial W}{\partial \nabla \boldsymbol{n}} \boldsymbol{\nu} \right] \cdot \boldsymbol{v} \right\} dA. \tag{3.17}$$

ERICKSEN's Identity

We show now that the invariance property for the mapping W stated in (3.12) entails a tensorial consequence that will play a role in Section 3.1.3 shortly below.

By differentiating both sides of equation (3.3), we easily see that a differentiable director field satisfies

$$(\nabla \boldsymbol{n})^{\mathsf{T}} \boldsymbol{n} \equiv \mathbf{0}, \tag{3.18}$$

which is a constraint for $\nabla \boldsymbol{n}$. Thus, letting $\mathbf{N}$ represent any admissible value of $\nabla \boldsymbol{n}$, for given $\boldsymbol{n}$, we realize that by (3.18) it lives in the linear subspace of $\mathsf{L}(\mathcal{V})$ defined by

$$\mathsf{L}(\boldsymbol{n}, \mathcal{V}) := \{\mathbf{L} \in \mathsf{L}(\mathcal{V}) : \mathbf{L}^{\mathsf{T}} \boldsymbol{n} = \mathbf{0}\}.$$

Properly, for $\boldsymbol{n} \in \mathbb{S}^2$, W is a real-valued mapping defined on $\mathsf{L}(\boldsymbol{n}, \mathcal{V})$, for which the requirement (3.12) acquires the following form:

$$W(\boldsymbol{n}, \mathbf{N}) = W(\mathbf{R}\boldsymbol{n}, \mathbf{R}\mathbf{N}\mathbf{R}^{\mathsf{T}}), \tag{3.19}$$

for all $\boldsymbol{n} \in \mathcal{V}$, $\mathbf{N} \in \mathsf{L}(\boldsymbol{n}, \mathcal{V})$, and all proper orthogonal transformations $\mathbf{R} \in \mathsf{SO}(3)$.

Let now $t \mapsto \mathbf{R}(t)$ be a differentiable trajectory in $\mathsf{SO}(3)$ such that $\mathbf{R}(0) = \mathbf{I}$. It follows from (3.19) that the mapping w defined by

$$w(t) := W(\mathbf{R}(t)\boldsymbol{n}, \mathbf{R}(t)\mathbf{N}\mathbf{R}^{\mathsf{T}}(t))$$

is constant. In particular, this implies that its first derivative vanishes at $t = 0$, that is, by the chain rule, that

$$\begin{aligned} \dot{w}(0) &= \frac{\partial W}{\partial \boldsymbol{n}} \cdot \mathbf{S}\boldsymbol{n} + \frac{\partial W}{\partial \mathbf{N}} \cdot (\mathbf{S}\mathbf{N} - \mathbf{N}\mathbf{S}) \\ &= \mathbf{S} \cdot \left[\frac{\partial W}{\partial \boldsymbol{n}} \otimes \boldsymbol{n} + \frac{\partial W}{\partial \mathbf{N}} \mathbf{N}^{\mathsf{T}} + \left(\frac{\partial W}{\partial \mathbf{N}} \right)^{\mathsf{T}} \mathbf{N} \right] = 0, \end{aligned} \tag{3.20}$$

where $\mathbf{S} := \dot{\mathbf{R}}(0)$. By differentiating with respect to the parameter t the identity $\mathbf{R}(t)\mathbf{R}^{\mathsf{T}}(t) \equiv \mathbf{I}$ and setting $t = 0$, since $\mathbf{R}(0) = \mathbf{I}$, we readily see that $\mathbf{S}$ is a skew-symmetric tensor. Since the trajectory $t \mapsto \mathbf{R}(t)$ is arbitrary, (3.20) must hold for all skew-symmetric tensors, which is the case, provided that

$$\left[\frac{\partial W}{\partial \boldsymbol{n}} \otimes \boldsymbol{n} + \frac{\partial W}{\partial \mathbf{N}} \mathbf{N}^{\mathsf{T}} + \left(\frac{\partial W}{\partial \mathbf{N}} \right)^{\mathsf{T}} \mathbf{N} \right] \in \mathsf{Sym}(\mathcal{V}), \tag{3.21}$$

where $\mathsf{Sym}(\mathcal{V})$ is the subspace of all symmetric tensors in $\mathsf{L}(\mathcal{V})$ (see Appendix A.1). In components, this condition reads as

$$\epsilon_{ijk} \left(n_j \frac{\partial W}{n_k} + N_{jl} \frac{\partial W}{N_{kl}} + N_{lj} \frac{\partial W}{N_{lk}} \right) = 0. \tag{3.22}$$

Equations (3.21) and (3.22) are equivalent forms of ERICKSEN's identity, which was first derived in [91].

Variation of the Working

After the preparation of writing the individual power contributions to the working as products of the velocities and generalized forces in the forms (3.17), (3.7), (3.10), and (3.9), the application of D'ALEMBERT's principle is straightforward. In these expressions, we merely have to replace the velocities $\boldsymbol{v}$ and $\dot{\boldsymbol{n}}$ by their variations $\delta\boldsymbol{v}$ and $\delta\dot{\boldsymbol{n}}$, so that (3.6) becomes

$$\begin{aligned}\delta\mathscr{W} = \int_{\mathcal{P}_t} \Bigg\{ &\left[\boldsymbol{b} - \nabla p - \varrho\dot{\boldsymbol{v}} - \operatorname{div}\left((\nabla\boldsymbol{n})^{\mathsf{T}}\frac{\partial W}{\partial\nabla\boldsymbol{n}}\right)\right]\cdot\delta\boldsymbol{v} \\ &+ \left[\boldsymbol{k}_{\mathsf{n}} + \gamma\boldsymbol{n} - \frac{\partial W}{\partial\boldsymbol{n}} + \operatorname{div}\frac{\partial W}{\partial\nabla\boldsymbol{n}}\right]\cdot\delta\dot{\boldsymbol{n}}\Bigg\}\,dV \\ + \int_{\partial^*\mathcal{P}_t}\Bigg\{&\left[\boldsymbol{t} + \left(p\mathbf{I} + (\nabla\boldsymbol{n})^{\mathsf{T}}\frac{\partial W}{\partial\nabla\boldsymbol{n}}\right)\boldsymbol{\nu}\right]\cdot\delta\boldsymbol{v} + \left[\boldsymbol{c}_{\mathsf{n}} - \frac{\partial W}{\partial\nabla\boldsymbol{n}}\boldsymbol{\nu}\right]\cdot\delta\dot{\boldsymbol{n}}\Bigg\}\,dA.\end{aligned} \tag{3.23}$$

This can vanish identically for arbitrary parts $\mathcal{P}_t$ only if the terms in the integrands multiplying the variations of the velocities vanish identically. This localization argument implies the equations

$$\varrho\dot{\boldsymbol{v}} = \boldsymbol{b} + \operatorname{div}\mathbf{T}, \tag{3.24}$$

$$\frac{\partial W}{\partial\boldsymbol{n}} - \operatorname{div}\frac{\partial W}{\partial\nabla\boldsymbol{n}} - \boldsymbol{k}_{\mathsf{n}} = \gamma\boldsymbol{n} \tag{3.25}$$

in $\mathcal{P}_t$ and

$$\boldsymbol{t} = \mathbf{T}\boldsymbol{\nu}, \tag{3.26}$$

$$\boldsymbol{c}_{\mathsf{n}} = \frac{\partial W}{\partial\nabla\boldsymbol{n}}\boldsymbol{\nu}$$

on $\partial^*\mathcal{P}_t$. Here, we have set

$$\mathbf{T} = -p\,\mathbf{I} - (\nabla\boldsymbol{n})^{\mathsf{T}}\frac{\partial W}{\partial\nabla\boldsymbol{n}}, \tag{3.27}$$

which, as shown by the traction condition (3.26), is CAUCHY's stress tensor. ERICKSEN [92] interprets the tensor

$$\mathbf{L}_{\mathsf{n}} := \frac{\partial W}{\partial\nabla\boldsymbol{n}} \tag{3.28}$$

as the *torque stress*[8] introduced by FRANK [109]. It should not be confused with the couple stress, which shall be identified in both equations (3.73) and (3.75) below. Often[9] the field

$$\boldsymbol{h}_{\mathsf{n}} := \operatorname{div}\mathbf{L}_{\mathsf{n}} - \frac{\partial W}{\partial\boldsymbol{n}} \tag{3.29}$$

is called the *molecular field*. By use of (3.28) and (3.29), equation (3.25) is also written as

$$\boldsymbol{h}_{\mathsf{n}} + \boldsymbol{k}_{\mathsf{n}} = -\gamma\boldsymbol{n}. \tag{3.30}$$

---

[8] In more recent literature (see, for example, [182]), it is also known as the *director stress*.

[9] See, for example, [59, p. 107].

Equation (3.24) is the linear momentum balance, and (3.25) is an equation that we shall soon relate to the balance of torques acting on the director.[10] The part of the stress that depends on the elastic free energy,

$$\mathbf{T}_{\mathrm{E}} := -(\nabla \boldsymbol{n})^{\mathsf{T}} \frac{\partial W}{\partial \nabla \boldsymbol{n}}, \tag{3.31}$$

is usually called the ERICKSEN stress. Depending on the actual form of W, it can fail to be symmetric, in which case the total stress $\mathbf{T}$ will be asymmetric as well.

As remarked by ERICKSEN [92], in the static limit, where $\boldsymbol{v} \equiv \mathbf{0}$, equations (3.24) and (3.25) become an overdetermined system for the equilibrium of the director $\boldsymbol{n}$. We shall show in Section 3.1.4 how they can be both made consistent with the stationarity requirement for an appropriate energy functional.

FRANK's Formula

FRANK [109] derived the most general form of $W(\boldsymbol{n}, \nabla \boldsymbol{n})$ at most quadratic in $\nabla \boldsymbol{n}$ that obeys the symmetry requirement (3.11) and is hemitropic, as prescribed by (3.12). He found the following formula, valid for cholesteric liquid crystals:

$$\begin{aligned} W_{\mathrm{F}}(\boldsymbol{n}, \nabla \boldsymbol{n}) :=& \frac{1}{2} K_1 (\operatorname{div} \boldsymbol{n})^2 + \frac{1}{2} K_2 \left(\boldsymbol{n} \cdot \operatorname{curl} \boldsymbol{n} + \tau_{\mathrm{c}}\right)^2 \\ &+ \frac{1}{2} K_3 |\boldsymbol{n} \times \operatorname{curl} \boldsymbol{n}|^2 + K_{24}[\operatorname{tr}(\nabla \boldsymbol{n})^2 - (\operatorname{div} \boldsymbol{n})^2], \end{aligned} \tag{3.32}$$

where K_1, K_2, and K_3 are FRANK's elastic constants and τ_{c} is the characteristic *twist* of the field

$$\boldsymbol{n}_{\mathrm{c}} = \cos(\tau_{\mathrm{c}} z)\boldsymbol{e}_x + \sin(\tau_{\mathrm{c}} z)\boldsymbol{e}_y, \tag{3.33}$$

which in the frame $(\boldsymbol{e}_x, \boldsymbol{e}_y, \boldsymbol{e}_z)$ represents the generic undistorted orientation of a cholesteric liquid crystal.[11] The field $\boldsymbol{n}_{\mathrm{c}}$ describes a helical texture in which the nematic director rotates uniformly along the $\boldsymbol{e}_z$ axis; the *pitch* p_{c} of the helix, which is defined as

$$p_{\mathrm{c}} := \frac{2\pi}{\tau_{\mathrm{c}}}, \tag{3.34}$$

represents the extension in space needed for $\boldsymbol{n}_{\mathrm{c}}$ to perform a complete turn. The elastic constants K_1, K_2, and K_3 are also referred to as the *splay*, *twist*, and *bend* constants, respectively, since they weight the contributions to the elastic energy density W_{F} arising from three distinct distortion modes almost pictorially described by these names.[12]

Since $\operatorname{curl} \boldsymbol{n}$ is a hemi-indifferent vector, that is, it transforms as in (2.93), W_{F} is isotropic only if $\tau_{\mathrm{c}} = 0$, in which case W_{F} represents the elastic free-energy density of a nematic liquid crystal.

---

[10] Cf. equation (3.74) below.

[11] See [353, p. 114] for more details on the derivation of (3.32).

[12] These distortion modes are described, for example, in [353, § 3.3].

The last term in (3.32) has a peculiar character: it is a *null Lagrangian*. It was first shown by ERICKSEN [93] that its integral over any shape $\mathcal{P}_t$ on the reduced boundary $\partial^*\mathcal{P}_t$ of which $\boldsymbol{n}$ is an assigned field $\boldsymbol{n}_*$ contributes an energy to $\mathscr{F}$ that depends only on $\boldsymbol{n}_*$, and so it does not affect the field $\boldsymbol{n}$ in $\mathcal{P}_t$, as long as $\boldsymbol{n}_*$ is kept fixed.[13] For this reason, this term is often omitted from (3.32), especially in the study of bulk properties.[14]

ERICKSEN determined in [94] the conditions that make W_F positive semidefinite on all admissible director fields, so as to measure the energy required to produce a local distortion starting from a natural, undistorted texture, characteristic of the phase. For nematic liquid crystals, the natural textures with zero energy are all uniform director fields, for which $\nabla\boldsymbol{n} \equiv \mathbf{0}$, whereas for cholesteric liquid crystals the natural textures are all like $\boldsymbol{n}_c$ in (3.33) in some appropriate frame.[15] ERICKSEN [94] proved that for nematic liquid crystals, W_F is positive semidefinite whenever

$$K_1 \geqq K_{24}, \quad K_2 \geqq K_{24} \geqq 0, \quad K_3 \geqq 0, \tag{3.35}$$

which are called ERICKSEN's inequalities.[16] For cholesteric liquid crystals, the positive semidefiniteness of W_F away from all fields $\boldsymbol{n}_c$ in the form (3.33) requires in addition that[17] $K_{24} = 0$.

3.1.2 Dissipative Dynamics

In a uniaxial nematic liquid crystal there are two different velocities that can lead to dissipation: the ordinary material velocity $\boldsymbol{v}$ and the rate of change of the director $\dot{\boldsymbol{n}}$. In our setting, outlined in general in Chapter 2, we look for a RAYLEIGH dissipation function that is a quadratic form in these velocities. At the same time, the dissipation function needs to be frame-indifferent, because it describes an objective quantity. In the case of an isotropic viscous fluid, we have seen in Section 2.3.2 how this requirement can be met by constructing the dissipation function as a quadractic form in the stretching $\mathbf{D} = \frac{1}{2}[\nabla\boldsymbol{v} + (\nabla\boldsymbol{v})^{\mathsf{T}}]$, which is the simplest indifferent tensorial quantity that is linear in the velocity.

To extend this idea to uniaxial nematics, we need an indifferent time derivative of the director. We use the simplest choice, the corotational time derivative already introduced in (2.83),

$$\overset{\circ}{\boldsymbol{n}} = \dot{\boldsymbol{n}} - \mathbf{W}\boldsymbol{n}, \tag{3.36}$$

where we recall that

$$\mathbf{W} = \frac{1}{2}[\nabla\boldsymbol{v} - (\nabla\boldsymbol{v})^{\mathsf{T}}] \tag{3.37}$$

is the vorticity tensor. If a different choice of indifferent time derivative is made, the procedure is exactly the same as that outlined below; in the end, this would merely lead to a different grouping of terms in the dissipation.

[13] See also [353, p. 159].

[14] When $\boldsymbol{n}$ is not prescribed on the whole of $\partial^*\mathcal{P}_t$, such an omission is fully unjustified.

[15] It is easily seen from (3.33) that $\operatorname{div}\boldsymbol{n}_c = 0$, $\operatorname{curl}\boldsymbol{n}_c + \tau_c\boldsymbol{n}_c = \mathbf{0}$, and $(\nabla\boldsymbol{n}_c)^2 = \mathbf{0}$.

[16] See also [353, p. 124].

[17] See also [353, § 3.4.2] and [153] for a criticism of this conclusion.

Generic Dissipation Function

The dissipation function R is constructed as a function of $\mathbf{D}$, $\mathring{\boldsymbol{n}}$, and of $\boldsymbol{n}$ itself in such a way that it is a quadratic form in $(\mathbf{D},\mathring{\boldsymbol{n}})$. In view of EULER's theorem on homogeneous functions, this implies that

$$\frac{\partial R}{\partial \mathbf{D}} \cdot \mathbf{D} + \frac{\partial R}{\partial \mathring{\boldsymbol{n}}} \cdot \mathring{\boldsymbol{n}} = 2R.$$

It is easily checked that the pair $(\nabla \boldsymbol{v}, \dot{\boldsymbol{n}})$ depends linearly on the pair $(\mathbf{D}, \mathring{\boldsymbol{n}})$, and so it follows that

$$\frac{\partial R}{\partial \nabla \boldsymbol{v}} \cdot \nabla \boldsymbol{v} + \frac{\partial R}{\partial \dot{\boldsymbol{n}}} \cdot \dot{\boldsymbol{n}} = 2R,$$

which shows that any R constructed as a quadratic form in $(\mathbf{D}, \mathring{\boldsymbol{n}})$ is indeed also a quadratic form in $(\nabla \boldsymbol{v}, \dot{\boldsymbol{n}})$, as required by the dissipation principle.

Before giving a specific form for R we formally perform the variation of the dissipation functional $\mathscr{R}$,

$$\mathscr{R}(\mathcal{P}_t, \chi) = \int_{\mathcal{P}_t} R(\boldsymbol{n}; \mathbf{D}, \mathring{\boldsymbol{n}}) dV. \tag{3.38}$$

Although R is given explicitly as a function of the indifferent rates $\mathring{\boldsymbol{n}}$ and $\mathbf{D}$, the variation needs to be performed with respect to $\boldsymbol{v}$ and $\dot{\boldsymbol{n}}$. Thus

$$\begin{aligned} \delta \mathscr{R} &= \int_{\mathcal{P}_t} \left[\frac{\partial R}{\partial \dot{\boldsymbol{n}}} \cdot \delta \dot{\boldsymbol{n}} + \frac{\partial R}{\partial \nabla \boldsymbol{v}} \cdot \delta \nabla \boldsymbol{v} \right] dV \\ &= \int_{\mathcal{P}_t} \left[\frac{\partial R}{\partial \dot{\boldsymbol{n}}} \cdot \delta \dot{\boldsymbol{n}} - \operatorname{div} \left(\frac{\partial R}{\partial \nabla \boldsymbol{v}} \right) \cdot \delta \boldsymbol{v} \right] dV + \int_{\partial^* \mathcal{P}_t} \left(\frac{\partial R}{\partial \nabla \boldsymbol{v}} \boldsymbol{\nu} \right) \cdot \delta \boldsymbol{v} \, dA. \end{aligned} \tag{3.39}$$

The partial derivatives of R in this expression can be found using the chain rule, which shows that

$$\frac{\partial R}{\partial \dot{\boldsymbol{n}}} = \frac{\partial R}{\partial \mathring{\boldsymbol{n}}} \tag{3.40}$$

and

$$\frac{\partial R}{\partial \nabla \boldsymbol{v}} = \frac{1}{2} \left(\boldsymbol{n} \otimes \frac{\partial R}{\partial \mathring{\boldsymbol{n}}} - \frac{\partial R}{\partial \mathring{\boldsymbol{n}}} \otimes \boldsymbol{n} \right) + \frac{\partial R}{\partial \mathbf{D}}. \tag{3.41}$$

Dynamic Equations

We are now in a position to give the general form of the dynamic equations for the evolution of a uniaxial liquid crystal. The dissipation principle amounts to the requirement (2.232):

$$\delta \mathscr{R} = \delta \mathscr{W}.$$

With the variation $\delta \mathscr{W}$ as in (3.23) and the variation $\delta \mathscr{R}$ as in (3.39) together with (3.40) and (3.41) we obtain, as before, the momentum balance in the bulk in the form

$$\varrho \dot{\boldsymbol{v}} = \boldsymbol{b} + \operatorname{div} \mathbf{T} \tag{3.42}$$

and the traction condition on the boundary as

$$\boldsymbol{t} = \mathbf{T}\boldsymbol{\nu}, \tag{3.43}$$

where now the stress is

$$\mathbf{T} = -p\,\mathbf{I} - (\nabla \boldsymbol{n})^{\mathsf{T}} \frac{\partial W}{\partial \nabla \boldsymbol{n}} + \frac{1}{2}\left(\boldsymbol{n} \otimes \frac{\partial R}{\partial \mathring{\boldsymbol{n}}} - \frac{\partial R}{\partial \mathring{\boldsymbol{n}}} \otimes \boldsymbol{n}\right) + \frac{\partial R}{\partial \mathbf{D}}. \tag{3.44}$$

Apart from a contribution proportional to the identity and the ERICKSEN stress $\mathbf{T}_{\mathrm{E}}$ as in (3.31), it contains a *viscous* or *dissipative* stress

$$\mathbf{T}_{\mathrm{dis}} := \frac{1}{2}\left(\boldsymbol{n} \otimes \frac{\partial R}{\partial \mathring{\boldsymbol{n}}} - \frac{\partial R}{\partial \mathring{\boldsymbol{n}}} \otimes \boldsymbol{n}\right) + \frac{\partial R}{\partial \mathbf{D}}. \tag{3.45}$$

The equation for the director in the bulk becomes

$$\frac{\partial R}{\partial \mathring{\boldsymbol{n}}} + \frac{\partial W}{\partial \boldsymbol{n}} - \operatorname{div} \frac{\partial W}{\partial \nabla \boldsymbol{n}} - \boldsymbol{k}_{\mathsf{n}} = \gamma \boldsymbol{n}, \tag{3.46}$$

and the condition on the boundary, which by (3.28) we now write as

$$\boldsymbol{c}_{\mathsf{n}} = \mathbf{L}_{\mathsf{n}} \boldsymbol{\nu}, \tag{3.47}$$

remains unchanged. This reflects the fact that there is no contribution proportional to $\dot{\boldsymbol{n}}$ in the boundary integral in (3.39), which in turn is due to the fact that we did not consider gradients of $\dot{\boldsymbol{n}}$ in the dissipation function.

General Dissipation Function

The dissipation function R must also obey the nematic symmetry in (3.4), and so it will be assumed that

$$R(-\boldsymbol{n}; \mathbf{D}, -\mathring{\boldsymbol{n}}) = R(\boldsymbol{n}; \mathbf{D}, \mathring{\boldsymbol{n}}). \tag{3.48}$$

The most general quadratic form in $\mathring{\boldsymbol{n}}$ and $\mathbf{D}$ that can be constructed in terms of these two rates and the director $\boldsymbol{n}$ and that obeys (3.48) has five different terms; we write it as[18]

$$R(\boldsymbol{n}; \mathbf{D}, \mathring{\boldsymbol{n}}) = \frac{1}{2}\gamma_1 \mathring{\boldsymbol{n}}^2 + \gamma_2 \mathring{\boldsymbol{n}} \cdot \mathbf{D}\boldsymbol{n} + \frac{1}{2}\gamma_3 (\mathbf{D}\boldsymbol{n})^2 + \frac{1}{2}\gamma_4 (\boldsymbol{n} \cdot \mathbf{D}\boldsymbol{n})^2 + \frac{1}{2}\gamma_5 \operatorname{tr} \mathbf{D}^2, \tag{3.49}$$

where the γ's are viscosity coefficients.[19] It is easily seen from (3.49) that R is an isotropic function, since the only hemitropic term quadratic in $(\mathbf{D}, \mathring{\boldsymbol{n}})$, namely

[18] See [98] and [184].

[19] These are constitutive constants as long as thermal effects are ignored. We refer the reader to [150] and [77] for studies on the thermal dependence of the viscosity coefficients.

$$R_c := \boldsymbol{n} \cdot \mathbf{D}\boldsymbol{n} \times \mathring{\boldsymbol{n}}, \tag{3.50}$$

is ruled out by (3.48). Thus, we conclude that both nematic and cholesteric liquid crystals are represented by a dissipation function R of one and the same form.

For the function R to be positive semidefinite, the viscosity coefficients must satisfy appropriate inequalities. We find these by introducing a general representation of $\boldsymbol{n}$, $\mathring{\boldsymbol{n}}$, and $\mathbf{D}$. Since $\boldsymbol{n}$ and $\mathring{\boldsymbol{n}}$ must be orthogonal to one another, we let $(\boldsymbol{e}_1, \boldsymbol{e}_2, \boldsymbol{e}_3)$ denote an orthonormal frame such that

$$\boldsymbol{n} = \boldsymbol{e}_1, \quad \mathring{\boldsymbol{n}} = N\boldsymbol{e}_2, \quad \text{and} \quad \mathbf{D} = A_{ij}\boldsymbol{e}_i \otimes \boldsymbol{e}_j, \tag{3.51}$$

where N is a scalar and the coefficients A_{ij} satisfy

$$A_{ij} = A_{ji} \quad \text{and} \quad A_{ii} = 0, \tag{3.52}$$

since $\mathbf{D}$ is a symmetric, traceless tensor. Using both (3.51) and (3.52) in (3.49), we write R as the sum of four independent quadratic forms:

$$\begin{aligned} R = {} & \left(\frac{1}{2}\gamma_3 + \gamma_5\right) A_{13}^2 + \gamma_5 A_{23}^2 \\ & + \frac{1}{2}\gamma_1 N^2 + \gamma_2 N A_{12} + \left(\frac{1}{2}\gamma_3 + \gamma_5\right) A_{12}^2 \\ & + \frac{1}{2}(\gamma_3 + \gamma_4 + 2\gamma_5) A_{11}^2 + \gamma_5 A_{11} A_{22} + \gamma_5 A_{22}^2. \end{aligned} \tag{3.53}$$

This function is positive semidefinite, provided that

$$\gamma_3 + 2\gamma_5 \geqq 0, \qquad \gamma_5 \geqq 0, \tag{3.54}$$

and the following symmetric matrices are positive semidefinite:

$$H_1 := \begin{bmatrix} \gamma_1 & \gamma_2 \\ \gamma_2 & \gamma_3 + 2\gamma_5 \end{bmatrix}, \quad H_2 := \begin{bmatrix} \gamma_3 + \gamma_4 + 2\gamma_5 & \gamma_5 \\ \gamma_5 & 2\gamma_5 \end{bmatrix}. \tag{3.55}$$

For a 2×2 symmetric matrix H to be positive semidefinite, both elements of its principal diagonal and its determinant must not be negative. For H_1 and H_2 in (3.55), such a criterion reduces to

$$\begin{aligned} \gamma_1 \geqq 0, \qquad & \gamma_1(\gamma_3 + 2\gamma_5) - \gamma_2^2 \geqq 0, \\ \gamma_3 + \gamma_4 + 2\gamma_5 \geqq 0, \qquad & \gamma_5(2\gamma_3 + 2\gamma_4 + 3\gamma_5) \geqq 0. \end{aligned} \tag{3.56}$$

Since $(3.54)_2$ and $(3.56)_4$ imply $(3.56)_3$, the independent inequalities that guarantee that R in (3.49) be positive semidefinite can be collected in the following list:

$$\gamma_1 \geqq 0, \tag{3.57a}$$

$$\gamma_3 + 2\gamma_5 \geqq 0, \tag{3.57b}$$

$$\gamma_5 \geqq 0, \tag{3.57c}$$

$$2(\gamma_3 + \gamma_4) + 3\gamma_5 \geqq 0, \tag{3.57d}$$

$$\gamma_1(\gamma_3 + 2\gamma_5) - \gamma_2^2 \geqq 0. \tag{3.57e}$$

It readily follows from (3.49) that[20]

$$\frac{\partial R}{\partial \mathbf{D}} = \gamma_2 \overline{\mathring{\boldsymbol{n}} \otimes \boldsymbol{n}} + \gamma_3 \overline{\boldsymbol{n} \otimes \mathbf{D}\boldsymbol{n}} + \gamma_4 (\boldsymbol{n} \cdot \mathbf{D}\boldsymbol{n}) \overline{\boldsymbol{n} \otimes \boldsymbol{n}} + \gamma_5 \mathbf{D}$$

and

$$\frac{\partial R}{\partial \mathring{\boldsymbol{n}}} = \gamma_1 \mathring{\boldsymbol{n}} + \gamma_2 [\mathbf{D}\boldsymbol{n} - (\mathbf{D}\boldsymbol{n} \cdot \boldsymbol{n})\boldsymbol{n}]. \tag{3.58}$$

Here, $\overline{\cdots}$ denotes the symmetric traceless part of a tensor (see also Appendix A.1),

$$\overline{\mathbf{A}} := \frac{1}{2}(\mathbf{A} + \mathbf{A}^{\mathsf{T}}) - \frac{1}{3}(\operatorname{tr}\mathbf{A})\mathbf{I}, \quad \forall \mathbf{A} \in \mathsf{L}(\mathcal{V}).$$

The viscous stress is thus

$$\begin{aligned} \mathbf{T}_{\text{dis}} = & \gamma_2 \overline{\boldsymbol{n} \otimes \mathring{\boldsymbol{n}}} + \gamma_3 \overline{\boldsymbol{n} \otimes \mathbf{D}\boldsymbol{n}} + \gamma_4 (\boldsymbol{n} \cdot \mathbf{D}\boldsymbol{n}) \overline{\boldsymbol{n} \otimes \boldsymbol{n}} + \gamma_5 \mathbf{D} \\ & + \gamma_1 \operatorname{skw}(\boldsymbol{n} \otimes \mathring{\boldsymbol{n}}) + \gamma_2 \operatorname{skw}(\boldsymbol{n} \otimes \mathbf{D}\boldsymbol{n}). \end{aligned} \tag{3.59}$$

With (3.58), the director evolution equation (3.46) takes the form

$$\gamma_1 \mathring{\boldsymbol{n}} + \gamma_2 \mathbf{D}\boldsymbol{n} + \frac{\partial W}{\partial \boldsymbol{n}} - \operatorname{div} \frac{\partial W}{\partial \nabla \boldsymbol{n}} - \boldsymbol{k}_{\mathsf{n}} = \gamma \boldsymbol{n}.$$

LESLIE Viscosity Coefficients

The form of the viscous stress commonly used is

$$\mathbf{T}_{\text{dis}} = \alpha_1 (\boldsymbol{n} \cdot \mathbf{D}\boldsymbol{n}) \boldsymbol{n} \otimes \boldsymbol{n} + \alpha_2 \mathring{\boldsymbol{n}} \otimes \boldsymbol{n} + \alpha_3 \boldsymbol{n} \otimes \mathring{\boldsymbol{n}} + \alpha_4 \mathbf{D} + \alpha_5 \mathbf{D}\boldsymbol{n} \otimes \boldsymbol{n} + \alpha_6 \boldsymbol{n} \otimes \mathbf{D}\boldsymbol{n},$$

where the α's are LESLIE's coefficients (cf. [185, eq. (4.6)]). This expression is the same as (3.59), provided that

$$\begin{aligned} \alpha_1 &= \gamma_4, & \alpha_2 &= \frac{1}{2}(\gamma_2 - \gamma_1), & \alpha_3 &= \frac{1}{2}(\gamma_2 + \gamma_1), \\ \alpha_4 &= \gamma_5, & \alpha_5 &= \frac{1}{2}(\gamma_3 - \gamma_2), & \alpha_6 &= \frac{1}{2}(\gamma_3 + \gamma_2), \end{aligned} \tag{3.60}$$

whence it follows that

$$\alpha_6 - \alpha_5 = \alpha_2 + \alpha_3, \tag{3.61}$$

which is a relation first derived by PARODI [265]. It is automatically satisfied in our setting because all generalized viscous forces derive here from a potential R.

For completeness, we also record here the formulas that express the γ's in terms of the α's, which are easily obtained from (3.60) and (3.61):

[20] The derivatives $\frac{\partial R}{\partial \mathbf{D}}$ and $\frac{\partial R}{\partial \mathring{\boldsymbol{n}}}$ are to be interpreted in the intrinsic sense (see, for example, [353, p. 133]): the former is a symmetric, traceless tensor, while the latter is a vector everywhere orthogonal to $\boldsymbol{n}$.

$$\gamma_1 = \alpha_3 - \alpha_2, \tag{3.62a}$$
$$\gamma_2 = \alpha_3 + \alpha_2, \tag{3.62b}$$
$$\gamma_3 = \alpha_2 + \alpha_3 + 2\alpha_5, \tag{3.62c}$$
$$\gamma_4 = \alpha_1, \tag{3.62d}$$
$$\gamma_5 = \alpha_4. \tag{3.62e}$$

Use of these formulas in inequalities (3.57) transforms them into the inequalities for the LESLIE viscosities that are reproduced below to ease the comparison with those derived in [319, p. 146]:

$$\alpha_3 \geqq \alpha_2, \tag{3.63a}$$
$$\alpha_4 \geqq 0, \tag{3.63b}$$
$$\alpha_2 + \alpha_3 + 2\alpha_4 + 2\alpha_5 \geqq 0, \tag{3.63c}$$
$$2(\alpha_1 + \alpha_2 + \alpha_3) + 3\alpha_4 + 4\alpha_5 \geqq 0, \tag{3.63d}$$
$$(\alpha_3 - \alpha_2)(\alpha_2 + \alpha_3 + 2\alpha_4 + 2\alpha_5) \geqq (\alpha_2 + \alpha_3)^2. \tag{3.63e}$$

3.1.3 Rotational Momentum and Couple Stress

We have derived the dynamic equations of nematic director theory using generalized forces acting on the generalized velocity $\dot{\boldsymbol{n}}$. We eventually obtained two coupled balance equations, (3.42) and (3.46), one for the linear momentum and the other for the orientational order. The format we used is quite general and can be applied in a similar manner to a wide variety of ordered media [311, 312]. In the present case, if one takes the naïve view that the director represents a rigid body capable both of being conveyed by the flow and of rotating relative to it, the director balance is equivalent to the balance of rotational momentum. Correspondingly, in such a kinematic interpretation, we may attribute two rotational velocities to $\boldsymbol{n}$ in one and the same frame: one is the spin vector $\boldsymbol{w}$ of the flow, that is, the axial vector associated with the vorticity tensor $\mathbf{W}$ in (3.37), and the other will be denoted by $\boldsymbol{w}_{\mathrm{n}}$. We write

$$\dot{\boldsymbol{n}} = \boldsymbol{w} \times \boldsymbol{n}$$

in the first case and

$$\dot{\boldsymbol{n}} = \boldsymbol{w}_{\mathrm{n}} \times \boldsymbol{n} \tag{3.64}$$

in the second case. In the former, by (3.36),

$$\mathring{\boldsymbol{n}} = \boldsymbol{0},$$

while in the latter,

$$\mathring{\boldsymbol{n}} = (\boldsymbol{w}_{\mathrm{n}} - \boldsymbol{w}) \times \boldsymbol{n}. \tag{3.65}$$

As LESLIE [185] has shown, it is indeed possible to formulate the theory based on the classical balances without resorting to generalized velocities and forces.

In general, the orientational balance cannot be derived from the balance of rotational momentum; this is obvious if the description of the orientation employs more than three degrees of freedom, as is the case, for example, in the order tensor theories treated in Chapter 4. However, in any event the balance of rotational momentum (2.141) must be satisfied, whether it is equivalent to the orientational balance or not. As shown in [311] for arbitrary tensorial order structure, this is indeed the case provided the kinetic and free energies satisfy appropriate invariance requirements.

We show now how the director balance can be interpreted as a balance of torques on the material element, and we identify the couple stress. Because we have neglected the kinetic energy associated with director rotation, the balance of rotational momentum (2.141) here takes the simpler form

$$2\boldsymbol{\tau} + \operatorname{div}\mathbf{L} + \boldsymbol{k} = 0, \tag{3.66}$$

where $\mathbf{L}$ is the couple stress, $\boldsymbol{k}$ is a body couple per unit volume, and $\boldsymbol{\tau}$ is the axial vector associated with the skew-symmetric part of the stress $\mathbf{T}$ via (2.132). In index notation,

$$2\tau_i = \epsilon_{ijk} T_{kj}. \tag{3.67}$$

Thus $\boldsymbol{\tau}$ is determined by our knowledge of the stress tensor (3.44), and we will now use this to identify $\mathbf{L}$ and $\boldsymbol{k}$ via the rotational momentum balance (3.66).

We start by giving the director balance a different form. Defining

$$\boldsymbol{g}_{\mathsf{n}} := -\frac{\partial R}{\partial \mathring{\boldsymbol{n}}} \tag{3.68}$$

and taking the vector product of $\boldsymbol{n}$ with (3.46), we find that

$$\boldsymbol{n} \times \left(\boldsymbol{g}_{\mathsf{n}} - \frac{\partial W}{\partial \boldsymbol{n}} + \operatorname{div} \frac{\partial W}{\partial \nabla \boldsymbol{n}} + \boldsymbol{k}_{\mathsf{n}} \right) = \mathbf{0}, \tag{3.69}$$

which, by (3.29), also acquires the more compact form

$$\boldsymbol{n} \times (\boldsymbol{g}_{\mathsf{n}} + \boldsymbol{h}_{\mathsf{n}} + \boldsymbol{k}_{\mathsf{n}}) = \mathbf{0}. \tag{3.70}$$

Computing explicitly (3.67) with the stress (3.44) yields

$$2\tau_i = \epsilon_{ijk} \left(n_j g_{\mathsf{n}k} + n_{l,j} \frac{\partial W}{\partial n_{l,k}} \right),$$

where we have denoted by $g_{\mathsf{n}k}$ the Cartesian components of $\boldsymbol{g}_{\mathsf{n}}$ and[21] by $\frac{\partial W}{\partial n_{l,k}}$ those of $\mathbf{L}_{\mathsf{n}}$. ERICKSEN's identity in the form (3.22) allows us to write $2\boldsymbol{\tau}$ as

$$\begin{aligned} 2\tau_i &= \epsilon_{ijk} \left(n_j g_{\mathsf{n}k} - n_j \frac{\partial W}{\partial n_k} - n_{j,l} \frac{\partial W}{\partial n_{k,l}} \right) \\ &= \epsilon_{ijk} \left(-n_j k_{\mathsf{n}k} - n_j \left[\frac{\partial W}{\partial n_{k,l}} \right]_{,l} - n_{j,l} \frac{\partial W}{\partial n_{k,l}} \right) \\ &= \epsilon_{ijk} \left(-n_j k_{\mathsf{n}k} - \left[n_j \frac{\partial W}{\partial n_{k,l}} \right]_{,l} \right), \end{aligned} \tag{3.71}$$

[21] With a common abuse of notation.

where the director balance (3.69) was used to obtain the second line. Comparing (3.71) with (3.66) shows that the latter is satisfied if we interpret

$$\boldsymbol{k} = \boldsymbol{n} \times \boldsymbol{k}_{\mathsf{n}} \tag{3.72}$$

as the body couple and

$$\mathbf{L} = \epsilon_{ijk} n_j \frac{\partial W}{\partial n_{k,l}} \boldsymbol{e}_i \otimes \boldsymbol{e}_l \tag{3.73}$$

as the couple stress, where $(\boldsymbol{e}_1, \boldsymbol{e}_2, \boldsymbol{e}_3)$ is any orthonormal positively oriented basis.[22] With (3.72) we can regard (3.69) as a balance of torques if we interpret

$$\boldsymbol{n} \times \boldsymbol{g}_{\mathsf{n}} = -\boldsymbol{n} \times \frac{\partial R}{\partial \mathring{\boldsymbol{n}}}$$

as a viscous torque and

$$\boldsymbol{n} \times \boldsymbol{h}_{\mathsf{n}} = \boldsymbol{n} \times \left(\operatorname{div} \frac{\partial W}{\partial \nabla \boldsymbol{n}} - \frac{\partial W}{\partial \boldsymbol{n}} \right)$$

as an elastic torque. Thus, in the inviscid limit, equation (3.25), which can equivalently be rewritten as

$$\boldsymbol{n} \times (\boldsymbol{k}_{\mathsf{n}} + \boldsymbol{h}_{\mathsf{n}}) = \boldsymbol{n} \times \left(\boldsymbol{k}_{\mathsf{n}} + \operatorname{div} \frac{\partial W}{\partial \nabla \boldsymbol{n}} - \frac{\partial W}{\partial \boldsymbol{n}} \right) = \mathbf{0}, \tag{3.74}$$

is interpreted as a balance of elastic and body couples.

Finally, the traction condition for the director (3.47) allows us to identify the surface couple as

$$\boldsymbol{c} = \boldsymbol{n} \times \boldsymbol{c}_{\mathsf{n}},$$

which is consistent with $\boldsymbol{c} = \mathbf{L}\boldsymbol{\nu}$. It readily follows from (3.73) that $\mathbf{L}$ can be characterized by its action on any vector $\boldsymbol{u} \in \mathcal{V}$ as

$$\mathbf{L}\boldsymbol{u} = \boldsymbol{n} \times \left(\frac{\partial W}{\partial \nabla \boldsymbol{n}} \right) \boldsymbol{u} = \boldsymbol{n} \times \mathbf{L}_{\mathsf{n}} \boldsymbol{u}. \tag{3.75}$$

Since by (2.92), the vector product of two indifferent vectors is hemi-indifferent, it follows from (3.75) that

$$\begin{aligned} \mathbf{L}^* \boldsymbol{u}^* &= \mathbf{L}^* \mathbf{R} \boldsymbol{u} = (\det \mathbf{R}) \mathbf{R} \mathbf{L} \boldsymbol{u} = (\det \mathbf{R}) \mathbf{R} \mathbf{L} \mathbf{R}^{\mathsf{T}} \mathbf{R} \boldsymbol{u} \\ &= (\det \mathbf{R}) \mathbf{R} \mathbf{L} \mathbf{R}^{\mathsf{T}} \boldsymbol{u}^*, \end{aligned} \tag{3.76}$$

where $\mathbf{R}$ is the orthogonal tensor representing a change of frame. By the arbitrariness of $\boldsymbol{u}^*$, it follows from (3.76) that the constitutive law (3.73) for the couple stress is hemitropic, as it should be.

[22] Strictly speaking, equation (3.44) determines $\mathbf{L}$ to within a divergence-free tensor. We shall see in Section 3.1.4 how this indeterminacy can be removed.

If the director is thought of as rotating independently from the flow, its time derivative is given by (3.64), where the rotational velocity $\boldsymbol{w}_{\mathsf{n}}$ may differ from the spin vector $\boldsymbol{w}$ of the flow and by (3.65), in general, $\mathring{\boldsymbol{n}} \neq \mathbf{0}$. With (3.72) and (3.64) we have

$$\boldsymbol{k}_{\mathsf{n}} \cdot \dot{\boldsymbol{n}} = \boldsymbol{k}_{\mathsf{n}} \cdot (\boldsymbol{w}_{\mathsf{n}} \times \boldsymbol{n}) = \boldsymbol{w}_{\mathsf{n}} \cdot (\boldsymbol{n} \times \boldsymbol{k}_{\mathsf{n}}) = \boldsymbol{k} \cdot \boldsymbol{w}_{\mathsf{n}}.$$

Similarly,

$$\boldsymbol{c}_{\mathsf{n}} \cdot \dot{\boldsymbol{n}} = \boldsymbol{c} \cdot \boldsymbol{w}_{\mathsf{n}},$$

and so the power of the external agents (3.7) can be equivalently written as

$$\mathscr{W}^{(\mathrm{a})}(\mathcal{P}_t, \chi) = \int_{\mathcal{P}_t} (\boldsymbol{b} \cdot \boldsymbol{v} + \boldsymbol{k} \cdot \boldsymbol{w}_{\mathsf{n}}) dV + \int_{\partial^* \mathcal{P}_t} (\boldsymbol{t} \cdot \boldsymbol{v} + \boldsymbol{c} \cdot \boldsymbol{w}_{\mathsf{n}}) dA,$$

which reduces to (2.142) only if $\boldsymbol{w}_{\mathsf{n}} = \boldsymbol{w}$, so that $\mathring{\boldsymbol{n}} = \mathbf{0}$.

3.1.4 Variational Compatibility

In this section, following essentially ERICKSEN [92], we establish the condition under which the static limit of the balance equations (3.66) and (3.25) can be related to the EULER–LAGRANGE equation for the stationarity of an energy functional, thus justifying the purely variational approach to the statics of liquid crystals, which was the first theory proposed by OSEEN [259] and FRANK [109].

Principle of Virtual Power

Our starting point here will be a principle of virtual power, which is actually the conceptual antecedent of D'ALEMBERT's principle, which in Section 3.1.1 served as a foundation for inviscid dynamics. Imagine any subbody $\mathcal{P}$ *carved* from the body $\mathcal{B}$, subject on its reduced boundary $\partial^*\mathcal{P}$ to a system of generalized tractions $(\boldsymbol{t}, \boldsymbol{c}_{\mathsf{n}})$ and, in its interior $\mathring{\mathcal{P}}$, to a system of generalized body forces $(\boldsymbol{b}, \boldsymbol{k}_{\mathsf{n}})$ expending power on any virtual motion described by the generalized velocities $(\boldsymbol{v}, \dot{\boldsymbol{n}})$, so that (3.7) still formally applies, though no real motion χ exists here:[23]

$$\mathscr{W}^{(\mathrm{a})}(\mathcal{P}) = \int_{\mathcal{P}} (\boldsymbol{b} \cdot \boldsymbol{v} + \boldsymbol{k}_{\mathsf{n}} \cdot \dot{\boldsymbol{n}}) dV + \int_{\partial^* \mathcal{P}} (\boldsymbol{t} \cdot \boldsymbol{v} + \boldsymbol{c}_{\mathsf{n}} \cdot \dot{\boldsymbol{n}}) dA. \tag{3.77}$$

As above, $\boldsymbol{b}$ and $\boldsymbol{k}_{\mathsf{n}}$ are assigned sources, whereas $\boldsymbol{t}$ and $\boldsymbol{c}_{\mathsf{n}}$ are shape functionals to be determined so as to comply with equilibrium. Like any real flow, the virtual flow $\boldsymbol{v}$ is also required to preserve the volume of any arbitrary subbody $\mathcal{P}$, since it is a means to mimic any admissible isochoric deformation of $\mathcal{P}$, possibly accompanied by an equally admissible change in the director distortion. Thus, the virtual flow $\boldsymbol{v}$ will also be subject to the kinematic constraint

[23] For notational coherence, we should denote the virtual flow by $\delta\boldsymbol{v}$, instead of $\boldsymbol{v}$, which is usually reserved for the real motion. However, no confusion is likely to arise here, since the only motion present is virtual.

$$\operatorname{div} \boldsymbol{v} = 0. \tag{3.78}$$

Similarly, denoting now by $\dot{\mathscr{F}}$ the virtual rate of change of the free energy $\mathscr{F}$ in (3.13), by the transport theorem (2.37) applied to a virtual motion, in complete analogy with (3.14), we write

$$\dot{\mathscr{F}}(\mathcal{P}) = \int_{\mathcal{P}} \dot{W}\, dV. \tag{3.79}$$

The principle of virtual power prescribes that at equilibrium an incompressible liquid crystal in $\mathcal{B}$ with elastic energy density (per unit volume) W and subject to the traction system $(\boldsymbol{t}, \boldsymbol{c}_{\mathsf{n}})$ and to the body force system $(\boldsymbol{b}, \boldsymbol{k}_{\mathsf{n}})$ satisfies the requirement

$$\dot{\mathscr{F}}(\mathcal{P}) = \mathscr{W}^{(\mathrm{a})}(\mathcal{P}) + \mathscr{W}^{(\mathrm{c})}(\mathcal{P}) \tag{3.80}$$

for all subbodies $\mathcal{P}$ of $\mathcal{B}$ and for all systems of virtual generalized velocities $(\boldsymbol{v}, \dot{\boldsymbol{n}})$, where $\mathscr{W}^{(\mathrm{c})}(\mathcal{P})$ is the power of the constraints[24] (3.78) and (3.8),

$$\mathscr{W}^{(\mathrm{c})}(\mathcal{P}) = \int_{\mathcal{P}} (p \operatorname{div} \boldsymbol{v} + \gamma \boldsymbol{n} \cdot \dot{\boldsymbol{n}}) dV.$$

Virtually with no change in the formal development of Section 3.1.1, we give $\dot{\mathscr{F}}$ in (3.79) the following form:

$$\dot{\mathscr{F}}(\mathcal{P}) = \int_{\partial^* \mathcal{P}} (\mathbf{T}\boldsymbol{\nu} \cdot \boldsymbol{v} + \mathbf{L}_{\mathsf{n}}\boldsymbol{\nu} \cdot \dot{\boldsymbol{n}}) dA - \int_{\mathcal{P}} (\operatorname{div} \mathbf{T} \cdot \boldsymbol{v} + \boldsymbol{h}_{\mathsf{n}} \cdot \dot{\boldsymbol{n}}) dV,$$

where $\boldsymbol{h}_{\mathsf{n}}$ is the molecular field defined in (3.29), $\mathbf{L}_{\mathsf{n}}$ is the torque stress (3.28), and $\mathbf{T}$ is the stress tensor as in (3.27). Thus, by (3.77) and (3.79), enforcing (3.80) requires the following equations to be satisfied:

$$\boldsymbol{b} + \operatorname{div} \mathbf{T} = \mathbf{0} \quad \text{and} \quad \boldsymbol{h}_{\mathsf{n}} + \boldsymbol{k}_{\mathsf{n}} + \gamma \boldsymbol{n} = \mathbf{0}, \tag{3.81a}$$

which hold in $\mathcal{B}$, and

$$\boldsymbol{t} = \mathbf{T}\boldsymbol{\nu} \quad \text{and} \quad \boldsymbol{c}_{\mathsf{n}} = \mathbf{L}_{\mathsf{n}}\boldsymbol{\nu}, \tag{3.81b}$$

which specify the generalized tractions as special functionals of $\partial^* \mathcal{P}$.

The first equation in (3.81a) is the static limit of the balance equation of linear momentum (3.66) in the inviscid case, while the second equation coincides with (3.30); both have been reobtained[25] within an alternative, independent formulation of statics. Since both these vectorial equations are in the single unknown field $\boldsymbol{n}$, they apparently overdetermine it. We shall see below how such an overdetermination can in general be resolved.

---

[24] Here, as in (3.9), p and γ are unknown LAGRANGE multipliers, with precisely the same mechanical meaning.

[25] There is more than a pedagogical reason in favor of such a derivation. The method illustrated here will guide us to obtain the appropriate equilibrium equations when in Section 5.2.1 we move in less traditional territories.

Now we pause briefly to consider other consequences of the principle of virtual power, which must hold whenever equations (3.81) do, though some might require more labor to be derived directly from these.[26]

Special Virtual Motions

The first of such consequences follows from taking as virtual motion a rigid motion, described by the system of generalized velocities $(\boldsymbol{v}_\mathrm{R}, \dot{\boldsymbol{n}}_\mathrm{R})$. Paraphrasing (2.71), we write $\boldsymbol{v}_\mathrm{R}$ as

$$\boldsymbol{v}_\mathrm{R}(x) = \boldsymbol{v}_o + \mathbf{W}(x - o), \tag{3.82}$$

where $\boldsymbol{v}_o$ is an arbitrary vector, representing the virtual velocity of the point o, and $\mathbf{W}$ is an arbitrary skew-symmetric tensor. Moreover, we set

$$\dot{\boldsymbol{n}}_\mathrm{R} := \mathbf{W}\boldsymbol{n}, \tag{3.83}$$

which, by (3.36), means that $\mathring{\boldsymbol{n}} = \mathbf{0}$, and so the director $\boldsymbol{n}$ is conveyed along with the fluid and remains at rest relative to it. By requiring $\dot{\mathscr{F}}(\mathcal{P})$ to vanish for all $\mathcal{P}$ along a rigid motion, which is a kinematic consequence of $\mathscr{F}(\mathcal{P})$ being frame-indifferent, with the aid of (3.79) and since $\boldsymbol{v}_\mathrm{R}$ satisfies (3.78), we reobtain ERICKSEN's identity (3.21), as the reader will easily verify.

Since $\dot{\mathscr{F}}(\mathcal{P})$ vanishes on all virtual rigid motions, (3.80) requires that $\mathscr{W}^{(a)}(\mathcal{P}) + \mathscr{W}^{(c)}(\mathcal{P})$ vanish as well. Use of (3.82) and (3.83), by the arbitrariness of $\boldsymbol{v}_o \in \mathcal{V}$ and $\mathbf{W} \in \mathsf{Skw}(\mathcal{V})$, yields

$$\int_{\mathcal{P}} \boldsymbol{b}\, dV + \int_{\partial^*\mathcal{P}} \boldsymbol{t}\, dA = \mathbf{0}, \tag{3.84a}$$

$$\int_{\mathcal{P}} [(x-o) \otimes \boldsymbol{b} + \boldsymbol{n} \otimes \boldsymbol{k}_\mathrm{n}] dV + \int_{\partial^*\mathcal{P}} [(x-o) \otimes \boldsymbol{t} + \boldsymbol{n} \otimes \boldsymbol{c}_\mathrm{n}] dA \in \mathsf{Sym}(\mathcal{V}). \tag{3.84b}$$

While (3.84a), easily recognized as the balance of all forces acting on the subbody $\mathcal{P}$ in equilibrium, is an immediate consequence of $(3.81a)_1$ and $(3.81b)_1$, (3.84b) can be employed to establish more directly formula (3.75) for the couple stress tensor $\mathbf{L}$.

By requiring the skew-symmetric part of the tensor in (3.84b) to vanish and using the fact that the axial vector of the tensor $(\boldsymbol{b} \otimes \boldsymbol{a} - \boldsymbol{a} \otimes \boldsymbol{b})$ is $\boldsymbol{a} \times \boldsymbol{b}$ (see also Appendix A.1), we give (3.84b) the following equivalent form:

$$\int_{\mathcal{P}} [(x - o) \times \boldsymbol{b} + \boldsymbol{n} \times \boldsymbol{k}_\mathrm{n}] dV + \int_{\partial^*\mathcal{P}} [(x - o) \times \mathbf{T}\boldsymbol{\nu} + \boldsymbol{n} \times \mathbf{L}_\mathrm{n}\boldsymbol{\nu}] dA = \mathbf{0}, \tag{3.85}$$

where use has also been made of (3.81b). Equation (3.85) clearly indicates that the tensor $\mathbf{L}$ defined as in (3.75) designates the couple stress. Use of (3.75) and the divergence theorem reduces (3.85), valid for all $\mathcal{P} \in \mathcal{B}$, to its equivalent local form (3.66), which we need not reproduce here.[27]

---

[26] This again serves more the purpose of illustrating a method than that of drawing new conclusions.

[27] We thus remove the indeterminacy on $\mathbf{L}$ signaled on page 182 above.

Compatibility Potential

Here we write explicitly the condition that must be met to ensure the compatibility of both equations in (3.84). We start from an identity that follows from (3.29) and (3.27),

$$\nabla W + \operatorname{div} \mathbf{T}_{\mathrm{E}} + (\nabla \boldsymbol{n})^{\mathsf{T}} \boldsymbol{h}_{\mathsf{n}} = \mathbf{0}, \tag{3.86}$$

and is more easily proved in indicial notation. By (3.31), we may write

$$w_{,i} - \left(n_{k,i} \frac{\partial W}{\partial n_{k,j}} \right)_{,j} = \frac{\partial W}{\partial n_k} n_{k,i} + \frac{\partial W}{\partial n_{k,j}} n_{k,ji} - n_{k,ij} \frac{\partial W}{\partial n_{k,j}} - n_{k,i} \left(\frac{\partial W}{\partial n_{k,j}} \right)_{,j},$$

whence, since for a smooth director field $n_{k,ji} = n_{k,ij}$, we obtain (3.86) in component form. Making use of both equations in (3.81a) and recalling (3.18), we arrive from (3.27) at the following compatibility equation:

$$\nabla W + \nabla p - \boldsymbol{b} - (\nabla \boldsymbol{n})^{\mathsf{T}} \boldsymbol{k}_{\mathsf{n}} = \mathbf{0}, \tag{3.87}$$

which is equivalently stated by requiring that there be a function $U = U(x, \boldsymbol{n})$ of position $x \in \mathcal{B}$ in space and orientation $\boldsymbol{n} \in \mathbb{S}^2$ such that

$$\boldsymbol{b} + (\nabla \boldsymbol{n})^{\mathsf{T}} \boldsymbol{k}_{\mathsf{n}} = \nabla U. \tag{3.88}$$

Once (3.88) is met, (3.87) delivers the pressure p to within a hydrostatic, uniform pressure p_0:

$$p = p_0 + U - W. \tag{3.89}$$

A way to satisfy (3.88), though presumably it is not the only one, is to assume that

$$\boldsymbol{b} = \frac{\partial U}{\partial x} \quad \text{and} \quad \boldsymbol{k}_{\mathsf{n}} = \frac{\partial U}{\partial \boldsymbol{n}}. \tag{3.90}$$

Here the derivative $\frac{\partial U}{\partial x}$ is meant to be computed ignoring the possible spatial dependence of $\boldsymbol{n}$. The derivative $\frac{\partial U}{\partial \boldsymbol{n}}$ is to be interpreted in the intrinsic sense, made clear for example by Lemma 3.6 of [353, p. 133]: it is by definition a vector orthogonal to $\boldsymbol{n}$. By (3.90) and (3.29), the second equation in (3.84), now compatible with the first, may also be written as

$$\operatorname{div} \left(\frac{\partial W}{\partial \nabla \boldsymbol{n}} \right) - \frac{\partial W}{\partial \boldsymbol{n}} + \frac{\partial U}{\partial \boldsymbol{n}} + \gamma \boldsymbol{n} = \mathbf{0}, \tag{3.91}$$

which is the EULER–LAGRANGE equation for the energy functional

$$\mathscr{E}[\boldsymbol{n}] := \int_{\mathcal{B}} (W - U) dV, \tag{3.92}$$

subject to the constraint (3.3).

Thus we have shown that the static limits of equations (3.24) and (3.25), which can also be obtained independently from a principle of virtual power, are compatible with a purely variational formulation with energy functional $\mathscr{E}$ as in (3.92), provided that compatibility condition (3.88) is satisfied. The stress tensor $\mathbf{T}$ in (3.27), computed with p as in (3.89) on a solution of (3.91), describes the distribution of internal forces in $\mathcal{B}$ at equilibrium. It has also been employed in [115] to compute elastic forces on defects of the director field, a topic that exceeds the scope of this book.[28]

We now illustrate a physically significant case in which both equations in (3.90) hold for a potential U that can easily be determined.[29] This case arises when the liquid crystal is subject to a magnetic field $\boldsymbol{H}$. According to the explicit computations of LEATHEM [176], $\boldsymbol{H}$ exerts both a magnetic body force $\boldsymbol{b}_{\mathrm{m}}$ and a body couple $\boldsymbol{k}_{\mathrm{m}}$ given by

$$\boldsymbol{b}_{\mathrm{m}} = (\nabla \boldsymbol{H})\boldsymbol{M} \quad \text{and} \quad \boldsymbol{k}_{\mathrm{m}} = \boldsymbol{M} \times \boldsymbol{H}, \tag{3.93}$$

where $\boldsymbol{M}$ is now the magnetization induced by $\boldsymbol{H}$. Under the assumption that $\boldsymbol{M}$ is linearly related to $\boldsymbol{H}$, a classical symmetry argument (see, for example, [307] and [353, p. 95]) shows that

$$\boldsymbol{M} = \chi_{\perp}\boldsymbol{H} + (\chi_{\parallel} - \chi_{\perp})(\boldsymbol{H}\cdot\boldsymbol{n})\boldsymbol{n}, \tag{3.94}$$

where $\chi_{\parallel}$ and $\chi_{\perp}$ are the magnetic susceptibilities of the material when $\boldsymbol{H}$ is parallel to $\boldsymbol{n}$ and when $\boldsymbol{H}$ is orthogonal to $\boldsymbol{n}$, respectively.[30] The *diamagnetic anisotropy*, defined by

$$\Delta\chi := \chi_{\parallel} - \chi_{\perp}, \tag{3.95}$$

can be either positive or negative, each sign characterizing a different interaction between $\boldsymbol{H}$ and $\boldsymbol{n}$, as we shall soon show.

By (3.94), (3.95), and (3.72), we obtain from (3.93) that

$$\boldsymbol{b}_{\mathrm{m}} = \chi_{\perp}(\nabla\boldsymbol{H})\boldsymbol{H} + \Delta\chi(\boldsymbol{H}\cdot\boldsymbol{n})(\nabla\boldsymbol{H})\boldsymbol{n}, \tag{3.96a}$$

$$\boldsymbol{k}_{\mathrm{m}} = \Delta\chi(\boldsymbol{H}\cdot\boldsymbol{n})\boldsymbol{n}\times\boldsymbol{H} = \boldsymbol{n}\times\boldsymbol{k}_{\mathrm{n}}. \tag{3.96b}$$

In particular, (3.96b) shows that for $\Delta\chi > 0$ the magnetic torque is *aligning*, since it tends to align $\boldsymbol{n}$ like $\boldsymbol{H}$, whereas for $\Delta\chi < 0$ it is *misaligning*, since it tends to align $\boldsymbol{n}$ at right angles with $\boldsymbol{H}$. Equation (3.96b) also implies that

$$\boldsymbol{k}_{\mathrm{n}} = \Delta\chi(\boldsymbol{H}\cdot\boldsymbol{n})(\mathbf{I} - \boldsymbol{n}\otimes\boldsymbol{n})\boldsymbol{H}, \tag{3.97}$$

since, by $(3.90)_2$, $\boldsymbol{k}_{\mathrm{n}}\cdot\boldsymbol{n} = 0$. We now show that both (3.96a) and (3.97) agree with (3.90), provided we set

---

28 Static defects are singularities in the solutions of equation (3.91). There is a vast literature concerned with them. We cite only the following works, which treat this topic at different levels of generality and rigor: [36, 37, 135, 136, 114, 166, 167, 168, 169, 172, 230, 286, 308, 355]. More recently, defect dynamics has also become the object of mathematical studies. We refer the reader to some papers that may be suggested as first readings: [25, 46, 47, 74, 151, 162, 261, 264, 269, 270, 285, 309, 314, 330, 336].

29 See also [92] and [183].

30 Here both $\chi_{\parallel}$ and $\chi_{\perp}$ are considered to be material constants, uniform in space.

$$U_{\mathrm{m}} = \frac{1}{2}\boldsymbol{M} \cdot \boldsymbol{H} = \frac{1}{2}[\chi_{\perp}(\boldsymbol{H} \cdot \boldsymbol{H}) + \Delta\chi(\boldsymbol{H} \cdot \boldsymbol{n})^2]. \tag{3.98}$$

While differentiating the magnetic potential U_{m} in (3.98) with respect to $\boldsymbol{n}$ readily delivers[31] (3.97), to obtain (3.96a) we note that

$$\frac{\partial U_{\mathrm{m}}}{\partial x} = \chi_{\perp}(\nabla \boldsymbol{H})^{\mathsf{T}}\boldsymbol{H} + \Delta\chi(\boldsymbol{H} \cdot \boldsymbol{n})(\nabla \boldsymbol{H})^{\mathsf{T}}\boldsymbol{n}, \tag{3.99}$$

since the field $\boldsymbol{n}$ is regarded as fixed in computing this derivative. To prove that (3.99) delivers the same body force $\boldsymbol{b}$ as (3.96a), we need only remark that $\boldsymbol{H}$ is irrotational in $\mathcal{B}$, and so its gradient $\nabla \boldsymbol{H}$ is a symmetric tensor.

By (3.92), $-U_{\mathrm{m}}$ is the magnetic energy that must be minimized together with the elastic energy W for $\boldsymbol{n}$ to attain a stable configuration. It readily follows from (3.98) that for $\Delta\chi > 0$ the energy $-U_{\mathrm{m}}$ is minimized when $\boldsymbol{n}$ is parallel to $\boldsymbol{H}$, whereas for $\Delta\chi < 0$ it is minimized for $\boldsymbol{n}$ orthogonal to $\boldsymbol{H}$. This shows how a magnetic field interacts with the nematic director. A completely analogous characterization can be derived for the action of an electric field in the linear regime.[32]

3.1.5 Thermal Effects

We have so far studied the isothermal dynamics of liquid crystals: all thermal effects have been deliberately neglected. On the other hand, in Section 2.3.3, we have shown already how heat conduction can be incorporated in the theory of isotropic, linearly viscous fluids, including the temperature gradient $\boldsymbol{g} := \nabla\theta$ among the local measures of dissipation. Here, following closely the line of thought presented in Section 2.3.3, we set the scene to study thermal effects in liquid crystals: we shall encounter a new torque affecting the director motion, which is imparted by a temperature gradient. We shall see how the polar symmetry of nematics prevents such an action from deploying, so that it will be effective only in cholesteric liquid crystals, proving itself a cause likely to explain an old effect, first observed by LEHMANN [178].

As we first did in (2.282), we allow $\boldsymbol{g}$ in the collection $\mathbf{d}$ of dissipation measures, along with $\mathbf{D}$ and $\mathring{\boldsymbol{n}}$. A quadratic form in $(\boldsymbol{g}, \mathbf{D}, \mathring{\boldsymbol{n}})$, which may also depend on $\boldsymbol{n}$ as a parameter, is both hemitropic and even under simultaneous inversion of $\boldsymbol{n}$ and $\mathring{\boldsymbol{n}}$, only if represented in the form[33]

---

[31] With the aid of the projection that makes $\boldsymbol{k}_{\mathsf{n}}$ obey the condition $\boldsymbol{k}_{\mathsf{n}} \cdot \boldsymbol{n} = 0$.

[32] MAXWELL's equations have here made only a partial appearance. We have subjected $\boldsymbol{H}$ to $\operatorname{curl} \boldsymbol{H} = \mathbf{0}$, but we neglected to enforce $\operatorname{div} \boldsymbol{B} = 0$, where the magnetic induction $\boldsymbol{B}$ is related to $\boldsymbol{H}$ in a fashion similar to (3.94). An approximation in which the equation for $\boldsymbol{B}$ can be ignored is discussed in [353, § 4.1.1], to which we also refer the reader for an analogous discussion on the dielectric interaction.

[33] The two hemitropic, but not isotropic, terms appearing in (3.100) are similar in structure to (3.50), which was ruled out by the nematic symmetry (3.4); the inclusion of $\boldsymbol{g}$ in the list $\mathbf{d}$ of dissipation measures has made these terms available. See also [279].

$$R(\boldsymbol{n};\boldsymbol{g},\mathbf{D},\mathring{\boldsymbol{n}}) = \frac{1}{2}\gamma_1\mathring{\boldsymbol{n}}^2 + \gamma_2\mathring{\boldsymbol{n}}\cdot\mathbf{D}\boldsymbol{n} + \frac{1}{2}\gamma_3(\mathbf{D}\boldsymbol{n})^2 + \frac{1}{2}\gamma_4(\boldsymbol{n}\cdot\mathbf{D}\boldsymbol{n})^2 + \frac{1}{2}\gamma_5\,\mathrm{tr}\,\mathbf{D}^2 + \frac{1}{2}\bar{\kappa}_1\boldsymbol{g}^2 + \frac{1}{2}\bar{\kappa}_2(\boldsymbol{g}\cdot\boldsymbol{n})^2 + \bar{\kappa}_3\boldsymbol{g}\cdot\boldsymbol{n}\times\mathring{\boldsymbol{n}} + \bar{\kappa}_4\boldsymbol{g}\cdot\mathbf{D}\boldsymbol{n}\times\boldsymbol{n}, \tag{3.100}$$

where the first five terms reproduce the dissipation function in (3.49), the sixth term is the analogue of the thermal term in (2.282), and only the last three terms are new, the first reflecting the anisotropy of the medium and the last two expressing thermal interactions with director relaxation and flow, respectively. Together with the γ's, the $\bar{\kappa}$'s are now functions of the temperature θ.

For R in (3.100) to be positive semidefinite, these functions cannot be arbitrary. Reasoning precisely as we did to obtain the inequalities (3.57), we prove that these must be supplemented by the following to ensure positive semidefiniteness to the extended dissipation in (3.100):

$$\begin{aligned} \bar{\kappa}_1 + \bar{\kappa}_2 &\geqq 0, \\ \bar{\kappa}_1(\gamma_3 + 2\gamma_5) - \bar{\kappa}_3^2 &\geqq 0, \\ \gamma_1\bar{\kappa}_1 - \bar{\kappa}_4^2 &\geqq 0, \\ \bar{\kappa}_1[\gamma_1(\gamma_3+2\gamma_5) - \gamma_2^2] - 2\gamma\bar{\kappa}_3\bar{\kappa}_4 - \gamma_1\bar{\kappa}_3^2 - \bar{\kappa}_4^2(\gamma_3+2\gamma_5) &\geqq 0. \end{aligned}$$

Requiring R in (3.100) to be an isotropic function, so as to comply with the symmetry of nematic liquid crystals, would set both $\bar{\kappa}_3$ and $\bar{\kappa}_4$ equal to zero. It will be apparent below how this would simply reduce all thermal effects to an anisotropic heat conduction. Thus, to avoid lessening the consequences of the theory, in this section we shall consider cholesteric liquid crystals.

Allowing for thermal effects, we also need to consider an elastic free-energy density W depending on the temperature θ. That this can be achieved by simply regarding the elastic constants in FRANK's formula (3.32) as functions of θ or by resorting to a more general functional dependence is here immaterial: we shall simply assume that $W = W(\theta, \boldsymbol{n}, \nabla\boldsymbol{n})$, so as to write the time rate of the free energy $\mathscr{F}$ as

$$\dot{\mathscr{F}}(\mathcal{P}_t,\chi) = \int_{\mathcal{P}_t}\left[\frac{\partial W}{\partial\theta}\dot{\theta} + \frac{\partial W}{\partial\boldsymbol{n}}\cdot\dot{\boldsymbol{n}} + \frac{\partial W}{\partial\nabla\boldsymbol{n}}\cdot(\nabla\boldsymbol{n})^{\cdot}\right]dV. \tag{3.101}$$

Echoing (2.284), we assume that the entropy per unit volume, $\varrho\eta$, is a function of $(\theta, \boldsymbol{n}, \nabla\boldsymbol{n})$:

$$\varrho\eta = \mathrm{H}(\theta, \boldsymbol{n}, \nabla\boldsymbol{n}), \tag{3.102}$$

so that, by (2.209), the thermal production $\mathscr{T}$ can be written as

$$\mathscr{T}(\mathcal{P}_t,\chi) = \int_{\mathcal{P}_t}\left(\mathrm{H}\dot{\theta} + \frac{1}{\theta}\boldsymbol{q}\cdot\boldsymbol{g}\right)dV, \tag{3.103}$$

where $\boldsymbol{q}$ is the heat flux. According to the general prescription (2.210), the total working $\mathscr{W}$ in (3.5) here becomes

$$\mathscr{W}^{(t)} = \mathscr{W}^{(a)} + \mathscr{W}^{(c)} - \dot{\mathscr{K}} - \dot{\mathscr{F}} - \mathscr{T},$$

where $\mathscr{W}^{(a)}$, $\mathscr{W}^{(c)}$, and $\dot{\mathscr{K}}$ are still given by (3.7), (3.9), and (3.10), respectively, while $\dot{\mathscr{F}}$ and $\mathscr{T}$ are as in (3.101) and (3.103) above.

The form of the principle of minimal reduced dissipation appropriate to the present setting is thus

$$\delta\mathscr{W}^{(t)} = \delta\mathscr{R}, \tag{3.104}$$

where $\mathscr{R}$ is the functional with density R in (3.100), formally defined as in (3.38). An easy computation shows that

$$\delta\mathscr{W}^{(t)} = \delta\mathscr{W} - \int_{\mathcal{P}_t} \left(\mathrm{H}\delta\dot{\theta} + \frac{1}{\theta}\boldsymbol{q}\cdot\boldsymbol{g} \right) dV - \int_{\mathcal{P}_t} \frac{\partial W}{\partial \theta}\delta\dot{\theta}\, dV,$$

which, by combining (3.23) with the reasoning that in Section 2.3.3 led us to (2.286a) and (2.289), here we conclude that

$$\mathrm{H} = -\frac{\partial W}{\partial \theta} \tag{3.105}$$

and

$$\boldsymbol{q} = -\theta\frac{\partial R}{\partial \boldsymbol{g}}. \tag{3.106}$$

While the former equation simply restates for liquid crystals the result proved in (2.247) for classical fluids, the latter extends the classical FOURIER's law. It is readily seen from (3.100) that $\boldsymbol{q}$ is explicitly given by

$$\boldsymbol{q} = -\kappa_1\boldsymbol{g} - \kappa_2(\boldsymbol{g}\cdot\boldsymbol{n})\boldsymbol{n} - \kappa_3\boldsymbol{n}\times\mathring{\boldsymbol{n}} - \kappa_4\mathbf{D}\boldsymbol{n}\times\boldsymbol{n},$$

where the functions κ_i of θ defined by

$$\kappa_i := \theta\bar{\kappa}_i, \quad i = 1,\dots,4,$$

represent generalized thermal conductivities.

Equations (3.105) and (3.106) followed from the thermal variations $(\delta\theta, \delta\boldsymbol{g})$ in (3.104); the mechanical variations $(\delta\boldsymbol{v}, \delta\dot{\boldsymbol{n}})$ are to be performed precisely as they were in Section 3.1.2 above, and they yield formally the same balance equations as in (3.42) and (3.46). However, new thermal contributions can be recognized in both the dissipative stress tensor $\mathbf{T}_{\mathrm{dis}}$ and the viscous generalized force $\boldsymbol{g}_{\mathrm{n}}$, still delivered by (3.45) and (3.68), but with R as in (3.100); these are given by

$$\mathbf{T}^{(t)}_{\mathrm{dis}} = \frac{1}{2}\bar{\kappa}_3\left[\boldsymbol{n}\otimes(\boldsymbol{g}\times\boldsymbol{n}) = (\boldsymbol{g}\times\boldsymbol{n})\otimes\boldsymbol{n}\right] + \bar{\kappa}_4\,\overline{(\boldsymbol{n}\times\boldsymbol{g})\otimes\boldsymbol{n}} \tag{3.107}$$

and

$$\boldsymbol{g}^{(t)}_{\mathrm{n}} = -\bar{\kappa}_3\boldsymbol{g}\times\boldsymbol{n},$$

respectively. Thus, the stress tensor $\mathbf{T}$ that enters the balance of linear momentum (3.42) now reads

$$\mathbf{T} = -p\mathbf{I} + \mathbf{T}_\mathrm{E} + \mathbf{T}^{(\mathrm{m})}_{\mathrm{dis}} + \mathbf{T}^{(\mathrm{t})}_{\mathrm{dis}},$$

where $\mathbf{T}_\mathrm{E}$ is the ERICKSEN stress (3.31), $\mathbf{T}^{(\mathrm{m})}_{\mathrm{dis}}$ is the mechanical dissipative stress tensor, still given by (3.59), and $\mathbf{T}^{(\mathrm{t})}_{\mathrm{dis}}$ is the thermal stress in (3.107). Likewise, the balance of torques expressed by (3.70) now becomes

$$\boldsymbol{n} \times \left(\boldsymbol{g}^{(\mathrm{m})}_{\mathsf{n}} + \boldsymbol{g}^{(\mathrm{t})}_{\mathsf{n}} + \boldsymbol{h}_{\mathsf{n}} + \boldsymbol{k}_{\mathsf{n}} \right) = \mathbf{0}, \tag{3.108}$$

where $\boldsymbol{g}^{(\mathrm{m})}_{\mathsf{n}}$ is the mechanical generalized force, which by (3.58) can still be written as

$$\boldsymbol{g}^{(\mathrm{m})}_{\mathsf{n}} = -\gamma_1 \mathring{\boldsymbol{n}} - \gamma_2[\mathbf{D}\boldsymbol{n} - (\mathbf{D}\boldsymbol{n} \cdot \boldsymbol{n})\boldsymbol{n}],$$

though the viscosities γ_1 and γ_2 are now functions of temperature. In (3.108),

$$\boldsymbol{k}^{(\mathrm{t})} := \boldsymbol{n} \times \boldsymbol{g}^{(\mathrm{t})}_{\mathsf{n}} = -\bar{\kappa}_3 \boldsymbol{n} \times (\boldsymbol{g} \times \boldsymbol{n}) = -\bar{\kappa}_3[\boldsymbol{g} - (\boldsymbol{g} \cdot \boldsymbol{n})\boldsymbol{n}] \tag{3.109}$$

is clearly to be interpreted as the *thermal torque* acting on the director.[34]

LEHMANN [178] first noted a thermal phenomenon in cholesteric liquid crystals that is now traditionally named after him. He observed that an undistorted cholesteric texture like the one represented by equation (3.33) would perform a precessional motion at uniform rotational speed, as though it were a rigid whole, once subjected to a temperature gradient $\boldsymbol{g}$ directed along the helical axis $\boldsymbol{e}_3$. No flow appears to sustain the motion, which is ultimately a uniform rotation of all directors $\boldsymbol{n}$ about the axis $\boldsymbol{e}_3$. LESLIE [181] showed how the thermal torque in equation (3.109) has the potential to explain LEHMANN's effect:[35] since $\boldsymbol{g}$ is orthogonal to $\boldsymbol{n}$, in this case $\boldsymbol{k}^{(\mathrm{t})}$ is parallel to $\boldsymbol{g}$; in general, as shown by (3.109), $\boldsymbol{k}^{(\mathrm{t})}$ is proportional to the component of $\boldsymbol{g}$ orthogonal to $\boldsymbol{n}$.

As also shown by LESLIE [181], the mechanical balance equations must be supplemented by an equation for the thermal field θ: in LEHMANN's problem this equation is solved by a constant $\boldsymbol{g}$, orthogonal to the plates bounding the sample, which

[34] Though $\boldsymbol{k}^{(\mathrm{t})}$ is clearly responsible for the thermal effects in cholesteric liquid crystals, the reader should not be induced to believe that it is the only one depending on the temperature: all other torques in (3.108) actually do.

[35] LEHMANN's effect was also referred to by OSEEN [259] as follows: "He [Lehmann] found that in certain cases a substance, spread out between two glass surfaces, would be put into motion, when influenced by a flow of heat coming from below, during which motion the different drops of liquid seemed to be in violent rotation. Further investigations convinced Lehmann that in this case it was not the drop itself, but the structure that moved." It is instructive to read how LESLIE [181] in turn comments about OSEEN's interpretation of LEHMANN's effect: "Oseen goes on to state that he considered that the motion was due to the molecules rotating with uniform speed around vertical axes drawn through their centres of gravity. He claims that his theory provided an explanation of the violent rotation, since his viscous terms vanished for such a motion. However, he offered no explanation of the forces creating the motion." The torques responsible for LEHMANN's effect have the thermal nature first illuminated by LESLIE [181]. We also refer the reader to [182], [34], and [75], where thermomechanical effects in cholesteric liquid crystals are further explored.

are taken at constant, different temperatures. As already shown in Section 2.3.3, such an equation stems from the first law of thermodynamics (2.195). Letting the internal energy per unit volume be denoted by W_e, by (2.204), (3.102), and (3.105), we set

$$W_e := W + \theta \mathrm{H} = W - \theta \frac{\partial W}{\partial \theta}. \tag{3.110}$$

Thus, (2.195) becomes

$$\left(\int_{\mathcal{P}_t} W_e dV \right)^{\cdot} = \mathscr{W}^{(e)}(\mathcal{P}_t, \chi) + \mathscr{Q}(\mathcal{P}_t, \chi), \tag{3.111}$$

where $\mathscr{W}^{(e)} = \mathscr{W}^{(a)} - \mathscr{K}$ is the external power,[36] and $\mathscr{Q}$ is the heating, which we recall from (2.194),

$$\mathscr{Q}(\mathcal{P}_t, \chi) = - \int_{\partial^* \mathcal{P}_t} \boldsymbol{q} \cdot \boldsymbol{\nu} \, dA + \int_{\mathcal{P}_t} \sigma \, dV, \tag{3.112}$$

where σ is the heat supply.

Before deriving from the local form of (3.111) the energy balance equation, we find it instructive to give $\mathscr{W}^{(e)}$ an equivalent form, valid along the solutions to the mechanical balance equations, thus extending a result obtained in (2.136) for the power stress of classical continuum mechanics. It readily follows from (3.42) and (3.46) that

$$\begin{aligned} \mathscr{W}^{(e)}(\mathcal{P}_t, \chi) &= \int_{\mathcal{P}_t} \left[-\operatorname{div} \mathbf{T} \cdot \boldsymbol{v} + \left(\frac{\partial R}{\partial \mathring{\boldsymbol{n}}} + \frac{\partial W}{\partial \boldsymbol{n}} - \operatorname{div} \frac{\partial W}{\partial \nabla \boldsymbol{n}} \right) \cdot \dot{\boldsymbol{n}} \right] dV \\ &\quad + \int_{\partial^* \mathcal{P}_t} \left(\mathbf{T}\boldsymbol{\nu} \cdot \boldsymbol{v} + \frac{\partial W}{\partial \nabla \boldsymbol{n}} \boldsymbol{\nu} \cdot \dot{\boldsymbol{n}} \right) dA \\ &= \int_{\mathcal{P}_t} \left[\mathbf{T} \cdot \nabla \boldsymbol{n} + \frac{\partial W}{\partial \nabla \boldsymbol{n}} \cdot \nabla \dot{\boldsymbol{n}} + \left(\frac{\partial R}{\partial \mathring{\boldsymbol{n}}} + \frac{\partial W}{\partial \boldsymbol{n}} \right) \cdot \dot{\boldsymbol{n}} \right] dV \\ &=: \mathscr{W}^{(i)}(\mathcal{P}_t, \chi), \end{aligned} \tag{3.113}$$

where, as in (2.136), $\mathscr{W}^{(i)}$ is interpreted as the power of all *internal actions*. Like (2.137), equation (3.113) expresses the balance of power: the power expended in an inertial frame by the external forces applied to a subbody $\mathcal{P}$ in a motion χ is balanced by the power expended by the internal forces and the rate of change of the kinetic energy. Clearly, in this case the viscous, dissipative actions, which also include thermal contributions, are to be reckoned among the internal forces.

Letting

$$W^{(i)} := \mathbf{T} \cdot \nabla \boldsymbol{v} + \frac{\partial W}{\partial \nabla \boldsymbol{n}} \cdot \nabla \dot{\boldsymbol{n}} + \left(\frac{\partial R}{\partial \mathring{\boldsymbol{n}}} + \frac{\partial W}{\partial \boldsymbol{n}} \right) \cdot \dot{\boldsymbol{n}} \tag{3.114}$$

denote the density per unit volume of the internal power $\mathscr{W}^{(i)}$, by (3.110) and (3.112), we give (3.111) the following local form:

---

[36] With $\mathscr{W}^{(e)}$ and $\mathscr{K}$ as in (3.7) and (3.10), respectively.

$$\left(W - \theta\frac{\partial W}{\partial \theta}\right)^{\cdot} = W^{(\mathrm{i})} - \operatorname{div}\boldsymbol{q} + \sigma. \tag{3.115}$$

We shall soon give an equivalent, more expressive variant of this equation. Here we pause to remark that, by (3.113), $\mathscr{W}^{(\mathrm{i})}$ must be frame-indifferent, since $\mathscr{W}^{(\mathrm{e})}$ is so. On the other hand, the single terms that constitute $W^{(\mathrm{i})}$ are not frame-indifferent. It is instructive to transform $W^{(\mathrm{i})}$ in (3.114) into a sum of indifferent terms. To this end, writing $\nabla\boldsymbol{v} = \mathbf{D} + \mathbf{W}$ and making use of (3.44), we easily prove that

$$\mathbf{T}\cdot\nabla\boldsymbol{v} = \mathbf{T}\cdot\mathbf{D} - \frac{\partial W}{\partial \nabla\boldsymbol{n}}\cdot(\nabla\boldsymbol{n})\mathbf{W} + \boldsymbol{n}\cdot\mathbf{W}\frac{\partial R}{\partial \mathring{\boldsymbol{n}}}. \tag{3.116}$$

Using (3.116), (3.36), and (3.21) in (3.114), we arrive at

$$W^{(\mathrm{i})} = \mathbf{T}\cdot\mathbf{D} + \frac{\partial W}{\partial \nabla\boldsymbol{n}}\cdot[\nabla\dot{\boldsymbol{n}} - \mathbf{W}(\nabla\boldsymbol{n})] + \left(\frac{\partial R}{\partial \mathring{\boldsymbol{n}}} + \frac{\partial W}{\partial \boldsymbol{n}}\right)\cdot\mathring{\boldsymbol{n}}. \tag{3.117}$$

While the first and last terms on the right-hand side of (3.117) are clearly frame-indifferent, proving that the tensor $[\nabla\dot{\boldsymbol{n}} - \mathbf{W}(\nabla\boldsymbol{n})]$ is also indifferent is an exercise that at this stage the reader should be able to do with little[37] or no guidance.

Yet another form of $W^{(\mathrm{i})}$ may be derived, which further simplifies (3.115). Since $\operatorname{tr}\mathbf{D} = 0$,

$$\mathbf{T}\cdot\mathbf{D} = \mathbf{T}_{\mathrm{dis}}\cdot\mathbf{D} - (\nabla\boldsymbol{n})^{\mathsf{T}}\frac{\partial W}{\partial \nabla\boldsymbol{n}}\cdot\mathbf{D}, \tag{3.118}$$

where $\mathbf{T}_{\mathrm{dis}} = \mathbf{T}^{(\mathrm{m})}_{\mathrm{dis}} + \mathbf{T}^{(\mathrm{t})}_{\mathrm{dis}}$ is the total dissipative stress tensor. It follows from inserting (3.118) into (3.117), from making again use of (3.36) and (3.21), and from applying the kinematic identity (3.16) that

$$W^{(\mathrm{i})} = \mathbf{T}_{\mathrm{dis}}\cdot\mathbf{D} + \frac{\partial W}{\partial \nabla\boldsymbol{n}}\cdot(\nabla\boldsymbol{n})^{\cdot} + \frac{\partial W}{\partial \boldsymbol{n}}\cdot\dot{\boldsymbol{n}}. \tag{3.119}$$

Computing $\dot{W}$ in (3.115) with the aid of the identity

$$\dot{W} = \frac{\partial W}{\partial \theta}\dot{\theta} + \frac{\partial W}{\partial \boldsymbol{n}}\cdot\dot{\boldsymbol{n}} + \frac{\partial W}{\partial \nabla\boldsymbol{n}}\cdot(\nabla\boldsymbol{n})^{\cdot},$$

already implicit in (3.101), we readily obtain from (3.119) that (3.115) reduces to

$$-\left(\frac{\partial W}{\partial \theta}\right)^{\cdot}\theta = \mathbf{T}_{\mathrm{dis}}\cdot\mathbf{D} + \mathring{\boldsymbol{n}}\cdot\frac{\partial R}{\partial \mathring{\boldsymbol{n}}} - \operatorname{div}\boldsymbol{q} + \sigma, \tag{3.120}$$

where all terms are frame-indifferent and $\boldsymbol{q}$ is given by (3.106). Equation (3.120) is the energy balance equation, which must be added to the mechanical balances to determine the evolution of the temperature θ along with that of the director $\boldsymbol{n}$ and the flow $\boldsymbol{v}$.

[37] The reader is advised to prove first that $(\nabla\dot{\boldsymbol{n}})^* = \boldsymbol{\Omega}(\nabla\boldsymbol{n})^* + \mathbf{R}\nabla\dot{\boldsymbol{n}}\mathbf{R}^{\mathsf{T}}$, where $\boldsymbol{\Omega}$ is the spin tensor in (2.50), and then employ (2.63) together with the frame-indifference of $\nabla\boldsymbol{n}$, that is, $(\nabla\boldsymbol{n})^* = \mathbf{R}\nabla\boldsymbol{n}\mathbf{R}^{\mathsf{T}}$.

One final consequence can be drawn from (3.117) when the director $\boldsymbol{n}$ is conveyed by the flow. Whenever this is case, $\mathring{\boldsymbol{n}}$ vanishes identically and (3.117) reduces to

$$W^{(\mathrm{i})} = \mathbf{T} \cdot \mathbf{D} + \frac{\partial W}{\partial \nabla \boldsymbol{n}} \cdot [\nabla \dot{\boldsymbol{n}} - \mathbf{W}(\nabla \boldsymbol{n})],$$

where now $\dot{\boldsymbol{n}} \equiv \mathbf{W}\boldsymbol{n}$. By differentiating in space both sides of this latter identity and applying (3.75), we arrive at

$$W^{(\mathrm{i})} = \mathbf{T} \cdot \mathbf{D} + \mathbf{L} \cdot \nabla \boldsymbol{w}, \tag{3.121}$$

where $\boldsymbol{w}$ is the spin vector. Equation (3.121) coincides with (2.146), thus showing that here the balance of power reduces to that contemplated by BEATTY's theory of general interactions presented in Section 2.1.6 only when the motion of $\boldsymbol{n}$ relative to the flow is artificially suppressed. An equivalent way of saying this is that BEATTY's theory is subsumed under ERICKSEN–LESLIE's.

3.2 Variable Degree of Orientation

We have seen in Section 1.3 that the MAIER–SAUPE mean field theory predicts uniaxial equilibrium states of the form

$$\mathbf{Q} = S\left(\boldsymbol{n} \otimes \boldsymbol{n} - \frac{1}{3}\mathbf{I}\right), \tag{3.122}$$

with a temperature-dependent scalar order parameter S that is given by

$$S = \langle P_2(\boldsymbol{n} \cdot \boldsymbol{u}) \rangle,$$

where the brackets $\langle \ldots \rangle$ indicate a local orientational average over the molecules and P_2 is the second Legendre polynomial in $\boldsymbol{n} \cdot \boldsymbol{u}$, the cosine of the angle between the molecular figure symmetry axis $\boldsymbol{u}$ and the nematic director $\boldsymbol{n}$. S can take values between $-\frac{1}{2}$ and 1. It can be expected and it is usually observed that the alignment is well approximated by the form (3.122) also away from equilibrium as long as distortions do not get too large. A theory for uniaxial nematics with variable order can naturally be obtained as a special case of a theory for the alignment tensor as treated in Chapter 4. Here, following mainly ERICKSEN [99], we directly derive such a theory by treating S and $\boldsymbol{n}$ as independent variables and extending the uniaxial director theory with constant scalar order presented in the preceding section. In this way, the new theory arises as an extension of the old one, and the differences between the two theories present themselves in a most transparent fashion.

The formal development of the theory is precisely the same as before, but now the ingredients of the power and dissipation can also depend on the variable scalar order parameter S and its spatial and time derivatives. D'ALEMBERT's principle and the principle of reduced minimum dissipation hold in the very same form. However, because $\boldsymbol{n}$ and S are treated as independent variables, a new independent generalized velocity $\dot{S}$ is present, and apart from the equations governing flow and director evolution, a third equation for the evolution of S arises.

3.2.1 Nondissipative Dynamics

Conceptually, our development here parallels rather closely that of Section 3.1, and so we can afford to be more concise, almost schematic.

External Agents

The external power expended on a uniaxial nematic with variable order is an extension of (3.7) that accounts for possible external actions on S,

$$\mathscr{W}^{(\mathrm{a})}(\mathcal{P}_t, \chi) = \int_{\mathcal{P}_t} \left(\boldsymbol{b} \cdot \boldsymbol{v} + \boldsymbol{k}_{\mathrm{n}} \cdot \dot{\boldsymbol{n}} + L\dot{S}\right) dV + \int_{\partial^* \mathcal{P}_t} \left(\boldsymbol{t} \cdot \boldsymbol{v} + \boldsymbol{c}_{\mathrm{n}} \cdot \dot{\boldsymbol{n}} + K\dot{S}\right) dA,$$

where L and K are generalized body and contact force densities acting on the order parameter S.

Power of the Constraints

The constraints on the velocity field and the unit director are the same as before. Although S is by its definition constrained[38] to be between $-\frac{1}{2}$ and 1, we treat it as a free variable, and so we have, as before,

$$\mathscr{W}^{(\mathrm{c})}(\mathcal{P}_t, \chi) = \int_{\mathcal{P}_t} (\gamma \boldsymbol{n} \cdot \dot{\boldsymbol{n}} - \nabla p \cdot \boldsymbol{v})\, dV + \int_{\partial^* \mathcal{P}_t} p\boldsymbol{\nu} \cdot \boldsymbol{v}\, dA.$$

Kinetic Energy

While the kinetic energy in principle contains a contribution related to $\dot{S}$, this is generally negligible for the same reasons cited before, namely, that molecular inertia is small. Thus we use as before

$$\dot{\mathscr{K}}(\mathcal{P}_t, \chi) = \int_{\mathcal{P}_t} \varrho \dot{\boldsymbol{v}} \cdot \boldsymbol{v}\, dV.$$

Free Energy

The free energy density is now a function $W = W(S, \nabla S, \boldsymbol{n}, \nabla \boldsymbol{n})$, and it will in general contain not only elastic terms, but also a LANDAU–DE GENNES potential in S, which is illustrated in detail in Section 4.1.1. In analogy to (3.22), W has to satisfy the invariance requirement

$$W(\boldsymbol{n}, \nabla \boldsymbol{n}) = W(S, \mathbf{R}\nabla S, \mathbf{R}\boldsymbol{n}, \mathbf{R}\nabla \boldsymbol{n}\mathbf{R}^{\mathsf{T}}) \tag{3.123}$$

[38] The unilateral bounds to which S is subject are different from the other constraints considered so far: they are to be valid on the solutions to the equations of the theory and may be favored by the action of some internal potential, such as the one considered in Section 4.1.1.

for any proper orthogonal transformation $\mathbf{R}$.

We again have

$$\mathscr{F}(\mathcal{P}_t, \chi) = \int_{\mathcal{P}_t} \dot{W}\, dV,$$

where now

$$\dot{W} = \frac{\partial W}{\partial \boldsymbol{n}} \cdot \dot{\boldsymbol{n}} + \frac{\partial W}{\partial \nabla \boldsymbol{n}} \cdot (\nabla \boldsymbol{n})^{\cdot} + \frac{\partial W}{\partial S} \dot{S} + \frac{\partial W}{\partial \nabla S} \cdot (\nabla S)^{\cdot}.$$

Computations analogous to those performed to arrive at (3.16) show that the material time derivative of the gradient of S can be written as

$$(\nabla S)^{\cdot} = \nabla \dot{S} - (\nabla \boldsymbol{v})^{\mathsf{T}} \nabla S.$$

With this, after an integration by parts we find that

$$\begin{aligned}
\dot{\mathscr{F}}(\mathcal{P}_t, \chi) = \int_{\mathcal{P}_t} \Bigg\{ & \operatorname{div}\left[(\nabla \boldsymbol{n})^{\mathsf{T}} \frac{\partial W}{\partial \nabla \boldsymbol{n}} + \nabla S \otimes \frac{\partial W}{\partial \nabla S} \right] \cdot \boldsymbol{v} \\
& + \left(\frac{\partial W}{\partial \boldsymbol{n}} - \operatorname{div} \frac{\partial W}{\partial \nabla \boldsymbol{n}} \right) \cdot \dot{\boldsymbol{n}} + \left(\frac{\partial W}{\partial S} - \operatorname{div} \frac{\partial W}{\partial \nabla S} \right) \cdot \dot{S} \Bigg\} dV \\
+ \int_{\partial^* \mathcal{P}_t} \Bigg\{ & \left[-(\nabla \boldsymbol{n})^{\mathsf{T}} \frac{\partial W}{\partial \nabla \boldsymbol{n}} \boldsymbol{\nu} - \nabla S \otimes \frac{\partial W}{\partial \nabla S} \boldsymbol{\nu} \right] \cdot \boldsymbol{v} \\
& + \left(\frac{\partial W}{\partial \nabla \boldsymbol{n}} \boldsymbol{\nu} \right) \cdot \dot{\boldsymbol{n}} + \left(\frac{\partial W}{\partial \nabla S} \cdot \boldsymbol{\nu} \right) \dot{S} \Bigg\} dA.
\end{aligned}$$

Variation of the Working

Requiring the variation $\delta \mathscr{W}$ of the total power to vanish shows that the equations for the director remain the same as before, while the balance of linear momentum and the associated traction condition hold with the stress tensor

$$\mathbf{T} = -p\mathbf{I} - (\nabla \boldsymbol{n})^{\mathsf{T}} \frac{\partial W}{\partial \nabla \boldsymbol{n}} - \nabla S \otimes \frac{\partial W}{\partial \nabla S}.$$

In addition, requiring the generalized forces multiplying the variations $\delta \dot{S}$ to vanish shows that the equation

$$\frac{\partial W}{\partial S} - \operatorname{div} \frac{\partial W}{\partial \nabla S} = L \tag{3.124}$$

must hold in $\mathcal{P}_t$ and that

$$\frac{\partial W}{\partial \nabla S} \cdot \boldsymbol{\nu} = K$$

on $\partial^* \mathcal{P}_t$.

3.2.2 Dissipative Dynamics

Since S is an indifferent scalar, its material time derivative $\dot{S}$ is frame-indifferent by (2.80), and we can construct the dissipation function as $R = R(\boldsymbol{n}, S; \mathbf{D}, \mathring{\boldsymbol{n}}, \dot{S})$, which is a quadratic form in $(\mathbf{D}, \mathring{\boldsymbol{n}}, \dot{S})$, where both $\boldsymbol{n}$ and S are regarded as parameters. The most general shape this can take that complies with the nematic symmetry (3.4) is

$$\begin{aligned} R(\boldsymbol{n}, S; \mathbf{D}, \mathring{\boldsymbol{n}}, \dot{S}) =& \beta_1 \dot{S}\, \boldsymbol{n} \cdot \mathbf{D}\boldsymbol{n} + \frac{1}{2}\beta_2 \dot{S}^2 + \frac{1}{2}\gamma_1 \mathring{\boldsymbol{n}}^2 + \gamma_2 \mathring{\boldsymbol{n}} \cdot \mathbf{D}\boldsymbol{n} \\ &+ \frac{1}{2}\gamma_3 (\mathbf{D}\boldsymbol{n})^2 + \frac{1}{2}\gamma_4 (\boldsymbol{n} \cdot \mathbf{D}\boldsymbol{n})^2 + \frac{1}{2}\gamma_5 \operatorname{tr} \mathbf{D}^2, \end{aligned} \tag{3.125}$$

where only two new viscosity coefficients β_1 and β_2 are needed to account for the dissipation due to changes in S. However, the viscosity coefficients are no longer constants, but all β's and γ's are functions of S subject to the requirement that R in (3.125) be positive semidefinite. Reasoning as to arrive at (3.57) above, we easily prove that the positive semidefiniteness of R is equivalent to the following list of inequalities:

$$\begin{aligned} \gamma_1 &\geqq 0, \\ \gamma_5 &\geqq 0, \\ \gamma_1(\gamma_3 + 2\gamma_5) - \gamma_2^2 &\geqq 0, \\ \beta_2 &\geqq 0, \\ \beta_2(2\gamma_4 + 2\gamma_3 + 3\gamma_5) - \beta_1^2 &\geqq 0, \end{aligned}$$

which were also obtained by ERICKSEN [99] by a different method.

It then follows that the equations of motion still hold in the form (3.24) and (3.25), where the stress instead of being given by (3.44) is now

$$\begin{aligned} \mathbf{T} = &- p\mathbf{I} - (\nabla \boldsymbol{n})^{\mathsf{T}} \frac{\partial W}{\partial \nabla \boldsymbol{n}} + \frac{1}{2}\left(\boldsymbol{n} \otimes \frac{\partial R}{\partial \mathring{\boldsymbol{n}}} - \frac{\partial R}{\partial \mathring{\boldsymbol{n}}} \otimes \boldsymbol{n}\right) + \frac{\partial R}{\partial \mathbf{D}} \\ &- \nabla S \otimes \frac{\partial W}{\partial \nabla S} + \beta_1 \dot{S}\, \overline{\boldsymbol{n} \otimes \boldsymbol{n}}\,. \end{aligned}$$

Furthermore, the additional equation (3.124) becomes

$$\frac{\partial W}{\partial S} - \operatorname{div} \frac{\partial W}{\partial \nabla S} + \frac{\partial R}{\partial \dot{S}} = L,$$

which with (3.125) is

$$\frac{\partial W}{\partial S} - \operatorname{div} \frac{\partial W}{\partial \nabla S} + \beta_1 \boldsymbol{n} \cdot \mathbf{D}\boldsymbol{n} + \beta_2 \dot{S} = L.$$

Apart from the microinertia, which we have neglected here, the evolution equations that we find are the same as ERICKSEN's [99], where again the extra PARODI-type relation he derived is automatically satisfied.

3.2.3 Rotational Momentum and Couple Stress

The invariance property (3.123) of the free energy here leads us to the equation

$$\left[\nabla S \otimes \frac{\partial W}{\partial \nabla S} + \frac{\partial W}{\partial \boldsymbol{n}} \otimes \boldsymbol{n} + \frac{\partial W}{\partial \nabla \boldsymbol{n}} (\nabla \boldsymbol{n})^{\mathsf{T}} + \left(\frac{\partial W}{\partial \nabla \boldsymbol{n}}\right)^{\mathsf{T}} (\nabla \boldsymbol{n})\right] \in \mathsf{Sym}(\mathcal{V}),$$

in analogy to (3.21). Performing exactly the same computations as in Section 3.1.3 for the standard director theory then shows that the balance of rotational momentum holds with body couple and couple stress that are formally precisely the same as in (3.72) and (3.73). No use is made in this derivation of the balance equation (3.124) for S, and the external actions on S do not contribute to the body couple. This shows that while once again the classical balance equation of rotational momentum is valid, other than in the case of the standard ERICKSEN–LESLIE theory, the continuum theory for a director with variable scalar order parameter cannot simply be reduced to the balance of rotational momentum.

3.3 Biaxial Nematics

As we have seen in Chapter 1, the average orientation of an ensemble of biaxial molecules can be represented by two order tensors **Q** and **B**, defined in (1.66) in terms of their microscopic counterparts **q** and **b**. If the simplifying assumption is made that these tensors share a common eigenframe $(\boldsymbol{n}, \boldsymbol{m}, \boldsymbol{l})$, they can be written in the form

$$\mathbf{Q} = S\left(\boldsymbol{n} \otimes \boldsymbol{n} - \frac{1}{3}\mathbf{I}\right) + T\,(\boldsymbol{m} \otimes \boldsymbol{m} - \boldsymbol{l} \otimes \boldsymbol{l}), \tag{3.126a}$$

$$\mathbf{B} = S'\left(\boldsymbol{n} \otimes \boldsymbol{n} - \frac{1}{3}\mathbf{I}\right) + T'\,(\boldsymbol{m} \otimes \boldsymbol{m} - \boldsymbol{l} \otimes \boldsymbol{l}) \tag{3.126b}$$

with four scalar order parameters S, T, S', and T'.

It is easy to confuse the microscopic origin of biaxiality and its macroscopic manifestion. Even experienced researchers have made the simplification of identifying the microscopic molecular axes with the macroscopic directors, for example in [43]. However, this is sensible and permissible only in special cases, such as for a perfectly oriented sample. In such a sample, the four order parameters have the values $T = S' = 0$ and $S = T' = 1$, and so **Q** and **B** take the form

$$\mathbf{Q} = \boldsymbol{n} \otimes \boldsymbol{n} - \frac{1}{3}\mathbf{I} \quad \text{and} \quad \mathbf{B} = \boldsymbol{m} \otimes \boldsymbol{m} - \boldsymbol{l} \otimes \boldsymbol{l}.$$

In this case, all the long molecular axes are aligned along the common direction $\boldsymbol{n}$ and all short molecular axes are aligned along $\boldsymbol{m}$. However, the more general case (3.126) can also be used to motivate a director theory. If all four order parameters are assumed to be constant, the orientation of the biaxial nematic can indeed be described by the triad of unit vectors $(\boldsymbol{n}, \boldsymbol{m}, \boldsymbol{l})$, where now no assumption is made about individual molecules.

The first theories for biaxial nematics were formulated in terms of three perpendicular directors [287, 165, 123]. While this approach has a certain appeal because of its symmetry with respect to the three directors, it ultimately leads to unnecessarily complicated equations. It also fails to account for the fact that the symmetry in the three directors is superficial: there is usually a dominant director that is the only one that survives in the transition from a biaxial to a uniaxial nematic phase.

A more transparent approach was presented in [186]: there the theory is phrased in terms of two unit vectors, the usual director $\boldsymbol{n}$ and a single secondary director $\boldsymbol{m}$ perpendicular to it.[39] In line with their interpretation as eigenvectors of $\boldsymbol{\epsilon}$, both directors have to obey the nematic symmetry, that is, one requires that both $\boldsymbol{n} \sim -\boldsymbol{n}$ and $\boldsymbol{m} \sim -\boldsymbol{m}$. In practice, this implies that scalar quantities like the elastic free energy have to be even expressions independently in both $\boldsymbol{n}$ and $\boldsymbol{m}$. The resulting equations for both directors are analogous to the one for the single director in the uniaxial case, and also many of the stress components are analogous to those found in the uniaxial case. We follow this approach and derive equations for two directors.

Two-Director Description of the Biaxial Phase

We want to phrase a continuum theory for a biaxial liquid crystal, assuming that the degree of orientational order is constant throughout. This means that we need to employ two orthogonal unit vectors, the directors $\boldsymbol{n}$ and $\boldsymbol{m}$. They satisfy the constraints

$$\boldsymbol{n} \cdot \boldsymbol{n} \equiv \boldsymbol{m} \cdot \boldsymbol{m} \equiv 1, \text{ and } \boldsymbol{n} \cdot \boldsymbol{m} \equiv 0. \tag{3.127}$$

We assume biaxial nematic symmetry of the material, that is,

$$\boldsymbol{n} \sim -\boldsymbol{n} \quad \text{and} \quad \boldsymbol{m} \sim -\boldsymbol{m}. \tag{3.128}$$

3.3.1 Nondissipative Dynamics

Here our development parallels that in Section 3.2.1: an extra order descriptor is added to $\boldsymbol{n}$, but instead of being the scalar S it is the other director $\boldsymbol{m}$.

External Agents

In the case of a biaxial nematic, external agents can expend power against both directors. Even though the directors themselves are not free to move independently, forces acting on them can in principle be completely independent. One could, for example, imagine molecules with two different axes of symmetry where each of the axes is susceptible to a different type of interaction with external fields. The macroscopic

[39] If desired, the third director $\boldsymbol{l}$ can always be recovered via the vector product $\boldsymbol{l} = \boldsymbol{n} \times \boldsymbol{m}$, but there is no need for $\boldsymbol{l}$ to enter explicitly the dynamical equations.

directors as averages of the molecular axes' orientations would in turn be susceptible to different external influences. The power of the external agents is thus given in the general form

$$\mathscr{W}^{(a)}(\mathcal{P}_t, \chi) = \int_{\mathcal{P}_t} (\boldsymbol{b} \cdot \boldsymbol{v} + \boldsymbol{k}_{\mathrm{n}} \cdot \dot{\boldsymbol{n}} + \boldsymbol{k}_{\mathrm{m}} \cdot \dot{\boldsymbol{m}})\, dV + \int_{\partial^* \mathcal{P}_t} (\boldsymbol{t} \cdot \boldsymbol{v} + \boldsymbol{c}_{\mathrm{n}} \cdot \dot{\boldsymbol{n}} + \boldsymbol{c}_{\mathrm{m}} \cdot \dot{\boldsymbol{m}})\, dA \tag{3.129}$$

with the body force $\boldsymbol{b}$ and traction $\boldsymbol{t}$, and generalized volume and surface forces $\boldsymbol{k}_{\mathrm{n}}$ and $\boldsymbol{c}_{\mathrm{n}}$ acting on $\boldsymbol{n}$ and $\boldsymbol{k}_{\mathrm{m}}$ and $\boldsymbol{c}_{\mathrm{m}}$ acting on $\boldsymbol{m}$.

Power of the Constraints

The power of the constraints has four parts. As usual, a pressure p serves as LAGRANGE multiplier for the incompressibilty constraint. Three further terms arise from the directors' constraints (3.127): they keep the directors at unit length and perpendicular to each other. We introduce three further LAGRANGE multiplier fields γ, τ, and κ that ensure that the power of the constraints, obtained by differentiating (3.127) with respect to time, vanishes. We obtain

$$\begin{aligned}
\mathscr{W}^{(c)}(\mathcal{P}_t, \chi) &= \int_{\mathcal{P}_t} [\gamma\, \boldsymbol{n} \cdot \dot{\boldsymbol{n}} + \tau\, \boldsymbol{m} \cdot \dot{\boldsymbol{m}} + \kappa\, (\boldsymbol{m} \cdot \dot{\boldsymbol{n}} + \boldsymbol{n} \cdot \dot{\boldsymbol{m}}) + p\, \mathrm{div}\, \boldsymbol{v}]\, dV \\
&= \int_{\mathcal{P}_t} [\gamma\, \boldsymbol{n} \cdot \dot{\boldsymbol{n}} + \tau\, \boldsymbol{m} \cdot \dot{\boldsymbol{m}} + \kappa\, (\boldsymbol{m} \cdot \dot{\boldsymbol{n}} + \boldsymbol{n} \cdot \dot{\boldsymbol{m}}) - \nabla p \cdot \boldsymbol{v}]\, dV \\
&\quad + \int_{\partial^* \mathcal{P}_t} p \boldsymbol{\nu} \cdot \boldsymbol{v}\, dA.
\end{aligned}$$

Kinetic Energy

While the kinetic energy in principle contains contributions related to $\dot{\boldsymbol{n}}$ and $\dot{\boldsymbol{m}}$, these can usually be considered negligible because of the small molecular inertia. Thus we use as before

$$\dot{\mathscr{K}}(\mathcal{P}_t, \chi) = \int_{\mathcal{P}_t} \varrho \dot{\boldsymbol{v}} \cdot \boldsymbol{v}\, dV.$$

Free Energy Density

We assume that the elastic free energy density is a function of the two directors and their first gradients, $W = W(\boldsymbol{n}, \nabla\boldsymbol{n}, \boldsymbol{m}, \nabla\boldsymbol{m})$. For such a function W to be compatible with the biaxial nematic symmetry (3.128), it has to satisfy

$$W(\boldsymbol{n}, \nabla\boldsymbol{n}, \boldsymbol{m}, \nabla\boldsymbol{m}) = W(-\boldsymbol{n}, -\nabla\boldsymbol{n}, \boldsymbol{m}, \nabla\boldsymbol{m}) = W(\boldsymbol{n}, \nabla\boldsymbol{n}, -\boldsymbol{m}, -\nabla\boldsymbol{m}).$$

Furthermore, in analogy to (3.12), W has to satisfy the invariance requirement

$$W(\boldsymbol{n}, \nabla\boldsymbol{n}, \boldsymbol{m}, \nabla\boldsymbol{m}) = W(\mathbf{R}\boldsymbol{n}, \mathbf{R}\nabla\boldsymbol{n}\mathbf{R}^{\mathsf{T}}, \mathbf{R}\boldsymbol{m}, \mathbf{R}\nabla\boldsymbol{m}\mathbf{R}^{\mathsf{T}}),$$

where $\mathbf{R}$ is an arbitrary orthogonal transformation, or, in the case of a chiral biaxial nematic, a proper orthogonal transformation. The general form of W is derived in [122], an equivalent expression also including surface terms is given in the appendix of [318].

The total free energy $\mathscr{F}$ then results as usual from the integral

$$\mathscr{F}(\mathcal{P}_t, \boldsymbol{\chi}) = \int_{\mathcal{P}_t} W(\boldsymbol{n}, \nabla\boldsymbol{n}, \boldsymbol{m}, \nabla\boldsymbol{m})\, dV.$$

Using exactly the same steps that led us to (3.17), we can cast its time rate $\dot{\mathscr{F}}$ in the form

$$\begin{aligned}
\dot{\mathscr{F}}(\mathcal{P}_t, \boldsymbol{\chi}) = \int_{\mathcal{P}_t} \Bigg\{ & \operatorname{div}\left[(\nabla\boldsymbol{n})^{\mathsf{T}}\frac{\partial W}{\partial \nabla\boldsymbol{n}} + (\nabla\boldsymbol{m})^{\mathsf{T}}\frac{\partial W}{\partial \nabla\boldsymbol{m}}\right]\cdot\boldsymbol{v} \\
& + \left(\frac{\partial W}{\partial \boldsymbol{n}} - \operatorname{div}\frac{\partial W}{\partial \nabla\boldsymbol{n}}\right)\cdot\dot{\boldsymbol{n}} + \left(\frac{\partial W}{\partial \boldsymbol{m}} - \operatorname{div}\frac{\partial W}{\partial \nabla\boldsymbol{m}}\right)\cdot\dot{\boldsymbol{m}}\Bigg\}\, dV \\
+ \int_{\partial^*\mathcal{P}_t} \Bigg\{ & \left[-(\nabla\boldsymbol{n})^{\mathsf{T}}\frac{\partial W}{\partial \nabla\boldsymbol{n}}\boldsymbol{\nu} + -(\nabla\boldsymbol{m})^{\mathsf{T}}\frac{\partial W}{\partial \nabla\boldsymbol{m}}\boldsymbol{\nu}\right]\cdot\boldsymbol{v} \\
& + \left(\frac{\partial W}{\partial \nabla\boldsymbol{n}}\boldsymbol{\nu}\right)\cdot\dot{\boldsymbol{n}} + \left(\frac{\partial W}{\partial \nabla\boldsymbol{m}}\boldsymbol{\nu}\right)\cdot\dot{\boldsymbol{m}}\Bigg\}\, dA.
\end{aligned}$$

Variation of the Working

The variation of the total power $\mathscr{W}$ now takes the form

$$\begin{aligned}
\delta\mathscr{W} = \int_{\mathcal{P}_t} \Bigg\{ & \left[\boldsymbol{b} - \nabla p - \rho\dot{\boldsymbol{v}} - \operatorname{div}\left((\nabla\boldsymbol{n})^{\mathsf{T}}\frac{\partial W}{\partial \nabla\boldsymbol{n}} + (\nabla\boldsymbol{m})^{\mathsf{T}}\frac{\partial W}{\partial \nabla\boldsymbol{m}}\right)\right]\cdot\delta\boldsymbol{v} \\
& + \left[\boldsymbol{k}_{\mathrm{n}} + \gamma\boldsymbol{n} + \kappa\boldsymbol{m} - \frac{\partial W}{\partial \boldsymbol{n}} + \operatorname{div}\frac{\partial W}{\partial \nabla\boldsymbol{n}}\right]\cdot\delta\dot{\boldsymbol{n}} \\
& + \left[\boldsymbol{k}_{\mathrm{m}} + \tau\boldsymbol{m} + \kappa\boldsymbol{n} - \frac{\partial W}{\partial \boldsymbol{m}} + \operatorname{div}\frac{\partial W}{\partial \nabla\boldsymbol{m}}\right]\cdot\delta\dot{\boldsymbol{m}}\Bigg\}\, dV \\
+ \int_{\partial^*\mathcal{P}_t} \Bigg\{ & \left[\boldsymbol{t} + \left(p\mathbf{I} + (\nabla\boldsymbol{n})^{\mathsf{T}}\frac{\partial W}{\partial \nabla\boldsymbol{n}} + (\nabla\boldsymbol{m})^{\mathsf{T}}\frac{\partial W}{\partial \nabla\boldsymbol{m}}\right)\boldsymbol{\nu}\right]\cdot\delta\boldsymbol{v} \\
& + \left[\boldsymbol{c}_{\mathrm{n}} - \frac{\partial W}{\partial \nabla\boldsymbol{n}}\boldsymbol{\nu}\right]\cdot\delta\dot{\boldsymbol{n}} + \left[\boldsymbol{c}_{\mathrm{m}} - \frac{\partial W}{\partial \nabla\boldsymbol{m}}\boldsymbol{\nu}\right]\cdot\delta\dot{\boldsymbol{m}}\Bigg\}\, dA.
\end{aligned}$$

The equations resulting from the requirement that this variation be identically zero are

$$\rho\dot{\boldsymbol{v}} = \boldsymbol{b} + \operatorname{div}\mathbf{T},$$

$$\frac{\partial W}{\partial \boldsymbol{n}} - \operatorname{div}\frac{\partial W}{\partial \nabla\boldsymbol{n}} - \boldsymbol{k}_{\mathrm{n}} = \gamma\boldsymbol{n} + \kappa\boldsymbol{m},$$

and

$$\frac{\partial W}{\partial \boldsymbol{m}} - \operatorname{div}\frac{\partial W}{\partial \nabla \boldsymbol{m}} - \boldsymbol{k}_{\mathrm{m}} = \tau \boldsymbol{m} + \kappa \boldsymbol{n}$$

in $\mathcal{P}_t$ and

$$\boldsymbol{t} = \mathbf{T}\boldsymbol{\nu},$$

$$\boldsymbol{c}_{\mathrm{n}} = \frac{\partial W}{\partial \nabla \boldsymbol{n}}\boldsymbol{\nu},$$

and

$$\boldsymbol{c}_{\mathrm{m}} = \frac{\partial W}{\partial \nabla \boldsymbol{m}}\boldsymbol{\nu},$$

on $\partial^* \mathcal{P}_t$. Here, the stress tensor is

$$\mathbf{T} = -p\,\mathbf{I} - (\nabla \boldsymbol{n})^{\mathsf{T}}\frac{\partial W}{\partial \nabla \boldsymbol{n}}, -(\nabla \boldsymbol{m})^{\mathsf{T}}\frac{\partial W}{\partial \nabla \boldsymbol{m}}.$$

The part of the stress that is analogous to the ERICKSEN stress (3.31) is

$$\mathbf{T}_{\mathrm{E}} = -(\nabla \boldsymbol{n})^{\mathsf{T}}\frac{\partial W}{\partial \nabla \boldsymbol{n}} - (\nabla \boldsymbol{m})^{\mathsf{T}}\frac{\partial W}{\partial \nabla \boldsymbol{m}}.$$

3.3.2 Dissipative Dynamics

According to our principle, the dissipation function depends on the stretching **D**, the directors $\boldsymbol{n}$ and $\boldsymbol{m}$, and indifferent time derivatives of the directors. We choose the corotational time derivatives

$$\mathring{\boldsymbol{n}} = \dot{\boldsymbol{n}} - \mathbf{W}\boldsymbol{n} \qquad \text{and} \qquad \mathring{\boldsymbol{m}} = \dot{\boldsymbol{m}} - \mathbf{W}\boldsymbol{m}.$$

The dissipation function needs to be a quadratic form in $(\mathbf{D}, \mathring{\boldsymbol{n}}, \mathring{\boldsymbol{m}})$; it is found by considering all scalar invariants in these three rates and the two directors. The biaxial nematic symmetry (3.128) requires that this function be even independently in both $(\boldsymbol{n}, \mathring{\boldsymbol{n}})$ and $(\boldsymbol{m}, \mathring{\boldsymbol{m}})$. The relevant invariants can be constructed, for example, from the table in [357]. However, the integrity basis given there is minimal only if all the vectors and tensors entering it are unrestrained and independent of one another, so care has to be taken to remove redundant terms.

We write the dissipation function as

$$\begin{aligned}
&R(\boldsymbol{n}, \boldsymbol{m}; \mathbf{D}, \mathring{\boldsymbol{n}}, \mathring{\boldsymbol{m}}) \\
&\quad = \frac{1}{2}\gamma_1 \mathring{\boldsymbol{n}}^2 + \gamma_2 \mathring{\boldsymbol{n}} \cdot \mathbf{D}\boldsymbol{n} + \frac{1}{2}\gamma_3 (\mathbf{D}\boldsymbol{n})^2 + \frac{1}{2}\gamma_4 (\boldsymbol{n} \cdot \mathbf{D}\boldsymbol{n})^2 + \frac{1}{2}\gamma_5 \operatorname{tr}\mathbf{D}^2 \\
&\quad + \frac{1}{2}\mu_1 \mathring{\boldsymbol{m}}^2 + \mu_2 \mathring{\boldsymbol{m}} \cdot \mathbf{D}\boldsymbol{m} + \frac{1}{2}\mu_3 (\mathbf{D}\boldsymbol{m})^2 + \frac{1}{2}\mu_4 (\boldsymbol{m} \cdot \mathbf{D}\boldsymbol{m})^2 \\
&\quad + \frac{1}{4}\lambda_1 \left[(\mathring{\boldsymbol{n}} \cdot \boldsymbol{m})^2 + (\mathring{\boldsymbol{m}} \cdot \boldsymbol{n})^2\right] + \frac{1}{2}\lambda_2 \left(\mathring{\boldsymbol{n}} \cdot \boldsymbol{m} - \mathring{\boldsymbol{m}} \cdot \boldsymbol{n}\right)(\boldsymbol{n} \cdot \mathbf{D}\boldsymbol{m}) \\
&\quad + \frac{1}{2}\lambda_3 (\boldsymbol{n} \cdot \mathbf{D}\boldsymbol{m})^2,
\end{aligned} \tag{3.130}$$

where the γ's are the viscosity coefficients already encountered in (3.49) for the single director theory, while the μ's and the λ's are additional viscosities introduced by the biaxial director theory.[40] The contributions to R in λ_1 and λ_2 are written in a symmetric fashion. Only one term is needed for each of these dissipation modes due to the identity $\mathring{\boldsymbol{n}} \cdot \boldsymbol{m} = -\mathring{\boldsymbol{m}} \cdot \boldsymbol{n}$, which is obtained by differentiating with respect to time $\boldsymbol{n} \cdot \boldsymbol{m} \equiv 0$. However, because we treat the constraints using LAGRANGE multipliers, the two directors should be considered independent in performing the variation, and hence both terms need to be retained. This also ensures that the two resulting director equations will have the same form; keeping only one of the two ultimately equivalent terms would destroy this formal symmetry.

Furthermore, a term proportional to $(\boldsymbol{n} \cdot \mathbf{D}\boldsymbol{n})(\boldsymbol{m} \cdot \mathbf{D}\boldsymbol{m})$ has been omitted because of the identity

$$\operatorname{tr}\mathbf{D}^2 = 2\left\{(\mathbf{D}\boldsymbol{n})^2 + (\mathbf{D}\boldsymbol{m})^2 + (\boldsymbol{m} \cdot \mathbf{D}\boldsymbol{m})(\boldsymbol{n} \cdot \mathbf{D}\boldsymbol{n}) - (\boldsymbol{m} \cdot \mathbf{D}\boldsymbol{n})^2\right\},$$

which holds for any traceless $\mathbf{D}$ and orthonormal vectors $\boldsymbol{n}$ and $\boldsymbol{m}$. Clearly, other terms could equivalently have been omitted. Our choice is the same as that made in [43].

Stress Tensor

In analogy to (3.45), the viscous stress takes the general form

$$\begin{aligned}\mathbf{T}_{\text{dis}} &= \frac{\partial R}{\partial \nabla \boldsymbol{v}} \\ &= \frac{\partial R}{\partial \mathbf{D}} + \frac{1}{2}\left(\boldsymbol{n} \otimes \frac{\partial R}{\partial \mathring{\boldsymbol{n}}} - \frac{\partial R}{\partial \mathring{\boldsymbol{n}}} \otimes \boldsymbol{n} + \boldsymbol{m} \otimes \frac{\partial R}{\partial \mathring{\boldsymbol{m}}} - \frac{\partial R}{\partial \mathring{\boldsymbol{m}}} \otimes \boldsymbol{m}\right).\end{aligned}$$

With $\mathscr{R}$ given by (3.130), the symmetric part of the viscous stress is

$$\begin{aligned}\overline{\mathbf{T}}_{\text{dis}} &= \frac{\partial R}{\partial \mathbf{D}} \\ &= \gamma_2\, \overline{\boldsymbol{n} \otimes \mathring{\boldsymbol{n}}} + \gamma_3\, \overline{\boldsymbol{n} \otimes \mathbf{D}\boldsymbol{n}} + \gamma_4 (\boldsymbol{n} \cdot \mathbf{D}\boldsymbol{n})\, \overline{\boldsymbol{n} \otimes \boldsymbol{n}} \\ &\quad + \mu_2\, \overline{\boldsymbol{m} \otimes \mathring{\boldsymbol{m}}} + \mu_3\, \overline{\boldsymbol{m} \otimes \mathbf{D}\boldsymbol{m}} + \mu_4 (\boldsymbol{m} \cdot \mathbf{D}\boldsymbol{m})\, \overline{\boldsymbol{m} \otimes \boldsymbol{m}} \\ &\quad + \left[\lambda_2 (\mathring{\boldsymbol{n}} \cdot \boldsymbol{m}) + \lambda_3 (\boldsymbol{n} \cdot \mathbf{D}\boldsymbol{m})\right] \overline{\boldsymbol{n} \otimes \boldsymbol{m}} + \lambda_4 \mathbf{D}.\end{aligned}$$

Furthermore, with

$$\frac{\partial R}{\partial \mathring{\boldsymbol{n}}} = \gamma_1 \mathring{\boldsymbol{n}} + \gamma_2 \mathbf{D}\boldsymbol{n} + \frac{1}{2}\left[\lambda_1 (\mathring{\boldsymbol{n}} \cdot \boldsymbol{m}) + \lambda_2 (\boldsymbol{n} \cdot \mathbf{D}\boldsymbol{m})\right] \boldsymbol{m} \tag{3.131}$$

---

[40] To our knowledge, a complete set of inequalities in all these 12 viscosities that would guarantee positive semidefiniteness of the dissipation function R in (3.130) as (3.57) guarantees positive semidefiniteness of the dissipation function in (3.49) has not yet been derived; some necessary conditions are mentioned in [43].

and

$$\frac{\partial R}{\partial \mathring{\boldsymbol{m}}} = \mu_1 \mathring{\boldsymbol{m}} + \mu_2 \mathbf{D}\boldsymbol{m} - \frac{1}{2}\left[\lambda_1(\mathring{\boldsymbol{n}} \cdot \boldsymbol{m}) + \lambda_2(\boldsymbol{n} \cdot \mathbf{D}\boldsymbol{m})\right]\boldsymbol{n} \tag{3.132}$$

we find the skew part of the viscous stress to be

$$\begin{aligned}
\operatorname{skw}(\mathbf{T}_{\text{dis}}) &= \operatorname{skw}\left(\boldsymbol{n} \otimes \frac{\partial R}{\partial \mathring{\boldsymbol{n}}}\right) + \operatorname{skw}\left(\boldsymbol{m} \otimes \frac{\partial R}{\partial \mathring{\boldsymbol{m}}}\right) \\
&= \gamma_1 \operatorname{skw}(\boldsymbol{n} \otimes \mathring{\boldsymbol{n}}) + \gamma_2 \operatorname{skw}(\boldsymbol{n} \otimes \mathbf{D}\boldsymbol{n}) \\
&\quad + \mu_1 \operatorname{skw}(\boldsymbol{m} \otimes \mathring{\boldsymbol{m}}) + \mu_2 \operatorname{skw}(\boldsymbol{m} \otimes \mathbf{D}\boldsymbol{m}) \\
&\quad + \left[\lambda_1(\mathring{\boldsymbol{n}} \cdot \boldsymbol{m}) + \lambda_2(\boldsymbol{n} \cdot \mathbf{D}\boldsymbol{m})\right] \operatorname{skw}(\boldsymbol{n} \otimes \boldsymbol{m}).
\end{aligned}$$

Director Evolution Equations

The equations for the two directors are formally given by

$$\frac{\partial R}{\partial \mathring{\boldsymbol{n}}} + \frac{\partial W}{\partial \boldsymbol{n}} - \operatorname{div} \frac{\partial W}{\partial \nabla \boldsymbol{n}} - \boldsymbol{k}_{\mathsf{n}} = \gamma \boldsymbol{n} + \kappa \boldsymbol{m} \tag{3.133}$$

and

$$\frac{\partial R}{\partial \mathring{\boldsymbol{m}}} + \frac{\partial W}{\partial \boldsymbol{m}} - \operatorname{div} \frac{\partial W}{\partial \nabla \boldsymbol{m}} - \boldsymbol{k}_{\mathsf{m}} = \tau \boldsymbol{m} + \kappa \boldsymbol{n}. \tag{3.134}$$

Equations (3.133) and (3.134) are basically those obtained in [186]. The only difference is that there the terms with λ_1 and λ_2 appear only in the equation for $\boldsymbol{n}$. However, this difference is merely formal, because it disappears once the LAGRANGE multipliers have been eliminated.

Formally, the LAGRANGE multipliers can be found by taking the scalar products of equations (3.133) and (3.134) with $\boldsymbol{n}$ and $\boldsymbol{m}$. The multipliers can then be inserted to find two scalar equations of motion: they turn out to be the $\boldsymbol{l}$-components of (3.133) and (3.134). A third equation arises from the requirement that the two different expressions obtained for the multiplier κ coincide.

An equivalent, more direct way of finding the equations of motion is by observing that the $\boldsymbol{l}$-components of the left sides of both (3.133) and (3.134) vanish, and that the $\boldsymbol{m}$-component of the left side of (3.133) equals the $\boldsymbol{n}$-component of the left-hand side of (3.134).

Using the abbreviations

$$\begin{aligned}
\boldsymbol{N} &:= \frac{\partial R}{\partial \mathring{\boldsymbol{n}}} + \frac{\partial W}{\partial \boldsymbol{n}} - \operatorname{div} \frac{\partial W}{\partial \nabla \boldsymbol{n}} - \boldsymbol{k}_{\mathsf{n}}, \\
\boldsymbol{M} &:= \frac{\partial R}{\partial \mathring{\boldsymbol{m}}} + \frac{\partial W}{\partial \boldsymbol{m}} - \operatorname{div} \frac{\partial W}{\partial \nabla \boldsymbol{m}} - \boldsymbol{k}_{\mathsf{m}},
\end{aligned}$$

the equations for the director orientation become

$$\boldsymbol{l} \cdot \boldsymbol{N} = 0, \tag{3.135a}$$

$$\boldsymbol{l} \cdot \boldsymbol{M} = 0, \tag{3.135b}$$

$$\boldsymbol{m} \cdot \boldsymbol{N} = \boldsymbol{n} \cdot \boldsymbol{M}. \tag{3.135c}$$

As can be seen from (3.131) and (3.132), this is a system of three equations that is implicit in the time derivatives of the directors $\boldsymbol{n}$ and $\boldsymbol{m}$. Explicit expressions for the time derivatives can be obtained by observing that

$$\mathring{\boldsymbol{n}} = \alpha \boldsymbol{l} + \beta \boldsymbol{m}, \tag{3.136a}$$

$$\mathring{\boldsymbol{m}} = \gamma \boldsymbol{l} - \beta \boldsymbol{n}, \tag{3.136b}$$

with some coefficients α, β, and γ. This is a consequence of the identities $\boldsymbol{n} \cdot \mathring{\boldsymbol{n}} = \boldsymbol{m} \cdot \mathring{\boldsymbol{m}} = 0$ and $\boldsymbol{n} \cdot \mathring{\boldsymbol{m}} = -\boldsymbol{m} \cdot \mathring{\boldsymbol{n}}$. Inserting (3.136a) and (3.136b) in (3.135a) to (3.135c) yields a system of three linear equations in the unknowns α, β, and γ. Its solution is[41]

$$\alpha = -\frac{\boldsymbol{l} \cdot \boldsymbol{A}}{\gamma_1},$$

$$\beta = \frac{\boldsymbol{n} \cdot \boldsymbol{B} - \boldsymbol{m} \cdot \boldsymbol{A}}{\gamma_1 + \mu_1 + \lambda_1},$$

$$\gamma = \frac{\boldsymbol{l} \cdot \boldsymbol{B}}{\mu_1}.$$

Here, we have introduced the abbreviations

$$\boldsymbol{A} := \frac{\partial W}{\partial \boldsymbol{n}} - \operatorname{div} \frac{\partial W}{\partial \nabla \boldsymbol{n}} - \boldsymbol{k}_{\mathrm{n}} + \gamma_2 \mathbf{D}\boldsymbol{n} + \frac{1}{2}\lambda_2 (\boldsymbol{n} \cdot \mathbf{D}\boldsymbol{m})\boldsymbol{m},$$

$$\boldsymbol{B} := \frac{\partial W}{\partial \boldsymbol{m}} - \operatorname{div} \frac{\partial W}{\partial \nabla \boldsymbol{m}} - \boldsymbol{k}_{\mathrm{m}} + \mu_2 \mathbf{D}\boldsymbol{m} - \frac{1}{2}\lambda_2 (\boldsymbol{n} \cdot \mathbf{D}\boldsymbol{m})\boldsymbol{n}.$$

Hence, we find that

$$\mathring{\boldsymbol{n}} = -\frac{\boldsymbol{l} \cdot \boldsymbol{A}}{\gamma_1}\boldsymbol{l} + \frac{\boldsymbol{n} \cdot \boldsymbol{B} - \boldsymbol{m} \cdot \boldsymbol{A}}{\gamma_1 + \mu_1 + \lambda_1}\boldsymbol{m},$$

$$\mathring{\boldsymbol{m}} = \frac{\boldsymbol{l} \cdot \boldsymbol{B}}{\mu_1}\boldsymbol{l} + \frac{\boldsymbol{m} \cdot \boldsymbol{A} - \boldsymbol{n} \cdot \boldsymbol{B}}{\gamma_1 + \mu_1 + \lambda_1}\boldsymbol{n}.$$

3.3.3 Rotational Momentum

Another form of the power has been used [186], and it is illuminating to compare the two different approaches. Because they ultimately stem from actions on different parts of a molecule, the generalized forces on the two directors are in principle independent. However, if one takes the naïve view that the director triad $(\boldsymbol{n}, \boldsymbol{m}, \boldsymbol{l})$ behaves as a rigid body, the external power $\mathscr{W}^{(\mathrm{a})}$ takes the form

$$\mathscr{W}^{(\mathrm{a})}(\mathcal{P}_t, \chi) = \int_{\mathcal{P}_t} (\boldsymbol{b} \cdot \boldsymbol{v} + \boldsymbol{k} \cdot \boldsymbol{w}_{\mathrm{n}}) dV + \int_{\partial^* \mathcal{P}_t} (\boldsymbol{t} \cdot \boldsymbol{v} + \boldsymbol{c} \cdot \boldsymbol{w}_{\mathrm{n}}) dA, \tag{3.137}$$

where $\boldsymbol{b}$ and $\boldsymbol{t}$ are body and surface forces, while $\boldsymbol{k}$ and $\boldsymbol{c}$ are body and surface moments, and $\boldsymbol{w}_{\mathrm{n}}$ is now the local rotational velocity of the director triad in the given frame.

[41] Under the assumption that $\gamma_1 \neq 0$, $\mu_1 \neq 0$, and $\gamma_1 + \mu_1 + \lambda_1 \neq 0$.

To see that (3.137) is just a special form of the more general power employed in (3.129), we note that if the directors are seen as a rigidly moving whole, then their time derivatives are given as functions of the rotational velocity $\boldsymbol{w}_{\mathsf{n}}$ by

$$\dot{\boldsymbol{n}} = \boldsymbol{w}_{\mathsf{n}} \times \boldsymbol{n} \quad \text{and} \quad \dot{\boldsymbol{m}} = \boldsymbol{w}_{\mathsf{n}} \times \boldsymbol{m}. \tag{3.138}$$

An elementary computation, using the orthonormality of the directors $\boldsymbol{n}$ and $\boldsymbol{m}$ along with (3.138), then shows that $\boldsymbol{w}_{\mathsf{n}}$ can in turn be expressed as a function of $\dot{\boldsymbol{n}}$ and $\dot{\boldsymbol{m}}$ as

$$\boldsymbol{w}_{\mathsf{n}} = \boldsymbol{n} \times \dot{\boldsymbol{n}} + \boldsymbol{m} \times \dot{\boldsymbol{m}} + (\dot{\boldsymbol{n}} \cdot \boldsymbol{m})\, \boldsymbol{m} \times \boldsymbol{n}.$$

Taking the scalar product with an arbitrary vector $\boldsymbol{l}$ yields

$$\begin{aligned} \boldsymbol{k} \cdot \boldsymbol{w}_{\mathsf{n}} &= \dot{\boldsymbol{n}} \cdot (\boldsymbol{k} \times \boldsymbol{n}) + \dot{\boldsymbol{m}} \cdot (\boldsymbol{k} \times \boldsymbol{m}) - (\dot{\boldsymbol{n}} \cdot \boldsymbol{m}) \left[\boldsymbol{m} \cdot (\boldsymbol{k} \times \boldsymbol{n})\right] \\ &= \dot{\boldsymbol{n}} \cdot \left(\mathbf{I} - \frac{1}{2}\boldsymbol{m} \otimes \boldsymbol{m}\right)(\boldsymbol{k} \times \boldsymbol{n}) + \dot{\boldsymbol{m}} \cdot \left(\mathbf{I} - \frac{1}{2}\boldsymbol{n} \otimes \boldsymbol{n}\right)(\boldsymbol{k} \times \boldsymbol{m}). \end{aligned} \tag{3.139}$$

To arrive at the second, symmetric, form of the above equation, we have used the identity

$$(\dot{\boldsymbol{n}} \cdot \boldsymbol{m}) \left[\boldsymbol{m} \cdot (\boldsymbol{k} \times \boldsymbol{n})\right] = (\dot{\boldsymbol{m}} \cdot \boldsymbol{n}) \left[\boldsymbol{n} \cdot (\boldsymbol{k} \times \boldsymbol{m})\right].$$

Equation (3.139) shows that a torque $\boldsymbol{k}$ on the material element can be interpreted in terms of generalized forces on the directors $\boldsymbol{n}$ and $\boldsymbol{m}$ if we set, for example,

$$\boldsymbol{k}_{\mathsf{n}} := \left(\mathbf{I} - \frac{1}{2}\boldsymbol{m} \otimes \boldsymbol{m}\right)(\boldsymbol{k} \times \boldsymbol{n})$$

and

$$\boldsymbol{k}_{\mathsf{m}} := \left(\mathbf{I} - \frac{1}{2}\boldsymbol{n} \otimes \boldsymbol{n}\right)(\boldsymbol{k} \times \boldsymbol{m}).$$

Corresponding relations can be used to define $\boldsymbol{c}_{\mathsf{n}}$ and $\boldsymbol{c}_{\mathsf{m}}$ in terms of the surface moment $\boldsymbol{c}$.

4

Order Tensor Theories

The history of order tensor theories is a long and winding one. Since DE GENNES introduced what he called the *tensor order parameter* in [57, 58] to phrase a LANDAU–GINZBURG-type theory for the nematic order, many steps have been taken toward a general continuum theory of nematics with tensorial order. Using standard methods of nonequilibrium thermodynamics, HESS [142, 143] and later OLMSTED and GOLDBART [254, 255] obtained constitutive theories for homogeneous alignments, later generalized by HESS and PARDOWITZ to include also spatial variations [145]. All these attempts were impaired by not yielding the full anisotropy of viscosities predicted by the ERICKSEN–LESLIE director theory. An extension using a codeformational model was proposed in [144], and while it recovered the complete anisotropy of viscosities, it failed to be otherwise fully consistent with the phenomenological ERICKSEN–LESLIE theory [267].

Perhaps more satisfactorily, QIAN and SHENG arrived in [280] at a system of evolution equations for both the velocity field and the order tensor that in the limit of uniaxial alignment reduces to the full director theory. However, their derivation is inspired by analogy with the balance of rotational momentum, which, as we have already learned in Section 3.2, is appropriate only for a director theory with fixed scalar order parameter, but which could be misleading for the more general tensorial order. Eventually, in [310] the dissipation principle of Section 2.2.3 was used to develop a general framework for a nematic described by a second-rank order tensor. This framework both unifies the older theories and provides an elegant way to phrase the general equations.

This chapter has two parts. In the first, we describe the general phenomenological theory for a single order tensor. We begin with the celebrated LANDAU–DE GENNES theory, which was originally conceived to describe a liquid crystal close to the nematic-to-isotropic phase transition. It employs a single order parameter tensor that can readily be related to a distribution of effectively uniaxial particles. We then develop a general continuum theory that describes the coupled evolution of flow and orientation.

This theory can, in principle, also be used to describe biaxial phases formed by uniaxial particles. However, there is as yet no experimental evidence of any nematic formed by such particles exhibiting phase biaxiality at equilibrium. As we have seen in Chapter 1, a biaxial phase formed by biaxial particles is most naturally described by a mean-field theory that employs two order tensors. We therefore devote the second part of this chapter to a phenomenological theory with two order tensors. As we shall see, this theory can be developed paralleling exactly the steps taken for the single-tensor theory, and it therefore offers surprisingly little technical difficulty. This emphasizes the power of the dissipation principle.

4.1 Uniaxial Nematics

The vast majority of nematic liquid crystals do not, at least in homogeneous equilibrium states, show any sign of biaxiality. This is consistent with the MAIER-SAUPE theory, which is based on a uniaxial interaction between molecules. As we have seen in Section 1.3, in all equilibrium states the order tensor $\mathbf{Q}$ is indeed uniaxial. This suggests that one regard the nematogenic molecules as effectively uniaxial and neglect any actual asymmetry and hence the related order tensor $\mathbf{B}$. Nevertheless, biaxiality can still occur, even in stationary states. When competing biaxial external agents such as electric or magnetic fields, boundary conditions, or a biaxial imposed flow break the uniaxial symmetry, the alignment can locally become biaxial [288, 7, 302, 207]. In chiral nematic liquid crystals, there can even be a small degree of biaxiality throughout a liquid crystal sample at equilibrium because of the interplay between chirality and biaxiality [275, 289, 76]. There is no spatially homogeneous uniaxial minimizer of the free energy. This state biaxiality is not a pure thermodynamical one but is brought about by the frustration of the liquid crystal that has to live in ordinary flat space when it could be at perfect ease only in a curved space [81].

All these manifestations of biaxiality, however, pertain exclusively to a biaxiality in the distribution of effectively uniaxial particles. If the particles are in fact biaxial, this might also encourage $\mathbf{B}$ to be different from zero. Describing effects of this type would, however, require a description of the liquid crystal order in terms of the two tensors, which we develop in Section 4.2 below. For now, we focus attention on a single tensor.

4.1.1 LANDAU–DE GENNES Free Energy

A phase transition is usually connected with a change in a certain type of order in the system. The idea behind the LANDAU theory of phase transitions is to write the free energy as a power series in an order parameter. This order parameter is a quantity that vanishes in one of the phases. In temperature-driven transitions, this phase is usually the one with higher symmetry. Upon decreasing the temperature, this symmetry is broken when the transition to the more ordered phase occurs.

Order Tensor

The application of LANDAU theory to the nematic-to-isotropic transition is due to DE GENNES [57, 58], who introduced what he called the tensor order parameter and what we call simply the *order tensor*. This is motivated by the optical properties of a liquid crystal. To find a suitable phenomenological order measure, we look again at (3.1), the relationship between the electric field $\boldsymbol{E}$ and the displacement vector $\boldsymbol{D}$:

$$\boldsymbol{D} = \boldsymbol{\epsilon} \boldsymbol{E} .$$

In the isotropic phase, $\boldsymbol{D}$ is simply parallel to $\boldsymbol{E}$, and so the dielectric tensor $\boldsymbol{\epsilon}$ is a multiple of the identity. By contrast, in the nematic phase the dielectric tensor is normally anisotropic. It can then be written in the form

$$\boldsymbol{\epsilon} = \frac{1}{3}(\operatorname{tr} \boldsymbol{\epsilon})\mathbf{I} + \overline{\boldsymbol{\epsilon}} \tag{4.1}$$

as the sum of two parts. The first, $\frac{1}{3}(\operatorname{tr} \boldsymbol{\epsilon})\mathbf{I}$, is the isotropic contribution, and the second, the symmetric, traceless tensor $\overline{\boldsymbol{\epsilon}} := \boldsymbol{\epsilon} - \frac{1}{3}(\operatorname{tr} \boldsymbol{\epsilon})\,\mathbf{I}$, is the deviation from isotropy. Because it vanishes in the isotropic phase, it can be used to introduce the desired order measure. We define the phenomenological order tensor $\mathbf{Q}^{\mathrm{p}}$ via

$$\mathbf{Q}^{\mathrm{p}} := \eta \, \overline{\boldsymbol{\epsilon}}, \tag{4.2}$$

where η is an arbitrary constant that can be chosen conveniently. By its definition, $\mathbf{Q}^{\mathrm{p}}$ is symmetric and traceless.

In Chapter 1 we had defined the order tensor $\mathbf{Q}$ in (1.66) as the ensemble average $\mathbf{Q} = \langle \mathbf{q} \rangle$ of the molecular tensor $\mathbf{q}$ defined in (1.7). We want to choose η so as to make the two definitions compatible. Recall that

$$\mathbf{q} = \overline{\boldsymbol{e}_1 \otimes \boldsymbol{e}_1} = \boldsymbol{e}_1 \otimes \boldsymbol{e}_1 - \frac{1}{3}\mathbf{I},$$

where $\boldsymbol{e}_1$ is a unit vector along the symmetry axis of the molecule. If $\epsilon_\parallel$ and $\epsilon_\perp$ denote the permittivities parallel and perpendicular to the symmetry axis, the permittivity tensor of a single molecule takes the form

$$\boldsymbol{\epsilon}^{\mathrm{m}} = \epsilon_\parallel \boldsymbol{e}_1 \otimes \boldsymbol{e}_1 + \epsilon_\perp (\boldsymbol{e}_2 \otimes \boldsymbol{e}_2 + \boldsymbol{e}_3 \otimes \boldsymbol{e}_3),$$

which can be written as

$$\boldsymbol{\epsilon}^{\mathrm{m}} = \bar{\epsilon}\,\mathbf{I} + \Delta\epsilon\,\mathbf{q}, \tag{4.3}$$

where, echoing (1.156), we have introduced the average permittivity

$$\bar{\epsilon} := \frac{1}{3} \operatorname{tr} \boldsymbol{\epsilon}^{\mathrm{m}} = \frac{1}{3}(\epsilon_\parallel + 2\epsilon_\perp)$$

and the dielectric anisotropy

$$\Delta\epsilon := \epsilon_\parallel - \epsilon_\perp. \tag{4.4}$$

Computing the effective permittivity tensor $\boldsymbol{\epsilon}$ as the ensemble average of the molecular permittivity tensor $\boldsymbol{\epsilon}^{\mathrm{m}}$ leads us to

$$\boldsymbol{\epsilon} = \langle \boldsymbol{\epsilon}^{\mathrm{m}} \rangle = \bar{\epsilon}\,\mathbf{I} + \Delta\epsilon \langle \mathbf{q} \rangle = \bar{\epsilon}\,\mathbf{I} + \Delta\epsilon\,\mathbf{Q}. \tag{4.5}$$

Comparing (4.5) with (4.1) shows that choosing $\eta = 1/\Delta\epsilon$ in (4.2) gives

$$\mathbf{Q} = \mathbf{Q}^{\mathrm{p}} = \frac{1}{\Delta\epsilon} \overset{\sqcap}{\boldsymbol{\epsilon}}.$$

From now on, we will use this definition, drop the superscript, and simply denote the order tensor by $\mathbf{Q}$.

By construction, $\mathbf{Q}$ and $\boldsymbol{\epsilon}$ have the same eigenvectors. Thus the single tensor $\mathbf{Q}$ describes both the degree of order and the orientation of the optical axes. Its eigenvalues, which add up to zero, are the scalar order parameters. When the latter are assumed to remain fixed, one can recover a director description such as those described in Sections 3.1 and 3.3. In this section, we begin by looking at the opposite case: we assume that the eigenframe of $\mathbf{Q}$ remains fixed, which leaves as variables two scalar order parameters for the description of the phase transition. We then allow the full tensor to vary and add an elastic contribution. Thus, we proceed to develop the full continuum theory for the order tensor in two stages, first without and then also including dissipation.

The free energy density proposed by DE GENNES has two contributions. The first is a LANDAU-type potential that is a polynomial in the invariants (1.83) of the order tensor $\mathbf{Q}$. The second is an elastic free-energy density in the form of a quadratic expression in the gradient of $\mathbf{Q}$. Because of these gradient terms, the resulting overall expression resembles the GINZBURG–LANDAU expression for the free-energy density of a superconductor, and the model is sometimes referred to as the LANDAU–GINZBURG–DE GENNES theory [194].

Bulk Free Energy

We want to write a polynomial in the invariants (1.83) of the order tensor. Because $\mathbf{Q}$ is traceless, its first invariant I_1 is zero, and only the two invariants $I_2 = \operatorname{tr}\mathbf{Q}^2$ and $I_3 = \operatorname{tr}\mathbf{Q}^3$ can be used in constructing a free energy. Up to fourth order, the most general form of the energy density is

$$U_{\mathrm{L}}(\mathbf{Q}) = U_0 + \frac{A}{2}\operatorname{tr}\mathbf{Q}^2 + \frac{B}{3}\operatorname{tr}\mathbf{Q}^3 + \frac{C}{4}\left(\operatorname{tr}\mathbf{Q}^2\right)^2 + O(\mathbf{Q}^5). \tag{4.6}$$

There is no linear term, and the constant term U_0 can be set equal to zero by an appropriate renormalization of the total free energy. In some older papers a further term of fourth order proportional to $\operatorname{tr}\mathbf{Q}^4$ appears, but this is redundant because the CAYLEY–HAMILTON theorem[1] guarantees that a second-order tensor has at most three independent invariants. In particular, its application to a traceless tensor shows

[1] See (A.4) in Appendix A.1 for a more general identity that implies this classical theorem.

that $2\,\mathrm{tr}\,\mathbf{Q}^4 = \left(\mathrm{tr}\,\mathbf{Q}^2\right)^2$. The fourth order is the lowest that is suitable for the description of a phase transition, because it allows the free-energy density to have two distinct minima. For the free energy to be bounded from below, the highest order of the expansion must be even. Higher orders than the fourth have been considered, and going to sixth order allows a stable biaxial phase to occur [59, 3]. We do not pursue this avenue here, because it is more natural to expect a biaxial phase to be formed by biaxial particles. These give rise to two molecular tensors and so to two order tensors, a case that is treated separately in Section 4.2 below.

The signs of the coefficients in (4.6) are as follows. For the energy to be bounded from below, C needs to be greater than zero. For a uniaxial state where the molecules are aligned on average along the nematic axis of symmetry, $\mathrm{tr}\,\mathbf{Q}^3 > 0$. By contrast, if the molecules are aligned on average perpendicular to the axis of symmetry, then $\mathrm{tr}\,\mathbf{Q}^3 < 0$. To be able to distinguish between those two states, we must have $B \neq 0$. For the free energy to favor the state of parallel alignment, $B < 0$ is required, which we henceforth assume. The coefficient A can have either sign. Normally, B and C are assumed to be temperature independent, in which case the value of A determines the phase transition. Usually, the additional assumption is made that, regarded as a function of temperature θ, A can be expanded in a TAYLOR series around a temperature θ^* close to the phase transition, and it is written in the form [59]

$$A(\theta) = A_0\left(\theta - \theta^*\right) \tag{4.7}$$

with a constant A_0. Here, θ^* marks the temperature at which $A = 0$ and, as we shall see below, at which the isotropic phase becomes unstable. Thus, it is the *supercooling* temperature of the isotropic phase at the isotropic-to-nematic phase transition.[2]

We do not use the specific form given in (4.7) but explore the properties of the LANDAU–DE GENNES potential in terms of a generic temperature-dependent function A. The first observation is that all minimizers of the free energy density (4.6) correspond to uniaxial states, which mirrors Theorem 1.9 of the MAIER–SAUPE theory.

Theorem 4.1. *If* $\mathbf{Q}$ *is a symmetric, traceless tensor that minimizes the free energy density in* (4.6) *with* $B \neq 0$*, then* $\mathbf{Q}$ *is either uniaxial or zero.*

Proof. While it is tempting to start from a representation of $\mathbf{Q}$ in the form (1.62) and perform directly a minimization with respect to the order parameters S and T, this is not as straightforward as it might appear. Apart from the obvious unixial states with $T = 0$, there are also uniaxial states with $T \neq 0$, as equations (1.92) show. We follow instead the indirect proof given in [212].

If the invariants (1.83) of $\mathbf{Q}$ are represented in terms of its eigenvalues as in (1.86), we have

$$\mathrm{tr}\,\mathbf{Q}^2 = \lambda_1^2 + \lambda_2^2 + \lambda_3^2 \quad \text{and} \quad \mathrm{tr}\,\mathbf{Q}^3 = \lambda_1^3 + \lambda_2^3 + \lambda_3^3,$$

[2] An alternative model for the temperature dependence of A that is sometimes used assumes that $A(\theta) = \tilde{A}_0\left(1 - \frac{\theta^*}{\theta}\right)$ with a constant $\tilde{A}_0$; see, for example, [142]. The special temperature θ^* plays essentially the same role in both models.

where the tracelessness of **Q** implies the constraint

$$\lambda_1 + \lambda_2 + \lambda_3 = 0. \tag{4.8}$$

The free energy (4.6) with $U_0 = 0$, when written as a function $\widetilde{U}_L$ of the eigenvalues of **Q**, thus takes the form

$$\widetilde{U}_L(\lambda_1, \lambda_2, \lambda_3) = \frac{A}{2}\sum_{i=1}^{3}\lambda_i^2 + \frac{B}{3}\sum_{i=1}^{3}\lambda_i^3 + \frac{C}{4}\left(\sum_{i=1}^{3}\lambda_i^2\right)^2.$$

Stationary points of $\widetilde{U}_L$ have to satisfy

$$A\lambda_i + B\lambda_i^2 + C\left(\sum_{k=1}^{3}\lambda_k^2\right)\lambda_i = \kappa, \quad i \in \{1,2,3\}, \tag{4.9}$$

where κ is a LAGRANGE multiplier that stems from the constraint (4.8). The system (4.9) can be rearranged in the form of the three equations

$$(\lambda_i - \lambda_j)\left[A + B(\lambda_i + \lambda_j) + C\sum_{k=1}^{3}\lambda_k^2\right] = 0, \quad 1 \leqq i < j \leqq 3. \tag{4.10}$$

Let us assume now that there is a solution $(\lambda_1, \lambda_2, \lambda_3)$ of this system with three distinct eigenvalues. Then the three expressions in the square brackets in (4.10) are all zero. Taking, for example, the difference of the two expressions with λ_1 then shows that

$$B(\lambda_2 - \lambda_3) = 0,$$

which, since $B \neq 0$, contradicts the assumption that $\lambda_2 \neq \lambda_3$. Hence any minimizer has at least two equal eigenvalues, that is, it is either uniaxial or zero.[3] □

Let now a uniaxial **Q** be given in the form

$$\mathbf{Q} = S\left(\boldsymbol{n} \otimes \boldsymbol{n} - \frac{1}{3}\mathbf{I}\right) \tag{4.11}$$

with a unit vector $\boldsymbol{n}$. An elementary computation shows that $\operatorname{tr}\mathbf{Q}^2 = \frac{2}{3}S^2$ and $\operatorname{tr}\mathbf{Q}^3 = \frac{2}{9}S^3$, so that the free energy $\widehat{U}_L$ as a function of S is

$$\widehat{U}_L(S) = \frac{A}{3}S^2 + \frac{2B}{27}S^3 + \frac{C}{9}S^4.$$

To find the minimizers, we set the derivative of $\widehat{U}_L$ equal to zero,

$$\widehat{U}_L'(S) = \frac{2A}{3}S + \frac{2B}{9}S^2 + \frac{4C}{9}S^3 = 0.$$

[3] We have actually proved that all stationary points of U_L in (4.6) other than **0** must be uniaxial tensors.

This equation has the three roots,

$$S_0 = 0 \quad \text{and} \quad S_\pm = \frac{-B \pm \sqrt{B^2 - 24AC}}{4C}.$$

The isotropic state with $S_0 = 0$ is always stationary. However, since

$$\widehat{U}_{\mathrm{L}}''(S) = \frac{2A}{3} + \frac{4B}{9}S + \frac{4C}{3}S^2,$$

and so $\widehat{U}_{\mathrm{L}}''(0) = \frac{2A}{3}$, S_0 is a minimizer of the free energy only when $A > 0 =: A^*$. Its free energy is $\widehat{U}_{\mathrm{L}}(S_0) = 0$. The other two solutions exist only when $A \leqq A^{**}$, where

$$A^{**} := \frac{B^2}{24C}.$$

In this case we find that always $S_+ > 0$, while $S_- < 0$ for $A < 0$ and $S_- > 0$ for $A > 0$. The former corresponds to an ordinary calamitic uniaxial state where the molecules are on average parallel to $\boldsymbol{n}$, the latter to a discotic uniaxial state where they tend to be perpendicular to $\boldsymbol{n}$; see also page 41 above. Their energies can be readily compared when written in the form [212]

$$\widehat{U}_{\mathrm{L}}(S_\pm) = \frac{S_\pm^2}{54}(9A + BS_\pm).$$

This shows that, with $B < 0$, the ordinary uniaxial state with S_+ always has the lower energy. The value A_c of A at the *clearing point*, where the nematic and the isotropic state both have the same (zero) energy, is given by

$$9A_c + BS_+ = 0,$$

which yields

$$A_c = \frac{B^2}{27C}.$$

To summarize, there are three crucial values of A. For $A < A^* = 0$ the isotropic phase is unstable. The value $A = A_c = \frac{B^2}{27C}$ marks the transition point at which the nematic and the isotropic phases have the same free energy. For $A > A^{**} = \frac{B^2}{24C}$ the nematic phase is unstable. Phase coexistence is possible for $A^* \leqq A \leqq A^{**}$. The isotropic phase is metastable for $A^* \leqq A < A_c$ and the nematic phase is metastable for $A_c < A \leqq A^{**}$. Furthermore, $A^{**} = \frac{9}{8}A_c$, so the temperature interval in which the isotropic phase is metastable is eight times bigger than that in which the nematic phase is metastable.

Elastic Free Energy

Up to second second order in $\mathbf{Q}$ and its gradient $\nabla\mathbf{Q}$ there are four scalar invariants:

$$Q_{ij,k}Q_{ij,k} = |\nabla \mathbf{Q}|^2, \tag{4.12a}$$
$$Q_{ij,j}Q_{ik,k} = |\operatorname{div} \mathbf{Q}|^2, \tag{4.12b}$$
$$Q_{ij,k}Q_{ik,j}, \tag{4.12c}$$
$$\epsilon_{ijk}Q_{jl,i}Q_{kl}. \tag{4.12d}$$

Under space inversion, the first three of these terms are invariant, while the last one changes sign. This latter is thus allowed only in the energy density of a chiral nematic liquid crystal.[4] Furthermore, the second and the third terms differ by only a null Lagrangian, as can be seen from the identity

$$q_{24} := Q_{ij,k}Q_{ik,j} - Q_{ij,j}Q_{ik,k} = (Q_{ij,k}Q_{ik} - Q_{ij}Q_{ik,k})_{,j} \tag{4.13}$$

and showing that the surface energy density $(Q_{ij,k}Q_{ik} - Q_{ij}Q_{ik,k})\nu_j$ to which (4.13) reduces upon integration over any subbody $\mathcal{P}$ depends only on the derivatives of $\mathbf{Q}$ along directions orthogonal to the normal $\boldsymbol{\nu}$ to $\partial^*\mathcal{P}$ [194]. Introducing a tensor-valued curl of a tensor via

$$\nabla \times \mathbf{Q} = \epsilon_{ikl}Q_{lj,k}\,\mathbf{e}_i \otimes \mathbf{e}_j,$$

the energy density can be written in compact form as[5]

$$W(\mathbf{Q}, \nabla\mathbf{Q}) = \frac{1}{2}L_1|\operatorname{div}\mathbf{Q}|^2 + \frac{1}{2}L_2\,|\nabla \times \mathbf{Q} + 2\tau_c\mathbf{Q}|^2 + L_{24}q_{24}, \tag{4.14}$$

where L_1, L_2, and L_{24} are elastic constants, while τ_c is related through (3.34) to the pitch p_c of the helix spontaneously induced in a chiral nematic. Inequalities akin to ERICKSEN's in (3.35) that guarantee that (4.14) is positive semidefinite are derived in [58, 194].[6]

In the purely (nonchiral) nematic case, for which $\tau_c = 0$, especially in view of numerical minimization of the total free energy $\mathscr{F}$ (see, for example [56, 114]), the elastic energy density W is often written in the following equivalent (indicial) form:

$$W = \frac{1}{2}L_1'Q_{ij,k}Q_{ij,k} + \frac{1}{2}L_2'Q_{ij,j}Q_{ik,k} + \frac{1}{2}L_3'Q_{ij,k}Q_{ik,j}. \tag{4.15}$$

It is easily seen by comparing (4.15) and (4.14) that the elastic constants L_1', L_2', and L_3' are related to L_1, L_2, and L_{24} through the relations

$$L_1' = L_2, \qquad L_2' = L_1 - 2L_{24}, \qquad L_3' = 2L_{24} - L_2.$$

The inequalities that guarantee that W in either (4.15) or (4.14) is positive semidefinite are expressed in terms of the elastic constants as follows:

---

[4] See the discussion of FRANK's formula (3.32) in Section 3.1.1.

[5] Cf. [59, p. 329].

[6] The elegant form of the energy given in [59] and reproduced here differs from that employed in [58, 194] by a reordering of terms and a different assignment of elastic constants.

$$L_1' \geqq 0, \qquad -L_1' \leqq L_3' \leqq 2L_1', \qquad 6L_1' + 10L_2' + L_3' \geqq 0,$$

$$-L_2 \leqq L_1 - 2L_{24} \leqq 2L_2, \qquad L_2 \geqq 0, \qquad 5(L_1 + L_2) - 8L_{24} \geqq 0.$$

In the chiral case when $\tau_c \neq 0$, the elastic energy written in the form (4.14) has a contribution $2L_2\tau_c^2 \operatorname{tr} \mathbf{Q}^2$. Thus, in view of (4.6), it is no longer the change of sign of A as in the pure bulk case, but the change of sign of $A + 4L_2\tau_c^2$ that triggers the phase transition.

Because there are only two independent constants in the elastic energy density (4.14), it is not possible to recover from it the full anisotropy of FRANK's elastic energy (3.32). The energy (4.14) does not distinguish between splay and bend deformations. It is possible to recover the full anisotropy by allowing energy terms that are quadratic in $\nabla\mathbf{Q}$ and linear in $\mathbf{Q}$. Though a single such term would suffice to recover the FRANK energy, there are seven of them, six for ordinary nematics and a further one for chiral nematics [15].

4.1.2 Nondissipative Dynamics

The total working of a fluid described by an order tensor theory is formally the same as that in a director theory; we write it as in (3.5) as

$$\mathscr{W} = \mathscr{W}^{(a)} + \mathscr{W}^{(c)} - \dot{\mathscr{K}} - \dot{\mathscr{F}},$$

where $\mathscr{W}^{(a)}$ is the power of the external agents, $\mathscr{W}^{(c)}$ the power of the constraints, $\mathscr{K}$ the kinetic energy, and $\mathscr{F}$ the free energy. We specify below the precise nature of these terms.

External Agents

As in the director case, the external power exerted on the nematic has two contributions. One is again the same power as that exerted by external forces acting on the material element of an isotropic fluid; see (2.261). In addition to this, power can also be expended on the order tensor, so that the total external power takes the form

$$\mathscr{W}^{(a)}(\mathcal{P}_t, \chi) = \int_{\mathcal{P}_t} \left(\boldsymbol{b} \cdot \boldsymbol{v} + \mathbf{K}^{\mathrm{Q}} \cdot \dot{\mathbf{Q}} \right) dV + \int_{\partial^*\mathcal{P}_t} \left(\boldsymbol{t} \cdot \boldsymbol{v} + \mathbf{C}^{\mathrm{Q}} \cdot \dot{\mathbf{Q}} \right) dA,$$

where $\mathbf{K}^{\mathrm{Q}}$ and $\mathbf{C}^{\mathrm{Q}}$ are generalized body and contact force densities acting on $\mathbf{Q}$.

Constraints

The order tensor is symmetric and traceless. In principle, these two constraints can be dealt with using appropriate LAGRANGE multipliers. However, it is simpler and more elegant to use intrinsic derivatives on the space of symmetric, traceless tensors. In practice, this means that we will regard any derivative taken with respect to a symmetric, traceless tensor or one of its derivatives as symmetric and traceless. Thus we will write, for example,

$$\frac{d\,\mathrm{tr}\,\mathbf{Q}^3}{d\mathbf{Q}} = 3\,\overline{\mathbf{Q}^2} = 3\mathbf{Q}^2 - (\mathrm{tr}\,\mathbf{Q}^2)\mathbf{I}.$$

Similarly, with an expression such as $\frac{dU_{\mathrm{L}}}{dQ_{ij}}$ we denote the components of the symmetric, traceless tensor $\frac{dU_{\mathrm{L}}}{d\mathbf{Q}}$ and *not* just the derivative of U_{L} with respect to the components of $\mathbf{Q}$.

We could have done systematically the same in Sections 3.1 and 3.2 on single-director theory, and this is indeed the way the theory is presented in [353, Chapter 3]. However, this does not extend easily to the two-director theory described in Section 3.3: the constraint that the two directors need to remain perpendicular to one another ultimately becomes a constraint coupling two equations. In this case, the use of LAGRANGE multipliers is both easier and more transparent because it allows us to treat the equations for the two directors independently and in a symmetric fashion.

In the case of two order parameter tensors in Section 4.2, no such complication arises. The two tensors are both symmetric and traceless, but independent of one another. Hence computing all derivatives intrinsically suffices.

The only constraint that we treat via its power is the one arising from incompressibility, so

$$\mathscr{W}^{(c)}(\mathcal{P}_t, \chi) = \int_{\mathcal{P}_t} p \,\mathrm{div}\, \boldsymbol{v}\, dV = -\int_{\mathcal{P}_t} \nabla p \cdot \boldsymbol{v}\, dV + \int_{\partial^* \mathcal{P}_t} p\boldsymbol{v} \cdot \boldsymbol{v}\, dA.$$

Kinetic Energy

As before, we neglect the kinetic energy connected with the orientation and have simply

$$\dot{\mathscr{K}}(\mathcal{P}_t, \chi) = \int_{\mathcal{P}_t} \rho \dot{\boldsymbol{v}} \cdot \boldsymbol{v}\, dV$$

for the rate of change of the kinetic energy.

Free Energy

The free energy has two contributions, the bulk or thermotropic energy given by the LANDAU–DE GENNES potential $U_{\mathrm{L}}(\mathbf{Q})$ as a function of $\mathbf{Q}$ and an elastic free energy density $W(\mathbf{Q}, \nabla\mathbf{Q})$ as a function of $\mathbf{Q}$ and its gradient $\nabla\mathbf{Q}$. Other than their dependence on these variables we here make no further assumptions on the specific form of either of these energies.

With the free energy in the form

$$\mathscr{F}(\mathcal{P}_t, \chi) = \int_{\mathcal{P}_t} (U_{\mathrm{L}} + W)\, dV$$

its time rate is simply

$$\dot{\mathscr{F}}(\mathcal{P}_t, \chi) = \int_{\mathcal{P}_t} \left(\dot{U}_{\mathrm{L}} + \dot{W} \right) dV,$$

where as usual the dot denotes the material time derivative. By the chain rule,

$$\dot{U}_{\mathrm{L}} = \frac{\partial U_{\mathrm{L}}}{\partial \mathbf{Q}} \cdot \dot{\mathbf{Q}}$$

and

$$\dot{W} = \frac{\partial W}{\partial \mathbf{Q}} \cdot \dot{\mathbf{Q}} + \frac{\partial W}{\partial \nabla \mathbf{Q}} \cdot (\nabla \mathbf{Q})^{\cdot}.$$

A computation analogous to that used to arrive at (3.16) here yields

$$(\nabla \mathbf{Q})^{\cdot} = \nabla \dot{\mathbf{Q}} - (\nabla \mathbf{Q}) \nabla \boldsymbol{v}, \tag{4.16}$$

where we have set

$$[(\nabla \mathbf{Q}) \nabla \boldsymbol{v}]_{ijk} = Q_{ij,l} v_{l,k}.$$

With this we find that

$$\begin{aligned} \dot{\mathscr{F}}(\mathcal{P}_t, \chi) &= \int_{\mathcal{P}_t} \left\{ \left(\frac{\partial U_{\mathrm{L}}}{\partial \mathbf{Q}} + \frac{\partial W}{\partial \mathbf{Q}} \right) \cdot \dot{\mathbf{Q}} + \frac{\partial W}{\partial \nabla \mathbf{Q}} \cdot (\nabla \mathbf{Q})^{\cdot} \right\} dV \\ &= \int_{\mathcal{P}_t} \left\{ \left(\frac{\partial U_{\mathrm{L}}}{\partial \mathbf{Q}} + \frac{\partial W}{\partial \mathbf{Q}} - \operatorname{div} \frac{\partial W}{\partial \nabla \mathbf{Q}} \right) \cdot \dot{\mathbf{Q}} \right. \\ &\qquad \left. + \left[\operatorname{div} \left(\nabla \mathbf{Q} \odot \frac{\partial W}{\partial \nabla \mathbf{Q}} \right) \right] \cdot \boldsymbol{v} \right\} dV \\ &\quad + \int_{\partial^{*} \mathcal{P}_t} \left\{ \left(\frac{\partial W}{\partial \nabla \mathbf{Q}} \boldsymbol{\nu} \right) \cdot \dot{\mathbf{Q}} - \left[\left(\nabla \mathbf{Q} \odot \frac{\partial W}{\partial \nabla \mathbf{Q}} \right) \boldsymbol{\nu} \right] \cdot \boldsymbol{v} \right\} dA, \end{aligned}$$

where we have performed integrations by parts to deal with the two terms from the right-hand side of (4.16) and have defined the second-rank tensor $\nabla \mathbf{Q} \odot \frac{\partial W}{\partial \nabla \mathbf{Q}}$ via its components in an arbitrary orthonormal positively oriented basis $(\boldsymbol{e}_1, \boldsymbol{e}_2, \boldsymbol{e}_3)$ as

$$\nabla \mathbf{Q} \odot \frac{\partial W}{\partial \nabla \mathbf{Q}} := Q_{kl,i} \frac{\partial W}{\partial Q_{kl,j}} \boldsymbol{e}_i \otimes \boldsymbol{e}_j. \tag{4.17}$$

Variation of the Working

With the individual power contributions to the working written as products of the velocities and generalized forces, the application of D'ALEMBERT's principle is again straightforward. We find the variation of the working to be

$$\delta\mathscr{W} = \int_{\mathcal{P}_t} \left\{ \left[\boldsymbol{b} - \nabla p - \rho\dot{\boldsymbol{v}} - \operatorname{div}\left(\nabla\mathbf{Q} \odot \frac{\partial W}{\partial \nabla\mathbf{Q}} \right) \right] \cdot \delta\boldsymbol{v} \right.$$
$$\left. + \left[\mathbf{K}^{\mathrm{Q}} - \frac{\partial W}{\partial \mathbf{Q}} + \operatorname{div} \frac{\partial W}{\partial \nabla\mathbf{Q}} \right] \cdot \delta\dot{\mathbf{Q}} \right\} dV$$
$$+ \int_{\partial^*\mathcal{P}_t} \left\{ \left[\boldsymbol{t} + \left(p\mathbf{I} + \nabla\mathbf{Q} \odot \frac{\partial W}{\partial \nabla\mathbf{Q}} \right) \boldsymbol{\nu} \right] \cdot \delta\boldsymbol{v} \right.$$
$$\left. + \left[\mathbf{C}^{\mathrm{Q}} - \frac{\partial W}{\partial \nabla\mathbf{Q}} \boldsymbol{\nu} \right] \cdot \delta\dot{\mathbf{Q}} \right\} dA. \tag{4.18}$$

For this to vanish identically, the integrands have to be zero. This yields the balance of linear momentum with corresponding traction condition in the classical form, (3.42) and (3.43), with the stress tensor

$$\mathbf{T} = -p\,\mathbf{I} - \nabla\mathbf{Q} \odot \frac{\partial W}{\partial \nabla\mathbf{Q}}.$$

The part of the stress that depends on the elastic free energy is

$$\mathbf{T}_{\mathrm{E}} := -\nabla\mathbf{Q} \odot \frac{\partial W}{\partial \nabla\mathbf{Q}}, \tag{4.19}$$

and it is analogous to the ERICKSEN stress (3.31) of the standard director theory.

The equations for the order tensor are

$$\frac{\partial W}{\partial \mathbf{Q}} - \operatorname{div} \frac{\partial W}{\partial \nabla\mathbf{Q}} - \mathbf{K}^{\mathrm{Q}} = \mathbf{0}$$

in $\mathcal{P}$ and

$$\mathbf{C}^{\mathrm{Q}} = \frac{\partial W}{\partial \nabla \boldsymbol{n}} \boldsymbol{\nu}$$

on $\partial^*\mathcal{P}$.

4.1.3 Dissipative Dynamics

The relevant quantities for constructing a frame-indifferent dissipation are the order tensor $\mathbf{Q}$, the symmetric part of the velocity gradient $\mathbf{D}$, and an invariant rate of change for the order tensor. The most general form of such a rate that also preserves the symmetry of the order tensor is the codeformational time derivative (2.89). While it has two parameters σ and τ that can be used as constitutive parameters [267], as remarked by ERICKSEN [99], choosing $\sigma \neq 0$ in general just amounts to a reordering of terms in the dissipation. The same is true for τ. In our formal development, we will thus simply use the corotational derivative (2.87) as the invariant rate $\mathring{\mathbf{Q}}$ of $\mathbf{Q}$. The RAYLEIGH dissipation density is thus a function $R(\mathbf{Q}; \mathbf{D}, \mathring{\mathbf{Q}})$ that is quadratic in the pair $(\mathbf{D}, \mathring{\mathbf{Q}})$ and depends on $\mathbf{Q}$ as a parameter. In the language of Section 2.2.3, the dissipation list d consists of the pair $(\mathbf{D}, \mathring{\mathbf{Q}})$.

Because we have chosen $\dot{\mathbf{Q}}$ and $\boldsymbol{v}$ as the measures of the rates against which the generalized forces expend power, to apply the variational principle (2.177) we need to compute the derivatives of the dissipation density R with respect to these quantities. By the chain rule, these are

$$\frac{\partial R}{\partial \dot{\mathbf{Q}}} = \frac{\partial R}{\partial \mathring{\mathbf{Q}}}$$

and

$$\frac{\partial R}{\partial \nabla \boldsymbol{v}} = \frac{\partial R}{\partial \mathbf{D}} + \mathbf{Q}\frac{\partial R}{\partial \mathring{\mathbf{Q}}} - \frac{\partial R}{\partial \mathring{\mathbf{Q}}}\mathbf{Q}.$$

The variation of the total dissipation then takes the form

$$\delta\mathscr{R} = \int_{\mathscr{P}_t}\left\{\frac{\partial R}{\partial \mathring{\mathbf{Q}}}\cdot\delta\dot{\mathbf{Q}} - \operatorname{div}\left(\mathbf{Q}\frac{\partial R}{\partial \mathring{\mathbf{Q}}} - \frac{\partial R}{\partial \mathring{\mathbf{Q}}}\mathbf{Q} + \frac{\partial R}{\partial \mathbf{D}}\right)\cdot\delta\boldsymbol{v}\right\}\mathrm{d}V$$
$$+\int_{\partial^*\mathscr{P}_t}\left\{\left(\mathbf{Q}\frac{\partial R}{\partial \mathring{\mathbf{Q}}} - \frac{\partial R}{\partial \mathring{\mathbf{Q}}}\mathbf{Q} + \frac{\partial R}{\partial \mathbf{D}}\right)\boldsymbol{\nu}\right\}\cdot\delta\boldsymbol{v}\,\mathrm{d}A. \tag{4.20}$$

Dynamic Equations

To obtain the dynamic equations we apply the dissipation principle (2.232) and require that

$$\delta\mathscr{R} = \delta\mathscr{W}.$$

With $\delta\mathscr{W}$ as in (4.18) and $\delta\mathscr{R}$ as in (4.20) this yields the momentum balance in its usual form with the stress

$$\mathbf{T} = -p\,\mathbf{I} - \nabla\mathbf{Q}\odot\frac{\partial W}{\partial \nabla\mathbf{Q}} + \mathbf{Q}\frac{\partial R}{\partial \mathring{\mathbf{Q}}} - \frac{\partial R}{\partial \mathring{\mathbf{Q}}}\mathbf{Q} + \frac{\partial R}{\partial \mathbf{D}}. \tag{4.21}$$

Apart from a contribution proportional to the identity and the ERICKSEN stress $\mathbf{T}_{\mathrm{E}}$ as in (4.19), it contains the viscous stress

$$\mathbf{T}_{\mathrm{dis}} := \mathbf{Q}\frac{\partial R}{\partial \mathring{\mathbf{Q}}} - \frac{\partial R}{\partial \mathring{\mathbf{Q}}}\mathbf{Q} + \frac{\partial R}{\partial \mathbf{D}}.$$

The equation for the order tensor in the bulk becomes

$$\frac{\partial R}{\partial \mathring{\mathbf{Q}}} + \frac{\partial W}{\partial \mathbf{Q}} - \operatorname{div}\frac{\partial W}{\partial \nabla\mathbf{Q}} - \mathbf{K}^{\mathrm{Q}} = \mathbf{0}, \tag{4.22}$$

and the condition on the boundary remains unchanged,

$$\mathbf{C}^{\mathrm{Q}} = \frac{\partial W}{\partial \nabla\mathbf{Q}}\boldsymbol{\nu}.$$

In general, the stress given by (4.21) contains a skew-symmetric part, at variance with the theory proposed by MACMILLAN [204], which provides a phenomenological setting for DOI's constitutive molecular theory [80], among others.

4.1.4 Specific Dissipation Functions

In contrast to in the case of the director theories, there is no limitation on the number of terms that can appear in the dissipation function for a nematic described by an order tensor $\mathbf{Q}$. While for the directors, because of their unit length, there is only a finite number of different scalars that can be constructed between them, their invariant time derivatives, and the stretching, here there is no limit on the number of times that $\mathbf{Q}$ itself can appear in the dissipation function. There are two different possible approaches that can be taken.

The tensor $\mathbf{Q}$, if defined as the average of the molecular tensor (1.7), has a bounded spectrum. Its eigenvalues lie in the interval $[-\frac{1}{3}, \frac{2}{3}]$, and in this sense it is a tensor with small norm. This motivates regarding the dissipation function as an expansion and limiting the order to which $\mathbf{Q}$ may occur in it.

A different approach can be taken that is reminiscent of the way we constructed the dissipation function (3.125) for the director theory with variable order parameter. In that case, while the number of terms remains finite, the viscosity coefficients are no longer constants but functions of the order parameter S. Similarly, the dissipation function for the order tensor can be written with a finite number of terms if the viscosity coefficients are allowed to be functions of the invariants I_2 and I_3 of the order tensor $\mathbf{Q}$.

For both ways of writing the dissipation function we need a list of all the scalar invariants that can be built with the three symmetric traceless tensors $\mathbf{Q}$, $\mathbf{D}$, and $\mathring{\mathbf{Q}}$.

Scalar Invariants

Lists of integrity bases for symmetric tensors can be found, for example, in [306] or [357]. Apart from the traces of the three tensors that are zero, we have omitted from the following list those invariants that are of order higher than two in the generalized velocities $\mathbf{D}$ and $\mathring{\mathbf{Q}}$.

1. Invariants of single tensors:

$$\operatorname{tr}\mathbf{Q}^2, \ \operatorname{tr}\mathbf{Q}^3, \ \operatorname{tr}\mathring{\mathbf{Q}}^2, \ \operatorname{tr}\mathbf{D}^2;$$

2. Invariants involving two tensors:

$$\begin{aligned}&\operatorname{tr}\mathbf{Q}\mathring{\mathbf{Q}}, \ \operatorname{tr}\mathbf{Q}\mathbf{D}, \ \operatorname{tr}\mathring{\mathbf{Q}}\mathbf{D},\\ &\operatorname{tr}\mathbf{Q}^2\mathring{\mathbf{Q}}, \ \operatorname{tr}\mathbf{Q}^2\mathbf{D}, \operatorname{tr}\mathring{\mathbf{Q}}^2\mathbf{Q}, \ \operatorname{tr}\mathbf{D}^2\mathbf{Q},\\ &\operatorname{tr}\mathbf{Q}^2\mathring{\mathbf{Q}}^2, \ \operatorname{tr}\mathbf{Q}^2\mathbf{D}^2;\end{aligned}$$

3. Invariants in all three tensors:

$$\operatorname{tr}\mathbf{Q}\mathring{\mathbf{Q}}\mathbf{D}, \ \operatorname{tr}\mathbf{Q}^2\mathring{\mathbf{Q}}\mathbf{D}.$$

Dissipation up to Second Order

From this list of invariants we choose all those terms that are at most of second order in $\mathbf{Q}$ *and* $\mathring{\mathbf{Q}}$. Using these terms and their combinations that are exactly of second order in $\mathbf{D}$ and $\mathring{\mathbf{Q}}$, we find that the general form of the dissipation in this limit is

$$\begin{aligned} R(\mathbf{Q};\mathbf{D},\mathring{\mathbf{Q}}) =& \frac{1}{2}\zeta_1 \operatorname{tr}\mathring{\mathbf{Q}}^2 + \zeta_2 \operatorname{tr}\mathbf{D}\mathring{\mathbf{Q}} + \zeta_3 \operatorname{tr}\mathbf{D}\mathring{\mathbf{Q}}\mathbf{Q} \\ &+ \zeta_4 \operatorname{tr}\mathbf{D}^2\mathbf{Q} + \frac{1}{2}\zeta_5 \operatorname{tr}\mathbf{D}^2\mathbf{Q}^2 + \frac{1}{2}\zeta_6(\operatorname{tr}\mathbf{D}\mathbf{Q})^2 \\ &+ \frac{1}{2}\zeta_7 \operatorname{tr}\mathbf{D}^2 \operatorname{tr}\mathbf{Q}^2 + \frac{1}{2}\zeta_8 \operatorname{tr}\mathbf{D}^2, \end{aligned} \tag{4.23}$$

where the ζ's are viscosity coefficients. Requiring the dissipation function in (4.23) to be positive semidefinite would result in a number of inequalities involving both the viscosity coefficients ζ_i and the order tensor $\mathbf{Q}$ that need to be satisfied for all admissible values of $\mathbf{Q}$. Sufficient consideration has not yet been given in the literature to the consequences of such inequalities.

We derived five algebraic inequalities that characterize all R in (4.23) that are *strictly*[7] positive definite, but they are too complicated and obscure to be reproduced in full here. The ones that can be printed in a single line are

$$\zeta_1 > 0 \tag{4.24a}$$

and

$$[\zeta_1(\zeta_5 + 6\zeta_7) - \zeta_3^2]S^2 + 6(\zeta_2\zeta_3 - \zeta_1\zeta_4)S + 9\zeta_1(\zeta_5 + 2\zeta_7)T^2 + 9(\zeta_1\zeta_8 - \zeta_2^2) > 0, \tag{4.24b}$$

where the pair of scalar order parameters (S, T) represents $\mathbf{Q}$ through equation (1.62). Inequality (4.24b) must hold for all (S, T) that satisfy the bounds in (1.95), which describe the admissible triangle in Figure 1.4. In particular, it follows from setting $S = T = 0$ in (4.24b) that

$$\zeta_1\zeta_8 - \zeta_2^2 > 0.$$

Other necessary conditions involving only the viscosities ζ_i can easily be obtained from (4.24b), but we shall not dwell any further on this issue, which is more technical than conceptual in character.

If terms of higher orders in $\mathbf{Q}$ are added to those in (4.23), the number of viscosity coefficients grows quickly, and the hope of determining them experimentally or obtaining inequalities restricting them diminishes correspondingly. On the other hand, it is possible to simplify (4.23) further. The dissipation function

---

[7] The quadratic form R in (4.23) is strictly positive definite whenever the symmetric matrix H associated to a representation of R akin to (3.53) has all positive eigenvalues. A criterion for the strict positive definiteness of H, which we seldom employ in this book, is recalled on page 263 below.

$$R(\mathbf{Q};\mathbf{D},\mathring{\mathbf{Q}}) = \frac{1}{2}\zeta_1 \operatorname{tr}\mathring{\mathbf{Q}}^2 + \zeta_2 \operatorname{tr}\mathbf{D}\mathring{\mathbf{Q}} + \zeta_4 \operatorname{tr}\mathbf{D}^2\mathbf{Q} + \frac{1}{2}\zeta_6(\operatorname{tr}\mathbf{D}\mathbf{Q})^2 + \frac{1}{2}\zeta_8 \operatorname{tr}\mathbf{D}^2 \quad (4.25)$$

has only five viscosity coefficients and leads to the model suggested by QIAN and SHENG [280], who in addition also consider an electric-field-induced MAXWELL stress.

While on first inspection the terms in (4.23) omitted in (4.25) appear to have been picked rather randomly, the particular choice made can be motivated by considering the order parameter dependence of the corresponding viscosity coefficients of the uniaxial limit. This is done by assuming the order tensor to remain uniaxial with constant scalar order parameter S and comparing coefficients of the dissipation function (4.23) in terms of a uniaxial order tensor and the dissipation function (3.49) in terms of a director. As shown in [310], if S remains constant in time, (4.11) implies that

$$\mathring{\mathbf{Q}} = S(\mathring{\boldsymbol{n}} \otimes \boldsymbol{n} + \boldsymbol{n} \otimes \mathring{\boldsymbol{n}}), \quad (4.26)$$

and inserting both (4.26) and (4.11) in (4.23), one concludes that the latter coincides with (3.49), provided that

$$\gamma_1 = 2\zeta_1 S^2, \quad (4.27a)$$

$$\gamma_2 = 2\zeta_2 S + \frac{1}{3}\zeta_3 S^2, \quad (4.27b)$$

$$\gamma_3 = \zeta_4 S + \frac{1}{2}\zeta_5 S^2, \quad (4.27c)$$

$$\gamma_4 = \zeta_6 S^2, \quad (4.27d)$$

$$\gamma_5 = \zeta_8 - \frac{1}{3}\zeta_4 S + \frac{1}{3}\left(\frac{1}{3}\zeta_5 + 2\zeta_7\right) S^2. \quad (4.27e)$$

Thus, setting ζ_3, ζ_5, and ζ_7 equal to zero merely amounts to neglecting corrections to the director viscosities that are of higher order in the scalar order parameter S.

Clearly, for R as in (4.25) to be positive semidefinite, the viscosities γ given by (4.27) as functions of S must satisfy inequalities (3.57) for all $S \in [-\frac{1}{2}, 1]$. Also, the simpler form of R in (4.25) encourages us to write the algebraic conditions ensuring that R is strictly positive definite for any $\mathbf{Q}$. Letting $\mathbf{Q}$ be represented as in (3.2), one finds that these conditions are given by (4.24a) and

$$\zeta_1(3\zeta_8 - 2\zeta_4 S) - 3\zeta_2^2 > 0,$$

$$\zeta_1[\zeta_4(S + 3T) + 3\zeta_8] - 3\zeta_2^2 > 0,$$

$$-4\zeta_1^2\zeta_4\zeta_6 S^3 + 2\zeta_1[\zeta_1(3\zeta_6\zeta_8 - \zeta_4^2) - 3\zeta_2^2\zeta_6]S^2 + 36\zeta_1^2\zeta_4\zeta_6 S T^2$$
$$+6\zeta_1[\zeta_1(\zeta_6\zeta_8 - 2\zeta_4^2) - 3\zeta_2^2\zeta_6]T^2 + 9(\zeta_2^2 - \zeta_1\zeta_8)^2 > 0,$$

$$\zeta_1\zeta_4(S - 3T) + 3(\zeta_1\zeta_2 - \zeta_2^2) > 0,$$

which must hold for all (S, T) subject to the bounds in (1.95).

General Dissipation Function

The most general dissipation function can be written as a sum of nineteen terms. These are products of viscosity coefficents that are regarded as functions of the invariants of $\mathbf{Q}$, namely, $\operatorname{tr}\mathbf{Q}^2$ and $\operatorname{tr}\mathbf{Q}^3$, and the following terms. The first nine terms are those in the list of invariants above that are already quadratic in the generalized velocities,

$$\operatorname{tr}\mathring{\mathbf{Q}}^2, \operatorname{tr}\mathbf{D}^2, \operatorname{tr}\mathring{\mathbf{Q}}\mathbf{D}, \operatorname{tr}\mathring{\mathbf{Q}}^2\mathbf{Q}, \operatorname{tr}\mathbf{D}^2\mathbf{Q}, \operatorname{tr}\mathbf{Q}^2\mathring{\mathbf{Q}}^2, \operatorname{tr}\mathbf{Q}^2\mathbf{D}^2, \operatorname{tr}\mathbf{Q}\mathring{\mathbf{Q}}\mathbf{D}, \operatorname{tr}\mathbf{Q}^2\mathring{\mathbf{Q}}\mathbf{D}.$$

A further ten terms are obtained as all possible products of two of the four terms in the set

$$\{\operatorname{tr}\mathbf{Q}\mathring{\mathbf{Q}}, \operatorname{tr}\mathbf{Q}\mathbf{D}, \operatorname{tr}\mathbf{Q}^2\mathring{\mathbf{Q}}, \operatorname{tr}\mathbf{Q}^2\mathbf{D}\}$$

of invariants that are linear in the velocities.

4.1.5 Rotational Momentum and Couple Stress

Whatever the specific form of the elastic energy density, for it to be objective it has to be invariant under rotations, which means that

$$W(\mathbf{Q}, \nabla\mathbf{Q}) = W(\mathbf{Q}^*, (\nabla\mathbf{Q})^*), \tag{4.28}$$

with

$$\mathbf{Q}^* = \mathbf{R}\mathbf{Q}\mathbf{R}^\mathsf{T} \quad \text{and} \quad (\nabla\mathbf{Q})^*_{ijk} = R_{ip}R_{jq}R_{kr}Q_{pq,r},$$

needs to hold for an arbitrary proper orthogonal transformation $\mathbf{R}$. Using the same reasoning that led us to (3.22) here shows that (4.28) implies

$$\epsilon_{ijk}\left(2Q_{jp}\frac{\partial W}{\partial Q_{pk}} + 2Q_{jp,q}\frac{\partial W}{\partial Q_{pk,q}} + Q_{pq,j}\frac{\partial W}{\partial Q_{pq,k}}\right) = 0. \tag{4.29}$$

Likewise, we compute the axial vector $\boldsymbol{\tau}$ associated with the skew-symmetric part of the stress tensor defined by (2.132). Starting from the stress tensor in (4.21), we find with (3.67) that

$$2\tau_i = \epsilon_{ijk}T_{kj} = \epsilon_{ijk}\left(2Q_{kl}\frac{\partial R}{\partial \mathring{Q}_{lj}} - Q_{lm,k}\frac{\partial W}{\partial Q_{lm,j}}\right).$$

By (4.29), this becomes

$$2\tau_i = 2\epsilon_{ijk}\left(Q_{kl}\frac{\partial R}{\partial \mathring{Q}_{lj}} + Q_{kl}\frac{\partial W}{\partial Q_{lj}} + Q_{kl,m}\frac{\partial W}{\partial Q_{lj,m}}\right),$$

and using the order balance (4.22) we find that

$$2\tau_i = 2\epsilon_{ijk}\left\{-Q_{kl}K_{lj} + Q_{kl}\left(\frac{\partial W}{\partial Q_{lj,m}}\right)_{,m} + Q_{kl,m}\frac{\partial W}{\partial Q_{lj,m}}\right\}$$
$$= 2\epsilon_{ijk}\left\{\left(Q_{kl}\frac{\partial W}{\partial Q_{lj,m}}\right)_{,m} - Q_{kl}K_{lj}\right\}.$$

The balance of rotational momentum (2.141) is indeed satisfied, provided we identify the torque per unit volume $\boldsymbol{k}$ and the couple stress tensor $\mathbf{L}$ as follows:[8]

$$\boldsymbol{k} = 2\epsilon_{ijk}Q_{kl}K_{lj}\,\boldsymbol{e}_i \tag{4.30}$$

and

$$\mathbf{L} = -2\epsilon_{ijk}Q_{kl}\frac{\partial W}{\partial Q_{lj,m}}\,\boldsymbol{e}_i \otimes \boldsymbol{e}_m,$$

where $(\boldsymbol{e}_1, \boldsymbol{e}_2, \boldsymbol{e}_3)$ is any orthonormal positively oriented basis.

It is clear from (2.140) that the couple stress depends only on the elastic energy. The absence of any viscous contribution to the couple stress is a consequence of $\nabla\dot{\mathbf{Q}}$ being excluded from the dissipation.[9]

To give a concrete example of a torque we consider an imposed electric field $\boldsymbol{E}$. This can be derived from a potential

$$W_{\mathrm{e}}(\mathbf{Q}) = -\frac{1}{2}\Delta\epsilon\boldsymbol{E}\cdot\mathbf{Q}\boldsymbol{E},$$

where $\Delta\epsilon$ is the dielectric anisotropy defined in (4.4). Hence a generalized force $\mathbf{K}_{\mathrm{e}}$ results from

$$\mathbf{K}_{\mathrm{e}} = \frac{\partial W_{\mathrm{e}}}{\partial\mathbf{Q}} = -\frac{1}{2}\Delta\epsilon\,\overline{\boldsymbol{E}\otimes\boldsymbol{E}},$$

and (4.30) implies that the electric torque per unit volume is then given by

$$\boldsymbol{k}_{\mathrm{e}} = \Delta\epsilon\boldsymbol{E}\times\mathbf{Q}\boldsymbol{E}.$$

Thus whenever the field $\boldsymbol{E}$ is an eigenvector of the order tensor $\mathbf{Q}$ it exerts no torque, and on average the molecules are either parallel or orthogonal to the field.

4.2 Biaxial Nematics

In this section we present a phenomenological theory for the description of a biaxial nematic liquid crystal in terms of two order tensors, $\mathbf{Q}$ and $\mathbf{B}$. These are the two tensors that arise naturally as ensemble averages of the molecular tensors $\mathbf{q}$ and $\mathbf{b}$ of STRALEY's interaction model for biaxial molecules, see the discussion in Section 1.2.6.

[8] The same conlusions as in (4.30) and (2.140) can also be reached by paralleling here the developments that in Section 3.1.4 led us to (3.84b).

[9] Cf. [185] for the analogous result in director theory.

The motivation for using two order tensors is primarily microscopic. At first glance, if only the macroscopic optical properties are considered, a single tensor appears to be sufficient to describe a nematic in its isotropic, uniaxial, and biaxial phases. It has indeed long been known that if the LANDAU–DE GENNES potential of a single order tensor is taken up to sixth order, it can have minima describing all three phases. However, from a microscopic point of view, it is more natural to expect the symmetry group of the phase to coincide with that of the molecules. In other words, it seems more likely to find a biaxial phase to be formed by biaxial rather than uniaxial molecules. Indeed, so far, all observed examples of biaxial phases stem from biaxial molecules [365, 205, 2, 298, 219].

After introducing the two order tensors and looking briefly at their macroscopic manifestations, we formulate the general theory in terms of such tensors first in terms of generic free energy and dissipation densities. We then briefly comment on the constitutive ingredients of the theory. Because not much is as yet known about the free energy and dissipation in their general forms, we have to restrict ourselves to giving an overview of the current state of play, pointing out what work still needs to be done to complete the theory.

As was the case for the director theories, where a second director was introduced to describe biaxiality, the advent of a new order tensor and its derivatives in the free energy and dissipation leads to a proliferation of new terms, which complicates matters considerably. This makes the need for simplifying assumptions even more pressing than in the uniaxial case. We therefore close the section by proposing what we regard as a minimal reduced model that can serve as a starting point for further explorations and practical applications.

4.2.1 Two-Tensor Theory

We have seen in Chapter 1 that the interaction of biaxial molecules can be conveniently described by the two symmetric traceless tensors

$$\mathbf{q} = \boldsymbol{e}_1 \otimes \boldsymbol{e}_1 - \frac{1}{3}\mathbf{I} \quad \text{and} \quad \mathbf{b} = \boldsymbol{e}_2 \otimes \boldsymbol{e}_2 - \boldsymbol{e}_3 \otimes \boldsymbol{e}_3, \tag{4.31}$$

where $(\boldsymbol{e}_1, \boldsymbol{e}_2, \boldsymbol{e}_3)$ is an orthonormal basis of eigenvectors of a molecular polarizability tensor. The ensemble averages of these molecular tensors are the order tensors

$$\mathbf{Q} = \langle \mathbf{q} \rangle \quad \text{and} \quad \mathbf{B} = \langle \mathbf{b} \rangle.$$

To see how these order tensors are related to the macroscopic properties of the liquid crystal, let us consider as in (4.3) a molecular permittivity tensor $\boldsymbol{\epsilon}^{\mathrm{m}}$. If ϵ_1, ϵ_2, and ϵ_3 are its eigenvalues, i.e., the permittivities in the directions of the respective eigenvectors, then we have

$$\boldsymbol{\epsilon}^{\mathrm{m}} = \epsilon_1 \boldsymbol{e}_1 \otimes \boldsymbol{e}_1 + \epsilon_2 \boldsymbol{e}_2 \otimes \boldsymbol{e}_2 + \epsilon_3 \boldsymbol{e}_3 \otimes \boldsymbol{e}_3.$$

This can be written in terms of the molecular tensors $\mathbf{q}$ and $\mathbf{b}$ as

$$\boldsymbol{\epsilon}^{\mathrm{m}} = \bar{\epsilon}\,\mathbf{I} + \Delta\epsilon\,\mathbf{q} + \frac{1}{2}\Delta\epsilon_{\perp}\,\mathbf{b},$$

where

$$\bar{\epsilon} := \frac{1}{3}\operatorname{tr}\boldsymbol{\epsilon}^{\mathrm{m}} = \frac{1}{3}(\epsilon_1 + \epsilon_2 + \epsilon_3)$$

is the average permittivity,

$$\Delta\epsilon := \epsilon_1 - \frac{1}{2}(\epsilon_2 + \epsilon_3)$$

is the anisotropy of the permittivity with respect to the main molecular axis $\boldsymbol{e}_1$, and

$$\Delta\epsilon_{\perp} := \epsilon_2 - \epsilon_3$$

is the transverse anistropy of the permittivity; see also (1.158). Computing the average shows that the macroscopic permittivity tensor $\boldsymbol{\epsilon}$ is given by

$$\boldsymbol{\epsilon} = \langle\boldsymbol{\epsilon}^{\mathrm{m}}\rangle = \bar{\epsilon}\,\mathbf{I} + \Delta\epsilon\,\mathbf{Q} + \frac{1}{2}\Delta\epsilon_{\perp}\,\mathbf{B} \tag{4.32}$$

as a function of $\mathbf{Q}$ and $\mathbf{B}$.

Equation (4.32) shows that in a biaxial phase of biaxial molecules both order tensors contribute to the macroscopic properties of the liquid crystal. While there is no simple way to reconstruct $\mathbf{Q}$ and $\mathbf{B}$ from the knowledge of $\boldsymbol{\epsilon}$, both play individual roles that can in principle also be distinguished experimentally, for example if $\Delta\epsilon$ and $\Delta\epsilon_{\perp}$ have different frequency dependencies. If either $\Delta\epsilon_{\perp}$ or $\mathbf{B}$ vanishes, the single order tensor case (4.5) is recovered.

4.2.2 Generic Dynamic Theory

The two tensors $\mathbf{Q}$ and $\mathbf{B}$, both symmetric and traceless, are in principle independent. The formal development of the theory is therefore very similar to that for the uniaxial nematic with a single order tensor as described in Section 4.1. Because with the material time derviative $\dot{\mathbf{B}}$ of $\mathbf{B}$ a third generalized velocity joins $\dot{\mathbf{Q}}$ and $\boldsymbol{v}$, there are now three bulk equations and three traction conditions. Owing to the identical tensorial nature of the two order tensors, the equations governing their dynamics are formally the same, and they are indeed carbon copies of the equation for $\mathbf{Q}$ in the uniaxial theory. It would be too repetitive and downright boring to give the full details of the derivation. We therefore content ourselves with a summary of the changes and additions that need to be made if a second tensor enters the stage and simply state the dynamic equations and the stress tensors.

External Agents

Because $\mathbf{Q}$ and $\mathbf{B}$ are independent, so are the generalized forces that could act on them. This is also apparent when we consider the microscopic origin of the order tensors: one is connected with the long molecular axis, and one with the plane perpendicular to it. In principle, different molecular axes can be susceptible to different agents that might act on them. Consequently, we write the power of the external agents as

$$\mathscr{W}^{(\mathrm{a})}(\mathcal{P}_t, \chi) = \int_{\mathcal{P}_t} \left(\boldsymbol{b} \cdot \boldsymbol{v} + \mathbf{K}^{\mathrm{Q}} \cdot \dot{\mathbf{Q}} + \mathbf{K}^{\mathrm{B}} \cdot \dot{\mathbf{B}} \right) dV + \int_{\partial^* \mathcal{P}_t} \left(\boldsymbol{t} \cdot \boldsymbol{v} + \mathbf{C}^{\mathrm{Q}} \cdot \dot{\mathbf{Q}} + \mathbf{C}^{\mathrm{B}} \cdot \dot{\mathbf{B}} \right) dA,$$

where $\mathbf{K}^{\mathrm{Q}}$ and $\mathbf{C}^{\mathrm{Q}}$ are generalized body and contact force densities acting on $\mathbf{Q}$, and $\mathbf{K}^{\mathrm{B}}$ and $\mathbf{C}^{\mathrm{B}}$ are generalized body and contact force densities acting on $\mathbf{B}$.

Free Energy

As in the case of a single tensor, the free energy has two contributions. There is a bulk or thermotropic energy density given by a LANDAU potential that here is a polynomial $U_{\mathrm{L}}(\mathbf{Q}, \mathbf{B})$ in both order tensors. Correspondingly, the elastic energy density is a function of both order tensors and their gradients, $W(\mathbf{Q}, \mathbf{B}, \nabla\mathbf{Q}, \nabla\mathbf{B})$. It is subject to the invariance requirement

$$W(\mathbf{Q}, \mathbf{B}, \nabla\mathbf{Q}, \nabla\mathbf{B}) = W(\mathbf{Q}^*, \mathbf{B}^*, (\nabla\mathbf{Q})^*, (\nabla\mathbf{B})^*),$$

where

$$\mathbf{Q}^* = \mathbf{R}\mathbf{Q}\mathbf{R}^{\top}, \qquad (\nabla\mathbf{Q})^*_{ijk} = R_{ip}R_{jq}R_{kr}Q_{pq,r},$$

$$\mathbf{B}^* = \mathbf{R}\mathbf{B}\mathbf{R}^{\top}, \quad \text{and} \quad (\nabla\mathbf{B})^*_{ijk} = R_{ip}R_{jq}R_{kr}B_{pq,r}$$

for an arbitrary rotation $\mathbf{R}$. In the usual way, it follows that

$$0 = \epsilon_{ijk} \left(2Q_{jp} \frac{\partial W}{\partial Q_{pk}} + 2Q_{jp,q} \frac{\partial W}{\partial Q_{pk,q}} + Q_{pq,j} \frac{\partial W}{\partial Q_{pq,k}} + 2B_{jp} \frac{\partial W}{\partial B_{pk}} + 2B_{jp,q} \frac{\partial W}{\partial B_{pk,q}} + B_{pq,j} \frac{\partial W}{\partial B_{pq,k}} \right).$$

Dissipation

We choose as an invariant rate of $\mathbf{B}$ its corotational derivative $\overset{\circ}{\mathbf{B}}$. The RAYLEIGH disspation density is then a function $R(\mathbf{Q}, \mathbf{B}; \mathbf{D}, \overset{\circ}{\mathbf{Q}}, \overset{\circ}{\mathbf{B}})$ quadratic in $(\mathbf{D}, \overset{\circ}{\mathbf{Q}}, \overset{\circ}{\mathbf{B}})$.

Dynamic Equations

The derivation of the evolution equations is precisely along the lines of that for a single order tensor. The variation of the additional velocity $\dot{\mathbf{B}}$ leads to a third equation of motion and to additional terms in the elastic and viscous stresses that are the double of the terms that stem from the variation of $\dot{\mathbf{Q}}$. We find the usual balance of linear momentum with the stress in the form

$$\mathbf{T} = -p\mathbf{I} + \mathbf{T}_{\mathrm{E}} + \mathbf{T}_{\mathrm{dis}}.$$

It has an elastic contribution $\mathbf{T}_{\mathrm{E}}$ given by

$$\mathbf{T}_{\mathrm{E}} := -\nabla\mathbf{Q} \odot \frac{\partial W}{\partial \nabla\mathbf{Q}} - \nabla\mathbf{B} \odot \frac{\partial W}{\partial \nabla\mathbf{B}},$$

where $\nabla\mathbf{B} \odot \frac{\partial W}{\partial \nabla\mathbf{B}}$ is defined in the same way as $\nabla\mathbf{Q} \odot \frac{\partial W}{\partial \nabla\mathbf{Q}}$ in equation (4.17). The viscous contribution to the stress is

$$\mathbf{T}_{\mathrm{dis}} := \mathbf{Q}\frac{\partial R}{\partial \overset{\circ}{\mathbf{Q}}} - \frac{\partial R}{\partial \overset{\circ}{\mathbf{Q}}}\mathbf{Q} + \mathbf{B}\frac{\partial R}{\partial \overset{\circ}{\mathbf{B}}} - \frac{\partial R}{\partial \overset{\circ}{\mathbf{B}}}\mathbf{B} + \frac{\partial R}{\partial \mathbf{D}}.$$

The equation governing the evolution of $\mathbf{Q}$ in the bulk is formally the same as (4.22),

$$\frac{\partial R}{\partial \overset{\circ}{\mathbf{Q}}} + \frac{\partial W}{\partial \mathbf{Q}} - \operatorname{div}\frac{\partial W}{\partial \nabla\mathbf{Q}} - \mathbf{K}^{\mathrm{Q}} = \mathbf{0}, \tag{4.33}$$

and the corresponding equation for $\mathbf{B}$ is

$$\frac{\partial R}{\partial \overset{\circ}{\mathbf{B}}} + \frac{\partial W}{\partial \mathbf{B}} - \operatorname{div}\frac{\partial W}{\partial \nabla\mathbf{B}} - \mathbf{K}^{\mathrm{B}} = \mathbf{0}. \tag{4.34}$$

The associated traction conditions on the reduced boundary are

$$\mathbf{C}^{\mathrm{Q}} = \frac{\partial W}{\partial \nabla\mathbf{Q}}\boldsymbol{\nu}$$

and

$$\mathbf{C}^{\mathrm{B}} = \frac{\partial W}{\partial \nabla\mathbf{B}}\boldsymbol{\nu}.$$

The couple stress is

$$\mathbf{L} = -2\epsilon_{ijk}\left(Q_{kl}\frac{\partial W}{\partial Q_{lj,m}} + B_{kl}\frac{\partial W}{\partial B_{lj,m}}\right)\boldsymbol{e}_i \otimes \boldsymbol{e}_m.$$

4.2.3 Constitutive Ingredients

The general theory presented in Section 4.2.2 springs to life only once the generic functions U_{L}, W, and R introduced above are made specific. We now show for each in turn the explicit form it can take.

LANDAU Potential

For two symmetric, traceless tensors in three-dimensional Euclidean space $\mathcal{E}$ there is a list $\mathcal{L}$ of eight basic scalars invariant under the full orthogonal group SO(3) [306]:

$$\mathcal{L} = \left\{\operatorname{tr}\mathbf{Q}^2, \operatorname{tr}\mathbf{B}^2, \operatorname{tr}\mathbf{Q}^3, \operatorname{tr}\mathbf{B}^3, \operatorname{tr}\mathbf{QB}, \operatorname{tr}\mathbf{Q}^2\mathbf{B}, \operatorname{tr}\mathbf{QB}^2, \operatorname{tr}\mathbf{Q}^2\mathbf{B}^2\right\}. \tag{4.35}$$

If $\mathbf{Q}$ and $\mathbf{B}$ have the same eigenframe, the number of basic invariants in $\mathcal{L}$ is reduced to seven, because in that case the identity

$$\operatorname{tr}\mathbf{B}^2 \operatorname{tr}\mathbf{Q}^2 = 6\operatorname{tr}(\mathbf{Q}^2\mathbf{B}^2) - 2(\operatorname{tr}\mathbf{QB})^2 \tag{4.36}$$

holds.[10]

The isotropic phase of a biaxial nematic is characterized by $\mathbf{Q} = \mathbf{B} = \mathbf{0}$. An appropriate LANDAU free-energy function is a rotationally invariant power-series expansion in the order tensors $\mathbf{Q}$ and $\mathbf{B}$. It can be constructed from the terms in (4.35). The lowest-order polynomial that allows the occurrence of isotropic, uniaxial, and biaxial nematic phases is of fourth order. The most general SO(3)-invariant polynomial up to fourth order in $\mathbf{Q}$ and $\mathbf{B}$ has fourteen terms. We write it in the form given to it in [65],

$$\begin{aligned} U_{\mathrm{L}}(\mathbf{Q},\mathbf{B}) =& a_1 \operatorname{tr}\mathbf{Q}^2 + a_2 \operatorname{tr}\mathbf{Q}^3 + a_3\left(\operatorname{tr}\mathbf{Q}^2\right)^2 + a_4 \operatorname{tr}\mathbf{B}^2 + a_5\left(\operatorname{tr}\mathbf{B}^2\right)^2 \\ &+ a_6 \operatorname{tr}\mathbf{QB}^2 + a_7 \operatorname{tr}\mathbf{Q}^2\mathbf{B}^2 + a_8 (\operatorname{tr}\mathbf{QB})^2 + a_9 \operatorname{tr}\mathbf{QB} + a_{10} \operatorname{tr}\mathbf{B}^3 \\ &+ a_{11} \operatorname{tr}\mathbf{Q}^2\mathbf{B} + a_{12} \operatorname{tr}\mathbf{Q}^3\mathbf{B} + a_{13} \operatorname{tr}\mathbf{QB}^3 + a_{14} \operatorname{tr}\mathbf{Q}^2 \operatorname{tr}\mathbf{B}^2, \end{aligned} \tag{4.37}$$

with phenomenological coefficients $a_1, \dots, a_{14}$. The terms $\operatorname{tr}\mathbf{Q}^3\mathbf{B}$ and $\operatorname{tr}\mathbf{B}^3\mathbf{Q}$ in (4.37) can be expressed as polynomial functions of the basic invariants:[11]

$$\operatorname{tr}\mathbf{Q}^3\mathbf{B} = \frac{1}{2}\operatorname{tr}\mathbf{Q}^2 \operatorname{tr}\mathbf{QB} \quad \text{and} \quad \operatorname{tr}\mathbf{B}^3\mathbf{Q} = \frac{1}{2}\operatorname{tr}\mathbf{B}^2 \operatorname{tr}\mathbf{QB}.$$

From (4.37) we easily recover the conventional single-tensor LANDAU–DE GENNES free energy (4.6) by setting $\mathbf{B} = \mathbf{0}$ and identifying a_1, a_2, and a_3 here with $\frac{1}{2}A$, $\frac{1}{3}B$, and $\frac{1}{4}C$ there.

Only a very limited analysis of the general potential (4.37) has so far been performed [65]. We will give a more detailed account of the properties of a simplified LANDAU potential below.

Elastic Energy

The most general elastic energy density quadratic in the two order tensors consists of the terms (4.12) in $\mathbf{Q}$ and $\nabla\mathbf{Q}$ and corresponding terms in $\mathbf{B}$ and $\nabla\mathbf{B}$,

---

10 The identity (4.36) follows from the generalized CAYLEY–HAMILTON theorem (A.4) and the fact that if $\mathbf{Q}$ and $\mathbf{B}$ have the same eigenframe, then they commute, so that $\mathbf{QB} = \mathbf{BQ}$, whence it follows that $\operatorname{tr}(\mathbf{Q}^2\mathbf{B}^2) = \operatorname{tr}(\mathbf{QB})^2$.

11 These identities are again consequences of (A.4).

$$B_{ij,k}\,B_{ij,k},$$
$$B_{ij,j}\,B_{ik,k},$$
$$B_{ij,k}\,B_{ik,j},$$
$$\epsilon_{ijk}\,B_{jl,i}\,B_{kl}.$$

In addition, there are a further five mixed terms,

$$Q_{ij,k}\,B_{ij,k},$$
$$Q_{ij,j}\,B_{ik,k},$$
$$Q_{ij,k}\,B_{ik,j},$$
$$\epsilon_{ijk}\,Q_{jl,i}\,B_{kl},$$
$$\epsilon_{ijk}\,B_{jl,i}\,Q_{kl}.$$

In analogy to (4.13), we can identify two new surface terms via

$$B_{ij,k}\,B_{ik,j} - B_{ij,j}\,B_{ik,k} = (B_{ij,k}\,B_{ik} - B_{ij}\,B_{ik,k})_{,j}$$

and

$$\begin{aligned} Q_{ij,k}\,B_{ik,j} - Q_{ij,j}\,B_{ik,k} &= (Q_{ij,k}\,B_{ik} - B_{ij}\,Q_{ik,k})_{,j} \\ &= (B_{ij,k}\,Q_{ik} - Q_{ij}\,B_{ik,k})_{,j}. \end{aligned}$$

Omitting the surface terms, the energy can be written as

$$\begin{aligned} W(\mathbf{Q}, \mathbf{B}, \nabla\mathbf{Q}, \nabla\mathbf{B}) = &\frac{1}{2}L_1|\operatorname{div}\mathbf{Q}|^2 + \frac{1}{2}L_2\,|\nabla\times\mathbf{Q} + 2(q_0\mathbf{Q} + q_1\mathbf{B})|^2 \\ &+\frac{1}{2}M_1|\operatorname{div}\mathbf{B}|^2 + \frac{1}{2}M_2\,|\nabla\times\mathbf{B} + 2(q_2\mathbf{Q} + q_3\mathbf{B})|^2 \\ &+N_1\operatorname{div}\mathbf{Q}\cdot\operatorname{div}\mathbf{B} + N_2\nabla\mathbf{Q}\cdot\nabla\mathbf{B}, \end{aligned}$$

where the dot in last term denotes the inner product between the two third-rank tensors defined by

$$\nabla\mathbf{Q}\cdot\nabla\mathbf{B} = Q_{ij,k}\,B_{ij,k}. \tag{4.38}$$

So far, neither have identities between these terms been ruled out nor has a general set of inequalities enforcing positive semidefiniteness of the elastic energy density been established.

Dissipation

If terms up to quadratic in the scalar order parameters and their time derivatives are taken into account, there are already 21 terms in the viscous dissipation. The RAYLEIGH function comprises the eight terms already listed in (4.23) for a single order tensor, seven parallel expressions relating to **B**, and a further six mixed terms featuring both order tensors. We write it as

$$
\begin{aligned}
&R(\mathbf{Q},\mathbf{B};\mathbf{D},\mathring{\mathbf{Q}},\mathring{\mathbf{B}})\\
&=\frac{1}{2}\zeta_1 \operatorname{tr}\mathring{\mathbf{Q}}^2+\zeta_2 \operatorname{tr}\mathbf{D}\mathring{\mathbf{Q}}+\zeta_3 \operatorname{tr}\mathbf{D}\mathring{\mathbf{Q}}\mathbf{Q}+\zeta_4 \operatorname{tr}\mathbf{D}^2\mathbf{Q}\\
&+\frac{1}{2}\zeta_5 \operatorname{tr}\mathbf{D}^2\mathbf{Q}^2+\frac{1}{2}\zeta_6(\operatorname{tr}\mathbf{D}\mathbf{Q})^2+\frac{1}{2}\zeta_7 \operatorname{tr}\mathbf{D}^2 \operatorname{tr}\mathbf{Q}^2+\frac{1}{2}\zeta_8 \operatorname{tr}\mathbf{D}^2\\
&+\frac{1}{2}\eta_1 \operatorname{tr}\mathring{\mathbf{B}}^2+\eta_2 \operatorname{tr}\mathbf{D}\mathring{\mathbf{B}}+\eta_3 \operatorname{tr}\mathbf{D}\mathring{\mathbf{B}}\mathbf{B}+\eta_4 \operatorname{tr}\mathbf{D}^2\mathbf{B}\\
&+\frac{1}{2}\eta_5 \operatorname{tr}\mathbf{D}^2\mathbf{B}^2+\frac{1}{2}\eta_6(\operatorname{tr}\mathbf{D}\mathbf{B})^2+\frac{1}{2}\eta_7 \operatorname{tr}\mathbf{D}^2 \operatorname{tr}\mathbf{B}^2\\
&+\xi_1 \operatorname{tr}\mathring{\mathbf{Q}}\mathring{\mathbf{B}}+\xi_2 \operatorname{tr}\mathbf{D}\mathring{\mathbf{Q}}\mathbf{B}+\xi_3 \operatorname{tr}\mathbf{D}\mathring{\mathbf{B}}\mathbf{Q}\\
&+\frac{1}{2}\xi_4 \operatorname{tr}\mathbf{D}^2\mathbf{Q}\mathbf{B}+\frac{1}{2}\xi_5(\operatorname{tr}\mathbf{D}\mathbf{Q})(\operatorname{tr}\mathbf{D}\mathbf{B})+\frac{1}{2}\xi_6 \operatorname{tr}\mathbf{D}^2 \operatorname{tr}\mathbf{Q}\mathbf{B}. \qquad (4.39)
\end{aligned}
$$

The new viscosity coefficients η_i pertain to $\mathbf{B}$ and the coefficients ξ_i to mixed terms in the two order tensors. This function needs to be positive semi-definite. Deriving neccessary and sufficient conditions on the viscosity coefficients for positive semi-definiteness of the dissipation density seems a formidable task.[12]

We derive below a set of inequalities (4.46) for a simplified model. In the general case, these inequalities are at least necessary conditions.

4.2.4 Simplified Models

We now consider plausible simplifications of the LANDAU potential, the elastic energy density, and the dissipation function. In principle, these are independent of one another, and our general format allows one to use as its constitutive ingredients any combination of the simplified and the more complex forms of these scalar functions. We conclude by giving the explicit form of the stress tensor and the dynamic equations for a simple model that might serve as a starting point for explorations into the realm of biaxial nematics described by two order tensors.

Symmetric LANDAU Potential

The general LANDAU potential (4.37) of fourth order consists of fourteen invariant terms. Here we describe a model that enjoys an additional symmetry property. We require the LANDAU potential to be invariant under the transformation $(\mathbf{Q},\mathbf{B}) \mapsto (\mathbf{Q},-\mathbf{B})$, so that (4.37) reduces to

$$
\begin{aligned}
U_{\mathrm{L}}(\mathbf{Q},\mathbf{B}) =& a_1 \operatorname{tr}\mathbf{Q}^2+a_2 \operatorname{tr}\mathbf{Q}^3+a_3\left(\operatorname{tr}\mathbf{Q}^2\right)^2+a_4 \operatorname{tr}\mathbf{B}^2+a_5\left(\operatorname{tr}\mathbf{B}^2\right)^2\\
&+a_6 \operatorname{tr}\mathbf{Q}\mathbf{B}^2+a_7 \operatorname{tr}\mathbf{Q}^2\mathbf{B}^2+a_8(\operatorname{tr}\mathbf{Q}\mathbf{B})^2\\
&+a_{14} \operatorname{tr}\mathbf{Q}^2 \operatorname{tr}\mathbf{B}^2. \qquad (4.40)
\end{aligned}
$$

12 Even establishing strict positive definiteness, usually an easier task, would be challenging.

Molecular Motivation

To motivate this choice of symmetry in the LANDAU potential, we need to digress from our line of purely phenomenological reasoning in this chapter and revert to an argument based on the molecular theories expounded in Chapter 1. A symmetry corresponding to that enforced in (4.40) has been proposed and its consequences examined in the pair-potential (1.10) of STRALEY's general biaxial molecular interaction [321]. To facilitate our discussion, we recall (1.10),

$$\widehat{H}_{\mathrm{b}} := -U_0\big[\mathbf{q}\cdot\mathbf{q}' + \gamma(\mathbf{q}\cdot\mathbf{b}' + \mathbf{b}\cdot\mathbf{q}') + \lambda\mathbf{b}\cdot\mathbf{b}'\big], \tag{4.41}$$

which specifies the pair-potential in terms of a typical interaction energy U_0 and two parameters γ and λ. Here $\mathbf{q}$ and $\mathbf{b}$ are the molecular tensors (4.31) of one molecule, and their primed versions are the corresponding tensors of a second molecule interacting with the first.

Setting $\gamma = 0$ in (4.41) makes the pair-potential invariant under the simultaneous changes $\mathbf{b} \mapsto -\mathbf{b}$ and $\mathbf{b}' \mapsto -\mathbf{b}'$. This specific choice of γ was introduced and its consequences explored in [315]: it models a nematogenic compound with a simple frequency-dependent biaxiality for which, for example, the low-frequency part would be purely uniaxial and the high-frequency part would add the required biaxiality. This amounts to the least possible coupling between uniaxial and biaxial susceptibilities. The mean-field phase diagram of this model exhibits first-order transition lines between the isotropic and nematic phases and both first- and second-order transition lines between the uniaxial and the biaxial nematic phase, meeting at a tricritical point. Furthermore, as shown in [24], there is a second tricritical point on the transition line between the biaxial and the isotropic phases at which the first-order transition becomes of second order. It was also shown in [24] that the topology of the phase diagram in the symmetric case with $\gamma = 0$ is the same as the topology of the universal phase diagram derived within the mean-field approximation of the general pair-potential (4.41).

This is not simply a coincidence, because there are indeed physical reasons to believe that the symmetric case embodies the essence of the general quadrupolar biaxial interaction. In the classical uniaxial interaction, only the long molecular axes interact with one another. If there is any type of interaction between them and the two short axes, a good first approximation would be to assume that this interaction is effectively the same for both short axes, which requires $\gamma = 0$, as exchanging the short axes in both molecules amounts to reversing the sign of both $\mathbf{b}$ and $\mathbf{b}'$ in (4.41). Since there the term multiplying λ aligns the short axes independently of the long axes, setting $\gamma = 0$ in (4.41) simply amounts to neglecting a small difference between two dominant interactions, likely with no qualitative consequence on the equilibrium phases. As shown in [315], even when $\gamma = 0$, the ground state of the pair-potential (4.41) is the one in which the interacting molecules are parallel with both long and short axes aligned, provided that $\lambda \geqq 0$. This discussion suggests that U_{L} in its symmetric form (4.40) should be sufficiently general to reproduce the universal phase diagram of biaxial nematics.

Symmetric Minimizers

It is natural to expect that in the absence of any external symmetry-breaking agent the minimizers of both the general LANDAU potential (4.37) and the symmetric one (4.40) would be pairs of order tensors $(\mathbf{Q},\mathbf{B})$ that share a common eigenframe. As shown in [65], any such pair would indeed be stationary at least with respect to a relative rotation of the tensors' eigenframes. Requiring those stationary pairs also to be minimizers of U_L leads to inequalities for the coefficients of the LANDAU potential.

Further inequalities arise when the critical points of U_L are required to reflect the symmetry property $U_\mathrm{L}(\mathbf{Q},\mathbf{B}) = U_\mathrm{L}(\mathbf{Q},-\mathbf{B})$. Precisely, one assumes that any tensorial macroscopic observable $\mathbf{A} = \alpha\mathbf{Q}+\beta\mathbf{B}$ resulting from the ensemble average of a microscopic observable $\mathbf{a} = \alpha\mathbf{q} + \beta\mathbf{b}$, with α and β scalars, has the same spectrum as the tensorial macroscopic observable $\mathbf{A}^* = \alpha\mathbf{Q}-\beta\mathbf{B}$ resulting from the ensemble average of $\mathbf{a}^* = \alpha\mathbf{q}-\beta\mathbf{b}$. It is shown in [63] that this requirement amounts to the following constraints on the two order tensors:

$$\operatorname{tr}\mathbf{B}^3 = 0, \quad \operatorname{tr}\mathbf{B}\mathbf{Q}^2 = 0, \quad \text{and} \quad \operatorname{tr}\mathbf{Q}\mathbf{B} = 0.$$

By combining these equations with the representations (1.62) and (1.70) of $\mathbf{Q}$ and $\mathbf{B}$ in one and the same eigenframe, it follows that both T and S' must vanish at equilibrium, and so the admissible critical points of U_L in (4.40) are represented by

$$\mathbf{Q} = S\left(\boldsymbol{e}_z \otimes \boldsymbol{e}_z - \frac{1}{3}\mathbf{I}\right), \quad \mathbf{B} = T'\left(\boldsymbol{e}_x \otimes \boldsymbol{e}_x - \boldsymbol{e}_y \otimes \boldsymbol{e}_y\right).$$

This effectively reduces the number of scalar order parameters at equilibrium from four to two. While this is a particular property of the symmetry-reduced potential, the scalar order parameters S and T' are also the dominant ones in a mean-field approximation to the general interaction (4.41) with $\gamma \neq 0$, as shown in [22].

Potential Properties and Phase Diagram

Normally, only the coefficients of the quadratic invariants, a_1 and a_4 in the case of (4.40), are taken to depend on temperature θ. If we assume this dependence to be linear, we can write

$$a_1 = a_0\left(\theta - \theta^*\right) \quad \text{and} \quad a_4 = b_0\left(\theta - \theta_b^*\right) \quad \text{with} \quad a_0, b_0 > 0. \tag{4.42}$$

The temperature θ^* is, as in the single-tensor case (4.7), the supercooling temperature of the isotropic phase at the isotropic-to-uniaxial phase transition, while θ_b^* is the supercooling temperature of the isotropic phase at the isotropic-to-biaxial phase transition. In principle, there could be liquid crystalline materials for which $\theta_b^* < \theta^*$, and others for which $\theta_b^* \geqq \theta^*$. In materials with $\theta_b^* < \theta^*$, the molecules interact in a predominantly uniaxial fashion. In materials with $\theta_b^* > \theta^*$, molecular interactions are predominantly biaxial, and, accordingly, the likelihood of a direct isotropic-to-biaxial phase transition is increased, thus possibly precluding an intermediate uniaxial phase.

Motivated by the desire that the LANDAU potential (4.40) lead to a phase diagram that shows the same topological features as the universal phase diagram predicted by the mean-field approximation to (4.41) in [24], a number of inequalities were derived in [65] that restrict the phenomenological coefficients in (4.40):

1. Stability of coincident eigenframes of the order tensors is guaranteed if
$$a_6 > 0 \quad \text{and} \quad a_7 > 0. \tag{4.43a}$$
2. Stability of symmetric minimizers with $T = S' = 0$ is guaranteed if
$$a_2 < 0, \quad a_6 > 0, \quad \text{and} \quad a_8 \geqq -\frac{1}{3}a_7. \tag{4.43b}$$
3. Positive definiteness of the LANDAU potential (4.40) when expressed in terms of the scalar order parameters (S, T') of a symmetric minimizer is guaranteed if
$$a_3 > 0, \quad a_5 > 0, \quad \text{and} \quad a_7 + 6a_{14} > -12\sqrt{a_3 a_5}. \tag{4.43c}$$
4. The existence in the phase diagram of a triple point and two separate tricritical points is guaranteed if
$$a_7 + 6a_{14} \leqq 0 \quad \text{and} \quad a_5 a_2^2 < a_3 a_6^2. \tag{4.43d}$$

The conditions (4.43) are quite general and not too restrictive. If they are met, a typical phase diagram in the a_1-a_4 plane would look like the one shown in Figure 4.1. The directions of the arrows indicate increasing temperature and hence potential phase sequences; different arrows correspond to different ratios of a_0 and b_0 and different values of the two supercooling temperatures θ^* and θ_b^* in (4.42).

Reduced Elasticity

Any elastic free energy density of a nonchiral biaxial nematic would have to encourage a homogeneous state in both order tensors **Q** and **B**. In the simplest case, inspired by the one-constant approximation to the elasticity of a uniaxial nematic, it would include only the terms $|\nabla \mathbf{Q}|^2$ and $|\nabla \mathbf{B}|^2$. Because **Q** is an average related to the long molecular axes while **B** is an average related to the short molecular axes, deformations in **Q** can be expected to have a different energetic cost from those in **B**. In any event, it is safe to assume that the ratio of the elastic constants pertaining to the two types of deformations is a positive number κ^2. If also a coupling between the gradients $\nabla\mathbf{Q}$ and $\nabla\mathbf{B}$ should be taken into account, it is natural to expect this to lie energetically between the contributions from the individual tensor gradients.

A simple model for an elastic energy density that meets these requirements is
$$W(\mathbf{Q}, \mathbf{B}, \nabla\mathbf{Q}, \nabla\mathbf{B}) = \frac{1}{2}L(\nabla\mathbf{Q} + \kappa\nabla\mathbf{B})^2 = \frac{1}{2}L|\nabla(\mathbf{Q} + \kappa\mathbf{B})|^2, \tag{4.44}$$
where L is a positive elastic constant and $\kappa \in \mathbb{R}$ can in principle be either positive or negative. The square in the first equality is to be interpreted using the inner product between third-rank tensors defined in (4.38). The form given on the far right shows that this model is a one-constant approximation for an elastic energy of a single effective order tensor $\mathbf{Q} + \kappa\mathbf{B}$.

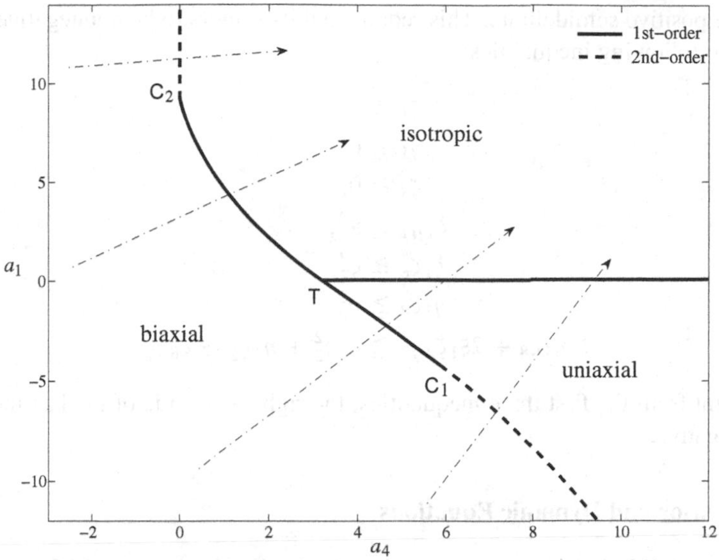

Fig. 4.1. A typical phase transition diagram predicted by the symmetric LANDAU potential U_L in (4.40). The coefficients are selected so as to comply with the inequalities (4.43). Both second- and first-order transitions between the isotropic and biaxial phases and between the uniaxial and biaxial phases are possible, with the appearance of two tricritical points, C_1 and C_2. The transition between the isotropic and uniaxial phases remains first-order for all parameters chosen according to (4.43). T is the triple point where the isotropic, uniaxial, and biaxial phases coexist in equilibrium. The arrows represent temperature axes with temperature increasing in the direction of the arrows. These straight lines are parameterized by the phenomenological coefficients a_0 and b_0 and by the difference between the two supercooling temperatures θ^* and θ_b^*.

Minimal Dissipation Function

A simple dissipation function can be constructed by considering only the minimum number of terms needed to account both for dissipation in the three individual rates and for dissipation arising from the three possible combinations of two different rates. This requirement is met if only the six quadratic terms in (4.39) are chosen. This means that we have

$$R(\mathbf{Q},\mathbf{B};\mathbf{D},\mathring{\mathbf{Q}},\mathring{\mathbf{B}}) = \frac{1}{2}\left(\zeta_1 \operatorname{tr}\mathring{\mathbf{Q}}^2 + \eta_1 \operatorname{tr}\mathring{\mathbf{B}}^2 + \zeta_8 \operatorname{tr}\mathbf{D}^2\right) + \xi_1 \operatorname{tr}\mathring{\mathbf{Q}}\mathring{\mathbf{B}} + \zeta_2 \operatorname{tr}\mathbf{D}\mathring{\mathbf{Q}} + \eta_2 \operatorname{tr}\mathbf{D}\mathring{\mathbf{B}}. \tag{4.45}$$

For this quadratic form to be positive semidefinite, its symmetric coefficient matrix

$$H = \begin{bmatrix} \zeta_1 & \xi_1 & \zeta_2 \\ \xi_1 & \eta_1 & \eta_2 \\ \zeta_2 & \eta_2 & \zeta_8 \end{bmatrix}$$

has to be positive semidefinite. This requires all its minors to be nonnegative, which yields the following inequalities:

$$\zeta_1 \geqq 0, \tag{4.46a}$$
$$\eta_1 \geqq 0, \tag{4.46b}$$
$$\zeta_8 \geqq 0, \tag{4.46c}$$
$$\zeta_1\eta_1 \geqq \xi_1^2, \tag{4.46d}$$
$$\zeta_1\zeta_8 \geqq \zeta_2^2, \tag{4.46e}$$
$$\eta_1\zeta_8 \geqq \eta_2^2, \tag{4.46f}$$
$$\zeta_1\eta_1\zeta_8 + 2\xi_1\zeta_2\eta_2 \geqq \zeta_1\eta_2^2 + \eta_1\zeta_2^2 + \zeta_8\xi_1^2. \tag{4.46g}$$

As evident from the first three inequalities, the right-hand side of the last inequality is nonnegative.

Stress Tensor and Dynamic Equations

With the simplified elastic energy density (4.44), the elastic stress (4.2.2) takes the form

$$\mathbf{T}_{\mathrm{E}} = -L\nabla(\mathbf{Q} + \kappa\mathbf{B}) \odot \nabla(\mathbf{Q} + \kappa\mathbf{B}),$$

which is symmetric.

The viscous stress (4.2.2) arising from the simplified dissipation function (4.45) is

$$\mathbf{T}_{\mathrm{dis}} = \zeta_8\mathbf{D} + \zeta_2\mathring{\mathbf{Q}} + \eta_2\mathring{\mathbf{B}} + 2\,\mathrm{skw}\left[\mathbf{Q}\left(\zeta_1\mathring{\mathbf{Q}} + \zeta_2\mathbf{D} + \xi_1\mathring{\mathbf{B}}\right) + \mathbf{B}\left(\eta_1\mathring{\mathbf{B}} + \eta_2\mathbf{D} + \xi_1\mathring{\mathbf{Q}}\right)\right].$$

When both the simplified elastic energy density (4.44) and the simplified dissipation function (4.45) are used, the dynamic equations (4.33) and (4.34) for **Q** and **B** take the explicit forms

$$\zeta_1\mathring{\mathbf{Q}} + \zeta_2\mathbf{D} + \xi_1\mathring{\mathbf{B}} + \frac{\partial U_{\mathrm{L}}}{\partial \mathbf{Q}} - L\Delta(\mathbf{Q} + \kappa\mathbf{B}) = \mathbf{0} \tag{4.47}$$

and

$$\eta_1\mathring{\mathbf{B}} + \eta_2\mathbf{D} + \xi_1\mathring{\mathbf{Q}} + \frac{\partial U_{\mathrm{L}}}{\partial \mathbf{B}} - \kappa L\Delta(\mathbf{Q} + \kappa\mathbf{B}) = \mathbf{0}. \tag{4.48}$$

Here, we have not made any assumptions about the specific form of the LANDAU potential U_{L} because it does not affect the structure of the equations: ultimately, the terms $\frac{\partial U_{\mathrm{L}}}{\partial \mathbf{Q}}$ and $\frac{\partial U_{\mathrm{L}}}{\partial \mathbf{B}}$ will be polynomials in the order tensors **Q** and **B**.

The viscosity coefficient ξ_1, which in the dissipation function couples $\mathring{\mathbf{Q}}$ and $\mathring{\mathbf{B}}$, also makes equations (4.47) and (4.48) coupled. Indeed, unless $\xi_1 = 0$, time derivatives of both order tensors are present in both equations. They cannot then be regarded as individual equations for the two order tensor evolutions, but they constitute a system of nonlinear partial differential equations for $\mathbf{Q}$ and $\mathbf{B}$, where $\overset{\circ}{\mathbf{Q}}$ and $\overset{\circ}{\mathbf{B}}$ enter linearly. To obtain explicit equations for the time evolution of the order tensors, we need to consider the determinant of the coefficient matrix of the linear system in $(\overset{\circ}{\mathbf{Q}}, \overset{\circ}{\mathbf{B}})$, which is $\zeta_1 \eta_1 - \xi_1^2$. This shows that the system of equations can be decoupled if we enforce strictly inequality (4.46d), that is, if we require $\zeta_1 \eta_1 - \xi_1^2 > 0$. Of course, once decoupled in $\overset{\circ}{\mathbf{Q}}$ and $\overset{\circ}{\mathbf{B}}$, the two resulting explicit equations will still be coupled together and with the flow.

5

Nematoacoustics

At this stage of our development we apply our general theory to a field that is still somewhat controversial. Though the orienting effect of an ultrasonic wave on the nematic texture has long been known,[1] its interpretation in terms of a coherent dynamical theory, widely if not universally accepted, has not yet been achieved. In this chapter, following [354] and [67], we phrase such a theory within the setting of this book; hopefully, this will serve as a further illustration of the generality of our method. Below, we first summarize the diverse theoretical approaches attempted in the past to describe the interaction between sound and molecular orientation in nematic liquid crystals. We then revive a theory for second-grade fluids, which we believe provides the most appropriate theoretical background to posit our nematoacoustic theory that elaborates on a proposal not new in its intuitive phrasing, but which has only recently found both a more precise theoretical formulation and its first experimental validation. We shall actually depart from the latest theoretical formulation in an effort to draw from it all its consequences. A closing explicit application to a simple computable case will also show some predictions of our theory, which are both qualitatively and quantitatively confirmed by a number of experimental results.

5.1 Overview

Experimental acoustic studies in nematic liquid crystals have a long history including early contributions from pioneers of liquid crystal science such as LEHMANN and ZOLINA.[2] Several reviews provide accounts of the effect of an acoustic field on the orientation of nematic molecules; we quote only [155], [156], and [160] among the most recent ones, which also report the still unappeased debate between the different theories that have attempted to explain the interaction between acoustic waves and nematic textures.

[1] Also with potential practical applications; see [134].

[2] See [155].

The main experimental findings that called for explanation were the anisotropy observed in both attenuation and speed of sound in the propagation of ultrasonic waves in nematic liquid crystals where the orientation of the director is kept fixed by an aligning magnetic field [196, 187, 164, 221, 231] and the reorientating action exerted on a uniformly aligned nematic cell by the propagation of ultrasonic waves in the absence of any other external action [211, 16, 10]. This evidence supported the hypothesis that a condensation wave can affect the director orientation in a way similar in its appearance, though not in its cause, to the action exerted by an external magnetic or electric field. Actually, the *acousto-optic* effect, as it is often called, produces an alteration of the birefringence in a nematic cell, which is easily detected and closely resembles the optic effect induced by an external field, as if the acoustic field could also impart a torque on the nematic director.

The theories so far proposed to explain the acoustic action on nematic liquid crystals can essentially be grouped into two wide categories: theories that explain the acoustic–nematic interaction by means of an intermediate hydrodynamic flow of one sort or another, and theories that explain the acoustic–nematic interaction through a direct coupling between acoustic field and nematic director, with its own associated elastic energy. The theories in the former category build essentially on the ERICKSEN–LESLIE theory presented in Chapter 3 and assume that an acoustic wave is capable of inducing a steady nonuniform flow, which in turn acts on the director field, thus distorting it, whereas the theories in the latter category posit an elastic interaction between an acoustic wave and the director field, which is also capable of inducing distortions in the absence of any induced flow.

The major hydrodynamic mechanism that has been imagined to transmit torque from the acoustic field to the nematic director is a nonlinear coupling relying on the occurrence of a variant of REYNOLDS stresses in the fluid. Related to these stresses is also the notion of *acoustic streaming*, which describes a phenomenon also known for dissipative isotropic fluids; a rather general description of these concepts and the mathematical techniques connected to them can be found in [189] (see, in particular, Section 4.7). Here, following in part [317], we shall be contented with outlining the general ideas underlining this method, to the extent that it may be applied to our context. Another, more recent application of these ideas is illustrated in [40].

Let u be any of the fields describing the flow: it may designate either the pressure or the density, a component of the velocity field or a component of the nematic director. We expand u in the form

$$u = u_0 + \varepsilon u_1 + \varepsilon^2 u_2 + o(\varepsilon^2),$$

where ε is a perturbation parameter, u_0 is the equilibrium value of u, and u_1 and u_2 are the first- and second-order corrections to u_0, respectively. In a plane-wave solution to the dynamical equations of the ERICKSEN–LESLIE theory, u_1 has zero average, whereas u_2 can in general be written as

$$u_2 = \overline{u}_2 + \widehat{u}_2,$$

where it is decomposed into a steady component, $\overline{u}_2$, and a varying component, $\widehat{u}_2$, oscillating at a frequency twice the frequency of u_1, which like this latter averages out to zero. The dynamical equations for the various fields like $\overline{u}_2$, which capture the slow, second-order evolution of the fluid, are derived by averaging in time the contributions to the general dynamical equations that are second-order in ε, as is typical in any perturbation method. Such second-order equations will invariably be affected by the time averages of terms quadratic in u_1, which will thus act as forces for the growth of inhomogeneities in $\overline{u}_2$. This is the essence of the acoustic streaming method applied in [317] to the ERICKSEN–LESLIE dynamic equations for nematic liquid crystals. The second-order character of the stresses responsible for the onset of the steady second-order flow makes them resemble REYNOLDS stresses of ordinary fluid dynamics (see, for example, pp. 328–330 of [189]). These stresses are responsible for making turbulent velocity perturbations about a mean flow interfere with the mean flow itself, thus generating sound. Conversely, waves propagating through a mean flow affect it through exactly the same mechanism (see p. 330 of [189]). This is indeed the conceptual connection between turbulence and acoustic streaming, also implied in the extension to nematic liquid crystals proposed in [317]. In summary, according to [317], acoustic streaming in nematic liquid crystals would be responsible for the hydrodynamic coupling that transfers torque from a traveling ultrasonic wave to the nematic director.

Essentially the same approach as in [317], though with some apparent variants, was more recently adopted in [171] and further applied in a series of other works [157, 158, 159] to explain nematic alignment produced by ultrasonic waves.

Within a slightly different category, though still postulating a hydrodynamic mediation, falls the explanation of the acoustic action on the nematic director proposed in [138]. In general, sound is known to produce a *radiation pressure* in the medium where it propagates (see, for example, Section 64 of [174]). Such a pressure is to be distinguished from the acoustic pressure, often also called the excess pressure; the latter is the difference between the pressure carrying an acoustic wave and the uniform pressure of the unperturbed medium; it averages out to zero in time, and so has no net mechanical effect. The radiation pressure is the time average of the second-order correction to the unperturbed pressure, and it is determined by the second-order components of the dynamical equation. In isotropic fluids, the radiation pressure can induce a force only along the direction of propagation, but in anisotropic fluids, such as nematic liquid crystals, the time average of second-order stresses may also induce transverse actions resulting in a torque on the nematic director. As for the acoustic streaming, such a torque would thus be of a viscous nature.

Here we shall follow a conceptual avenue that essentially differs from those already outlined in the nature of the postulated aligning torque, which will be elastic rather than viscous. Thus no flow will be needed for an acoustic field to act upon the nematic director. This line of thought first arose in [225], whose experimental results suggested that the elastic energy density be supplemented with the following acoustic contribution:

$$W_a = c_1 k^2 + c_2 (\boldsymbol{n} \cdot \boldsymbol{k})^2, \tag{5.1}$$

where c_1 and c_2 are constitutive constants, $\boldsymbol{k}$ is the acoustic wave vector, and $\boldsymbol{n}$ is the nematic director. A similar interaction, even if not explicitly formulated as in (5.1), was also adduced in [10] to interpret some acousto-optical observations.

DION [78] is often credited with having first proposed a direct interaction between acoustic propagation and molecular alignment. However, the further interpretation of this interaction within the general principle of minimum entropy production [79] has obscured its elastic character, thus bringing it into the realm of dissipation, to which it does not really belong. In DION and JACON's own words [79], according to their hypothesis, "in a medium with acoustical anisotropy, the molecules tend to reorient so as to minimize propagation losses." Such an interpretation of the acoustic–molecular interaction has fueled controversies and caused misunderstanding (exemplary to this effect is the comment on DION's work on p. 184 of [160]).

As proposed independently and almost simultaneously in [294] and [31], we hold that the acoustic–nematic interaction is of an elastic nature and results from the coupling between the density gradient induced by the acoustic wave and the average molecular orientation represented by the nematic director. Since the typical characteristic times of acoustic waves are much shorter than the director's relaxation time, it is actually the time-averaged interaction energy that will affect the nematic elastic energy. Both papers [294] and [31] were followed by further extensions of the original assumption along with the first experimental confirmations of that theory; in particular, we refer the reader to the series of works [126, 213, 124, 125, 214] and [32, 30, 268]. Here, we shall indeed posit a slight variant of this assumption and we shall interpret through the ensuing theory experiments long published in the literature, though never completely explained.

At the time scale of the acoustic oscillations, at which the director texture can be regarded as prescribed and immobile, a nematic liquid crystal behaves like an anisotropic KORTEWEG *fluid*, that is, an elastic fluid whose free energy density also depends on the density gradient. KORTEWEG [170] first considered an isotropic fluid with the elastic stress tensor depending on both the first and second gradients of the density field; he built his capillarity theory on such a constitutive assumption, as also recalled in [345] (see, in particular, pp. 513–515). Under appropriate assumptions, a KORTEWEG stress tensor is *hyperelastic*, that is, it can be derived from a potential that depends on the density and its first gradient (see also Section 18 of [41]).

In the following section, mainly following [354], we shall present a general variational theory for KORTEWEG fluids, which will be further adapted to nematoacoustics in Section 5.3, where we show how the time-averaged elastic actions associated with acoustic propagation affects the dynamics of nematic liquid crystals, in both its components, director relaxation and hydrodynamic flow. In particular, we shall draw the consequences of our general theory for the propagation of acoustic plane waves in a uniformly aligned nematic liquid crystal: we shall compute both the speed of propagation and the wave attenuation as functions of frequency, propagation direction, and nematic viscosities. For simplicity, these conclusions are reached under the assumption that the director is kept fixed in a uniform alignment by some external agent, as was indeed the case in most of the early experiments. In the closing Section 5.4, we relax this assumption by allowing oscillations of the director around a fixed uniform orientation to be excited by an incoming acoustic wave. Such a motion, which we call a *libration*, will affect the attenuation of the wave, but not its speed of propagation.

5.2 KORTEWEG Fluids

In this section we digress slightly from our natural development and consider both equilibrium and motion of a perfect second-grade fluid, whose elastic energy density is a function of both the mass density ϱ and its spatial gradient $\nabla\varrho$. Our objective is identifying both stresses and traction laws relevant to this class of fluids. The extension to nematoacoustics of the balance laws derived here will be the object of the following section, where the time scale at which a nematic liquid crystal behaves like a KORTEWEG fluid will be separated from the time scale at which only the average effects of such a behavior survive.

5.2.1 Principle of Virtual Power

Here, as in the classical treatment of second-grade materials of TOUPIN [338, 339] (see also [112] for a more recent application of the same method), we start by deriving both balance equations and traction laws of statics from a principle of virtual power. In the hierarchy of variational principles encountered in this book, the principle of virtual power is the most basic one: it is subsumed under D'ALEMBERT's principle of inviscid dynamics, which is in turn subsumed under the principle of minimum reduced dissipation of dissipative dynamics illustrated in Section 2.2.3. Formally, by setting equal to zero both the dissipation functional $\mathscr{D}$ and the thermal production of energy $\mathscr{T}$, we learn from (2.206) that the net working $\mathscr{W}^{(e)}(\mathcal{P}_t, \chi)$ and the time rate $\dot{\mathscr{F}}(\mathcal{P}_t, \chi)$ of the free energy functional $\mathscr{F}(\mathcal{P}_t, \chi)$ must be equal for all shapes $\mathcal{P}_t$ in any motion χ. The principle of virtual power ignores inertia and requires $\dot{\mathscr{F}}$ to equal $\mathscr{W}^{(a)}$, the power expended by all applied external agencies[3] in all *virtual* motions that would attempt to deform any subbody $\mathcal{P}$ of $\mathcal{B}$ at equilibrium.[4]

Let the free energy $\mathscr{F}_K(\mathcal{P})$ of the fluid occupying the subbody $\mathcal{P}$ of the body $\mathcal{B}$ be given by

$$\mathscr{F}_K(\mathcal{P}) := \int_{\mathcal{P}} \varrho\sigma_K(\varrho, \nabla\varrho)dV,$$

where σ_K is the internal energy per unit mass and V denotes as above the volume measure. Following [339, 112], we posit for $\mathscr{W}^{(a)}(\mathcal{P})$ the following form:

$$\mathscr{W}^{(a)}(\mathcal{P}) := \int_{\mathcal{P}} \boldsymbol{b}\cdot\boldsymbol{v}\,dV + \int_{\partial^*\mathcal{P}} \left(\boldsymbol{t}\cdot\boldsymbol{v} + \boldsymbol{m}\cdot\frac{\partial\boldsymbol{v}}{\partial\nu}\right)dA, \tag{5.2}$$

[3] If inertia is neglected, $\mathscr{K} \equiv 0$, and so, by (2.104), $\mathscr{W}^{(e)}$ reduces to $\mathscr{W}^{(a)}$.

[4] A similar study, though in a simpler context, was presented in Section 3.1.4 above. Here we extend the method illustrated there.

where A is the area measure and $\boldsymbol{v}$ is the velocity field inducing a *virtual* flow of the subbody $\mathcal{P}$, thought of as *carved* out of the whole body $\mathcal{B}$, while the actions exerted both in its bulk and on its boundary are held fixed, and the subbody $\mathcal{B} \setminus \mathcal{P}$ surrounding it is equally *frozen*.[5] Since $\mathcal{P}$ is an arbitrary subbody of $\mathcal{B}$, it complies with the requirements of a fit region listed on page 73. In particular, $\partial^* \mathcal{P}$ denotes in (5.2) the reduced boundary of $\mathcal{P}$, where the unit outer normal $\boldsymbol{\nu}$ to $\partial \mathcal{P}$ is defined.

In (5.2), $\boldsymbol{b}$ is the external *body force* defined in the whole of $\mathcal{B}$, while $\boldsymbol{t}$ and $\boldsymbol{m}$ are surface *contact* actions, the former expending power against $\boldsymbol{v}$, and so identifiable as a force, the latter expending power against the normal derivative of $\boldsymbol{v}$,

$$\frac{\partial \boldsymbol{v}}{\partial \nu} := (\nabla \boldsymbol{v})\boldsymbol{\nu}, \tag{5.3}$$

with $\boldsymbol{\nu}$ the outer unit normal to $\partial^* \mathcal{P}$, and so identifiable as a *hypertraction*, according to TOUPIN [339] (see also Section 98 of [345]). The hypertraction $\boldsymbol{m}$ would not be present in a classical simple fluid, for which the elastic energy density is independent of $\nabla \varrho$; as is soon to be shown, its presence in (5.2) is needed to counterbalance the internal power associated with the dependence of σ_{K} on $\nabla \varrho$. While, as already illustrated in Chapter 2, the body force $\boldsymbol{b}$ is a prescribed source, both surface actions $\boldsymbol{t}$ and $\boldsymbol{m}$ should be considered as unknown functionals of the boundary $\partial^* \mathcal{P}$, to be determined so as to comply with the variational principle posited by the theory. For statics, this principle is illustrated below; it is intended to provide both the balance equations valid within the body at equilibrium and the traction laws revealing how contact actions are transmitted through the boundary of internal subbodies.

We shall require the equilibrium configurations of the body $\mathcal{B}$ to be such that, for every subbody $\mathcal{P} \subset \mathcal{B}$,

$$\dot{\mathscr{F}}_{\mathrm{K}}(\mathcal{P}_t)\Big|_{t=0} = \mathscr{W}^{(\mathrm{a})}(\mathcal{P}), \tag{5.4}$$

where the time derivative of $\mathscr{F}_{\mathrm{K}}$ is meant to be computed along the virtual incipient flow $\boldsymbol{v}$ that at time t brings $\mathcal{P}$ into $\mathcal{P}_t$.

A virtual flow of $\mathcal{P}$ is described by a velocity field $\boldsymbol{v}(\cdot, t)$ defined for every $t \in [0, T]$ with $T > 0$ on the evolved subbody $\mathcal{P}_t$. Formally, for every $t \in [0, T]$, $p(t) \in \mathcal{P}_t$ whenever the trajectory $t \mapsto p(t)$ solves the evolution problem

$$\dot{p}(t) = \boldsymbol{v}(p(t), t), \quad \text{with} \quad p(0) \in \mathcal{P},$$

so that $\mathcal{P}_0 = \mathcal{P}$.

5.2.2 KORTEWEG Stress

The time derivative of $\mathscr{F}_{\mathrm{K}}$ in (5.4) is to be computed with the aid of REYNOLD's transport theorem in the Eulerian formalism, which we now recall from page 82. For a functional Φ defined on the evolving subbody $\mathcal{P}_t$ as

[5] For all intents and purposes, in the notation of Section 2.2.3, the virtual velocity in (5.2) is properly a variation $\delta \boldsymbol{v}$ of the still field $\boldsymbol{v} \equiv \boldsymbol{0}$. We prefer denoting it simply by $\boldsymbol{v}$ to avoid unnecessary complication.

$$\Phi(\mathcal{P}_t) := \int_{\mathcal{P}_t} \varphi(x,t)dV, \tag{5.5}$$

where $\varphi(\cdot, t)$ is a smooth scalar field on $\mathcal{P}_t$, REYNOLD's transport theorem says that

$$\dot{\Phi}(\mathcal{P}_t) = \int_{\mathcal{P}_t} (\varphi \operatorname{div} \boldsymbol{v} + \dot{\varphi})\, dV, \tag{5.6}$$

where $\dot{\varphi}$ is the *material* time derivative of φ defined in (2.38).

A mass evolution is associated with the virtual flow $\boldsymbol{v}$; it is described by a mass density function $\varrho(\cdot, t)$ defined on $\mathcal{P}_t$ for every $t \in [0, T]$. In particular, the functional

$$M(\mathcal{P}_t) := \int_{\mathcal{P}_t} \varrho(x,t)dV,$$

which represents the mass stored in $\mathcal{P}_t$, is a special form of Φ in (5.5). By (5.6), requiring

$$\dot{M}(\mathcal{P}_t) \equiv 0 \qquad \text{for all } \mathcal{P} \subset \mathcal{B},$$

which translates the conservation of mass along the virtual motion of any subbody, is equivalent to the continuity equation (2.108), which we now recall for the reader's ease:

$$\dot{\varrho} + \varrho \operatorname{div} \boldsymbol{v} = 0. \tag{5.7}$$

This equation must hold identically along all virtual motions of $\mathcal{P}$, and so the transport theorem in the form (2.111) will also apply to virtual motions. Thus we obtain that

$$\dot{\mathscr{F}}_{\mathrm{K}}(\mathcal{P}_t) = \int_{\mathcal{P}_t} \varrho \dot{\sigma}_{\mathrm{K}}\, dV, \tag{5.8}$$

where, by the chain rule,

$$\dot{\sigma}_{\mathrm{K}} = \frac{\partial \sigma_{\mathrm{K}}}{\partial \varrho}\dot{\varrho} + \frac{\partial \sigma_{\mathrm{K}}}{\partial \nabla \varrho}(\nabla \varrho)^{\cdot}. \tag{5.9}$$

Applying (2.12) to the vector field $\nabla \varrho$, we readily arrive at

$$(\nabla \varrho)^{\cdot} = (\nabla^2 \varrho)\boldsymbol{v} + \frac{\partial}{\partial t}(\nabla \varrho). \tag{5.10}$$

Under the assumption that ϱ is a sufficiently smooth function, also by (5.7) we see that

$$\begin{aligned} \frac{\partial}{\partial t}(\nabla \varrho) = \nabla\left(\frac{\partial \varrho}{\partial t}\right) &= -\nabla(\nabla \varrho \cdot \boldsymbol{v}) - \nabla(\varrho \operatorname{div} \boldsymbol{v}) \\ &= -(\nabla^2 \varrho)\boldsymbol{v} - (\nabla \boldsymbol{v})^{\mathsf{T}} \nabla \varrho - \nabla(\varrho \operatorname{div} \boldsymbol{v}), \end{aligned}$$

where the superscript ^{T} denotes transposition, and thus (5.10) becomes

$$(\nabla \varrho)^{\cdot} = -\nabla(\varrho \operatorname{div} \boldsymbol{v}) - (\nabla \boldsymbol{v})^{\mathsf{T}}(\nabla \varrho).$$

By this latter equation, using (5.7), from (5.8) and (5.9) we finally arrive at

$$\dot{\mathscr{F}}_{\mathrm{K}}(\mathcal{P}_t)\Big|_{t=0} = -\int_{\mathcal{P}} \varrho \left\{ \frac{\partial \sigma_{\mathrm{K}}}{\partial \varrho} \varrho \operatorname{div} \boldsymbol{v} + \frac{\partial \sigma_{\mathrm{K}}}{\partial \nabla \varrho} \cdot \left[\nabla(\varrho \operatorname{div} \boldsymbol{v}) + (\nabla \boldsymbol{v})^{\mathsf{T}} \nabla \varrho \right] \right\} dV, \tag{5.11}$$

since $\mathcal{P}_0 = \mathcal{P}$. Integrations by parts and repeated use of the divergence theorem allow us to give (5.11) the following form:

$$\dot{\mathscr{F}}_{\mathrm{K}}(\mathcal{P}_t)\Big|_{t=0} = -\int_{\mathcal{P}} \operatorname{div} \mathbf{T}_{\mathrm{K}} \cdot \boldsymbol{v}\, dV + \int_{\partial^* \mathcal{P}} \mathbf{T}_{\mathrm{K}} \boldsymbol{\nu} \cdot \boldsymbol{v}\, dA - \int_{\partial^* \mathcal{P}} \varrho^2 \frac{\partial \sigma_{\mathrm{K}}}{\partial \nabla \varrho} \cdot \boldsymbol{\nu} \operatorname{div} \boldsymbol{v}\, dA, \tag{5.12}$$

where

$$\mathbf{T}_{\mathrm{K}} := -p_{\mathrm{K}} \mathbf{I} - \varrho \nabla \varrho \otimes \frac{\partial \sigma_{\mathrm{K}}}{\partial \nabla \varrho} \tag{5.13}$$

is the KORTEWEG *stress tensor* and

$$p_{\mathrm{K}} := \varrho^2 \frac{\partial \sigma_{\mathrm{K}}}{\partial \varrho} - \varrho \operatorname{div} \left(\varrho \frac{\partial \sigma_{\mathrm{K}}}{\partial \nabla \varrho} \right) \tag{5.14}$$

is the associated KORTEWEG *pressure*.

5.2.3 Surface Calculus

The second surface integral in (5.12) must be further transformed to give (5.12) a form compatible with (5.2). To this end, we recall from Section 2.3.6 of [354] the surface-divergence theorem.

Let $\mathcal{S} \subset \mathcal{E}$ be a smooth, orientable, closed surface in the three-dimensional Euclidean space $\mathcal{E}$ and let $\boldsymbol{u}$ be a differentiable vector field on $\mathcal{S}$. The *surface* divergence of $\boldsymbol{u}$ is defined by

$$\operatorname{div}_{\mathrm{s}} \boldsymbol{u} := \operatorname{tr} \nabla_{\mathrm{s}} \boldsymbol{u},$$

where $\nabla_{\mathrm{s}} \boldsymbol{u}$ is the *surface* gradient of $\boldsymbol{u}$.

It can be shown that

$$\nabla_{\mathrm{s}} \boldsymbol{u} = (\nabla \hat{\boldsymbol{u}}) \mathbf{P}(\boldsymbol{\nu}), \tag{5.15}$$

where

$$\mathbf{P}(\boldsymbol{\nu}) := \mathbf{I} - \boldsymbol{\nu} \otimes \boldsymbol{\nu} \tag{5.16}$$

is the projection onto the plane orthogonal to a unit normal $\boldsymbol{\nu}$ to $\mathcal{S}$, and $\hat{\boldsymbol{u}}$ is any smooth extension of $\boldsymbol{u}$ to a three-dimensional neighborood of $\mathcal{S}$. It follows from (5.15) and (5.16) that

$$\nabla_{\mathrm{s}} \boldsymbol{u} = \nabla \hat{\boldsymbol{u}} - (\nabla \hat{\boldsymbol{u}}) \boldsymbol{\nu} \otimes \boldsymbol{\nu} = \nabla \hat{\boldsymbol{u}} - \frac{\partial \hat{\boldsymbol{u}}}{\partial \nu} \otimes \boldsymbol{\nu},$$

which, letting

$$\nabla_{\nu} \hat{\boldsymbol{u}} := \frac{\partial \hat{\boldsymbol{u}}}{\partial \nu} \otimes \boldsymbol{\nu}$$

and noting that $\nabla_s u = \nabla_s \hat{u}$, we can also rewrite as

$$\nabla \hat{u} = \nabla_s \hat{u} + \nabla_\nu \hat{u}, \tag{5.17}$$

whence we interpret $\nabla_\nu \hat{u}$ as the *normal* gradient of $\hat{u}$. By computing the trace of the tensors on both sides of (5.17), we conclude that

$$\operatorname{div} \hat{u} = \operatorname{div}_s \hat{u} + \operatorname{div}_\nu \hat{u}, \tag{5.18}$$

where

$$\operatorname{div}_\nu \hat{u} := \operatorname{tr} \nabla_\nu \hat{u} = \frac{\partial \hat{u}}{\partial \nu} \cdot \nu \tag{5.19}$$

is the *normal* divergence of $\hat{u}$.

The surface-divergence theorem states that

$$\int_{\mathcal{S}} \operatorname{div}_s u \, dA = \int_{\mathcal{S}} u \cdot \nu \operatorname{div}_s \nu \, dA, \tag{5.20}$$

for all smooth vector fields u on $\mathcal{S}$. In (5.20), $\operatorname{div}_s \nu$ embodies the differential properties of the surface $\mathcal{S}$,

$$\operatorname{div}_s \nu = \operatorname{tr} \nabla_s \nu = 2H, \tag{5.21}$$

where $\nabla_s \nu$ is the *curvature tensor*, which enjoys the properties

$$(\nabla_s \nu)^{\mathsf{T}} = \nabla_s \nu \quad \text{and} \quad (\nabla_s \nu)\nu \equiv \mathbf{0},$$

and H is the *mean curvature* of $\mathcal{S}$.

Similarly, for a smooth scalar field χ on $\mathcal{S}$, the surface gradient-integral theorem says that

$$\int_{\mathcal{S}} \nabla_s \chi \, dA = \int_{\mathcal{S}} \chi (\operatorname{div}_s \nu) \nu \, dA, \tag{5.22}$$

which is the analogue of (2.1).

5.2.4 Traction and Hypertraction

By (5.18) and (5.19), we have that

$$\int_{\partial^* \mathcal{P}} \varrho^2 \frac{\partial \sigma_K}{\partial \nabla \varrho} \cdot \nu \operatorname{div} v \, dA = \int_{\partial^* \mathcal{P}} \varrho^2 \frac{\partial \sigma_K}{\partial \nabla \varrho} \cdot \nu \left(\operatorname{div}_s v + \frac{\partial v}{\partial \nu} \cdot \nu \right) dA, \tag{5.23}$$

and using the identity

$$\varrho^2 \frac{\partial \sigma_K}{\partial \nabla \varrho} \cdot \nu \operatorname{div}_s v = \operatorname{div}_s \left[\left(\varrho^2 \frac{\partial \sigma_K}{\partial \nabla \varrho} \cdot \nu \right) v \right] - \nabla_s \left(\varrho^2 \frac{\partial \sigma_K}{\partial \nabla \varrho} \right) \cdot v$$

and the surface-divergence theorem in (5.20), we can also write

$$\int_{\partial^*\mathcal{P}} \varrho^2 \frac{\partial \sigma_K}{\partial \nabla \varrho} \cdot \boldsymbol{\nu} \operatorname{div}_s \boldsymbol{v}\, dA = \int_{\partial^*\mathcal{P}} \left(\varrho^2 \frac{\partial \sigma_K}{\partial \nabla \varrho} \cdot \boldsymbol{\nu}\right) 2H(\boldsymbol{v}\cdot\boldsymbol{\nu})dA - \int_{\partial^*\mathcal{P}} \nabla_s \left(\varrho^2 \frac{\partial \sigma_K}{\partial \nabla \varrho}\right) \cdot \boldsymbol{v}\, dA.$$

Making use of both this equation and (5.23), we finally arrive at

$$\begin{aligned} \dot{\mathscr{F}}_K(\mathcal{P}_t)\Big|_{t=0} &= -\int_{\mathcal{P}} \operatorname{div} \mathbf{T}_K \cdot \boldsymbol{v}\, dA - \int_{\partial^*\mathcal{P}} \varrho^2 \left(\frac{\partial \sigma_K}{\partial \nabla \varrho} \cdot \boldsymbol{\nu}\right) \frac{\partial \boldsymbol{v}}{\partial \nu} \cdot \boldsymbol{\nu}\, dA \\ &+ \int_{\partial^*\mathcal{P}} \left[\mathbf{T}_K \boldsymbol{\nu} + \nabla_s \left(\varrho^2 \frac{\partial \sigma_K}{\partial \nabla \varrho} \cdot \boldsymbol{\nu}\right) - \left(\varrho^2 \frac{\partial \sigma_K}{\partial \nabla \varrho} \cdot \boldsymbol{\nu}\right) 2H\boldsymbol{\nu}\right] \cdot \boldsymbol{v}\, dA. \end{aligned} \tag{5.24}$$

Inserting both (5.24) and (5.2) into (5.4), and requiring the latter to be valid for every virtual flow $\boldsymbol{v}$ of $\mathcal{P}$ and for every subbody $\mathcal{P}$ of $\mathcal{B}$, we derive the equation

$$\boldsymbol{b} + \operatorname{div} \mathbf{T}_K = \mathbf{0}, \tag{5.25}$$

expressing the balance of external and internal forces at equilibrium in $\mathcal{B}$, and the traction laws

$$\boldsymbol{t} = \mathbf{T}_K \boldsymbol{\nu} + \nabla_s \left(\varrho^2 \frac{\partial \sigma_K}{\partial \nabla \varrho} \cdot \boldsymbol{\nu}\right) - \varrho^2 \left(\frac{\partial \sigma_K}{\partial \nabla \varrho} \cdot \boldsymbol{\nu}\right) 2H\boldsymbol{\nu} \tag{5.26}$$

and

$$\boldsymbol{m} = -\varrho^2 \left(\frac{\partial \sigma_K}{\partial \nabla \varrho} \cdot \boldsymbol{\nu}\right) \boldsymbol{\nu}, \tag{5.27}$$

valid on the boundary $\partial^*\mathcal{P}$ of every subbody $\mathcal{P}$ of $\mathcal{B}$.

Equation (5.26) illustrates a notable variance from the linear dependence of the traction $\boldsymbol{t}$ on the outer unit normal $\boldsymbol{\nu}$ established by CAUCHY's classical theorem (see Theorem 2.5 above), a deviation typical of second-grade fluids. It should be noted, however, that by (5.21) $\boldsymbol{t}$ is still an odd function of $\boldsymbol{\nu}$, thus complying with NEWTON's action and reaction principle (see also p. 164 of [349]).

Equation (5.27) represents the hypertraction $\boldsymbol{m}$ as a function of $\boldsymbol{\nu}$; unlike $\boldsymbol{t}$, $\boldsymbol{m}$ is even in $\boldsymbol{\nu}$; it is given by a third-rank tensor M, which TOUPIN [338, 339] suggested be called a *hyperstress*: in Cartesian components,

$$m_i = M_{jik} \nu_j \nu_k, \tag{5.28}$$

with repeated indices denoting summation and

$$M_{jik} := -\varrho^2 \frac{\partial \sigma_K}{\partial \varrho_{,j}} \delta_{ik}, \tag{5.29}$$

where δ_{ik} is KRONECKER's symbol and a comma denotes differentiation with respect to Cartesian coordinates (x_1, x_2, x_3).

Clearly, were σ_K independent of $\nabla\varrho$, $\boldsymbol{m}$ would vanish identically, while $\boldsymbol{t}$ would be given the classical CAUCHY's form with a stress tensor purely spherical and the KORTEWEG pressure p_K in (5.14) reduced to[6]

$$p_K = p_0(\varrho) := \varrho^2\sigma_K'(\varrho), \tag{5.30}$$

characteristic of a compressible perfect fluid.[7] Since the deviations from the behavior of such a fluid introduced in equations (5.13), (5.26), and (5.27) all stem from σ_K also being a function of $\nabla\varrho$, we often refer to these equations as being *nonlocal* extensions of the classical equations.

As also reported in Section 124 of [345], in his original work, KORTEWEG [170] assumed the following form for $\mathbf{T}_K$:

$$\mathbf{T}_K = -(\alpha|\nabla\varrho|^2 - \gamma\Delta\varrho)\mathbf{I} - \beta\nabla\varrho\otimes\nabla\varrho + \delta\nabla^2\varrho, \tag{5.31}$$

where α, β, γ, and δ are constitutive functions of ϱ, and Δ denotes the Laplacian. It is easy to verify that setting

$$\varrho\sigma_K = \frac{1}{2}\beta|\nabla\varrho|^2, \qquad \alpha = -\frac{d}{d\varrho}(\varrho\beta), \quad \gamma = \varrho\beta, \quad \text{and} \quad \delta = 0 \tag{5.32}$$

would make (5.13) agree with (5.31). Under these assumptions, KORTEWEG's original fluid would then become *hyperelastic*, since $\mathbf{T}_K$ could be derived from the energy density σ_K in (5.32) within the theory presented here.

5.2.5 Symmetry of the KORTEWEG Stress

Another property of the KORTEWEG stress in (5.13) is worth mentioning: it is necessarily symmetric.This is indeed a direct consequence of σ_K being frame-indifferent.

Let $\mathbf{R} \in \mathrm{O}(3)$ be any orthogonal tensor representing a change of observer through the equations (2.43). In particular, we recall that

$$\boldsymbol{x}^* = \mathbf{R}\boldsymbol{x}, \tag{5.33}$$

where $\boldsymbol{x} := x - o$ is the position vector of a body point at $x \in \mathcal{E}$ relative to the origin o selected by one observer and $\boldsymbol{x}^* := x^* - o$ is the position vector of the same body point seen at $x^* \in \mathcal{E}$ by a different observer selecting the same origin. As already discussed on page 84, x and x^* refer one and the same body point to two frames.

It readily follows from (5.33) that

$$\nabla\boldsymbol{x}^* = \mathbf{R}.$$

Letting the density fields ϱ and ϱ^* relative to the two observers be functions of x and x^*, respectively, we require them to satisfy

---

6 In (5.30), σ_K' would then denote the derivative of σ_K with respect to ϱ.

7 See also (2.266) above.

$$\varrho^*(x^*) = \varrho(x), \tag{5.34}$$

since both x^* and x refer to the same body point seen against two different frames in space. That is, we require ϱ to be a frame-indifferent scalar, as defined in (2.44). By differentiating both sides of (5.34) with respect to x, by (5.33), the chain rule, and the property $\mathbf{R}^{-1} = \mathbf{R}^{\mathsf{T}}$, we show that

$$\nabla\varrho^* = \mathbf{R}\nabla\varrho,$$

where the gradient of ϱ^* is meant to be computed at x^*. Thus, for σ_{K} to retain its meaning for the new observer, in complete analogy with (5.34), it must obey the invariance property

$$\sigma_{\mathrm{K}}(\varrho^*, \nabla\varrho^*) = \sigma_{\mathrm{K}}(\varrho, \mathbf{R}\nabla\varrho) = \sigma_{\mathrm{K}}(\varrho, \nabla\varrho) \qquad \forall\, \mathbf{R} \in \mathrm{O}(3). \tag{5.35}$$

A classical theorem of CAUCHY (see also [357]) ensures that (5.35) is satisfied if and only if σ_{K} can be expressed as a function $\hat{\sigma}_{\mathrm{K}}$ of ϱ and $|\nabla\varrho|$:

$$\sigma_{\mathrm{K}}(\varrho, \nabla\varrho) = \hat{\sigma}_{\mathrm{K}}(\varrho, |\nabla\varrho|). \tag{5.36}$$

Since

$$\frac{\partial|\nabla\varrho|}{\partial\nabla\varrho} = \frac{\partial}{\partial\nabla\varrho}\sqrt{\nabla\varrho\cdot\nabla\varrho} = \frac{\nabla\varrho}{|\nabla\varrho|},$$

by the chain rule,

$$\frac{\partial\sigma_{\mathrm{K}}}{\partial\nabla\varrho} = \frac{\partial\hat{\sigma}_{\mathrm{K}}}{\partial|\nabla\varrho|}\frac{\nabla\varrho}{|\nabla\varrho|},$$

and so $\mathbf{T}_{\mathrm{K}}$ in (5.13) is symmetric.

There is another, perhaps less direct, proof of this property, which follows from a variant of the principle of frame-indifference, that is, the requirement that the free energy time rate $\dot{\mathscr{F}}_{\mathrm{K}}(\mathcal{P}_t)\big|_{t=0}$ be zero for every subbody $\mathcal{P} \subset \mathcal{B}$ along any rigid motion. To prove this, we begin by representing a rigid motion through the flow

$$\boldsymbol{v}_{\mathrm{R}}(x) = \boldsymbol{v}_{\mathrm{R}}(o) + \mathbf{W}x, \tag{5.37}$$

where $\mathbf{W}$ is a skew tensor, also called the *spin* tensor of $\boldsymbol{v}_{\mathrm{R}}$.[8] It readily follows from (5.37) that $\nabla\boldsymbol{v}_{\mathrm{R}} = \mathbf{W}$, and so, for a rigid motion, $\operatorname{div}\boldsymbol{v}_{\mathrm{R}} = 0$. Thus (5.11) becomes

$$\dot{\mathscr{F}}_{\mathrm{K}}(\mathcal{P}_t)\Big|_{t=0} = \int_{\mathcal{P}} \varrho\frac{\partial\sigma_{\mathrm{K}}}{\partial\nabla\varrho}\cdot\mathbf{W}\nabla\varrho\, dV = -\mathbf{W}\cdot\int_{\mathcal{P}} \varrho\nabla\varrho\otimes\frac{\partial\sigma_{\mathrm{K}}}{\partial\nabla\varrho}\, dV.$$

Hence requiring $\dot{\mathscr{F}}_{\mathrm{K}}(\mathcal{P}_t)\big|_{t=0}$ to vanish along any rigid flow and for every $\mathcal{P}$ amounts to requiring that the tensor

$$\nabla\varrho\otimes\frac{\partial\sigma_{\mathrm{K}}}{\partial\nabla\varrho}$$

be symmetric, thus proving the symmetry of $\mathbf{T}_{\mathrm{K}}$.

---

[8] By comparing (5.37) and (2.71), the reader will easily realize that they are identical, provided that $\mathbf{W}$ is identified with $\boldsymbol{\Omega}^*$.

5.2.6 Balances of Forces and Torques

A second-grade material can in general convey internal torques by means of a couple stress deriving from the hyperstress [338, 339] (see also Section 94 of [345]). We show now that the couple stress associated with the hyperstress **M** in (5.29) vanishes identically. To this end, we consider again a rigid virtual flow like (5.37). Since along it the left-hand side of (5.4) vanishes, so must also its right-hand side, provided that the balance equation (5.25) and the traction laws (5.26) and (5.27) are satisfied.

By inserting (5.26) and (5.27) into (5.2) evaluated along the flow (5.37), we readily obtain that

$$\begin{aligned}\mathscr{W}^{(a)}(\mathcal{P}) =& \boldsymbol{v}(o)\cdot\left[\int_{\mathcal{P}}\boldsymbol{b}\,dV+\int_{\partial^*\mathcal{P}}(\mathbf{T}_{\mathrm{K}}\boldsymbol{\nu}+\boldsymbol{t}_{\mathrm{K}})\,dA\right]\\ &+\mathbf{W}\cdot\left[\int_{\mathcal{P}}\boldsymbol{b}\otimes\boldsymbol{x}\,dV+\int_{\partial^*\mathcal{P}}(\mathbf{T}_{\mathrm{K}}\boldsymbol{\nu}\otimes\boldsymbol{x}+\boldsymbol{t}_{\mathrm{K}}\otimes\boldsymbol{x}+\boldsymbol{m}\otimes\boldsymbol{\nu})\,dA\right],\end{aligned}$$

where we have introduced the KORTEWEG *traction*

$$\boldsymbol{t}_{\mathrm{K}} := \nabla_{\mathrm{s}}\left(\varrho^2\frac{\partial\sigma_{\mathrm{K}}}{\partial\nabla\varrho}\cdot\boldsymbol{\nu}\right)-\varrho^2\left(\frac{\partial\sigma_{\mathrm{K}}}{\partial\nabla\varrho}\cdot\boldsymbol{\nu}\right)(\mathrm{div}_{\mathrm{s}}\,\boldsymbol{\nu})\boldsymbol{\nu}. \tag{5.38}$$

$\mathscr{W}^{(a)}(\mathcal{P})$ vanishes identically for all choices of $\boldsymbol{v}(o)$ and $\mathbf{W}$ if and only if

$$\int_{\mathcal{P}}\boldsymbol{b}\,dV+\int_{\partial^*\mathcal{P}}(\mathbf{T}_{\mathrm{K}}\boldsymbol{\nu}+\boldsymbol{t}_{\mathrm{K}})\,dA=\mathbf{0}\qquad\forall\,\mathcal{P}\subset\mathcal{B} \tag{5.39}$$

and

$$\int_{\mathcal{P}}\boldsymbol{x}\times\boldsymbol{b}\,dV+\int_{\partial^*\mathcal{P}}[\boldsymbol{x}\times(\mathbf{T}_{\mathrm{K}}\boldsymbol{\nu}+\boldsymbol{t}_{\mathrm{K}})+\boldsymbol{\nu}\times\boldsymbol{m}]\,dA=\mathbf{0}\quad\forall\,\mathcal{P}\subset\mathcal{B}. \tag{5.40}$$

These equations have a transparent mechanical interpretation; the former represents the balance of all forces acting on $\mathcal{P}$ and the latter represents the balance of all torques exerted by both forces and couples. By applying the divergence theorem, use of (5.25) reduces (5.39) to

$$\int_{\partial^*\mathcal{P}}\boldsymbol{t}_{\mathrm{K}}\,dA=\mathbf{0}\qquad\forall\mathcal{P}\subset\mathcal{B}, \tag{5.41}$$

while (5.27) and the symmetry of $\mathbf{T}_{\mathrm{K}}$ reduce (5.40) to

$$\int_{\partial^*\mathcal{P}}\boldsymbol{x}\times\boldsymbol{t}_{\mathrm{K}}\,dA=\mathbf{0}\qquad\forall\mathcal{P}\subset\mathcal{B}. \tag{5.42}$$

This latter equation shows that, by its specific structure, the hypertraction $\boldsymbol{m}$ in (5.27) does not convey torque, and so the couple stress associated with the hyperstress **M** in (5.29) vanishes identically.

We now prove directly that both equations (5.41) and (5.42) are identically satisfied as a consequence of (5.38), as they should, having been obtained by applying the principle of virtual power to a specific virtual flow, whereas both the balance equation (5.25) and the traction laws (5.26) and (5.27) were established by that very principle in its full generality.

Let $\boldsymbol{e}$ be any given unit vector. Then, by (5.38), (5.41) is equivalent to

$$\int_{\partial^*\mathcal{P}} \left[\boldsymbol{e} \cdot \nabla_{\mathrm{s}} \left(\varrho^2 \frac{\partial \sigma_{\mathrm{K}}}{\partial \nabla \varrho} \cdot \boldsymbol{\nu} \right) - \varrho^2 \left(\frac{\partial \sigma_{\mathrm{K}}}{\partial \nabla \varrho} \cdot \boldsymbol{\nu} \right) \operatorname{div}_{\mathrm{s}} \boldsymbol{\nu} (\boldsymbol{e} \cdot \boldsymbol{\nu}) \right] dA = 0,$$

which, since $\nabla \boldsymbol{e} \equiv \mathbf{0}$, can also be written as

$$\int_{\partial^*\mathcal{P}} \left\{ \operatorname{div}_{\mathrm{s}} \left[\left(\varrho^2 \frac{\partial \sigma_{\mathrm{K}}}{\partial \nabla \varrho} \cdot \boldsymbol{\nu} \right) \boldsymbol{e} \right] - \varrho^2 \left(\frac{\partial \sigma_{\mathrm{K}}}{\partial \nabla \varrho} \cdot \boldsymbol{\nu} \right) \operatorname{div}_{\mathrm{s}} \boldsymbol{\nu} (\boldsymbol{e} \cdot \boldsymbol{\nu}) \right\} dA = 0. \tag{5.43}$$

By applying to (5.43) the surface-divergence theorem, we conclude that this equation is identically satisfied for all $\boldsymbol{e} \in \mathbb{S}^2$ and $\mathcal{P} \subset \mathcal{B}$.

We find it convenient to rephrase (5.42) in Cartesian components:

$$\int_{\partial^*\mathcal{P}} \epsilon_{ijk} \left[x_j \chi_{;k} - x_j \nu_{h;h} \chi \nu_k \right] dA = 0, \tag{5.44}$$

where

$$\chi := \varrho^2 \frac{\partial \sigma_{\mathrm{K}}}{\partial \varrho_{,i}} \nu_i,$$

ϵ_{ijk} is RICCI's alternator, and a semicolon denotes surface differentiation. Integration by parts and use of the surface gradient-integral theorem in (5.22) allow us to rewrite the left-hand side of (5.44) as follows:

$$\int_{\partial^*\mathcal{P}} \epsilon_{ijk} \left[x_j \chi \nu_k \nu_{h;h} - \chi x_{j;k} - x_j \nu_{h;h} \chi \nu_k \right] dA = - \int_{\partial^*\mathcal{P}} \epsilon_{ijk} \chi P_{jk} \, dA, \tag{5.45}$$

where P_{jk} are the Cartesian components of the projection $\mathbf{P}(\boldsymbol{\nu})$ in (5.16). Since $\mathbf{P}(\boldsymbol{\nu})$ is symmetric, the integral on the right-hand side of (5.45) vanishes, and so (5.42) is identically satisfied for all $\mathcal{P}$.

We thus conclude that the KORTEWEG traction $\boldsymbol{t}_{\mathrm{K}}$ defined in (5.38) represents a system of self-equilibrated contact forces, which, in particular, would not affect the motion of any submerged rigid body. By contrast, in general, the hypertraction $\boldsymbol{m}$ in (5.27) is not self-equilibrated. However, according to (5.2) and (5.3), the power expended by $\boldsymbol{m}$ against a rigid motion vanishes identically, since, by (5.37),

$$\boldsymbol{m} \cdot \frac{\partial \boldsymbol{v}_{\mathrm{R}}}{\partial \nu} = \boldsymbol{m} \cdot \mathbf{W} \boldsymbol{\nu} = -\varrho^2 \left(\frac{\partial \sigma_{\mathrm{K}}}{\partial \nabla \varrho} \cdot \boldsymbol{\nu} \right) \boldsymbol{\nu} \cdot \mathbf{W} \boldsymbol{\nu} = 0,$$

since $\mathbf{W}$ is a skew tensor.

5.2.7 Dissipative Dynamics

The foregoing discussion on the equilibrium of isotropic KORTEWEG fluids served the purpose of identifying KORTEWEG stress, traction, and hypertraction. Our main interest in this book lies with dissipative fluid dynamics. To derive the basic equations of motion for a dissipative isotropic KORTEWEG fluid, we may replace the principle of virtual power with the principle of minimum reduced dissipation formulated for deformable bodies on page 137 of Chapter 2. Our treatment of both inertial and viscous forces would here be the same as that in Section 2.3.2 for a compressible NAVIER–STOKES fluid, and need not be repeated at this stage of our study. Thus the extension of (5.25) into the balance equation for linear momentum is as in (2.279),

$$\varrho \dot{\boldsymbol{v}} = \operatorname{div} \mathbf{T} + \boldsymbol{b}, \tag{5.46}$$

where now

$$\mathbf{T} = \mathbf{T}_{\mathrm{K}} + 2\mu \mathbf{D} + \lambda (\operatorname{div} \boldsymbol{v}) \mathbf{I} \tag{5.47}$$

is the complete stress tensor, including both elastic and viscous stresses.[9] The traction law (5.26) is accordingly modified,

$$\boldsymbol{t} = \mathbf{T}\boldsymbol{\nu} + \boldsymbol{t}_{\mathrm{K}},$$

while the expression for $\boldsymbol{m}$ in (5.27) remains unchanged. These equations should further be supplemented by the continuity equation (5.7).

Constitutive Assumption

We shall consider in this section a simple constitutive assumption for σ_{K}, which complies with the frame-indifference requirement (5.36):

$$\sigma_{\mathrm{K}} = \sigma_0(\varrho) + \frac{1}{2} u_1 |\nabla \varrho|^2, \tag{5.48}$$

where σ_0 is an increasing convex function of ϱ and u_1 is the *acoustic susceptibility*, which we take independent of ϱ. With this choice of σ_{K}, the stress tensor $\mathbf{T}_{\mathrm{K}}$ and the traction and hypertraction $\boldsymbol{t}_{\mathrm{K}}$ and $\boldsymbol{m}$ are given the following explicit forms:

$$\mathbf{T}_{\mathrm{K}} = -p_{\mathrm{K}} \mathbf{I} - u_1 \varrho \nabla \varrho \otimes \nabla \varrho \quad \text{with} \quad p_{\mathrm{K}} = \varrho^2 \sigma_0' - u_1 \varrho \operatorname{div}(\varrho \nabla \varrho), \tag{5.49}$$

$$\boldsymbol{t}_{\mathrm{K}} = u_1 \left[\nabla_{\mathrm{s}} \left(\varrho^2 \nabla_{\nu} \varrho \right) - (\operatorname{div}_{\mathrm{s}} \boldsymbol{\nu}) \varrho^2 (\nabla_{\nu} \varrho) \boldsymbol{\nu} \right],$$

$$\boldsymbol{m} = -u_1 \varrho^2 (\nabla_{\nu} \varrho) \boldsymbol{\nu},$$

where σ_0' denotes the derivative of σ_0 with respect to ϱ, and $\nabla_{\nu} \varrho = \nabla \varrho \cdot \boldsymbol{\nu}$. We further assume that $\boldsymbol{b} = \mathbf{0}$, so that no body force is applied to the medium.

It is easily seen that, under these assumptions, any uniform density field ϱ_0 would be compatible with equations (5.46) and (5.7) with $\boldsymbol{v} \equiv \mathbf{0}$, the equilibrium value ϱ_0 being selected by the total mass present in the body when the body is confined in space, or being treated as a parameter when the body is indefinite.

[9] Cf. (5.47) with (2.280).

5.2.8 Acoustic Plane Waves

We now seek a special solution to the equation of motion (5.46) under the specific constitutive assumption (5.48).

We imagine that an acoustic plane wave is being forced in the fluid by the vibration of a rigid plane at the angular frequency ω, which produces a disturbance in ϱ represented as

$$\varrho = \varrho_0(1 + s), \tag{5.50}$$

where the *condensation* s is given the form

$$s(x,t) = s_0 \Re\left(e^{\mathrm{i}(\boldsymbol{k}\cdot\boldsymbol{x}-\omega t)}\right), \tag{5.51}$$

where $\Re$ denotes the real part of a complex number, $\boldsymbol{x} := x - o$ with o a given origin, s_0 is a small dimensionless parameter, and $\boldsymbol{k}$ is the *complex* wave vector to be determined in terms of ω. Correspondingly, the velocity field $\boldsymbol{v}$ is taken as

$$\boldsymbol{v}(x,t) = s_0 \Re\left(e^{\mathrm{i}(\boldsymbol{k}\cdot\boldsymbol{x}-\omega t)}\right)\boldsymbol{a}, \tag{5.52}$$

where the amplitude $\boldsymbol{a}$ is an unknown complex vector.

As customary in acoustics, equations (5.7) and (5.46) for the perturbed mass density and the oscillating flow are linearized in s_0; in them the fields ϱ and $\boldsymbol{v}$ will be represented as complex exponentials, with the proviso that only their real parts bear a physical meaning. We shall represent $\boldsymbol{k}$ as

$$\boldsymbol{k} = k\boldsymbol{e} \quad \text{with} \quad k = k_1 + \mathrm{i}k_2, \tag{5.53}$$

where $\boldsymbol{e} \in \mathbb{S}^2$ represents the propagation direction. Similarly, $\boldsymbol{a}$ will be represented as

$$\boldsymbol{a} = a\boldsymbol{f} \quad \text{with} \quad a = a_1 + \mathrm{i}a_2, \tag{5.54}$$

where $\boldsymbol{f} \in \mathbb{S}^2$. The imaginary part k_2 of k will be associated with the *attenuation* of the wave: when $k_2 > 0$, its reciprocal represents the length over which the wave amplitude is reduced by the factor $1/e$; such a length is also called the *attenuation length*. The imaginary part a_2 of a is related to the phase shift between the condensation wave and the velocity field it carries along.

Our program is now to seek solutions in the form (5.51) and (5.52) to the continuity equation (5.7) and the balance equation of linear momentum (5.46) with the stress tensor $\mathbf{T}$ expressed as in (5.47), under the assumption that only linear terms in the perturbation parameter s_0 are to be retained.

To this end, we set

$$E := e^{\mathrm{i}(\boldsymbol{k}\cdot\boldsymbol{x}-\omega t)}, \tag{5.55}$$

for brevity, and we compute

$$\nabla\boldsymbol{v} = s_0 \mathrm{i} E \boldsymbol{a} \otimes \boldsymbol{k},$$

whence it follows that

$$\mathbf{D} = \frac{1}{2} s_0 \mathrm{i} E (\boldsymbol{a} \otimes \boldsymbol{k} + \boldsymbol{k} \otimes \boldsymbol{a}) \tag{5.56}$$

and

$$\operatorname{div} \boldsymbol{v} = s_0 \mathrm{i} E \boldsymbol{a} \cdot \boldsymbol{k}.$$

Similarly, we obtain

$$\nabla \varrho = s_0 \mathrm{i} \varrho_0 E \boldsymbol{k}.$$

It easily follows from (2.38), (5.50), and (5.51) that

$$\dot{\varrho} = -s_0 \mathrm{i} \varrho_0 E k + o(s_0).$$

Thus, up to first order in s_0, equation (5.7) becomes

$$\omega = \boldsymbol{a} \cdot \boldsymbol{k}. \tag{5.57}$$

Similarly, also by (5.52) and (5.56),

$$\varrho \dot{\boldsymbol{v}} = -s_0 \mathrm{i} \varrho_0 \omega E \boldsymbol{a} + o(s_0) \tag{5.58}$$

and

$$\mathbf{T} = -p_{\mathrm{K}} \mathbf{I} + s_0 \mathrm{i} \mu E (\boldsymbol{a} \otimes \boldsymbol{k} + \boldsymbol{k} \otimes \boldsymbol{a}) + s_0 \mathrm{i} \lambda E (\boldsymbol{a} \cdot \boldsymbol{k}) \mathbf{I}. \tag{5.59}$$

By (5.49), p_{K} can here be given the following expression:

$$p_{\mathrm{K}} = p_0(\varrho_0) + s_0 \varrho_0 c_0^2 E + s_0 \varrho_0^3 u_1 k^2 E + o(s_0), \tag{5.60}$$

where we have set

$$p_0(\varrho_0) := \varrho_0^2 \sigma_0'(\varrho_0)$$

and

$$c_0 := \sqrt{p_0'}, \tag{5.61}$$

where p_0', the derivative of p_0 with respect to ϱ, is the velocity of sound in a classical isotropic compressible viscous fluid described by (5.47) and (5.49) with $u_1 = 0$.

It readily follows from (5.58) and (5.59) that for ϱ and $\boldsymbol{v}$ as in (5.50), (5.51), and (5.52), up to first order in s_0, the balance equation of linear momentum (5.46) becomes

$$\mathrm{i} \varrho_0 \omega \boldsymbol{a} = \mathrm{i} \varrho_0 c_0^2 \boldsymbol{k} + \mathrm{i} u_1 \varrho_0^3 k^2 \boldsymbol{k} + \mu [k^2 \boldsymbol{a} + (\boldsymbol{a} \cdot \boldsymbol{k}) \boldsymbol{k}] + \lambda (\boldsymbol{a} \cdot \boldsymbol{k}) \boldsymbol{k}. \tag{5.62}$$

By (5.53) and (5.54), this equation implies that $\boldsymbol{f} = \boldsymbol{e}$, and so the wave is purely longitudinal. This, in turn, transforms (5.57) into

$$\left(\frac{\omega}{c} + \mathrm{i} k_2 \right) (a_1 + \mathrm{i} a_2) = \omega, \tag{5.63}$$

where we have set

$$k_1 =: \frac{\omega}{c}, \tag{5.64}$$

with c representing the still undetermined velocity of sound in the KORTEWEG fluid being studied. Equation (5.63) has the following solution:

$$a_1 = \frac{\omega^2 c}{\omega^2 + c^2 k_2^2}, \qquad a_2 = -\frac{\omega c^2 k_2}{\omega^2 + c^2 k_2^2}, \tag{5.65}$$

showing that for $k_2 > 0$ the velocity wave has a negative phase shift relative to the condensation wave, so that the latter wave precedes the former. Moreover, taking the inner product with $\boldsymbol{k}$ of both sides of equation (5.62), we arrive at

$$\mathrm{i}\omega^2 = \mathrm{i}c_0^2 k^2 + \mathrm{i}u_1 \varrho_0^2 k^4 + \nu\omega k^2, \tag{5.66}$$

where

$$\nu := \frac{1}{\varrho_0}(2\mu + \lambda).$$

It follows from inequalities (2.255), which ensure positive semidefiniteness to the RAYLEIGH dissipation function for a compressible NAVIER–STOKES fluid, that

$$\nu \geqq \frac{4\mu}{3\varrho_0} \geqq 0.$$

By (5.64) and (5.53), (5.66) is equivalent to the pair of equations corresponding for any $\omega > 0$ to its real and imaginary parts:

$$\left[2\left(\frac{c_0}{c}\right) + \omega^2\tau_1^2\left(\frac{c_0}{c}\right)^3\right]k_2' - \omega^2\tau_1^2\left(\frac{c_0}{c}\right)k_2'^3 - \nu'\left[\left(\frac{c_0}{c}\right)^2 - k_2'^2\right] = 0, \tag{5.67}$$

$$\left(\frac{c_0}{c}\right)^2\left[1 + \frac{1}{4}\omega^2\tau_1^2\left(\frac{c_0}{c}\right)^2\right] - 1 - \left[1 + \frac{3}{2}\omega^2\tau_1^2\left(\frac{c_0}{c}\right)^2\right]k_2'^2 + \frac{1}{4}\omega^2\tau_1^2 k_2'^4 + 2\left(\frac{c_0}{c}\right)\nu' k_2' = 0. \tag{5.68}$$

Here we have set

$$\tau_1 := 2\frac{\varrho_0}{c_0^2}\sqrt{u_1}, \tag{5.69}$$

which defines a characteristic time related to the acoustic susceptibility u_1, and we have introduced the following dimensionless quantities:

$$k_2' := \frac{c_0}{\omega}k_2 \quad \text{and} \quad \nu' := \frac{\omega}{c_0^2}\nu. \tag{5.70}$$

For $\nu' = 0$, that is, in the inviscid limit, equation (5.67) possesses three roots, namely, $k_2' = 0$ and

$$k_2' = \pm\sqrt{\frac{2}{\omega^2\tau_1^2} + \left(\frac{c_0}{c}\right)^2}.$$

Inserting these latter into (5.68) with $\nu' = 0$, we find the following equation for c:

$$\omega^2\tau_1^2 + 1 = -\omega^2\tau_1^2\left(\frac{c_0}{c}\right)^2\left[\omega^2\tau_1^2\left(\frac{c_0}{c}\right)^2 + 2\right],$$

which clearly fails to possess a real root. Hence, if $\nu' = 0$, then $k_2' = 0$ and c is determined by the *dispersion* equation

$$\frac{1}{4}\omega^2\tau_1^2\left(\frac{c_0}{c}\right)^4 + \left(\frac{c_0}{c}\right)^2 - 1 = 0, \tag{5.71}$$

obtained from setting $k_2' = 0$ in (5.68). The only positive real root of (5.71) is

$$\frac{c}{c_0} = \frac{\omega\tau_1}{\sqrt{2\left(\sqrt{1+\omega^2\tau_1^2} - 1\right)}}. \tag{5.72}$$

Thus, as expected, in the inviscid limit the wave is not attenuated, and, as also shown in Figure 5.1, $c \geqq c_0$, for all $\omega \geqq 0$, where $c = c_0$ only for $u_1 = 0$. Moreover,

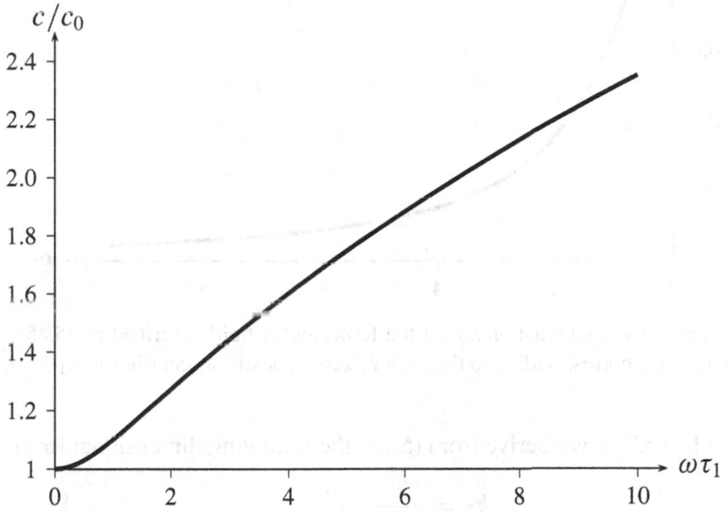

Fig. 5.1. The speed of sound c in the KORTEWEG fluid described by (5.48), scaled to the speed c_0 corresponding to the limit of zero acoustic susceptibility, $u_1 = 0$.

asymptotically,

$$c \approx \frac{c_0}{\sqrt{2}}\sqrt{\omega\tau_1} \quad \text{for} \quad \omega\tau_1 \gg 1.$$

Attenuation

For $\nu' > 0$, we now determine both k_2' and c in a perturbation limit by continuing the solution just found for $\nu' = 0$. In particular, we assume that both ν' and k_2' are

$O(s_0)$ and neglect in equations (5.67) and (5.68) all terms of order higher than 1 in s_0. Doing so, we readily obtain from (5.67) that

$$k_2' = \frac{\nu'}{2} \frac{1}{\frac{c}{c_0} + \frac{1}{2}\omega^2 \tau_1^2 \frac{c_0}{c}}, \tag{5.73}$$

while, to this order of approximation, (5.68) still reduces to (5.71). By (5.70), assuming k_2' to be small amounts to assuming that the attenuation length associated with k_2 is much larger than the wavelength corresponding to the limiting case of zero acoustic susceptibility, $u_1 \to 0$.

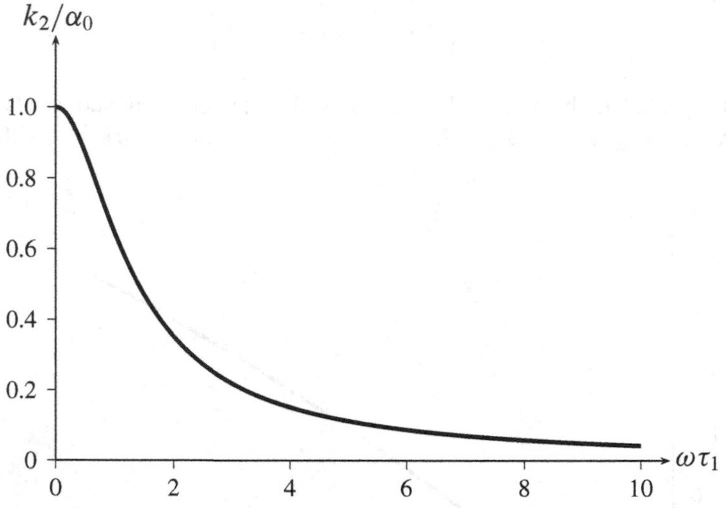

Fig. 5.2. The acoustic attenuation k_2 for the KORTEWEG fluid described by (5.48), scaled to the attenuation α_0 corresponding to the limit of zero acoustic susceptibility, $u_1 = 0$.

Again by (5.70), we derive from (5.73) the following dimensional form:

$$k_2 = \frac{\alpha_0}{\frac{c}{c_0} + \frac{1}{2}\omega^2 \tau_1^2 \frac{c_0}{c}}, \tag{5.74}$$

where

$$\alpha_0 := \frac{\nu\omega^2}{2c_0^3}$$

is the attenuation in the limit of zero acoustic susceptibility and c is expressed by (5.72) as a function of ω. As also shown in Figure 5.2, $k_2 < \alpha_0$ for all $\omega > 0$, provided that $u_1 > 0$; asymptotically,

$$k_2 \approx \frac{\sqrt{2}\alpha_0}{(\omega\tau_1)^{3/2}} \quad \text{for} \quad \omega\tau_1 \gg 1.$$

Thus, for any given frequency ω, the wave propagating in the viscous KORTEWEG fluid considered here is quicker and less attenuated than the corresponding wave propagating in the limit of zero acoustic susceptibility.

Within the same approximation that led us to (5.74), we obtain from (5.65) that

$$a_1 = c \quad \text{and} \quad a_2 = -\frac{c^2}{\omega} k_2.$$

Acoustic Intensity

The *acoustic intensity* carried by the wave is defined as

$$I_a := \langle p_K \boldsymbol{v} \cdot \boldsymbol{e} \rangle, \tag{5.75}$$

where $\langle \cdot \rangle$ now denotes the time average over a period and $\boldsymbol{e} \in \mathbb{S}^2$ designates the direction of propagation. To within second order in s_0, by (5.60), (5.64), and (5.75),

$$I_a = \varrho_0 s_0^2 \left(c_0^2 c + u_1 \varrho_0^2 \frac{\omega^2}{c} \right) \left\langle (\Re E)^2 \right\rangle = I_0 \left[\left(\frac{c}{c_0} \right) + \frac{1}{4} \omega^2 \tau_1^2 \left(\frac{c_0}{c} \right) \right], \tag{5.76}$$

where

$$I_0 := \frac{1}{2} \varrho_0 s_0^2 c_0^3 e^{-2k_2 \boldsymbol{x} \cdot \boldsymbol{e}} \tag{5.77}$$

is the acoustic intensity of the wave in the limit of zero acoustic susceptibility and c is given by (5.72) as a function of ω.

The graph of I_a scaled to I_0 is shown in Figure 5.3 against $\omega\tau_1$; it reveals how the acoustic susceptibility u_1 increases the acoustic intensity. The asymptotic behavior of I_a for large frequencies compensates exactly the attenuation k_2, since

$$I_a k_2 \approx I_0 \alpha_0 \quad \text{for} \quad \omega\tau_1 \gg 1.$$

The viscous KORTEWEG fluid studied in in this section was isotropic. We shall examine in the following section what consequences relate especially to the propagation of a plane acoustic wave in a nematic liquid crystal by adding to the elastic free energy density in (5.48) an anisotropic acoustic susceptibility coupling the density gradient with the director. Such a formal alteration will open the way to regard a nematic liquid crystal as an anisotropic KORTEWEG fluid at the time scales comparable to the period of ultrasonic waves.

5.3 Nematoacoustic Theory

Having developed in the preceding section the general theory for dissipative isotropic KORTEWEG fluids, in this section we base our nematoacoustic theory on the postulation that at sufficiently high frequencies a nematic liquid crystal behaves like a particular anisotropic KORTEWEG fluid, symmetric about the local director $\boldsymbol{n}$. The

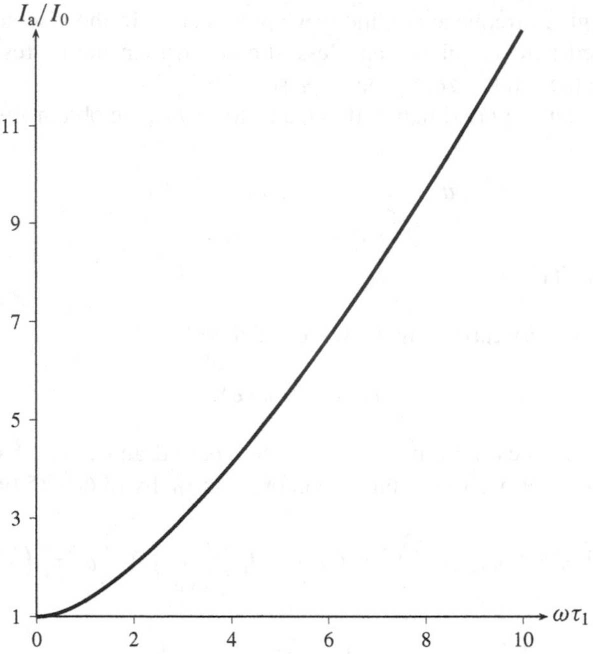

Fig. 5.3. The acoustic intensity I_a for the KORTEWEG fluid described by (5.48), scaled to the acoustic intensity I_0 corresponding to the limit of zero acoustic susceptibility, $u_1 = 0$.

behavior at longer time scales remains as already described in the preceding chapters of this book; in the presence of a fast phenomenon, such as the propagation of an ultrasonic wave, what survives at the longer time scales is the average of whatever fast variable bears a mechanical meaning. We imagine distinguishing fast and slow dynamics, the former evolving as if the latter were not, this latter being influenced only by the time average of the other.

In the fast dynamics, a nematic liquid crystal may reveal features that do not generally characterize its slow dynamics. For example, the very possibility of sound propagation in liquid crystals resides in their being compressible, a property that is generally denied to the slow dynamics. Fast and slow dynamics mutually interfere with one another: the fast dynamics interfere with the slow dynamics by providing time-averaged sources; the slow dynamics in turn drive the background against which the fast dynamics are taking place. Such an interplay will in particular be illuminated by the propagation of ultrasonic waves: they produce an acoustic torque on the nematic director, which later affects the slow director dynamics; this will eventually alter the wave propagation and with it the acoustic torque. Bridging rigorously the different time scales of fast and slow dynamics for ultrasonic wave propagation in nematic liquid crystals will be the primary object of this section. We begin by considering the RAYLEIGH dissipation function for a compressible nematic liquid crystal.

5.3.1 Acoustic Dissipation Function

At the acoustic time scale, a nematic liquid crystal is regarded as being compressible, and so the velocity field $\boldsymbol{v}$ is no longer solenoidal, though its time average is so. This point of view is not unprecedented in the literature: for example, in the hydrodynamic theory of liquid crystals proposed in [106] and [149], liquid crystals are compressible fluids. Thus, the *acoustic* dissipation function R_a, which like R in (3.49) above depends on the director $\boldsymbol{n}$, its corotational time derivative $\mathring{\boldsymbol{n}}$, and the stretching tensor $\mathbf{D}$, being quadratic in the pair $(\mathbf{D}, \mathring{\boldsymbol{n}})$, may also depend on $\operatorname{tr}\mathbf{D}$, the new invariant introduced by removing the constraint on the divergence of $\boldsymbol{v}$. Only two quadratic terms in $\mathbf{D}$ containing $\operatorname{tr}\mathbf{D}$ may be added to R in (3.49), namely, $(\operatorname{tr}\mathbf{D})^2$ and $(\operatorname{tr}\mathbf{D})\boldsymbol{n}\cdot\mathbf{D}\boldsymbol{n}$. Therefore, R_a is defined as

$$
\begin{aligned}
R_a(\boldsymbol{n};\mathbf{D},\mathring{\boldsymbol{n}}) :=& \frac{1}{2}\gamma_1\mathring{\boldsymbol{n}}^2 + \gamma_2\mathring{\boldsymbol{n}}\cdot\mathbf{D}\boldsymbol{n} + \frac{1}{2}\gamma_3(\mathbf{D}\boldsymbol{n})^2 + \frac{1}{2}\gamma_4(\boldsymbol{n}\cdot\mathbf{D}\boldsymbol{n})^2 + \frac{1}{2}\gamma_5\operatorname{tr}\mathbf{D}^2 \\
&+ \frac{1}{2}\gamma_6(\operatorname{tr}\mathbf{D})^2 + \gamma_7(\operatorname{tr}\mathbf{D})\boldsymbol{n}\cdot\mathbf{D}\boldsymbol{n},
\end{aligned} \tag{5.78}
$$

where $\gamma_1,\dots,\gamma_5$ are viscosity coefficients related to LESLIE's viscosities $\alpha_1,\dots,\alpha_5$ through equations (3.62)[10] and γ_6, γ_7 are new viscosity coefficients arising from the material compressibility in the fast acoustic propagation. Here all coefficients $\gamma_1,\dots,\gamma_7$ will be considered as functions of the mass density ϱ.

We already described in (3.57) the conditions on the viscosity coefficients γ_1, $\dots,\gamma_5$ that make R in (3.49) positive semidefinite in all admissible motions. That discussion must now be extended to R_a; the additional viscosities γ_6 and γ_7 are likely to play a role in the positive semidefiniteness of R_a. For given $\boldsymbol{n}$, $\mathring{\boldsymbol{n}}$ is subject only to the condition of being orthogonal to $\boldsymbol{n}$, while $\mathbf{D}$ is here an arbitrary symmetric tensor. As we did in (3.53) for R, with no loss in generality, we can now represent them in the following form:

$$
\mathring{\boldsymbol{n}} = N\boldsymbol{e}_2 \quad \text{and} \quad \mathbf{D} = \sum_{i,j=1}^{3} A_{ij}\boldsymbol{e}_i\otimes\boldsymbol{e}_j \quad \text{with} \quad A_{ij} = A_{ji}, \tag{5.79}
$$

where $(\boldsymbol{e}_1,\boldsymbol{e}_2,\boldsymbol{e}_3)$ is an orthonormal frame such that $\boldsymbol{n} = \boldsymbol{e}_1$. By inserting (5.79) into (5.78), we transform R_a into the sum of four quadratic forms in the independent variables A_{13}, A_{23}, (N, A_{12}), and (A_{11}, A_{22}, A_{33}), respectively:

$$
\begin{aligned}
R_a =& \left(\frac{1}{2}\gamma_3 + \gamma_5\right)A_{13}^2 + \gamma_5 A_{23}^2 \\
&+ \frac{1}{2}\gamma_1 N^2 + \gamma_2 N A_{12} + \left(\frac{1}{2}\gamma_3 + \gamma_5\right)A_{12}^2 \\
&+ \frac{1}{2}(\gamma_3+\gamma_4+\gamma_5+\gamma_6+2\gamma_7)A_{11}^2 + (\gamma_6+\gamma_7)A_{11}A_{22} + \frac{1}{2}(\gamma_5+\gamma_6)A_{22}^2 \\
&+ \gamma_6 A_{22}A_{33} + (\gamma_6+\gamma_7)A_{11}A_{33} + \frac{1}{2}(\gamma_5+\gamma_6)A_{33}^2.
\end{aligned}
$$

[10] Conversely, equations (3.60) express the α's in terms of the γ's.

Necessary and sufficient conditions for R_a to be positive semidefinite are the inequalities

$$\gamma_5 \geqq 0 \quad \text{and} \quad \gamma_3 + 2\gamma_5 \geqq 0, \tag{5.80}$$

and the positive semidefiniteness of the symmetric matrices

$$H_1 := \begin{bmatrix} \gamma_1 & \gamma_2 \\ \gamma_2 & \gamma_3 + 2\gamma_5 \end{bmatrix}$$

and

$$H_2 := \begin{bmatrix} \gamma_3 + \gamma_4 + \gamma_5 + \gamma_6 + 2\gamma_7 & \gamma_6 + \gamma_7 & \gamma_6 + \gamma_7 \\ \gamma_6 + \gamma_7 & \gamma_5 + \gamma_6 & \gamma_6 \\ \gamma_6 + \gamma_7 & \gamma_6 & \gamma_5 + \gamma_6 \end{bmatrix}.$$

Following in part [319] (see, in particular, p. 146), we recall that both H_1 and H_2 are positive semidefinite whenever all their principal minors are nonnegative[11] (see, for example, p. 7 of [18] for this positive semidefiniteness criterion). The principal minors of H_1 are its determinant and the entries γ_1 and $\gamma_3+2\gamma_5$, and so H_1 is positive semidefinite whenever

$$\gamma_1 \geqq 0, \quad \gamma_3 + 2\gamma_5 \geqq 0, \quad \text{and} \quad \gamma_1(\gamma_3 + 2\gamma_5) - \gamma_2^2 \geqq 0. \tag{5.81}$$

Clearly, $(5.81)_2$ reproduces $(5.80)_2$, which will henceforth be redundant.

To ensure that H_2 is positive semidefinite, we begin by requiring that all its leading principal minors[12] be nonnegative:

$$\gamma_3 + \gamma_4 + \gamma_5 + \gamma_6 + 2\gamma_7 \geqq 0, \tag{5.82a}$$

$$\gamma_3\gamma_5 + \gamma_3\gamma_6 + \gamma_5^2 + 2\gamma_5\gamma_6 + \gamma_5\gamma_4 + \gamma_4\gamma_6 + 2\gamma_5\gamma_7 - \gamma_7^2 \geqq 0, \tag{5.82b}$$

$$\gamma_5[\gamma_3\gamma_5 + 2\gamma_3\gamma_6 + \gamma_5^2 + 3\gamma_5\gamma_6 + \gamma_5\gamma_4 + 2\gamma_4\gamma_6 + 2\gamma_5\gamma_7 - 2\gamma_7^2] \geqq 0. \tag{5.82c}$$

It is easily seen, also with the aid of $(5.80)_1$, that inequalities (5.82b) and (5.82c) are equivalent to

$$\gamma_5 a + b \geqq 0 \quad \text{and} \quad \gamma_5 a + 2b \geqq 0, \tag{5.83}$$

respectively, with

$$a := \gamma_3 + \gamma_4 + \gamma_5 + \gamma_6 + 2\gamma_7 \quad \text{and} \quad b := \gamma_6(\gamma_3 + \gamma_4 + \gamma_5) - \gamma_7^2.$$

Since, by $(5.80)_1$ and (5.82a), both γ_5 and a are nonnegative, the second inequality in (5.83) is more stringent than the first, and so (5.82c) implies (5.82b).

Three extra inequalities are derived by also requiring the remaining principal minors of H_2 to be nonnegative. The first coincides with $(5.83)_1$, while the others are

---

[11] A principal submatrix of an $n \times n$ matrix M is a submatrix of M whose principal diagonal is part of the principal diagonal of M. A principal minor of M is the determinant of a principal submatrix of M. We suggest [17] and [18] as general references on these matters.

[12] A leading principal minor of an $n \times n$ matrix M is the determinant of a principal submatrix of M identified by the first j rows and the first j columns of M, with $j \leqq n$.

$$\gamma_5 + \gamma_6 \geqq 0 \quad \text{and} \quad \gamma_5(\gamma_5 + 2\gamma_6) \geqq 0.$$

Again by $(5.80)_1$, the latter inequality implies the former.[13]

In summary, R_a is positive semidefinite whenever

$$\gamma_1 \geqq 0, \tag{5.84a}$$

$$\gamma_3 + 2\gamma_5 \geqq 0, \tag{5.84b}$$

$$\gamma_5 \geqq 0, \tag{5.84c}$$

$$\gamma_5 + 2\gamma_6 \geqq 0, \tag{5.84d}$$

$$\gamma_3 + \gamma_4 + \gamma_5 + \gamma_6 + 2\gamma_7 \geqq 0, \tag{5.84e}$$

$$\gamma_1\gamma_3 + 2\gamma_1\gamma_5 - \gamma_2^2 \geqq 0, \tag{5.84f}$$

$$\gamma_5(\gamma_3 + \gamma_4 + \gamma_5 + \gamma_6 + 2\gamma_7) + 2[\gamma_6(\gamma_3 + \gamma_4 + \gamma_5) - \gamma_7^2] \geqq 0. \tag{5.84g}$$

These inequalities should be compared to those already collected in (3.57) for the positive semidefiniteness of the RAYLEIGH dissipation function R in the incompressible limit and to the corresponding inequalities (3.63), expressed in terms of LESLIE's viscosities, which we reproduce below for the reader's ease:

$$\alpha_3 \geqq \alpha_2, \tag{5.85a}$$

$$\alpha_4 \geqq 0, \tag{5.85b}$$

$$\alpha_2 + \alpha_3 + 2\alpha_4 + 2\alpha_5 \geqq 0, \tag{5.85c}$$

$$2(\alpha_1 + \alpha_2 + \alpha_3) + 3\alpha_4 + 4\alpha_5 \geqq 0, \tag{5.85d}$$

$$(\alpha_3 - \alpha_2)(\alpha_2 + \alpha_3 + 2\alpha_4 + 2\alpha_5) \geqq (\alpha_2 + \alpha_3)^2. \tag{5.85e}$$

By letting $\gamma_6 = \gamma_7 = 0$ in (5.84), which amounts to silencing the extra acoustic dissipation terms introduced in (5.78), and by using (3.62), we see that (5.84d) collapses into (5.84c), which is equivalent to (5.85b), and that, by (5.84c), (5.84g) reduces to (5.84e), and together they become

$$\alpha_1 + \alpha_2 + \alpha_3 + \alpha_4 + 2\alpha_5 \geqq 0. \tag{5.86}$$

While (5.84a), (5.84b), and (5.84f) reproduce (5.85a), (5.85c), and (5.85e), respectively, (5.86) and (5.85b) imply (5.85d), as is easily seen by multiplying both sides of (5.86) by 2 and then adding α_4 to the left-hand side. This shows that inequalities (5.84) and (5.85) are not equivalent in the limit as both γ_6 and γ_7 vanish, but that

---

[13] We note in passing that were we to require a real symmetric matrix M to be strictly positive, we could be contented with requiring all its leading principal minors to be positive; positivity of all other principal minors would follow as a consequence. The simple matrix

$$M = \begin{bmatrix} 0 & 0 \\ 0 & -1 \end{bmatrix}$$

illustrates how nonnegativity of all leading principal minors of M does not imply positive semidefiniteness of M (see again p. 7 of [18]).

the former inequalities imply the latter. That inequalities (5.84) with $\gamma_6 = \gamma_7 = 0$ are more stringent than (5.85) should not surprise us, for in deriving (5.84) the variables A_{11}, A_{22}, and A_{33} were not subject to the constraint $A_{11} + A_{22} + A_{33} = 0$ in (3.52), which was instead enforced to obtain (3.57), and so (5.85), which simply transliterate the former into the alphabet of the α's.

5.3.2 Nematoacoustic Equations

Here we derive the equations that govern acoustic propagation in nematic liquid crystals, assuming that a nematic liquid crystal, as seen from an acoustic wave propagating through it, behaves like an anisotropic, compressible KORTEWEG fluid with RAYLEIGH dissipation function R_a as in (5.78). More specifically, we assume that at the acoustic time scale (comparable with the wave period) the nematic director $\boldsymbol{n}$ is immobile, so that its dynamics can be appreciated only over much longer time scales. Similarly, we assume that at the acoustic length scale (comparable with the wavelength) $\boldsymbol{n}$ is undistorted, that is, $\nabla\boldsymbol{n} \equiv \mathbf{0}$, so that nematic distortions can appear only over much larger length scales. In particular, this latter assumption will imply that the nematoacoustic equations may be derived by neglecting the elastic energy density W introduced in (3.13) on page 170. At the acoustic length scale, the role of W is to be played by the KORTEWEG energy density σ_K introduced in Section 5.2 above.

Balance Laws

Following the general theory presented in Section 3.1, in the absence of body forces, the balance of linear momentum is expressed by the equation

$$\varrho\dot{\boldsymbol{v}} = \operatorname{div}(\mathbf{T}_E + \mathbf{T}_K + \mathbf{T}_{dis}), \tag{5.87}$$

where, as shown in in Section 3.1.4, the ERICKSEN stress has the form

$$\mathbf{T}_E = -W\mathbf{I} - (\nabla\boldsymbol{n})^{\mathsf{T}}\frac{\partial W}{\partial\nabla\boldsymbol{n}},$$

the KORTEWEG stress $\mathbf{T}_K$ is defined as in (5.13), and the dissipative stress $\mathbf{T}_{dis}$ is given by

$$\mathbf{T}_{dis} = \frac{1}{2}\left(\boldsymbol{n}\otimes\frac{\partial R_a}{\partial\mathring{\boldsymbol{n}}} - \frac{\partial R_a}{\partial\mathring{\boldsymbol{n}}}\otimes\boldsymbol{n}\right) + \frac{\partial R_a}{\partial\mathbf{D}}. \tag{5.88}$$

It is worth noting that by (5.78),

$$\frac{\partial R_a}{\partial\mathring{\boldsymbol{n}}} = \frac{\partial R}{\partial\mathring{\boldsymbol{n}}},$$

where R is the RAYLEIGH dissipation function in the incompressible limit (3.49), whereas

$$\frac{\partial R_a}{\partial\mathbf{D}} = \frac{\partial R}{\partial\mathbf{D}} + [\gamma_6\operatorname{tr}\mathbf{D} + \gamma_7(\boldsymbol{n}\cdot\mathbf{D}\boldsymbol{n})]\mathbf{I} + \gamma_7(\operatorname{tr}\mathbf{D})\boldsymbol{n}\otimes\boldsymbol{n}.$$

Similarly, by (3.68) and (3.69), the balance of torques is expressed by the equation

$$-\boldsymbol{n} \times \left(\varrho \frac{\partial \sigma_{\mathrm{K}}}{\partial \boldsymbol{n}} + \frac{\partial R_{\mathrm{a}}}{\partial \mathring{\boldsymbol{n}}} + \frac{\partial W}{\partial \boldsymbol{n}} - \operatorname{div} \frac{\partial W}{\partial \nabla \boldsymbol{n}} \right) = \mathbf{0}, \tag{5.89}$$

where the external body couple $\boldsymbol{n} \times \boldsymbol{k}^{n}$ has been set equal to zero. As customary in nematodynamics, no inertial torque appears in (5.89), since the *microkinetic* energy κ associated with the motion of $\boldsymbol{n}$ is systematically neglected relative to the predominant macroscopic kinetic energy $\kappa_0 = \frac{1}{2}\varrho v^2$ of the flow. At the acoustic frequencies, however, such an assumption must be subject to scrutiny. Here we write κ as

$$\kappa := \frac{1}{2} \varrho \delta^2 \dot{\boldsymbol{n}}^2, \tag{5.90}$$

with δ a molecular *radius of gyration*. We shall estimate in Section 5.4.2 below the acoustic frequency that cannot be exceeded for this energy to be safely neglected with respect to κ_0.

Letting W be a quadratic form[14] of $\nabla \boldsymbol{n}$, we write

$$W(\boldsymbol{n}, \nabla \boldsymbol{n}) := \frac{1}{2} K_1 (\operatorname{div} \boldsymbol{n})^2 + \frac{1}{2} K_2 (\boldsymbol{n} \cdot \operatorname{curl} \boldsymbol{n})^2 + \frac{1}{2} K_3 |\boldsymbol{n} \times \operatorname{curl} \boldsymbol{n}|^2, \tag{5.91}$$

where the *elastic constants* K_1, K_2, and K_3 are material parameters, which we assume to be independent of the mass density ϱ. We easily conclude from (5.91) that for a uniform director alignment, for which $\nabla \boldsymbol{n} \equiv \mathbf{0}$, $\mathbf{T}_{\mathrm{E}}$ vanishes identically, and so also does the elastic torque, since

$$\frac{\partial W}{\partial \boldsymbol{n}} = \mathbf{0} \quad \text{and} \quad \frac{\partial W}{\partial \nabla \boldsymbol{n}} = \mathbf{0}.$$

Both these equations must be inserted into (5.89).

At the acoustic time and length scales, equation (5.89) does *not* govern the director evolution: as shown below, its time average over an acoustic period will provide the acoustic torque unbalance, responsible for linking the fast acoustic dynamics with the slow director relaxation.

As usual, the balance equations (5.87) and (5.89) are to be supplemented by the mass continuity equation (5.7).

As explained in Section 5.2.4, the total traction $\boldsymbol{t}$ transmitted through a surface $\mathcal{S}$ within the fluid is given by

$$\boldsymbol{t} = (\mathbf{T}_{\mathrm{K}} + \mathbf{T}_{\mathrm{dis}})\, \boldsymbol{\nu} + \boldsymbol{t}_{\mathrm{K}},$$

where $\boldsymbol{t}_{\mathrm{K}}$ is as in (5.38) and $\boldsymbol{\nu}$ is the outer normal to $\mathcal{S}$. Likewise, the hypertraction $\boldsymbol{m}$ is given by (5.27).

[14] As originally proposed by FRANK [109] (see also Chapter 3 of [353]).

Constitutive Assumption

Elaborating on (5.48), here we write σ_K as a function of ϱ, $\nabla\varrho$, and $\boldsymbol{n}$,

$$\sigma_K(\varrho, \nabla\varrho, \boldsymbol{n}) = \sigma_0(\varrho) + \frac{1}{2}\left[u_1|\nabla\varrho|^2 + u_2\,(\nabla\varrho\cdot\boldsymbol{n})^2\right], \tag{5.92}$$

where the *acoustic susceptibilities* u_1 and u_2 are assumed to be constitutive parameters independent of ϱ. Clearly, the KORTEWEG behavior of a nematic liquid crystal at the acoustic time and length scales described by (5.92) is anisotropic about $\boldsymbol{n}$. In (5.92), the terms in square brackets represent the most general addition to σ_0 that depends on $\boldsymbol{n}$ and is both quadratic in $\nabla\varrho$ and frame-indifferent. It is easily seen that for such an additional energy to be positive semidefinite, it is necessary and sufficient that u_1 and u_2 obey the inequalities

$$u_1 \geqq 0 \quad \text{and} \quad u_1 + u_2 \geqq 0.$$

By (5.13) and (5.14), the associated KORTEWEG stress tensor $\mathbf{T}_K$ is then

$$\mathbf{T}_K = -p_K\mathbf{I} - \varrho\,[u_1\nabla\varrho\otimes\nabla\varrho + u_2(\nabla\varrho\cdot\boldsymbol{n})\nabla\varrho\otimes\boldsymbol{n}], \tag{5.93}$$

where

$$p_K = \varrho^2\sigma_0' - \varrho\,\mathrm{div}\,[\varrho\,(u_1\nabla\varrho + u_2(\nabla\varrho\cdot\boldsymbol{n})\boldsymbol{n})]. \tag{5.94}$$

For completeness, we record here the form given by (5.27) and (5.38) to the hypertraction $\boldsymbol{m}$ and to the KORTEWEG traction $\boldsymbol{t}_K$, respectively, under the constitutive assumption (5.92):

$$\begin{aligned}\boldsymbol{m} &= -\varrho^2\,[u_1(\nabla\varrho\cdot\boldsymbol{\nu}) + u_2(\nabla\varrho\cdot\boldsymbol{n})\boldsymbol{n}\cdot\boldsymbol{\nu}]\,\boldsymbol{\nu},\\ \boldsymbol{t}_K &= \nabla_s\left\{\varrho^2\,[u_1(\nabla\varrho\cdot\boldsymbol{\nu}) + u_2(\nabla\varrho\cdot\boldsymbol{n})\boldsymbol{n}\cdot\boldsymbol{\nu}]\right\}\\ &\quad - \varrho^2\,[u_1(\nabla\varrho\cdot\boldsymbol{\nu}) + u_2(\nabla\varrho\cdot\boldsymbol{n})\boldsymbol{n}\cdot\boldsymbol{\nu}]\,(\mathrm{div}_s\,\boldsymbol{\nu})\boldsymbol{\nu}.\end{aligned}$$

Finally, it follows from (5.88) and (5.78) that

$$\begin{aligned}\mathbf{T}_{\mathrm{dis}} =& \frac{1}{2}\gamma_1(\boldsymbol{n}\otimes\mathring{\boldsymbol{n}} - \mathring{\boldsymbol{n}}\otimes\boldsymbol{n}) + \frac{1}{2}\gamma_2(\boldsymbol{n}\otimes\mathbf{D}\boldsymbol{n} - \mathbf{D}\boldsymbol{n}\otimes\boldsymbol{n}) + \frac{1}{2}\gamma_2(\mathring{\boldsymbol{n}}\otimes\boldsymbol{n} + \boldsymbol{n}\otimes\mathring{\boldsymbol{n}})\\ &+ \frac{1}{2}\gamma_3(\boldsymbol{n}\otimes\mathbf{D}\boldsymbol{n} + \mathbf{D}\boldsymbol{n}\otimes\boldsymbol{n}) + \gamma_5\mathbf{D} + (\gamma_4\boldsymbol{n}\cdot\mathbf{D}\boldsymbol{n} + \gamma_7\,\mathrm{tr}\,\mathbf{D})\,\boldsymbol{n}\otimes\boldsymbol{n}\\ &+ (\gamma_6\,\mathrm{tr}\,\mathbf{D} + \gamma_7\boldsymbol{n}\cdot\mathbf{D}\boldsymbol{n})\,\mathbf{I}.\end{aligned} \tag{5.95}$$

In the following section, we shall seek plane wave solutions to equation (5.87) with $\mathbf{T}_K$ and $\mathbf{T}_{\mathrm{dis}}$ given as in (5.93) and (5.95).

5.3.3 Propagation Equations

Our postulation[15] here is that, at the acoustic time scale, $\dot{\boldsymbol{n}} \equiv \mathbf{0}$, since $\boldsymbol{n}$ is thought of as being immobile; thus, by (3.36), the corotational time derivative $\mathring{\boldsymbol{n}}$ reduces to

[15] This assumption will be relaxed in Section 5.4 below, where we shall allow for a librational motion of $\boldsymbol{n}$ around a uniform immobile direction. We found it instructive to proceed by degrees of increasing complexity in presenting this subject.

$\overset{\circ}{\boldsymbol{n}} = -\mathbf{W}\boldsymbol{n}$, where $\mathbf{W}$ is the vorticity tensor, and the dissipative stress tensor $\mathbf{T}_{\text{dis}}$ becomes

$$\begin{aligned}\mathbf{T}_{\text{dis}} =& \frac{1}{2}(\gamma_1 - \gamma_2)\mathbf{W}\boldsymbol{n} \otimes \boldsymbol{n} - \frac{1}{2}(\gamma_1 + \gamma_2)\boldsymbol{n} \otimes \mathbf{W}\boldsymbol{n} + \frac{1}{2}(\gamma_2 + \gamma_3)\boldsymbol{n} \otimes \mathbf{D}\boldsymbol{n} \\ &+ \frac{1}{2}(\gamma_3 - \gamma_2)\mathbf{D}\boldsymbol{n} \otimes \boldsymbol{n} + \gamma_5\mathbf{D} + (\gamma_4\boldsymbol{n} \cdot \mathbf{D}\boldsymbol{n} + \gamma_7 \operatorname{tr}\mathbf{D})\, \boldsymbol{n} \otimes \boldsymbol{n} \\ &+ (\gamma_6 \operatorname{tr}\mathbf{D} + \gamma_7\boldsymbol{n} \cdot \mathbf{D}\boldsymbol{n})\,\mathbf{I}. \end{aligned} \tag{5.96}$$

With the acoustic fields ϱ and $\boldsymbol{v}$ represented as in (5.50), (5.51), and (5.52) above, the stretching tensor $\mathbf{D}$ reads as in (5.56), while the vorticity tensor becomes

$$\mathbf{W} = \frac{1}{2}s_0 \mathrm{i} E(\boldsymbol{a} \otimes \boldsymbol{k} - \boldsymbol{k} \otimes \boldsymbol{a}), \tag{5.97}$$

where E is the complex exponential function defined in (5.55), $\boldsymbol{k}$ is the wave vector, and $\boldsymbol{a}$ is the amplitude vector. Both $\boldsymbol{k}$ and $\boldsymbol{a}$ are represented here as in (5.53) and (5.54) above. By use of (5.56) and (5.97) in (5.96), we readily arrive at

$$\begin{aligned}\operatorname{div}\mathbf{T}_{\text{dis}} =& -\frac{1}{2}s_0 E\left\{\left[\frac{1}{2}(\gamma_1 - 2\gamma_2 + \gamma_3)(\boldsymbol{k}\cdot\boldsymbol{n})^2 + \gamma_5 k^2\right]\boldsymbol{a}\right. \\ &+ \left[\frac{1}{2}(\gamma_3 - \gamma_1 + 4\gamma_7)(\boldsymbol{a}\cdot\boldsymbol{n})(\boldsymbol{k}\cdot\boldsymbol{n}) + (\gamma_5 + 2\gamma_6)(\boldsymbol{a}\cdot\boldsymbol{k})\right]\boldsymbol{k} \\ &+ \left[\frac{1}{2}(\gamma_3 - \gamma_1 + 4\gamma_7)(\boldsymbol{k}\cdot\boldsymbol{n})(\boldsymbol{k}\cdot\boldsymbol{a})\right. \\ &\left.\left. + \frac{1}{2}(\gamma_1 + 2\gamma_2 + \gamma_3)(\boldsymbol{a}\cdot\boldsymbol{n})k^2 + 2\gamma_4(\boldsymbol{a}\cdot\boldsymbol{n})(\boldsymbol{k}\cdot\boldsymbol{n})^2\right]\boldsymbol{n}\right\}.\end{aligned}$$

On the other hand, reasoning precisely as in Section 5.2.8, by (5.93) and (5.94), we show that

$$\begin{aligned}\operatorname{div}\mathbf{T}_{\mathrm{K}} &= -\nabla p_{\mathrm{K}} + o(s_0) \\ &= -s_0\varrho_0 \mathrm{i}E\left[c_0^2 + \varrho_0^2\left(u_1 k^2 + u_2(\boldsymbol{k}\cdot\boldsymbol{n})^2\right)\right]\boldsymbol{k} + o(s_0),\end{aligned}$$

where ϱ_0 is the unperturbed density and c_0 is the velocity of sound in the limit of zero acoustic susceptibilities, $u_1 = u_2 = 0$, defined as in (5.61).

Up to the first order in s_0, the balance equation of linear momentum (5.87) then reduces to the purely geometric form

$$\begin{aligned}2\mathrm{i}\omega\boldsymbol{a} =& 2\mathrm{i}\left[c_0^2 + \varrho_0^2\left(u_1 k^2 + u_2(\boldsymbol{k}\cdot\boldsymbol{n})^2\right)\right]\boldsymbol{k} \\ &+ \left[\frac{1}{2}(\nu_1 - 2\nu_2 + \nu_3)(\boldsymbol{k}\cdot\boldsymbol{n})^2 + \nu_4 k^2\right]\boldsymbol{a} \\ &+ \left[\frac{1}{2}(\nu_3 - \nu_1 + 4\nu_7)(\boldsymbol{a}\cdot\boldsymbol{n})(\boldsymbol{k}\cdot\boldsymbol{n}) + (\nu_4 + 2\nu_6)(\boldsymbol{a}\cdot\boldsymbol{k})\right]\boldsymbol{k} \\ &+ \left[\frac{1}{2}(\nu_3 - \nu_1 + 4\nu_7)(\boldsymbol{k}\cdot\boldsymbol{n})(\boldsymbol{k}\cdot\boldsymbol{a}) + \frac{1}{2}(\nu_1 + 2\nu_2 + \nu_3)(\boldsymbol{a}\cdot\boldsymbol{n})k^2\right. \\ &\left. + 2\nu_5(\boldsymbol{a}\cdot\boldsymbol{n})(\boldsymbol{k}\cdot\boldsymbol{n})^2\right]\boldsymbol{n},\end{aligned} \tag{5.98}$$

where we have set

$$\nu_i := \frac{\gamma_i}{\varrho_0} \quad \text{for} \quad i = 1, \ldots, 7. \tag{5.99}$$

Equation (5.98) must be supplemented with the mass continuity equation (5.7), which for the acoustic fields still takes the form (5.57); this latter, however, does not reduce to the form (5.63), since by (5.98), in the nematic case, acoustic waves are no longer longitudinal. Letting $\boldsymbol{k}$ and $\boldsymbol{a}$ be represented as

$$\boldsymbol{k} = k\boldsymbol{e} \quad \text{and} \quad \boldsymbol{a} = a_e\boldsymbol{e} + a_n\boldsymbol{n}, \tag{5.100}$$

with $\boldsymbol{e} \in \mathbb{S}^2$ designating the propagation direction and k, a_e, and a_n all complex numbers to be determined, we write (5.57) in the form

$$ka_e + ka_n\cos\beta = \omega, \quad \text{with} \quad \cos\beta := \boldsymbol{e}\cdot\boldsymbol{n}. \tag{5.101}$$

It follows from (5.100) and (5.101) that whenever $\sin\beta = 0$, a_e and a_n are not uniquely defined; we resolve this ambiguity by setting $a_n = 0$ for $\sin\beta = 0$.

Before solving equations (5.98) and (5.101), we introduce new dimensionless variables defined as

$$k' := \frac{c_0}{\omega}k := \left(\frac{c_0}{c} + \mathrm{i}k_2'\right), \quad a_e' := \frac{a_e}{c_0}, \quad a_n' := \frac{a_n}{c_0}, \quad \text{and} \quad \nu_i' := \frac{\omega}{c_0^2}\nu_i, \tag{5.102}$$

for $i = 1, \ldots, 7$, where c is the velocity of sound in the nematic medium, still to be determined. Written in the new variables, equation (5.101) readily yields

$$a_e' = \frac{1}{k'} - a_n'\cos\beta. \tag{5.103}$$

Similarly, by (5.57), taking the inner product of both sides of (5.98) with $\boldsymbol{k}$, we obtain the scalar equation

$$\begin{aligned}
2\mathrm{i} =& 2\mathrm{i}\left(1 + \frac{1}{4}\omega^2\tau^2k'^2\right)k'^2 \\
&+ \left[\frac{1}{2}\left(\nu_1' - 2\nu_2' + \nu_3'\right)\cos^2\beta \right.\\
&\quad \left. + \frac{1}{2}\left(\nu_3' - \nu_1' + 4\nu_7'\right)\left(\cos\beta + a_n'k'\sin^2\beta\right)\cos\beta + 2\left(\nu_4' + \nu_6'\right)\right]k'^2 \\
&+ \left[\frac{1}{2}\left(\nu_3' - \nu_1' + 4\nu_7'\right)\cos\beta + \frac{1}{2}\left(\nu_1' + 2\nu_2' + \nu_3'\right)\left(\cos\beta + a_n'k'\sin^2\beta\right)\right. \\
&\quad \left. + 2\nu_5'\left(\cos\beta + a_n'k'\sin^2\beta\right)\cos^2\beta\right]k'^2\cos\beta,
\end{aligned} \tag{5.104}$$

where use has been made of (5.103) and τ is the *anisotropic* characteristic time defined by

$$\tau := 2\frac{\varrho_0}{c_0^2}\sqrt{u_1 + u_2\cos^2\beta}. \tag{5.105}$$

Moreover, taking the inner product of both sides of (5.98) with $\boldsymbol{n}$ and using again (5.103), we arrive at

$$\begin{aligned}
2\mathrm{i}(\cos\beta + a_n' k' \sin^2\beta) \\
= 2\mathrm{i}\left(1 + \frac{1}{4}\omega^2\tau^2 k'^2\right) k'^2 \cos\beta \\
+ \left[\frac{1}{2}\left(\nu_1' - 2\nu_2' + \nu_3'\right)\cos^2\beta + \nu_4'\right]\left(\cos\beta + a_n' k' \sin^2\beta\right) k'^2 \\
+ \left[\frac{1}{2}\left(\nu_3' - \nu_1' + 4\nu_7'\right)\left(\cos\beta + a_n' k' \sin^2\beta\right)\cos\beta + \left(\nu_4' + 2\nu_6'\right)\right] k'^2 \cos\beta \\
+ \left[\frac{1}{2}\left(\nu_3' - \nu_1' + 4\nu_7'\right)\cos\beta + \frac{1}{2}\left(\nu_1' + 2\nu_2' + \nu_3'\right)\left(\cos\beta + a_n' k' \sin^2\beta\right)\right. \\
\left. + \nu_5'\left(\cos\beta + a_n' k' \sin^2\beta\right)\cos^2\beta\right] k'^2.
\end{aligned} \tag{5.106}$$

Equations (5.104) and (5.106) are algebraic in k' and a_n'. They determine all propagation modes allowed by our theory. We begin by considering special instances of these equations that are easier to solve. Let us first set $\sin\beta = 0$, so that $a_n' = 0$ and the wave is longitudinal. Then equation (5.104) becomes

$$\mathrm{i} = \mathrm{i}\left(1 + \frac{1}{4}\omega^2\tau^2 k'^2\right) k'^2 + \left(\nu_3' + \nu_4' + \nu_5' + \nu_6' + 2\nu_7'\right) k'^2, \tag{5.107}$$

and (5.106) reduces to the same equation (5.107) with both sides multiplied by $\cos\beta$. It also follows from (5.103) that

$$a_e' = \frac{1}{k'}. \tag{5.108}$$

Let us now set $\cos\beta = 0$. Then (5.103) implies that a_e' is still given by (5.108). Moreover, equations (5.104) and (5.106) become

$$\mathrm{i} = \mathrm{i}\left(1 + \frac{1}{4}\omega^2\tau_1^2 k'^2\right) k'^2 + \left(\nu_4' + \nu_6'\right) k'^2 \tag{5.109}$$

and

$$a_n' k' \left[2\mathrm{i} - \frac{1}{2}\left(\nu_1' + 2\nu_2' + \nu_3' + 2\nu_4'\right) k'^2\right] = 0,$$

whence it follows that $a_n' = 0$. Then the wave propagating at right angles to the nematic director is also longitudinal.

Though both equations (5.107) and (5.109) can easily be solved explicitly, their solutions are given by rather cumbersome expressions, which do not make their interpretation transparent. As in Section 5.2.8, we prefer to study the limit of small viscosities, where all ν_i' are treated as perturbation parameters of the same order. To this end, we first consider the inviscid limit, in which all viscosities are set equal to zero. Equation (5.104) then becomes

$$\left(1+\frac{1}{4}\omega^2\tau^2k'^2\right)k'^2-1=0, \tag{5.110}$$

which, together with (5.106), also requires $a'_n=0$, and correspondingly implies that a'_e is as in (5.108). The only solution of (5.110) with positive real part has $k'_2=0$ and

$$\frac{c}{c_0}=\frac{\omega\tau}{\sqrt{2\left(\sqrt{1+\omega^2\tau^2}-1\right)}}. \tag{5.111}$$

We drop the solution with negative real part, since it represents the same wave propagating in the opposite direction. We also drop the other two purely imaginary solutions, since they do not represent traveling waves. Were τ replaced by τ_1 in (5.69), equation (5.111) would exactly coincide with (5.72). However, since τ depends on β, the dispersion described by (5.111) is anisotropic.

We now assume that all dimensionless viscosities ν'_i are $O(\varepsilon_0)$, where ε_0 is a small dimensionless parameter, and we continue in ε_0 the solution to the propagation equations already found in the inviscid limit. In particular, we write k' as

$$k'=\frac{c_0}{c}+h'_1+\mathrm{i}k'_2, \tag{5.112}$$

and we assume that both h'_1 and k'_2 are $O(\varepsilon_0)$. By (5.102), assuming $k'_2\ll 1$ amounts to assuming that the attenuation length of the propagating wave is expected to be much smaller that the wavelength. Intuitively, this is grounded in the assumption that all viscosities are small, in the sense made precise by requiring that $\nu'_i\ll 1$, for all i.

Inserting (5.112) into (5.104), (5.106), and (5.103), at the lowest order of approximation in s_0, we obtain that

$$h'_1=0, \tag{5.113a}$$

$$k'_2=\frac{1}{2\frac{c}{c_0}+\omega^2\tau^2\frac{c_0}{c}}\left[\nu'_4+\nu'_6+(\nu'_3+2\nu'_7)\cos^2\beta+\nu'_5\cos^4\beta\right], \tag{5.113b}$$

$$a'_n=-\mathrm{i}\frac{1}{2}\left(\frac{c_0}{c}\right)\frac{\cos\beta}{\sin^2\beta}\left[\nu'_2+\nu'_3+2\nu'_7-(\nu'_2+\nu'_3-2\nu'_5+2\nu'_7)\cos^2\beta\right.$$
$$\left.-2\nu'_5\cos^4\beta\right]\quad\text{for}\quad\sin\beta\neq 0, \tag{5.113c}$$

$$a'_e=\frac{c}{c_0}-\mathrm{i}\left(\frac{c}{c_0}\right)^2k'_2-a'_n\cos\beta, \tag{5.113d}$$

where c is expressed by (5.111) as a function of both ω and β.

The solutions to equations (5.104) and (5.106) for which the real part vanishes in the inviscid limit can also be continued as all dimensionless viscosities ν'_i move away from zero. The continued solution with positive real part of k' can be represented as

$$k'=h'_1+\mathrm{i}h'_2,$$

where

$$h_2' = \frac{\sqrt{2\left(1\sqrt{1+\omega^2\tau^2}\right)}}{\omega\tau}$$

and $h_1' = O(\varepsilon_0)$. It turns out that, at the lowest order of approximation,

$$\frac{h_2'}{h_1'} = \frac{2\sqrt{1+\omega^2\tau^2}}{\nu_4' + \nu_6' + (\nu_3' + 2\nu_7')\cos^2\beta + \nu_5'\cos^4\beta} = O\left(\frac{1}{\varepsilon_0}\right).$$

This would thus correspond to a wave propagating with a speed c much larger than c_0 and with an attenuation length much shorter than the wavelength. Such a wave could not indeed propagate, and so it will henceforth be disregarded, though it might rise and compete with the wave that propagates in the asymptotic limit of small viscosities in the complete nonlinear analysis of propagation equations (5.104) and (5.106).

Anisotropic Dispersion

The same Figure 5.1 that above represented the dispersion of c in the case of isotropic acoustic susceptibility studied in Section 5.2.8 also represents equation (5.111) for a given propagation direction specified by the angle β: it suffices to replace τ_1 with τ. What the graph in Figure 5.1 fails to represent is the anisotropy in the speed of sound. To capture this feature of (5.111), we define the *relative* sound speed *anisotropy* Δc as

$$\Delta c := \frac{c - c|_{\beta=\frac{\pi}{2}}}{c|_{\beta=0}}. \tag{5.114}$$

Here Δc is a function of both ω and β, which vanishes for $\beta = \frac{\pi}{2}$. To distinguish in Δc the dependence on ω from the dependence on β, we find it convenient to let

$$\varepsilon := \frac{u_2}{u_1} \tag{5.115}$$

and to assume that ε is also a small parameter. Then, by (5.111), (5.114) yields

$$\Delta c = \varepsilon f(\omega\tau_1)\cos^2\beta + O(\varepsilon^2), \tag{5.116}$$

where τ_1 is defined as in (5.69) and

$$f(x) := \frac{1}{4}\frac{x^2 - 2\left(\sqrt{1+x^2}-1\right)}{\sqrt{1+x^2}\left(\sqrt{1+x^2}-1\right)}. \tag{5.117}$$

It readily follows from (5.117) that

$$f(x) = \frac{1}{8}x^2 + O(x^4) \quad \text{for} \quad x \ll 1, \quad \text{and} \quad \lim_{x\to\infty} f(x) = \frac{1}{4}.$$

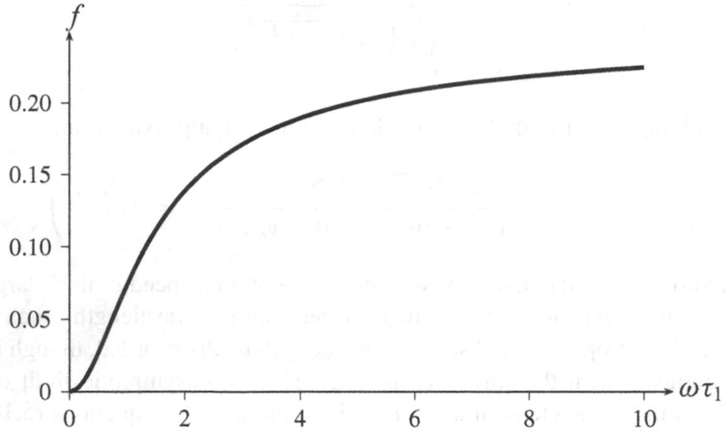

Fig. 5.4. In the limit where the susceptibility u_2 is much smaller than the susceptibility u_1, the frequency dependence of the speed anisotropy Δc in (5.114) is represented by the function f in (5.117), here plotted against $\omega\tau_1$. At small frequencies, f is quadratic; at large frequencies, it saturates to $\frac{1}{4}$.

As shown by Figure 5.4, f is a positive, strictly increasing function, so that, in particular, the speed of propagation along the nematic director is larger than the speed of propagation at right angles to it. The prediction in (5.116) also agrees with the observations of [225] for *p-n*-butyl-aniline (MBBA) at 21°C and wave frequency 10 MHz under the action of an aligning magnetic field with strength 5 Oe. The data for Δc were represented in Figure 2 of [225] as $\Delta c = A\cos^2\beta$; they will be further discussed in Section 5.4.3 below in this chapter.

Anisotropic Attenuation

As already remarked in Section 5.2.8 above, dispersion in wave propagation disrupts the simple quadratic dependence in ω of the wave attenuation k_2. Here we shall further explore the dependence of k_2 on the propagation direction. By (5.99) and (5.102), we readily derive from (5.113b) the dimensional form of the attenuation:

$$k_2 = \frac{\omega^2}{2\varrho_0 c_0^3} \frac{1}{\dfrac{c}{c_0} + \dfrac{1}{2}\omega^2\tau^2\dfrac{c_0}{c}} \left[\gamma_5 + \gamma_6 + (\gamma_3 + 2\gamma_7)\cos^2\beta + \gamma_4\cos^4\beta\right]. \tag{5.118}$$

It is worth noting that $k_2 \geqq 0$ for both $\beta = 0$ and $\beta = \frac{\pi}{2}$, as a consequence of inequalities (5.84c) and (5.84e). The proof that $k_2 \geqq 0$ also for all $\beta \in [0, \pi]$ will be given shortly below, on page 284; it will follow as a special case of a more general result.

The angular dependence exhibited by (5.118) coincides with that predicted by LEE & ERINGEN [177] in their theory for wave propagation in nematic liquid crystals phrased within the general *micromorphic* theory of continuum mechanics first put forward by ERINGEN & SUHUBI [103] and later extended by ERINGEN [100, 101, 102]. However, as pointed out in [361] and [223], at the lowest order in the condensation, this theory does not predict dispersion of sound, and consequently the frequency dependence of the attenuation is classically quadratic. In particular, it is shown in [223] that this is indeed a feature common to both the theories presented in [106] and [149] and the theory of LESLIE [180].[16] While the dependence on β of k_2 in (5.118) has been widely confirmed [361, 139], a purely quadratic dependence of k_2 on ω has no experimental basis [361, 223]. Since in our theory c depends on ω and τ does not vanish, equation (5.118) exhibits indeed a nonquadratic dependence on ω, which we now explore more closely, introducing an appropriate measure of attenuation anisotropy.

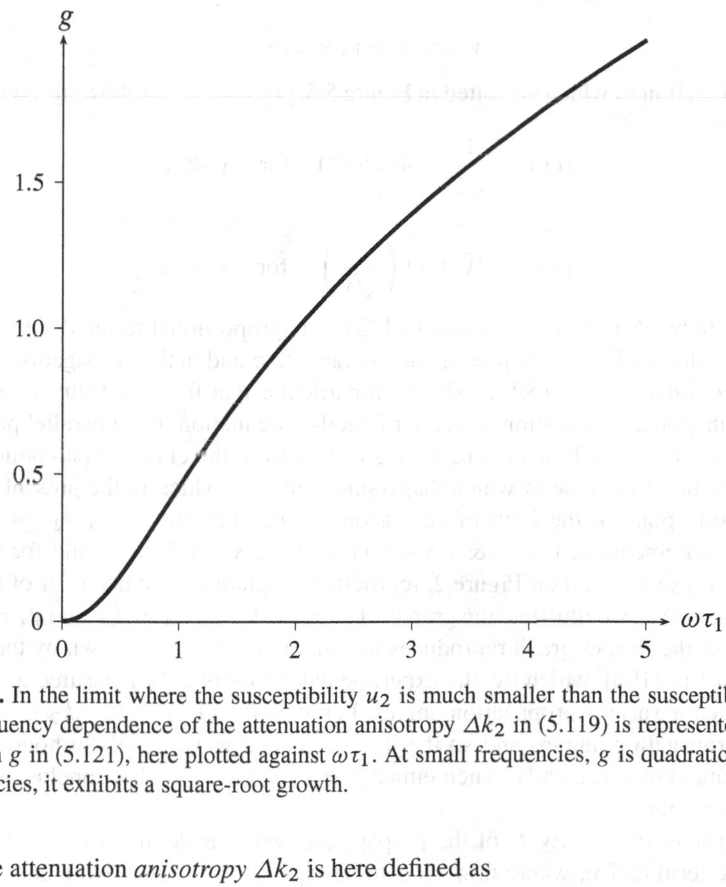

Fig. 5.5. In the limit where the susceptibility u_2 is much smaller than the susceptibility u_1, the frequency dependence of the attenuation anisotropy Δk_2 in (5.119) is represented by the function g in (5.121), here plotted against $\omega\tau_1$. At small frequencies, g is quadratic; at large frequencies, it exhibits a square-root growth.

The attenuation *anisotropy* Δk_2 is here defined as

[16] Properly speaking, to be applicable to the propagation of condensation waves, LESLIE'S theory should be extended to encompass also compressible liquid crystals, as done in [223] essentially along the lines followed in Section 5.3.1 above.

$$\Delta k_2 := k_2 - k_2|_{\beta=0}, \tag{5.119}$$

which like Δc is a function of both ω and β. By assuming again that ε in (5.115) is a small parameter, we easily give (5.119) the following form:

$$\Delta k_2 = \frac{c_0 \gamma}{4\sqrt{2}\varrho_0^3 u_1} g(\omega\tau_1) G(\beta) + O(\varepsilon^2), \tag{5.120}$$

where

$$g(x) := \frac{x\sqrt{\sqrt{1+x^2}-1}}{\sqrt{1+x^2}}, \tag{5.121}$$

$$G(\beta) := -\sin^2\beta \left(1 + \frac{\gamma_4}{\gamma}\cos^2\beta\right), \tag{5.122}$$

and

$$\gamma := \gamma_3 + \gamma_4 + 2\gamma_7.$$

The function g, which is plotted in Figure 5.5, possesses the following asymptotic behaviors:

$$g(x) = \frac{1}{\sqrt{2}} x^2 + O(x^4) \quad \text{for} \quad x \ll 1$$

and

$$g(x) = \sqrt{x} + O\left(\frac{1}{\sqrt{x}}\right) \quad \text{for} \quad x \gg 1.$$

For $\gamma > 0$, by (5.119), (5.120), and (5.122), g is proportional to the difference between the attenuations in the propagation parallel to $\boldsymbol{n}$ and in the propagation orthogonal to $\boldsymbol{n}$. Since $g \geqq 0$, (5.122) shows in particular that for $\gamma > 0$ the attenuation in the orthogonal propagation is smaller than the attenuation in the parallel propagation. It is to be noted how the graph of g differs from the classical parabolic form, characteristic of the case in which dispersion is absent, which in the present setting would correspond to the limit of zero acoustic susceptibilities, u_1, $u_2 \to 0$. The early measurements of LORD & LABES [196] for Δk_2 at $\beta = \frac{\pi}{2}$ and for various frequencies, shown in their Figure 2, reproduce qualitatively the behavior of g.

In Figure 5.6, we illustrate the graphs of G against β for $\gamma_4 = \frac{1}{2}\gamma$ and $\gamma_4 = -2\gamma$. For $\gamma > 0$, the former graph reproduces the qualitative features shown by the graph in Figure 1 of [196], which fits the experimental values of Δk_2 measured at a given frequency for various propagation angles. For $\gamma > 0$ and $\gamma_4 \geqq 0$, Δk_2 is negative for all propagation angles, and so the less-attenuated wave travels orthogonally to the nematic director; clearly, when either γ or γ_4 is negative, this conclusion is not necessarily true.

The acoustic intensity I_a of the propagating wave, as defined by (5.75), is still given the form (5.76), where now τ_1 is to be replaced by τ in (5.105), and I_0 is as in (5.77), but with k_2 now given by (5.118). For a prescribed value of β, the graph in Figure 5.3 is still appropriate for representing I_a scaled to I_0, provided we read it against $\omega\tau$. One should, however, keep in mind that two independent sources of anisotropy are now hidden in I_a, namely, τ and k_2.

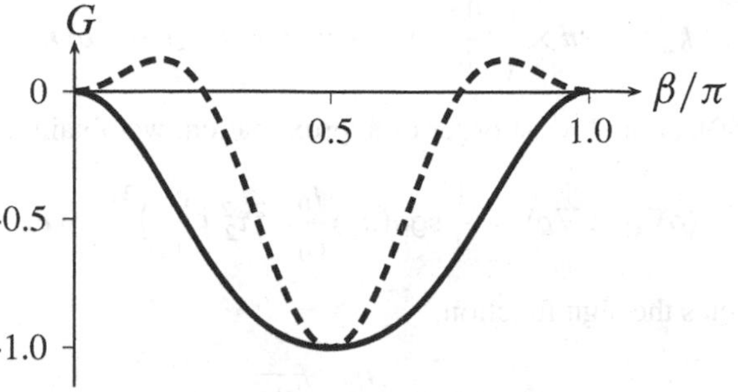

Fig. 5.6. In the limit where the susceptibility u_2 is much smaller than the susceptibility u_1, the angular dependence of the attenuation anisotropy Δk_2 in (5.119), for $\gamma > 0$, is represented by the function G in (5.122), here plotted against β for $\gamma_4 = \frac{1}{2}\gamma$ (solid line) and $\gamma_4 = -2\gamma$ (dashed line).

Acoustic Torque

As already pointed out, one major issue related to ultrasonic wave propagation in nematic liquid crystals is to explain the ability of ultrasound to act on the nematic director.[17] The KORTEWEG nature of nematic liquid crystals at the time and length scales characteristic of ultrasound propagation has been here the main idea to explain the acoustic interaction with the molecular alignment.

We read through the balance equation of torques (5.89) the action exerted by the acoustic field on the nematic director. It follows from (5.92) and (5.78) that

$$\frac{\partial \sigma_{\mathrm{K}}}{\partial \boldsymbol{n}} = u_2 \left[(\nabla_{\varrho} \cdot \boldsymbol{n})\, \nabla_{\varrho} - (\nabla_{\varrho} \cdot \boldsymbol{n})^2\, \boldsymbol{n} \right]$$

and

$$\frac{\partial R_{\mathrm{a}}}{\partial \mathring{\boldsymbol{n}}} = \gamma_1 \mathring{\boldsymbol{n}} + \gamma_2 \left[\mathbf{D}\boldsymbol{n} - (\boldsymbol{n} \cdot \mathbf{D}\boldsymbol{n})\, \boldsymbol{n} \right]. \qquad (5.123)$$

In particular, for the acoustic flow considered here, where the director does not librate, $\dot{\boldsymbol{n}} \equiv \mathbf{0}$ and equation (5.123) becomes

$$\frac{\partial R_{\mathrm{a}}}{\partial \mathring{\boldsymbol{n}}} = -\gamma_1 \mathbf{W} + \gamma_2 \left[\mathbf{D}\boldsymbol{n} - (\boldsymbol{n} \cdot \mathbf{D}\boldsymbol{n})\, \boldsymbol{n} \right],$$

which, by (5.97), (5.56), and (5.55), implies that

$$\left\langle \frac{\partial R_{\mathrm{a}}}{\partial \mathring{\boldsymbol{n}}} \right\rangle = \mathbf{0},$$

where $\langle \cdot \rangle$ denotes time average over an acoustic period. Thus, at time scales longer than the acoustic period, equation (5.89) reveals an unbalanced acoustic torque $\boldsymbol{k}_{\mathrm{a}}$ that has its origin in the KORTEWEG coupling we have postulated; $\boldsymbol{k}_{\mathrm{a}}$ is defined as

[17] Also, the role of director–ultrasound coupling in the order–disorder transitions in nematic liquid crystals has recently been illuminated in [200].

$$\boldsymbol{k}_{\mathrm{a}} := -\boldsymbol{n} \times \left\langle \varrho \frac{\partial \sigma_{\mathrm{K}}}{\partial \boldsymbol{n}} \right\rangle = -u_2 \boldsymbol{n} \times \langle \varrho \nabla \varrho \otimes \nabla \varrho \rangle \, \boldsymbol{n}. \tag{5.124}$$

For ϱ as in (5.50), at the lowest order of approximation, we obtain that

$$\langle \varrho \nabla \varrho \otimes \nabla \varrho \rangle = \frac{1}{4} \operatorname{sgn}(u_2) \frac{I_0}{c_0} \omega^2 \tau_2^2 \left(\frac{c_0}{c} \right)^2 \boldsymbol{e} \otimes \boldsymbol{e}, \tag{5.125}$$

where sgn denotes the sign function,

$$\tau_2 := 2 \frac{\varrho_0}{c_0^2} \sqrt{|u_2|},$$

c is as in (5.111), and I_0 is given by (5.77) with k_2 as in (5.118). By inserting (5.125) into (5.124), we arrive at

$$\boldsymbol{k}_{\mathrm{a}} = -\operatorname{sgn}(u_2) K_0 (\boldsymbol{n} \cdot \boldsymbol{e}) \boldsymbol{n} \times \boldsymbol{e}, \tag{5.126}$$

where

$$K_0 := \frac{1}{4} \frac{I_0}{c_0} \omega^2 \tau_2^2 \left(\frac{c_0}{c} \right)^2.$$

A couple of remarks are suggested by (5.126). First, since $K_0 \geqq 0$, $\boldsymbol{k}_{\mathrm{a}}$ is an *aligning* torque, that is, it tends to bring $\boldsymbol{n}$ along the propagation direction $\boldsymbol{e}$ only if $u_2 < 0$; otherwise, it is a *misaligning* torque, which tends to make $\boldsymbol{n}$ orthogonal to $\boldsymbol{e}$. Second, it may appear that $\boldsymbol{k}_{\mathrm{a}}$ behaves essentially like the magnetic torque $\boldsymbol{k}_{\mathrm{m}}$ encountered in Section 3.1.4, the case with positive diamagnetic anisotropy being the analogue of the case with negative acoustic susceptibility u_2, and, conversely, the case with negative diamagnetic anisotropy being the analogue of the case with positive acoustic susceptibility u_2.[18] This analogy, however, is only formal, since the dependence of K_0 on the propagation direction makes the dependence of $\boldsymbol{k}_{\mathrm{a}}$ on the angle between $\boldsymbol{n}$ and $\boldsymbol{e}$ more complicated than it appears from (5.126). In case of pure acoustic relaxation of the nematic director, such a dependence might result in a relaxation law more complicated than a simple exponential decay.

Synopsis

The nematoacoustic theory presented in this chapter is variational in that it retraces the source of the interaction between the acoustic field and the nematic molecular alignment in an elastic coupling of capillary type. It remains a phenomenological theory, since the acoustic susceptibilities u_1 and u_2 introduced in (5.92) need to be determined experimentally by exploring the consequences of the theory. Among these, some appear particularly promising, namely, the anisotropy and dispersion in sound speed, and the unconventional frequency-dependence of wave attenuation. These features, which other theories do not possess, stem from the assumed KORTEWEG nature of the acoustic coupling.

[18] The reader is advised to compare (5.126) and (3.96b).

Strictly speaking, our propagation equations in Section 5.3.3 were derived under the assumption that the director $\boldsymbol{n}$ is uniform and immobile, as if it were held fixed by some external action, such as an applied magnetic field. This is indeed the situation envisaged in the wealth of experimental studies recalled above. We shall see in the following Section 5.4 the consequences of relaxing in part such an assumption by allowing the nematic director to vibrate around a fixed orientation.

More generally, in the absence of any external cause, the director is free to vary in time and be distorted in space. These variations take place at time and length scales much larger than the acoustic characteristic times and lengths, so that especially an ultrasonic wave propagates locally in an undistorted medium, where our equations still apply. Such reasoning might suggest that the evolution of the director, which is governed by the balance of torques where now $\boldsymbol{k}_{\mathrm{a}}$ results from the acoustic interaction, would interfere with the wave propagation only marginally, by affecting locally its anisotropic character. This would indeed be correct, were the sound speed independent of the propagation direction. On the contrary, we have shown above that this is not the case in our theory. Such an *acoustic birefringence* causes the director texture to alter the ultrasound propagation: the director, which can be distorted by an acoustic wave, in turn causes the refringence of the distorting wave. Studying the ultrasound propagation in a moderately distorted nematic medium is a challenge our theory should next face. We might also learn from it how to steer an acoustic wave by acting on the nematic texture through controllable external actions.

5.4 Director Libration

In Section 5.3, at the time scale of the acoustic vibrations, the director texture is still regarded as immobile, in accordance with previous experimental work and in line with experimental studies where the director is kept fixed by external magnetic fields. Here, following [67], we shall relax such an assumption: the director is set free to vibrate in time and be possibly distorted in space around a constant and uniform orientation, this latter possibly held fixed by an external (magnetic) field. We refer to such a motion as the director *libration*, and we imagine it to arise as a consequence of the sound wave that propagates through the nematic liquid crystal. The theoretical outcomes of our analysis will allow us to interpret quantitatively the experimental results published long ago in the literature and to estimate some phenomenological parameters involved in the theory.

5.4.1 Dynamical Balance Equations

We begin by recapitulating the balance equations that govern the motion of the fluid. Here care must be used in treating elastic forces and torques, since $\boldsymbol{n}$, vibrating with the wave, is no longer uniform in space. To simplify matters, we shall assume that in (5.91) all elastic constants are equal to K, so that (5.91) reduces to

$$W = \frac{1}{2}K|\nabla \boldsymbol{n}|^2. \tag{5.127}$$

The balance of mass has the classical form of the continuity equation (5.7), which we find it convenient to write in the equivalent form

$$\frac{\partial \varrho}{\partial t} + \operatorname{div}(\varrho \boldsymbol{v}) = 0. \tag{5.128}$$

In the absence of body forces, the balance of linear momentum is expressed by equation (5.87), which we recall for the reader's ease:

$$\varrho \dot{\boldsymbol{v}} = \operatorname{div}(\mathbf{T}_{\mathrm{E}} + \mathbf{T}_{\mathrm{K}} + \mathbf{T}_{\mathrm{dis}}). \tag{5.129}$$

With the aid of (5.92), (5.127), and (5.78), the ERICKSEN elastic stress $\mathbf{T}_{\mathrm{E}}$, the KORTEWEG elastic stress $\mathbf{T}_{\mathrm{K}}$, and the dissipative stress $\mathbf{T}_{\mathrm{dis}}$ are given by

$$\mathbf{T}_{\mathrm{E}} = -K\left[W\mathbf{I} + (\nabla \boldsymbol{n})^{\mathsf{T}}(\nabla \boldsymbol{n})\right], \tag{5.130}$$

$$\mathbf{T}_{\mathrm{K}} = -p_{\mathrm{K}}\mathbf{I} - \varrho\left[u_1 \nabla\varrho \otimes \nabla\varrho + u_2(\nabla\varrho \cdot \boldsymbol{n})\nabla\varrho \otimes \boldsymbol{n}\right], \tag{5.131}$$

where

$$p_{\mathrm{K}} := p_0(\varrho) - \varrho \operatorname{div}\left[\varrho\left(u_1 \nabla\varrho + u_2(\nabla\varrho \cdot \boldsymbol{n})\boldsymbol{n}\right)\right] \tag{5.132}$$

with

$$p_0(\varrho) := \varrho^2 \sigma_0'(\varrho)$$

assumed to be an *increasing* function of ϱ, and

$$\begin{aligned}
\mathbf{T}_{\mathrm{dis}} =& \frac{1}{2}\gamma_1(\boldsymbol{n} \otimes \mathring{\boldsymbol{n}} - \mathring{\boldsymbol{n}} \otimes \boldsymbol{n}) + \frac{1}{2}\gamma_2(\boldsymbol{n} \otimes \mathbf{D}\boldsymbol{n} - \mathbf{D}\boldsymbol{n} \otimes \boldsymbol{n}) \\
&+ \frac{1}{2}\gamma_2(\mathring{\boldsymbol{n}} \otimes \boldsymbol{n} + \boldsymbol{n} \otimes \mathring{\boldsymbol{n}}) + \frac{1}{2}\gamma_3(\boldsymbol{n} \otimes \mathbf{D}\boldsymbol{n} + \mathbf{D}\boldsymbol{n} \otimes \boldsymbol{n}) \\
&+ \gamma_5 \mathbf{D} + (\gamma_4 \boldsymbol{n} \cdot \mathbf{D}\boldsymbol{n} + \gamma_7 \operatorname{tr}\mathbf{D})\, \boldsymbol{n} \otimes \boldsymbol{n} \\
&+ (\gamma_6 \operatorname{tr}\mathbf{D} + \gamma_7 \boldsymbol{n} \cdot \mathbf{D}\boldsymbol{n})\,\mathbf{I}.
\end{aligned} \tag{5.133}$$

Similarly, using again the constitutive equations (5.92), (5.127), and (5.78) for σ_{K}, W, and R_{a}, we give the balance of torques in equation (5.89) the following form:

$$K \operatorname{div}(\nabla \boldsymbol{n}) \times \boldsymbol{n} - u_2 \varrho\,(\nabla\varrho \cdot \boldsymbol{n})\,\nabla\varrho \times \boldsymbol{n} - \gamma_1 \mathring{\boldsymbol{n}} \times \boldsymbol{n} - (\gamma_2 \mathbf{D}\boldsymbol{n} - \gamma_1 \mathbf{W}\boldsymbol{n}) \times \boldsymbol{n} = \mathbf{0}. \tag{5.134}$$

Equations (5.128), (5.129), and (5.134), where $\mathbf{T}_{\mathrm{E}}$, $\mathbf{T}_{\mathrm{K}}$, and $\mathbf{T}_{\mathrm{dis}}$ are as in (5.130), (5.131), and (5.133) above, represent the basic balances of the theory. They will be solved below in a special setting.

5.4.2 Plane Wave Solutions

Here, as in Section 5.3, we study the propagation of forced plane waves of condensation,[19] but we abandon the simplifying assumption that $\boldsymbol{n}$ is held fixed by a compliant external action, such as a magnetic field. In such an approach, the elastic torque

[19] Not to be confused with either the shear acoustic waves studied in [216] and [9] or the weak and twist waves studied in [95] and [96].

in (5.134) vanishes, as does the viscous torque opposing the tumbling of $\boldsymbol{n}$, while the viscous torque opposing the acoustic flow vanishes once averaged in time. The only nonvanishing torque is the time-averaged acoustic torque that must be imagined as balanced by a reactive torque exerted by the external constraint keeping $\boldsymbol{n}$ fixed. Here, allowing the director to vibrate, we need to solve the balance equation of torques (5.134): we can no longer be contented with reading off from it the average unbalanced acoustic torque. Thus, our attention will now turn to the director motion and to its consequences on the balances of linear momentum and torque.

The linearized balance laws resulting from (5.128), (5.129), and (5.134) and the constitutive relations in (5.130), (5.131), and (5.133) are solved and used to find the anisotropic dispersion of waves and to study the relationship between energy dissipation and wave attenuation. Solutions are sought as above in the plane wave form

$$\varrho(\boldsymbol{x},t) = \varrho_0 \left(1 + s_0 \Re E\right), \quad \boldsymbol{v}(\boldsymbol{x},t) = s_0 \Re(E\boldsymbol{a}), \tag{5.135}$$

where $E := \mathrm{e}^{\mathrm{i}(\boldsymbol{k}\cdot\boldsymbol{x}-\omega t)}$, $\Re$ denotes the real part of a complex number, $\boldsymbol{x}$ is the position vector, ϱ_0 is the unperturbed mass density, s_0 is a small dimensionless parameter measuring the scale of perturbation, $\boldsymbol{k}$ is the *complex* wave vector to be determined in terms of the angular frequency ω, and $\boldsymbol{a}$ is an unknown complex amplitude vector. We also allow for a director *libration*[20] described by

$$\boldsymbol{n} = [\mathbf{I} + s_0 \Re(E\mathbf{A})]\,\boldsymbol{n}_0, \tag{5.136}$$

where $\mathbf{A}$ is a complex skew-symmetric tensor and $\boldsymbol{n}_0$ is a uniform unperturbed director field.

The basic governing equations are solved within the above class of flows, in the limit where s_0 is a small perturbation parameter. In particular, it follows from (5.135) that

$$\mathbf{D} = \frac{1}{2} s_0 \mathrm{i} E(\boldsymbol{a} \otimes \boldsymbol{k} + \boldsymbol{k} \otimes \boldsymbol{a}), \tag{5.137a}$$

$$\mathbf{W} = \frac{1}{2} s_0 \mathrm{i} E(\boldsymbol{a} \otimes \boldsymbol{k} - \boldsymbol{k} \otimes \boldsymbol{a}). \tag{5.137b}$$

Here and in what follows we drop $\Re$ from equations like (5.137), while keeping in mind that as in (5.135) and (5.136) only their real parts bear a physical meaning. Up to first order in s_0, equation (5.128) becomes

$$\omega = \boldsymbol{a} \cdot \boldsymbol{k},$$

which is the same as (5.57). Using (5.135) and (5.56), one readily arrives at

$$\varrho \dot{\boldsymbol{v}} = -s_0 \mathrm{i} \varrho_0 \omega E \boldsymbol{a} + o(s_0). \tag{5.138}$$

Since $\boldsymbol{n}_0$ is uniform in space, it follows from (5.130) that $\mathbf{T}_\mathrm{E}$ is $o(s_0)$, and so at the lowest approximation in s_0 the elastic stress does not contribute to the balance of linear momentum. On the other hand, by (5.131) and (5.132), one obtains that

[20] By *libration* we mean here a motion in which the director keeps a nearly uniform orientation $\boldsymbol{n}_0$ and vibrates about it as a consequence of the flow perturbation.

$$\operatorname{div}\mathbf{T}_{\mathrm{K}} = -s_0\varrho_0 \mathrm{i}E\left\{c_0^2 + \varrho_0^2\left[u_1k^2 + u_2(\boldsymbol{k}\cdot\boldsymbol{n}_0)^2\right]\right\}\boldsymbol{k} + o(s_0), \tag{5.139}$$

where

$$c_0(\varrho_0) := \sqrt{p_0'(\varrho_0)} \tag{5.140}$$

is the velocity of sound in the *isotropic* limit where $u_1 = u_2 = 0$. Finally, by (5.56), (5.97), and (5.57), one also arrives at

$$\begin{aligned}\operatorname{div}\mathbf{T}_{\mathrm{dis}} =& \frac{1}{2}(\gamma_1+\gamma_2)s_0\omega E\,(\boldsymbol{k}\cdot\mathbf{A}\boldsymbol{n}_0)\,\boldsymbol{n}_0 + \frac{1}{2}(\gamma_2-\gamma_1)s_0\omega E(\boldsymbol{k}\cdot\boldsymbol{n}_0)\mathbf{A}\boldsymbol{n}_0\\ &-\frac{1}{2}s_0E\left\{\left[\frac{1}{2}(\gamma_1-2\gamma_2+\gamma_3)(\boldsymbol{k}\cdot\boldsymbol{n}_0)^2+\gamma_5k^2\right]\boldsymbol{a}\right.\\ &+\left[\frac{1}{2}(\gamma_3-\gamma_1+4\gamma_7)(\boldsymbol{a}\cdot\boldsymbol{n}_0)(\boldsymbol{k}\cdot\boldsymbol{n}_0)+(\gamma_5+2\gamma_6)(\boldsymbol{a}\cdot\boldsymbol{k})\right]\boldsymbol{k}\\ &+\left[\frac{1}{2}(\gamma_3-\gamma_1+4\gamma_7)(\boldsymbol{k}\cdot\boldsymbol{n}_0)(\boldsymbol{k}\cdot\boldsymbol{a})\right.\\ &\left.\left.+\frac{1}{2}(\gamma_1+2\gamma_2+\gamma_3)(\boldsymbol{a}\cdot\boldsymbol{n}_0)k^2+2\gamma_4(\boldsymbol{a}\cdot\boldsymbol{n}_0)(\boldsymbol{k}\cdot\boldsymbol{n}_0)^2\right]\boldsymbol{n}_0\right\},\end{aligned} \tag{5.141}$$

where $k^2 := \boldsymbol{k}\cdot\boldsymbol{k}$. Equations (5.138), (5.139), and (5.141) will be used in Section 5.4.2 to derive the propagation equation that applies in the presence of director libration. To this end we also need to solve the balance equation of torques (5.134).

Libration Equation

At the lowest order in s_0 and under the assumption that $\boldsymbol{n}_0$ is uniform in space and constant in time, equation (5.134) becomes

$$\begin{aligned}\left[Kk^2 - \mathrm{i}\gamma_1\omega\right](\mathbf{A}\boldsymbol{n}_0)\times\boldsymbol{n}_0 + \frac{1}{2}\mathrm{i}\,(\gamma_2-\gamma_1)\,(\boldsymbol{k}\cdot\boldsymbol{n}_0)\,\boldsymbol{a}\times\boldsymbol{n}_0&\\ +\frac{1}{2}\mathrm{i}\,(\gamma_2+\gamma_1)\,(\boldsymbol{a}\cdot\boldsymbol{n}_0)\,\boldsymbol{k}\times\boldsymbol{n}_0 = \mathbf{0}.&\end{aligned} \tag{5.142}$$

By the constraint set in Section 5.3 on the director motion, the balance of torque was there satisfied only on average, since, as shown by (5.142), for a proper instantaneous balance, the viscous torque exerted by the acoustic flow entrains a director vibration. It is also worth noting that at the lowest order of approximation in the acoustic condensation parameter s_0, the acoustic torque in (5.134) does not contribute to the libration equation (5.142). As already shown in Section 5.3.3, the acoustic torque is of second order in s_0 and manifests itself at times longer than the acoustic period through a time-averaged action similar in character to an *acoustic streaming*. We shall further elaborate on this in Section 5.3.3; here we only remark that among the second-order effects that characterize the slow (streaming) dynamics taking place at time and length scales larger than the acoustic ones, we should also include ERICKSEN's elastic stress in the balance equation of linear momentum.

Since $\mathbf{A}$ is skew-symmetric, $\mathbf{A}\boldsymbol{n}_0$ is orthogonal to $\boldsymbol{n}_0$. Letting $\mathbf{A}\boldsymbol{n}_0 = \boldsymbol{d} \times \boldsymbol{n}_0$, with $\boldsymbol{d} \perp \boldsymbol{n}_0$, we can easily solve (5.142) for $\boldsymbol{d}$ and then arrive at

$$\mathbf{A}\boldsymbol{n}_0 = -\Sigma\Big\{\gamma_2\,(\boldsymbol{k}\cdot\boldsymbol{n}_0)\,(\boldsymbol{a}\cdot\boldsymbol{n}_0)\,\boldsymbol{n}_0 + \frac{1}{2}\,(\gamma_1-\gamma_2)\,(\boldsymbol{k}\cdot\boldsymbol{n}_0)\,\boldsymbol{a} - \frac{1}{2}\,(\gamma_1+\gamma_2)\,(\boldsymbol{a}\cdot\boldsymbol{n}_0)\,\boldsymbol{k}\Big\}, \tag{5.143}$$

where

$$\Sigma := \frac{1}{\gamma_1\omega + \mathrm{i}Kk^2}.$$

A consequence of (5.143) is especially worthy of notice. In the limit as both viscosities γ_1 and γ_2 vanish, so does $\mathbf{A}\boldsymbol{n}_0$, this implying by (5.136) that no director libration occurs in that limit and a wave can propagate while $\boldsymbol{n}$ remains immobile, even in the absence of any external restraining field. In brief, one could also say that the director libration is a motion fed by dissipation. Henceforth we shall assume that $\gamma_1 > 0$.

We now use the explicit solution (5.143) for the director libration to derive the equation that governs the wave propagation from the balance of linear momentum (5.129).

Wave Propagation

By employing (5.139) and (5.141), with the aid of (5.143), up to first order in s_0, equation (5.129) is reduced to the purely kinematic form

$$\begin{aligned}
2\mathrm{i}\varrho_0\omega\boldsymbol{a} =& 2\mathrm{i}\varrho_0\left[c_0^2 + \varrho_0^2\left(u_1k^2 + u_2(\boldsymbol{k}\cdot\boldsymbol{n}_0)^2\right)\right]\boldsymbol{k} \\
&+ \left[\frac{1}{2}(\gamma_1 - 2\gamma_2 + \gamma_3)(\boldsymbol{k}\cdot\boldsymbol{n}_0)^2 + \gamma_5k^2 - \frac{1}{2}(\gamma_2-\gamma_1)^2\Sigma\omega(\boldsymbol{k}\cdot\boldsymbol{n}_0)^2\right]\boldsymbol{a} \\
&+ \left[\frac{1}{2}(\gamma_3 - \gamma_1 + 4\gamma_7)(\boldsymbol{a}\cdot\boldsymbol{n}_0)(\boldsymbol{k}\cdot\boldsymbol{n}_0) + (\gamma_5 + 2\gamma_6)(\boldsymbol{a}\cdot\boldsymbol{k}) \right.\\
&\qquad \left. + \frac{1}{2}(\gamma_1^2 - \gamma_2^2)\Sigma\omega(\boldsymbol{k}\cdot\boldsymbol{n}_0)(\boldsymbol{a}\cdot\boldsymbol{n}_0)\right]\boldsymbol{k} \\
&+ \left[\frac{1}{2}(\gamma_3 - \gamma_1 + 4\gamma_7)(\boldsymbol{k}\cdot\boldsymbol{n}_0)(\boldsymbol{k}\cdot\boldsymbol{a}) \right.\\
&\qquad + \frac{1}{2}(\gamma_1 + 2\gamma_2 + \gamma_3)(\boldsymbol{a}\cdot\boldsymbol{n}_0)k^2 + 2\gamma_4(\boldsymbol{a}\cdot\boldsymbol{n}_0)(\boldsymbol{k}\cdot\boldsymbol{n}_0)^2 \\
&\qquad + 2\gamma_2^2\Sigma\omega(\boldsymbol{k}\cdot\boldsymbol{n}_0)^2(\boldsymbol{a}\cdot\boldsymbol{n}_0) + \frac{1}{2}(\gamma_1^2 - \gamma_2^2)\Sigma\omega(\boldsymbol{k}\cdot\boldsymbol{n}_0)(\boldsymbol{a}\cdot\boldsymbol{k}) \\
&\qquad \left. - \frac{1}{2}(\gamma_1 + \gamma_2)^2\Sigma\omega(\boldsymbol{a}\cdot\boldsymbol{n}_0)k^2\right]\boldsymbol{n}_0,
\end{aligned} \tag{5.144}$$

where all γ_i's are evaluated at the unperturbed density ϱ_0. We let $\boldsymbol{k}$ and $\boldsymbol{a}$ be represented as

$$\boldsymbol{k} = k\boldsymbol{e} \quad \text{and} \quad \boldsymbol{a} = a_e\boldsymbol{e} + a_n\boldsymbol{n}_0, \quad a_n,\ a_e,\ k \in \mathbb{C}, \tag{5.145}$$

where the unit vector $\boldsymbol{e}$ designates the propagation direction and k, a_e and a_n are all complex numbers to be determined. The imaginary part k_2 of k is associated with the attenuation of the wave: $1/k_2$ represents the attenuation length,[21] that is, the length over which the wave amplitude is reduced by the factor 1/e. Equation (5.144) must be supplemented with the mass continuity equation (5.57), which by (5.145) takes the form

$$ka_e + ka_n \cos\beta = \omega, \quad \text{with} \quad \cos\beta := \boldsymbol{e}\cdot\boldsymbol{n}_0. \tag{5.146}$$

It follows from (5.145) and (5.146) that whenever $\sin\beta = 0$, a_e and a_n are not uniquely defined; we resolve this ambiguity by setting $a_n = 0$ for $\sin\beta = 0$.

Were $\Sigma = 0$, equation (5.144) would reduce to the propagation equation (5.98) found in the absence of director libration. What makes (5.144) more difficult to solve than that equation is the way Σ depends on the unknown k^2. Here we assume that

$$\gamma_1\omega \gg K|k^2|, \tag{5.147}$$

so that, in (5.144), $\Sigma\omega$ can be approximated by $1/\gamma_1$. Physically, this approximation amounts to disregarding the elastic torque in the balance equation (5.134); elastic effects thus disappear from the balances of both linear momentum and torque, though for different reasons. In Section 5.4.2, we shall derive the *upper* bound to be imposed on ω to make (5.147) compatible with the solution to the propagation equation obtained here. As in Section 5.3.3, we also consider the limit of (5.144) where all viscosities are small. More precisely, we assume that there is a small dimensionless parameter ε_0 such that

$$\gamma_i = \varrho_0\frac{c_0^2}{\omega}\,O(\varepsilon_0), \qquad i = 1,\dots,7, \tag{5.148}$$

and we further seek solutions of (5.144) such that

$$k_2 = \frac{\omega}{c_0}\,O(\varepsilon_0), \quad a_n = c_0\,O(\varepsilon_0), \tag{5.149}$$

where c_0 is as in (5.140). The validity of (5.148) requires that ω not exceed an upper bound that will be discussed in Section 5.4.2 along with the one that makes (5.147) compatible.

Under assumptions (5.147), (5.148), and (5.149), proceeding exactly as in Section 5.3.3 above, we finally arrive at the following solution of (5.144) and (5.146) at the lowest order of approximation in ε_0:

---

[21] We shall show in Section 5.4.2 that inequalities (5.84) imply $k_2 \geqq 0$.

$$k = \frac{\omega}{c} + \mathrm{i}k_2, \tag{5.150a}$$

$$k_2 = \frac{\omega^2}{2\varrho_0 c_0^3} \frac{1}{\frac{c}{c_0} + \frac{1}{2}\omega^2\tau^2\frac{c_0}{c}} \left[\gamma_5 + \gamma_6 + \left(\gamma_3 + 2\gamma_7 - \frac{\gamma_2^2}{\gamma_1}\right)\cos^2\beta + \left(\gamma_4 + \frac{\gamma_2^2}{\gamma_1}\right)\cos^4\beta \right], \tag{5.150b}$$

$$a_n = -\frac{\mathrm{i}}{2}\frac{\cos\beta}{\sin^2\beta}\left\{\frac{\omega}{c\varrho_0}\left[\gamma_2 + \gamma_3 + 2\gamma_7 - \left(\gamma_3 - 2\gamma_4 + 2\gamma_7 - 3\frac{\gamma_2^2}{\gamma_1}\right)\cos^2\beta - 2\left(\gamma_4 + \frac{\gamma_2^2}{\gamma_1}\right)\cos^4\beta\right] - \frac{c_0^2}{c}\frac{\gamma_2(\gamma_1+\gamma_2)}{\gamma_1^2}\right\}$$
$$\text{for} \quad \sin\beta \neq 0, \tag{5.150c}$$

$$a_e = c - \mathrm{i}\frac{c^2}{\omega}k_2 - a_n\cos\beta. \tag{5.150d}$$

Here c is the velocity of sound along $\boldsymbol{e}$, which depends on both ω and β through the same equation (5.111) obtained in Section 5.3.3 in the absence of director libration, while the expressions for k_2, a_n, and a_e above would correspondingly reduce to the equations (5.118), (5.113c), and (5.113d) found in Section 5.3.3 in the limit as $\gamma_1/\gamma_2 \to \infty$, where the director libration would be hampered by an arbitrarily large rotational viscosity γ_1. We record here for future reference the limiting expression of k_2 in the absence of libration, which here we denote by k_2^∞,

$$k_2^\infty = \frac{\omega^2}{2\varrho_0 c_0^3} \frac{1}{\frac{c}{c_0} + \frac{1}{2}\omega^2\tau^2\frac{c_0}{c}} [\gamma_5 + \gamma_6 + (\gamma_3 + 2\gamma_7)\cos^2\beta + \gamma_4\cos^4\beta]. \tag{5.151}$$

As already pointed out in Section 5.3.3 for k_2^∞, since c in (5.111) is a function of ω, k_2 in (5.150b) does not depend on ω in a purely quadratic fashion, as in earlier theoretical studies on wave propagation in nematic liquid crystals [361, 223, 106, 149, 139]. Such a nonquadratic dependence is a characteristic signature of our assumption on the KORTEWEG nature of the acoustic coupling; it will be quantitatively compared in Section 5.4.3 with the available experimental data.

Wave Attenuation

We now proceed to show that both k_2 and k_2^∞ are not negative whenever the dissipation inequalities (5.84) are satisfied. To this end, we set $z := \cos^2\beta$ and denote by $h(z)$ the function defined by the expression enclosed in brackets on the right side of (5.150b):

$$h(z) := \gamma_5 + \gamma_6 + (\gamma_3 + 2\gamma_7)z + \gamma_4 z^2 - \frac{\gamma_2^2}{\gamma_1}[-(z^2 - z)].$$

Clearly, the sign of k_2 is the same as the sign of h. Since we assumed that $\gamma_1 > 0$, by (5.84f) we also have that

$$\frac{\gamma_2^2}{\gamma_1} \leqq \gamma_3 + 2\gamma_5,$$

and so,

$$h(z) \geqq \gamma_5+\gamma_6+2(\gamma_7-\gamma_5)z+(\gamma_4+\gamma_3+2\gamma_5)z^2 =: h_0(z) \quad \forall\, z \in [0,1]. \tag{5.152}$$

It readily follows from (5.84c), (5.84d), and (5.84e)) that both $h_0(0) \geqq 0$ and $h_0(1) \geqq 0$. Thus, if h_0 is either linear or concave, that is, if $\gamma_4 + \gamma_3 + 2\gamma_5 \leqq 0$, then $h_0(z) \geqq 0\ \forall z \in [0,1]$, which by (5.152) is the desired conclusion. If, on the other hand, h_0 is convex, that is, if $\gamma_4 + \gamma_3 + 2\gamma_5 > 0$, the desired conclusion follows from the inequality $h_0(z_m) \geqq 0$, where z_m is the minimizer of h_0 in $\mathbb{R}$. Indeed, an easy computation shows that

$$h_0(z_m) = \frac{(\gamma_5+\gamma_6)(\gamma_3+\gamma_4+2\gamma_5) - (\gamma_7-\gamma_5)^2}{\gamma_3+\gamma_4+2\gamma_5}.$$

While the denominator of this ratio is positive by assumption, the numerator can be shown to be nonnegative by taking the product of the left sides of (5.84c) and (5.84e) and adding the result to the left side of (5.84g).

Similarly, we show that $k_2^\infty \geqq 0$. By (5.151), h is then replaced by

$$h^\infty(z) := \gamma_5 + \gamma_6 + (\gamma_3 + 2\gamma_7)z + \gamma_4 z^2,$$

which can also be written as

$$h^\infty(z) = h(z) + \frac{\gamma_2^2}{\gamma_1}\left[-(z^2 - z)\right] \geqq h(z) \geqq 0 \quad \forall z \in [0,1], \tag{5.153}$$

since $\gamma_1 > 0$.

Admissible Frequency Ranges

Several simplifying assumptions have been made to arrive at (5.150); here we identify the ranges in which to choose the angular frequency ω of the propagating wave to make these assumptions admissible.

First, we identify the values of ω that make ε_0 in (5.148) a small parameter. Estimating from [4] the velocity of sound $c_0 = 1.3\times10^3\ \mathrm{m\,s^{-1}}$, from [59] (p. 231) the average viscosity $\gamma = 10^{-1}$ Pa s, and from [369] the mass density $\varrho_0 = 10^3\ \mathrm{Kg\,m^{-3}}$, one easily sees from (5.148) that $\varepsilon_0 \ll 1$ whenever $\omega \ll \omega_\gamma$, with

$$\omega_\gamma := \frac{c_0^2 \varrho_0}{\gamma} = 0.8 \times 10^{10}\ \mathrm{s^{-1}} \sim 10^4\ \mathrm{MHz}.$$

Second, we identify the angular frequencies that make the solution (5.150) of the propagation equation (5.144) compatible with the assumption (5.147). To this end, we note from Figure 5.1 that $c \sim c_0$ for $\omega\tau < 10$, and so, since by (5.150a) $k \sim \omega/c$, estimating γ_1 as 2γ and taking $K \sim 10^{-11}$ N from [59] (p. 103), we see that (5.147) is satisfied for the solution (5.150) whenever $\omega \ll \omega_K$, with

$$\omega_K := \frac{c_0^2\gamma_1}{K} = 3.38 \times 10^{16}\,\mathrm{s}^{-1} \sim 10^{10}\,\mathrm{MHz}.$$

Finally, as remarked in Section 5.4.1 above, our theory has neglected the director inertia. Such an approximation is valid if κ in (5.90) is much smaller than the kinetic energy density associated with the acoustic flow. By (5.135), (5.145), and (5.150d), this latter can be estimated as $\frac{1}{2}\varrho_0 s_0^2 c_0^2$. On the other hand, by (5.136), $|\dot{\boldsymbol{n}}| = s_0\omega|\mathbf{A}\boldsymbol{n}_0| + O(s_0^2)$, and by (5.143), for the solution (5.150), $|\mathbf{A}\boldsymbol{n}_0| = O(1)$, so that κ can be neglected whenever $\omega \ll \omega_\delta$, with[22]

$$\omega_\delta := \frac{c_0}{\delta} = 6.5 \times 10^{12}\,\mathrm{s}^{-1} \sim 10^{6}\,\mathrm{MHz}.$$

It is clear from the estimates above that the largest upper bound on ω for the validity of our theory is ω_K: such a bound of an elastic origin can still not be exceeded when the rotational kinetic energy κ comes into play. For $\omega \sim \omega_\delta$, however, our solution (5.150) ceases to be valid, since the upper bound ω_γ is violated and so (5.148) no longer applies. The most stringent bound on ω is thus ω_γ; it will follow from the estimate of τ in Section 5.4.3 that this easily complies with the requirement that $\omega\tau < 10$.

5.4.3 Phenomenological Parameters

Using data published in the literature for *N*-(*p*-methoxybenzylidene)-*p*-butylaniline (MBBA), numerical evaluations for both dispersion and attenuation are made in this section and compared to acoustic experiments. Our objective is to estimate the phenomenological parameters introduced by our theory, namely, u_1, u_2, γ_6, and γ_7.

To account for the experimental data available in the literature [196, 225], we use the measure of anisotropy Δc for the speed of sound c introduced in (5.114). Δc is a function of both ω and β, which vanishes for $\beta = \frac{\pi}{2}$. By assuming that ε defined as in (5.115) is a small parameter, we approximate Δc as in (5.116). By (5.115), we can write the characteristic time τ in (5.105) as

$$\tau = \tau_1\left(1 + \frac{1}{2}\varepsilon\cos^2\beta\right) + O(\varepsilon^2),$$

where τ_1 is the same as in (5.69).

Similarly, we introduce the following measure of anisotropy for the attenuation k_2:

$$\Delta_1 k_2 := \frac{\Delta k_2}{\Delta_\perp k_2},$$

---

22 Here we take the molecular radius of gyration δ as a typical molecular length, and following [59] (p. 98) we estimate $\delta \approx 2\,\mathrm{nm}$.

where $\Delta k_2 := k_2 - k_2|_{\beta=0}$ expresses the change in attenuation and $\Delta_\perp k_2 := \Delta k_2|_{\beta=\frac{\pi}{2}}$ is the value of Δk_2 when the wave propagates at right angles to $\boldsymbol{n}_0$. It follows from (5.150b) that, in the limit of small ε,

$$\Delta_1 k_2 = G(\beta) + O(\varepsilon), \tag{5.154}$$

where

$$G(\beta) := \sin^2\beta\left(1 + \frac{\gamma_1\gamma_4 + \gamma_2^2}{\gamma_1(\gamma_3 + \gamma_4 + 2\gamma_7)}\cos^2\beta\right).$$

To represent how $\Delta_\perp k_2$ depends on ω relative to a reference angular frequency ω_0, we introduce the quantity

$$\Delta_2 k_2 := \frac{\Delta_\perp k_2}{\Delta_\perp k_2|_{\omega=\omega_0}},$$

which in the limit of small ε becomes

$$\Delta_2 k_2 = \frac{\omega}{\omega_0}\sqrt{\frac{\sqrt{1+\tau_1^2\omega^2}-1}{\sqrt{1+\tau_1^2\omega_0^2}-1}\sqrt{\frac{1+\tau_1^2\omega_0^2}{1+\tau_1^2\omega^2}}} + O(\varepsilon). \tag{5.155}$$

In this approximation, $\Delta_2 k_2$, unlike $\Delta_1 k_2$, is independent of the viscosities: it depends on a single phenomenological parameter, that is, τ_1.

We used formulas (5.114), (5.116), (5.154), and (5.155) to fit the data measured for Δc in [225] for MBBA at the wave frequency $\omega/2\pi = 10\,\text{MHz}$ and the data measured for Δk_2 in [196] in the range 2–6 MHz for $\omega/2\pi$. The former data were taken from Figure 2 of [225] and the latter data from Figures 1 and 2 of [196].

We started from Figure 2 of [196], which exhibits the dependence of Δk_2 on the wave frequency $\omega/2\pi$. We fitted these data with formula (5.155) using $\omega_0/2\pi = 6\,\text{MHz}$ as reference frequency. By employing the built-in function `FindFit` in *Mathematica* [363] for least-squares fit, we obtained $\tau_1 = 3.47 \times 10^{-8}\,\text{s}$ for the only fitting parameter. From this value, we computed $\tau_1\omega_1 = 2.18$ at the frequency $\omega_1/2\pi = 10\,\text{MHz}$ used in [225]. We inserted this value of $\omega_1\tau_1$ in the exact formula (5.114) for Δc to validate our assumption about the smallness of ε, which was then found to be $\varepsilon = 7.74 \times 10^{-3}$ by fitting the data in Figure 2 of [225] for $u_2 = \varepsilon u_1$. Thus, our using (5.155) to fit the data in [196] was fully justified. Moreover, the data in Figure 2 of [225] for Δc could also be fitted directly with the function $A\cos^2\beta$, which gave $A = 11.25 \times 10^{-4}$, consistent with the value $A = 12 \times 10^{-4}$ found in [225] from the raw data. On the other hand, using the approximate formula (5.116) for Δc, we found instead $A = \varepsilon f(\tau_1\omega_1) = 11.28 \times 10^{-4}$, which is in very good agreement with the value found by the direct fit and thus further confirmed the validity of our assumption on ε. Finally, we used (5.154) to find the value of the viscosity γ_7, taking for $\gamma_1, \dots, \gamma_4$ in G the standard values for MBBA (see p. 231 of [59]). We thus arrived at $\gamma_7 = 1.58 \times 10^{-1}\,\text{Pa s}$. Since both $\Delta_1 k_2$ and $\Delta_2 k_2$ are independent of γ_6, this latter viscosity would be determined only with the aid of (5.150b) from direct measurements of k_2.

Synopsis

We extended the variational theory for nematoacoustics presented in Section 5.3 to the case in which the nematic director can freely librate around an average orientation. According to this theory, the acoustic field interacts with the nematic texture through an additional elastic energy of KORTEWEG type, characterized by the phenomenological susceptibilities u_1 and u_2 introduced in (5.92). We estimated these constitutive parameters with the aid of experimental data available in the literature for the anisotropy of both sound speed and wave attenuation. We also estimated one additional viscosity introduced in the RAYLEIGH dissipation function (5.78) by the relaxation of the incompressibility constraint.

We solved the balance equations of the theory in the linear approximation, appropriate for acoustic propagation. In particular, we sought plane wave solutions also involving the director libration and we proved that the wave attenuation is not negative as a consequence of the semipositive definiteness of the RAYLEIGH dissipation function.

We showed that in the (fast) acoustical regime, elastic stresses and torques do not affect the motion: the ERICKSEN stress tensor is of second order in the acoustical perturbation parameter, and the elastic torque is negligible with respect to viscous torques as long as the angular frequency of the propagating wave does not exceed an upper bound much larger than the frequency at which the rotational inertia of the director—neglected here as usual—should also be taken into account.

Also the torque imparted to the director by the acoustic wave is of second order in the acoustical perturbation parameter. Acoustic torques and elastic stresses thus act at time and length scales larger than the acoustic ones. They should be regarded as *streaming* sources that affect the flow through their time averages. Other steady motions of the director texture can then take place at time and length scales much larger than the acoustic characteristic times and lengths. Correspondingly, though no net hydrodynamic flow takes place at the time scale of the acoustic vibrations, at longer time scales even an initially stagnant fluid may develop steady flows by *acoustic* streaming, as already proved for isotropic viscous fluids in a vast literature [87, 360, 251, 188], originated from RAYLEIGH's work [325, 326, 328]. For nematic liquid crystals, elastic stresses and acoustic torques are new streaming sources: they both affect the slow director and flow dynamics, mutually interwoven, for which the appropriate balance has still to be derived and studied.

A

Notation and Basic Concepts

This appendix is meant to introduce the reader to the notation employed throughout the book and to illustrate some basic concepts in possibly a less formal style, though with the same rigor as in the main text. Only on occasion does it serve the purpose of a repository of technical results to be invoked wherever the need arises, since we strove to make the text as self-contained as possible, preferring moderate digressions to excessive referencing.

A.1 Points, Vectors, and Tensors

In this book, we consider fluids in the three-dimensional Euclidean space $\mathcal{E}$. The basic elements of this space are its *points*.[1] Associated with $\mathcal{E}$ is the linear space of *translations* $\mathcal{V}$ that are mappings of $\mathcal{E}$ into itself. The elements of $\boldsymbol{v} \in \mathcal{V}$ are called *vectors*:

$$\boldsymbol{v} : \mathcal{E} \to \mathcal{E}, \quad p \mapsto \boldsymbol{v}(p).$$

For any two points $p, q \in \mathcal{E}$ there is exactly one vector $\boldsymbol{v}$ that takes p into q. We denote it by $\boldsymbol{v} = q - p$ and write $\boldsymbol{v}(p) = q$, or sometimes $q = p + \boldsymbol{v}$. We note that while the difference of two points is a well-defined unique vector, the sum of two points is *not* defined. If an origin $o \in \mathcal{E}$ is chosen, points can be identified by their *position vectors* relative to that origin, $\boldsymbol{p} = p - o$. Care needs to be taken not to confuse position vectors with ordinary vectors. A given point has different position vectors with respect to different origins. A vector $\boldsymbol{v} = p - q$, however, is independent of the choice of origin. By contrast, $\boldsymbol{p} = p - o$ and $\boldsymbol{p}^* = p - o^*$ are different position vectors of the same point p with respect to different origins o and o^*. We also write $\boldsymbol{v}^2$ for $\boldsymbol{v} \cdot \boldsymbol{v}$.

The *inner product* of two vectors $\boldsymbol{u}$ and $\boldsymbol{v}$ is denoted by $\boldsymbol{u} \cdot \boldsymbol{v}$, and the *cross product* by $\boldsymbol{u} \times \boldsymbol{v}$. The *mixed* product of three vectors $\boldsymbol{u}$, $\boldsymbol{v}$, and $\boldsymbol{w}$ is the scalar

---

[1] For simplicity, we identify material points with their positions in Euclidean space, even though they may be richer in mechanical structure than a mere geometric point can suggest.

$$\boldsymbol{u} \cdot \boldsymbol{v} \times \boldsymbol{w}.$$

It is invariant under cyclic permutations of the three vectors, that is,

$$\boldsymbol{u} \cdot \boldsymbol{v} \times \boldsymbol{w} = \boldsymbol{v} \cdot \boldsymbol{w} \times \boldsymbol{u} = \boldsymbol{w} \cdot \boldsymbol{u} \times \boldsymbol{v} \quad \forall\, \boldsymbol{u}, \boldsymbol{v}, \boldsymbol{w} \in \mathcal{V}.$$

Two vectors $\boldsymbol{u}$ and $\boldsymbol{v}$ are *orthogonal* if $\boldsymbol{u} \cdot \boldsymbol{v} = 0$ The length of a vector $\boldsymbol{v}$ is given by $v = |\boldsymbol{v}| = \sqrt{\boldsymbol{v} \cdot \boldsymbol{v}}$.

A *tensor* of rank two, usually simply called a tensor, is an element of the linear space $\mathsf{L}(\mathcal{V})$ of all linear mappings of $\mathcal{V}$ into itself:

$$\mathbf{T} : \mathcal{V} \to \mathcal{V}, \quad \boldsymbol{v} \mapsto \mathbf{T}(\boldsymbol{v}).$$

We write $\mathbf{T}\boldsymbol{v}$ for $\mathbf{T}(\boldsymbol{v})$. The *transpose* $\mathbf{T}^{\mathsf{T}}$ of a tensor $\mathbf{T}$ is defined via

$$\boldsymbol{u} \cdot \mathbf{T}^{\mathsf{T}}\boldsymbol{v} = \boldsymbol{v} \cdot \mathbf{T}\boldsymbol{u} \quad \text{for all } \boldsymbol{u}, \boldsymbol{v} \in \mathcal{V}.$$

We say that a tensor $\mathbf{T}$ is *symmetric* if $\mathbf{T} = \mathbf{T}^{\mathsf{T}}$ and *skew-symmetric* if $\mathbf{T} = -\mathbf{T}^{\mathsf{T}}$. All symmetric tensors constitute a linear subspace of $\mathsf{L}(\mathcal{V})$, which we denote by $\mathsf{Sym}(\mathcal{V})$. Likewise, all skew-symmetric tensors constitute a linear subspace of $\mathsf{L}(\mathcal{V})$, which we denote by $\mathsf{Skw}(\mathcal{V})$. Given any tensor $\mathbf{T}$, it can be uniquely written as the sum of a tensor in $\mathsf{Sym}(\mathcal{V})$ and a tensor in $\mathsf{Skw}(\mathcal{V})$:

$$\mathbf{T} = \frac{1}{2}(\mathbf{T} + \mathbf{T}^{\mathsf{T}}) + \frac{1}{2}(\mathbf{T} - \mathbf{T}^{\mathsf{T}}).$$

We set

$$\operatorname{sym}(\mathbf{T}) := \frac{1}{2}(\mathbf{T} + \mathbf{T}^{\mathsf{T}}) \quad \text{and} \quad \operatorname{skw}(\mathbf{T}) := \frac{1}{2}(\mathbf{T} - \mathbf{T}^{\mathsf{T}}),$$

and we call the former the *symmetric part* of $\mathbf{T}$ and the latter the *skew-symmetric* part of $\mathbf{T}$. The identity tensor $\mathbf{I}$ is defined by $\mathbf{I}\boldsymbol{v} := \boldsymbol{v}$ for all $\boldsymbol{v} \in \mathcal{V}$, and it is clearly symmetric. The trace and determinant of a tensor $\mathbf{T}$ are denoted by $\operatorname{tr}\mathbf{T}$ and $\det\mathbf{T}$. It is sometimes convenient to denote in a concise manner the symmetric, traceless part of a tensor $\mathbf{T}$; we shall employ the following notation:

$$\overline{\mathbf{T}} := \operatorname{sym}(\mathbf{T}) - \frac{1}{3}(\operatorname{tr}\mathbf{T})\mathbf{I}.$$

For a tensor $\mathbf{L} \in \mathsf{L}(\mathcal{V})$, we say that $\lambda \in \mathbb{R}$ and $\boldsymbol{e} \in \mathbb{S}^2$ are an *eigenvalue* and the corresponding (normalized) *eigenvector* of $\mathbf{L}$ if

$$\mathbf{L}\boldsymbol{e} = \lambda \boldsymbol{e}.$$

Symmetric tensors are peculiar in regard to eigenvalues and eigenvectors. The spectral theorem states that for every tensor $\mathbf{S} \in \mathsf{Sym}(\mathcal{V})$ there is a basis $(\boldsymbol{e}_1, \boldsymbol{e}_2, \boldsymbol{e}_3)$ of $\mathcal{V}$ such that

$$\mathbf{S} = \lambda_1 \boldsymbol{e}_1 \otimes \boldsymbol{e}_1 + \lambda_2 \boldsymbol{e}_2 \otimes \boldsymbol{e}_2 + \lambda_3 \boldsymbol{e}_3 \otimes \boldsymbol{e}_3.$$

Thus, clearly, e_1, e_2, and e_3 are eigenvectors of $\mathbf{S}$, and λ_1, λ_2, and λ_3 are the corresponding eigenvalues. Given any skew-symmetric tensor $\mathbf{W}$ in $\mathsf{Skw}(\mathcal{V})$, there is precisely one vector $w \in \mathcal{V}$ such that

$$\mathbf{W}v = w \times v \qquad \forall\, v \in \mathcal{V}. \tag{A.1}$$

We say that w is the *axial vector* associated with $\mathbf{W}$. The linear mapping established by (A.1) between the spaces $\mathsf{Skw}(\mathcal{V})$ and $\mathcal{V}$ is indeed invertible. The axial vector w of $\mathbf{W}$ belongs the *null space* of $\mathbf{W}$, that is, it is a (nonnormalized) eigenvector of $\mathbf{W}$ with zero eigenvalue. Actually, all (nonnormalized) eigenvectors of $\mathbf{W}$ are parallel to w, which amounts to saying that the null space of $\mathbf{W}$ is one-dimensional. For the special skew-symmetric tensor $\mathbf{W}$ in the form

$$\mathbf{W} = b \otimes a - a \otimes b,$$

the associated axial vector w is

$$w = a \times b.$$

The *dyadic* product $u \otimes v$ of two vectors u and v is a tensor that is defined by its action on an arbitrary third vector w:

$$(u \otimes v)w := (v \cdot w)u \quad \text{for all } w \in \mathcal{V}.$$

The composition of two tensors $\mathbf{C} = \mathbf{AB}$ is defined via

$$\mathbf{C}v = \mathbf{AB}v \quad \text{for all } v \in \mathcal{V}.$$

An inner product between two tensors $\mathbf{S}$ and $\mathbf{T}$ is defined by

$$\mathbf{S} \cdot \mathbf{T} := \operatorname{tr} \mathbf{S}\mathbf{T}^{\mathsf{T}}. \tag{A.2}$$

With it, we have $\operatorname{tr} \mathbf{T} = \mathbf{T} \cdot \mathbf{I}$ for all tensors $\mathbf{T}$. The norm of a tensor $\mathbf{T}$ is $|\mathbf{T}| = \sqrt{\mathbf{T} \cdot \mathbf{T}}$. In particular, $\mathbf{I} \cdot \mathbf{I} = \operatorname{tr} \mathbf{I} = 3$.

The inner product in $\mathsf{L}(\mathcal{V})$ defined by (A.2) enjoys the following properties:

$$\mathbf{S} \cdot \mathbf{T} = \mathbf{S}^{\mathsf{T}} \cdot \mathbf{T}^{\mathsf{T}} \quad \forall\, \mathbf{S}, \mathbf{T} \in \mathsf{L}(\mathcal{V}),$$

$$\mathbf{AB} \cdot \mathbf{C} = \mathbf{A} \cdot \mathbf{C}\mathbf{B}^{\mathsf{T}} = \mathbf{B} \cdot \mathbf{A}^{\mathsf{T}}\mathbf{C} \quad \forall\, \mathbf{A}, \mathbf{B}, \mathbf{C} \in \mathsf{L}(\mathcal{V}).$$

Moreover, given two skew-symmetric tensors $\mathbf{W}_1$ and $\mathbf{W}_2$ with associated axial vectors w_1 and w_2, respectively, the inner product of $\mathbf{W}_1$ and $\mathbf{W}_2$ is related to the inner product of w_1 and w_2 through the equation

$$\mathbf{W}_1 \cdot \mathbf{W}_2 = 2 w_1 \cdot w_2.$$

The linear subspaces $\mathsf{Sym}(\mathcal{V})$ and $\mathsf{Skw}(\mathcal{V})$ are orthogonal complements of $\mathsf{L}(\mathcal{V})$ with respect to the inner product in (A.2), and so any symmetric tensor is orthogonal to any skew-symmetric tensor.

The *principal invariants* $\widehat{I}_i(\mathbf{T})$, $i = 1, 2, 3$, of a tensor $\mathbf{T} \in \mathsf{L}(\mathcal{V})$ are defined as the coefficients of the following polynomial in λ:

$$\det(\mathbf{T} + \lambda \mathbf{I}) = \lambda^3 + \widehat{I}_1(\mathbf{T})\lambda^2 + \widehat{I}_2(\mathbf{T})\lambda + \widehat{I}_3(\mathbf{T}).$$

We have that

$$\widehat{I}_1(\mathbf{T}) = \operatorname{tr} \mathbf{T}, \quad \widehat{I}_2(\mathbf{T}) = \frac{1}{2}\left[(\operatorname{tr} \mathbf{T})^2 - \operatorname{tr} \mathbf{T}^2\right], \quad \widehat{I}_3(\mathbf{T}) = \det \mathbf{T}.$$

While the principal invariant $\widehat{I}_1(\mathbf{T})$ coincides with the invariant $I_1(\mathbf{T}) := \operatorname{tr} \mathbf{T}$ introduced in (1.83), $\widehat{I}_2(\mathbf{T})$ and $\widehat{I}_3(\mathbf{T})$ are not the same as the corresponding invariants $I_2(\mathbf{T}) := \operatorname{tr} \mathbf{T}^2$ and $I_3(\mathbf{T}) := \operatorname{tr} \mathbf{T}^3$. They are related as follows:

$$\widehat{I}_2 = \frac{1}{2}\left(I_1^2 - I_2\right), \qquad \widehat{I}_3 = \frac{1}{3}\left(I_3 - \frac{3}{2} I_1 I_2 + \frac{1}{2} I_1^3\right).$$

The latter equation in particular follows from the CAYLEY–HAMILTON theorem, which states that every tensor $\mathbf{T}$ obeys the equation

$$\mathbf{T}^3 - \widehat{I}_1 \mathbf{T}^2 + \widehat{I}_2 \mathbf{T} - \widehat{I}_3 \mathbf{I} = \mathbf{0}. \tag{A.3}$$

Equation (A.3) also follows from a more general identity of RIVLIN [281] valid for any triple of tensors $\mathbf{A}$, $\mathbf{B}$, $\mathbf{C}$:

$$\begin{aligned}
&\mathbf{ABC} + \mathbf{ACB} + \mathbf{BCA} + \mathbf{BAC} + \mathbf{CAB} + \mathbf{CBA} \\
&- (\operatorname{tr} \mathbf{BC} - \operatorname{tr} \mathbf{B} \operatorname{tr} \mathbf{C})\mathbf{A} - (\operatorname{tr} \mathbf{CA} - \operatorname{tr} \mathbf{C} \operatorname{tr} \mathbf{A})\mathbf{B} - (\operatorname{tr} \mathbf{AB} - \operatorname{tr} \mathbf{A} \operatorname{tr} \mathbf{B})\mathbf{C} \\
&- \operatorname{tr} \mathbf{A}(\mathbf{BC} + \mathbf{CB}) - \operatorname{tr} \mathbf{B}(\mathbf{CA} + \mathbf{AC}) - \operatorname{tr} \mathbf{C}(\mathbf{AB} + \mathbf{BA}) \\
&- (\operatorname{tr} \mathbf{A} \operatorname{tr} \mathbf{B} \operatorname{tr} \mathbf{C} - \operatorname{tr} \mathbf{A} \operatorname{tr} \mathbf{BC} - \operatorname{tr} \mathbf{B} \operatorname{tr} \mathbf{CA} \\
&\qquad - \operatorname{tr} \mathbf{C} \operatorname{tr} \mathbf{AB} + \operatorname{tr} \mathbf{ABC} + \operatorname{tr} \mathbf{CBA})\mathbf{I} \\
&= \mathbf{0},
\end{aligned} \tag{A.4}$$

by setting $\mathbf{A} = \mathbf{B} = \mathbf{C} = \mathbf{T}$.

For an invertible tensor $\mathbf{A}$ and every tensor $\mathbf{C}$,

$$\det(\mathbf{A} + s\mathbf{C}) = (\det \mathbf{A})(1 + \widehat{I}_1(\mathbf{A}^{-1}\mathbf{C})s + \widehat{I}_2(\mathbf{A}^{-1}\mathbf{C})s^2 + \widehat{I}_3(\mathbf{A}^{-1}\mathbf{C})s^3),$$

whence it follows that

$$\left.\frac{d}{ds} \det(\mathbf{A} + s\mathbf{C})\right|_{s=0} = (\det \mathbf{A})\widehat{I}_1(\mathbf{A}^{-1}\mathbf{C}) = \operatorname{tr}[(\det \mathbf{A})\mathbf{A}^{-1}\mathbf{C}]. \tag{A.5}$$

A tensor of rank three is a linear mapping a from the vector space $\mathcal{V}$ into the space $\mathsf{L}(\mathcal{V})$ of second-rank tensors:

$$\mathsf{a} : \mathcal{V} \to \mathsf{L}(\mathcal{V}), \quad \boldsymbol{v} \mapsto \mathsf{a}(\boldsymbol{v}).$$

We write $\mathsf{a}\boldsymbol{v}$ for the second-rank tensor $\mathsf{a}(\boldsymbol{v})$. More generally, a tensor of rank $n \geqq 1$ is a linear mapping that takes a vector into a tensor of order $n - 1$. A continuous mapping from a region $\mathcal{B}$ of $\mathcal{E}$ into the space of tensors of rank n that assigns a tensor to every point $p \in \mathcal{B}$ is called a *tensor field* of rank n.

A.2 Bases and Coordinates

An ordered set of three mutually orthogonal unit vectors $(\boldsymbol{e}_1, \boldsymbol{e}_2, \boldsymbol{e}_3)$ forms a *basis* for $\mathcal{V}$. We normally choose a *positively oriented* basis, that is, one for which $\boldsymbol{e}_1 \times \boldsymbol{e}_2 = \boldsymbol{e}_3$.[2] Every vector $\boldsymbol{v} \in \mathcal{V}$ can be expressed in terms of its *components* $v_i := \boldsymbol{v} \cdot \boldsymbol{e}_i$ as

$$\boldsymbol{v} = \sum_{i=1}^{3} v_i \boldsymbol{e}_i.$$

To simplify notation, we apply the *summation convention*: a term with a repeated index is to be summed over all three values of that index from 1 to 3. Hence, we simply write $\boldsymbol{v} = v_i \boldsymbol{e}_i$.

The components of a second-rank tensor are defined as $T_{ij} := \boldsymbol{e}_i \cdot \mathbf{T}\boldsymbol{e}_j$. For a dyadic product this implies $(\boldsymbol{u} \otimes \boldsymbol{v})_{ij} = u_i v_j$. The components of the identity $\mathbf{I}$ in any basis are $\boldsymbol{e}_i \cdot \mathbf{I}\boldsymbol{e}_j = \boldsymbol{e}_i \cdot \boldsymbol{e}_j = \delta_{ij}$, with

$$\delta_{ij} := \begin{cases} 1 & \text{if } i = j, \\ 0 & \text{if } i \neq j, \end{cases}$$

the KRONECKER delta. The tensor $\mathbf{T}$ can be represented in terms of its components T_{ij} as

$$\mathbf{T} = T_{ij} \boldsymbol{e}_i \otimes \boldsymbol{e}_j.$$

Indeed,

$$\boldsymbol{e}_i \cdot \mathbf{T}\boldsymbol{e}_j = \boldsymbol{e}_i \cdot (T_{ab}\boldsymbol{e}_a \otimes \boldsymbol{e}_b)\boldsymbol{e}_j = (\boldsymbol{e}_i \cdot \boldsymbol{e}_a) T_{ab} (\boldsymbol{e}_b \cdot \boldsymbol{e}_j) = \delta_{ia} T_{ab} \delta_{bj} = T_{ij}.$$

Thus, if $(\boldsymbol{e}_1, \boldsymbol{e}_2, \boldsymbol{e}_3)$ is a basis of $\mathcal{V}$, then $(\boldsymbol{e}_i \otimes \boldsymbol{e}_j, i, j = 1, 2, 3)$ is a basis of $\mathsf{L}(\mathcal{V})$. In general, if v_i are the components of a vector $\boldsymbol{v}$ in $(\boldsymbol{e}_1, \boldsymbol{e}_2, \boldsymbol{e}_3)$, then $T_{ij} v_j$ are the components of $\mathbf{T}\boldsymbol{v}$ in the same basis. It follows immediately from the definition of the transpose tensor that $(\mathbf{T}^\top)_{ij} = T_{ji}$. The trace of a tensor $\mathbf{T}$ is given by $\operatorname{tr} \mathbf{T} = T_{ii}$, and the components of the product $\mathbf{C} = \mathbf{AB}$ of two tensors are given by $C_{ik} = A_{ij} B_{jk}$. The inner product of two tensors is thus $\operatorname{tr} \mathbf{S}\mathbf{T}^\top = S_{ij} T_{ij}$. Specifically for dyadic products, we have

$$(\boldsymbol{u} \otimes \boldsymbol{v}) \cdot \mathbf{I} = \operatorname{tr}(\boldsymbol{u} \otimes \boldsymbol{v}) = \boldsymbol{v} \cdot \boldsymbol{u}$$

and

$$(\boldsymbol{s} \otimes \boldsymbol{t}) \cdot (\boldsymbol{u} \otimes \boldsymbol{v}) = (\boldsymbol{s} \cdot \boldsymbol{u})(\boldsymbol{t} \cdot \boldsymbol{v}).$$

For the components of a vector product we have

$$\boldsymbol{e}_i \cdot (\boldsymbol{u} \times \boldsymbol{v}) = \epsilon_{ijk} u_j v_k,$$

where

$$\epsilon_{ijk} := \begin{cases} 1 & \text{if } (i, j, k) \text{ is an even permutation of } (1, 2, 3), \\ -1 & \text{if } (i, j, k) \text{ is an odd permutation of } (1, 2, 3), \\ 0 & \text{in all other cases,} \end{cases}$$

is the RICCI alternator.

[2] A *negatively oriented* basis $(\boldsymbol{e}_1, \boldsymbol{e}_2, \boldsymbol{e}_3)$ is one for which $\boldsymbol{e}_1 \times \boldsymbol{e}_2 \cdot \boldsymbol{e}_3 = -1$. We shall always use positively oriented bases.

A.3 Rotations

An *orthogonal transformation* is a linear transformation that preserves the length of all vectors and their relative angles. Hence, it takes a set of basis vectors $(\boldsymbol{e}_1, \boldsymbol{e}_2, \boldsymbol{e}_3)$ into a new set of basis vectors, say $(\boldsymbol{e}_1^*, \boldsymbol{e}_2^*, \boldsymbol{e}_3^*)$. More generally, given any two vectors $\boldsymbol{u}$ and $\boldsymbol{v}$ in $\mathcal{V}$,

$$\mathbf{R}\boldsymbol{u} \cdot \mathbf{R}\boldsymbol{v} = \boldsymbol{u} \cdot \boldsymbol{v},$$

whence it follows that

$$(\mathbf{R}^{\mathsf{T}}\mathbf{R} - \mathbf{I})\boldsymbol{u} \cdot \boldsymbol{v} = 0 \qquad \forall\, \boldsymbol{u},\, \boldsymbol{v} \in \mathcal{V}.$$

This means that $\mathbf{R}^{\mathsf{T}}\mathbf{R} = \mathbf{I}$. It follows that $1 = \det \mathbf{I} = \det(\mathbf{R}^{\mathsf{T}}\mathbf{R}) = (\det \mathbf{R})^2$, and so $\det \mathbf{R} = \pm 1$. Two consecutive orthogonal transformations $\mathbf{R}_1\mathbf{R}_2$ result in a new orthogonal transformation $\mathbf{R} = \mathbf{R}_1\mathbf{R}_2$. The set of orthogonal transformations in three-dimensional Euclidean space forms the group $\mathsf{O}(3)$.

For any two vectors $\boldsymbol{u}$ and $\boldsymbol{v}$, an orthogonal tensor $\mathbf{R}$ is such that

$$\mathbf{R}\boldsymbol{u} \times \mathbf{R}\boldsymbol{v} = (\det \mathbf{R})\mathbf{R}(\boldsymbol{u} \times \boldsymbol{v}).$$

A *rotation* is an orthogonal transformation that transforms a positively oriented basis into a positively oriented basis.[3] It is an orthogonal transformation $\mathbf{R} \in \mathsf{O}(3)$ with $\det \mathbf{R} = +1$. Rotations form the group of *special* orthogonal transformations $\mathsf{SO}(3)$.

Both $\mathsf{SO}(3)$ and $\mathsf{O}(3)$ are proper subgroups of the *unimodular* group $\mathsf{U}(3)$ defined by

$$\mathsf{U}(3) := \{\mathbf{U} \in \mathsf{L}(\mathcal{V}) \,|\, |\det \mathbf{U}| = 1\}.$$

As proved by BRAUER and NOLL [35, 239], $\mathsf{O}(3)$ is the maximal subgroup of $\mathsf{U}(3)$, meaning that no subgroup of $\mathsf{U}(3)$ exists that is not included in $\mathsf{O}(3)$.

A.4 Time Derivatives

The position of a point p in motion is a function of time $p(t)$. The *velocity* $\boldsymbol{v}$ of the point p is defined as

$$\boldsymbol{v}(p,t) = \dot{p} = \frac{dp}{dt} = \lim_{\epsilon \to 0} \frac{p(t+\epsilon) - p(t)}{\epsilon}. \tag{A.6}$$

The field $\boldsymbol{v}(p,t)$ thus gives the velocity of the point that at time t occupies the position p.

For a scalar quantity $\phi(p,t)$ that depends on time and position we write $\frac{\partial p}{\partial t}$ for the *local* time derivative. It is the rate of change of ϕ at the fixed position that p attains at time t. The dependence of ϕ at fixed time on position is given by the gradient $\nabla\phi = \frac{\partial \phi}{\partial p}$. The *material* time derivative is the rate of change of ϕ at the moving

[3] Actually, it also transforms a negatively oriented basis into a negatively oriented one.

material point p. We denote it by $\dot{\phi}$, and it is the total time derivative of $\phi(p(t), t)$, $\dot{\phi} = \frac{d\phi}{dt} = \frac{\partial \phi}{\partial p}\frac{dp}{dt} + \frac{\partial \phi}{\partial t}$. With (A.6), we obtain

$$\dot{\phi} = \frac{\partial \phi}{\partial t} + \nabla \phi \cdot \boldsymbol{v}.$$

Time derivatives of vectors and tensors are defined similarily. We use the convention that

$$(\nabla \boldsymbol{v})_{ij} = v_{i,j}.$$

For example, the acceleration of a point p is $\boldsymbol{a} = \dot{\boldsymbol{v}} = \ddot{p}$, and we have

$$\dot{\boldsymbol{v}} = \frac{\partial \boldsymbol{v}}{\partial t} + (\nabla \boldsymbol{v})\boldsymbol{v}.$$

Elaborating on these basic kinematic concepts, in Section 2.1.2 we define the motion of a continuous body.

Gradients of tensors of rank two and higher are defined in a similar way, for example,

$$(\nabla \mathbf{T})_{ijk} = T_{ij,k},$$

so that

$$\dot{\mathbf{T}} = \frac{\partial \mathbf{T}}{\partial t} + (\nabla \mathbf{T})\boldsymbol{v}.$$

The divergence of a vector field $\boldsymbol{v}$ is defined by

$$\operatorname{div} \boldsymbol{v} := \operatorname{tr}(\nabla \boldsymbol{v}) = v_{i,i}.$$

We define the divergence of a tensor field $\mathbf{T}$ to be the vector $\operatorname{div} \mathbf{T}$ such that

$$\boldsymbol{a} \cdot \operatorname{div} \mathbf{T} = \operatorname{div}(\mathbf{T}^{\mathsf{T}} \boldsymbol{a}) \quad \text{for all } \boldsymbol{a} \in \mathcal{V}.$$

To find an expression for the components of $\operatorname{div} \mathbf{T}$ we first compute

$$\mathbf{T}^{\mathsf{T}} \boldsymbol{e}_i = T_{ba}(\boldsymbol{e}_a \otimes \boldsymbol{e}_b)\boldsymbol{e}_i = T_{ba}(\boldsymbol{e}_i \cdot \boldsymbol{e}_b)\boldsymbol{e}_a = T_{ba}\delta_{ib}\boldsymbol{e}_a = T_{ia}\boldsymbol{e}_a.$$

Now

$$\operatorname{div}(T_{ia}\boldsymbol{e}_a) = \nabla(T_{ia}) \cdot \boldsymbol{e}_a = T_{ia,j}\boldsymbol{e}_j \cdot \boldsymbol{e}_a = T_{ia,j}\delta_{ja} = T_{ij,j},$$

and so

$$\boldsymbol{e}_i \cdot \operatorname{div} \mathbf{T} = T_{ij,j}.$$

The divergence of a higher-rank tensor is defined similarly as the contraction of the gradient of that tensor over its last two indices.

A.5 Divergence Theorems

For a smooth vector field $\boldsymbol{v} : \mathcal{P} \to \mathcal{V}$ on a fit region[4] $\mathcal{P} \subset \mathcal{E}$, the divergence theorem states that

$$\int_{\mathcal{P}} \operatorname{div} \boldsymbol{v} \, dV = \int_{\partial^* \mathcal{P}} \boldsymbol{v} \cdot \boldsymbol{\nu} \, dA,$$

where $\partial^* \mathcal{P}$ is the *reduced* boundary of $\mathcal{P}$, and $\boldsymbol{\nu}$ is its outer unit normal.

An immediate consequence of the divergence theorem is the integral-gradient theorem. By applying the divergence theorem to a vector field $\boldsymbol{v}(p) = f(p)\boldsymbol{a}$ that is the product of a scalar function $f(p)$ and an arbitrary vector $\boldsymbol{a}$ uniform in space, one finds that

$$\int_{\mathcal{P}} \nabla f \, dV = \int_{\partial^* \mathcal{P}} f \boldsymbol{\nu} \, dA.$$

Considering the divergence theorem for the vector field $\mathbf{T}^{\mathsf{T}}\boldsymbol{a}$, where $\boldsymbol{a}$ is again an arbitrary uniform vector, leads us to the following identities:

$$\int_{\mathcal{P}} \operatorname{div}(\mathbf{T}^{\mathsf{T}}\boldsymbol{a}) dV = \int_{\partial^* \mathcal{P}} \mathbf{T}^{\mathsf{T}}\boldsymbol{a} \cdot \boldsymbol{\nu} \, dA = \boldsymbol{a} \cdot \int_{\partial^* \mathcal{P}} \mathbf{T}\boldsymbol{\nu} \, dA =: \boldsymbol{a} \cdot \int_{\mathcal{P}} \operatorname{div} \mathbf{T} \, dV,$$

the last of which actually defines $\operatorname{div} \mathbf{T}$ in such a way that the divergence theorem holds in the form

$$\int_{\mathcal{P}} \operatorname{div} \mathbf{T} \, dV = \int_{\partial^* \mathcal{P}} \mathbf{T}\boldsymbol{\nu} \, dA.$$

Analogous theorems hold for higher-rank tensors.

[4] The definition of a *fit region* is given in Section 2.1.1, where also the use of these regions to describe the shapes of a body in continuum mechanics is justified.

References

1. M. ABRAMOWITZ & I. A. STEGUN (Eds.), *Handbook of Mathematical Functions with Formulas, Graphs, and Mathematical Tables*. Dover, New York, 1965. 9th Dover printing, 10th GPO printing.
2. B. R. ACHARYA, A. PRIMAK, & S. KUMAR, "Biaxial nematic phase in bent-core thermotropic mesogens." *Phys. Rev. Lett.* **92**(2004): 145506.
3. D. ALLENDER & L. LONGA, "Landau-de Gennes theory of biaxial nematics reexamined." *Phys. Rev. E* **78**(2008): 011704.
4. N. H. AYACHIT, S. T. VASAN, F. M. SANNANINGANNAVAR, & D. K. DESHPANDE, "Thermodynamic and acoustical parameters of some nematic liquid crystals." *J. Molecular Liq.* **133**(2007): 134–138.
5. C. BANFI & M. FABRIZIO, "Sul concetto di sottocorpo nella meccanica dei continui." *Rend. Accad. Naz. Lincei* **66**(1979): 136–142.
6. E. BARBERA, "On the principle of minimal entropy production for Navier-Stokes-Fourier fluids." *Continuum Mech. Thermodyn.* **11**(1999): 327–330. 10.1007/s001610050127.
7. R. BARBERI, F. CIUCHI, G. E. DURAND, M. IOVANE, D. SIKHARULIDZE, A. M. SONNET, & E. G. VIRGA, "Electric field induced order reconstruction in a nematic cell." *Eur. Phys. J. E* **13**(2004): 61–71.
8. J. B. BARBOUR, *The Discovery of Dynamics*. Oxford University Press, New York, 2001. Originally published as *Absolute or Relative Motion ?* Volume 1, *The Discovery of Dynamics*, Cambridge University Press 1989.
9. P. J. BARRATT & F. M. LESLIE, "Reflection and refraction of an obliquely incident shear wave at a solid-nematic interface." *J. Phys. (Paris) Colloque* **C3**(1979): 73–77. Supplément au n° 4, Tome 40, Avril 1979.
10. R. BARTOLINO, M. BERTOLOTTI, F. SCUDIERI, & D. SETTE, "Ultrasonic modulation of light with a liquid crystal in the smectic-A and nematic phases." *J. Appl. Phys.* **46**(1975): 1928–1933.
11. H. BATEMAN, "On dissipative systems and related variational principles." *Phys. Rev.* **38**(1931): 815–819.
12. M. A. BATES & G. R. LUCKHURST, "Biaxial nematic phases and v-shaped molecules: A Monte Carlo simulation study." *Phys. Rev. E* **72**(2005): 051702.
13. M. F. BEATTY, "On the foundation principles of general classical mechanics." *Arch. Rational Mech. Anal.* **24**(1967): 264–273.

14. B. BERGERSEN, P. PALFFY-MUHORAY, & D. A. DUNMUR, "Uniaxial nematic phase in fluids of biaxial particles." *Liq. Cryst.* **3**(1988): 347–352.
15. D. W. BERREMAN & S. MEIBOOM, "Tensor representation of Oseen-Frank strain energy in uniaxial cholesterics." *Phys. Rev. A* **30**(1984): 1955–1959.
16. M. BERTOLOTTI, S. MARTELLUCCI, F. SCUDIERI, & D. SETTE, "Acoustic modulation of light by nematic liquid crystals." *Appl. Phys. Lett.* **21**(1972): 74–75.
17. R. BHATIA, *Matrix Analysis*. Graduate Texts in Mathematics 169. Springer, New York, 1997.
18. R. BHATIA, *Positive Definite Matrices*. Princeton Series in Applied Mathematics 169. Princeton University Press, Princeton, 2007.
19. M. A. BIOT, "Variational principles in irreversible thermodynamics with application to viscoelasticity." *Phys. Rev.* **97**(1955): 1463–1469.
20. M. A. BIOT, *Variational Principles in Heat Transfer: Unified Lagrangian Analysis of Dissipative Phenomena*. Oxford University Press, Oxford, 1970.
21. M. A. BIOT, "A virtual dissipation principle and Lagrangian equations in non-linear irreversible thermodynamics." *Bulletin de l'Académie royale de Belgique (Classe des Sciences)* **61**(1975): 6–30. Also available at `http://www.pmi.ou.edu/Biot2005/biotConferenceBiotsPapers.htm`.
22. F. BISI, G. R. LUCKHURST, & E. G. VIRGA, "Dominant biaxial quadrupolar contribution to the nematic potential of mean torque." *Phys. Rev. E* **78**(2008): 021710.
23. F. BISI, S. ROMANO, & E. G. VIRGA, "Uniaxial rebound at the nematic biaxial transition." *Phys. Rev. E* **75**(2007): 041705.
24. F. BISI, E. G. VIRGA, E. C. GARTLAND, JR.,, G. DE MATTEIS, A. M. SONNET, & G. E. DURAND, "Universal mean-field phase diagram for biaxial nematics obtained from a minimax principle." *Phys. Rev. E* **73**(2006): 051709.
25. C. BLANC, D. SVENŠEK, S. ŽUMER, & M. NOBILI, "Dynamics of nematic liquid crystal disclinations: The role of the backflow." *Phys. Rev. Lett.* **95**(2005): 097802.
26. N. BOCCARA, R. MEJDANI, & L. DE SEZE, "Solvable model exhibiting a first-order phase transition." *J. Phys. (Paris)* **38**(1977): 149–151.
27. N. N. BOGOLIUBOV, JR.,, *A Method for Studying Model Hamiltonians*. Clarendon Press, Oxford, 1972.
28. N. N. BOGOLIUBOV, JR.,, B. I. SADOVNIKOV, & A. S. SHUMOVSKY, *Mathematical Methods of Statistical Mechanics of Model Systems*. CRC Press, Boca Raton, 1994.
29. H. BOLDER, "Deformation of tensor fields described by time-dependent mappings." *Arch. Rational Mech. Anal.* **35**(1969): 321–341.
30. F. BONETTO & E. ANOARDO, "Spin-lattice dispersion in nematc and smectic-A mesophases in the presence of ultrasonic waves: A theoretical approach." *Phys. Rev. E* **68**(2003): 021703.
31. F. BONETTO, E. ANOARDO, & R. KIMMICH, "Enhancement of order fluctuations in a nematic liquid crystal by sonication." *Chem. Phys. Lett.* **361**(2002): 237–244.
32. F. BONETTO, E. ANOARDO, & R. KIMMICH, "Ultrasound-order director fluctuations interaction in nematic liquid crystals: A nuclear magnetic resonance relaxometry study." *J. Chem. Phys.* **118**(2003): 9037–9043.
33. M. BORN & E. WOLF, *Principles of Optics*. Cambridge University Press, Cambridge, 7th edn., 1999.
34. H. R. BRAND & H. PLEINER, "New theoretical results for the Lehmann effect in cholesteric liquid crystals." *Phys. Rev. A* **37**(1988): 2736–2738.
35. R. BRAUER, "On the relation between the orthogonal group and the unimodular group." *Arch. Rational Mech. Anal.* **18**(1965): 97–99.

36. H. BREZIS, "S^k-valued maps with singularities." In M. GIAQUINTA (Ed.), *Topics in Calculus of Variations*, vol. 1365 of *Lecture Notes in Mathematics*, 1–30. Springer, Berlin, 1989. CIME Course, 1987.
37. H. BREZIS, J. M. CORON, & E. H. LIEB, "Harmonic maps with defects." *Commun. Math. Phys.* **107**(1986): 649–705.
38. A. D. BUCKINGHAM, "Permanent and induced molecular moments and long-range intermolecular forces." In J. O. HIRSCHFELDER (Ed.), *Advances in Chemical Physics: Intermolecular Forces, Volume 12*, 107–142. John Wiley & Sons, New York, 1967.
39. H. B. CALLEN, *Thermodynamics and an Introduction to Thermostatistics*. John Wiley & Sons, New York, 2nd edn., 1985.
40. S. CANDAU, A. FERRE, A. PETERS, G. WATON, & P. PIERANSKI, "Acoustical streaming in a film of nematic liquid crystal." *Mol. Cryst. Liq. Cryst.* **61**(1980): 7–30.
41. G. CAPRIZ, *Continua with Microstructure*. Springer-Verlag, New York, 1989. Springer Tracts in Natural Philosophy, Vol. 35.
42. T. CARLSSON & F. M. LESLIE, "The development of theory for flow and dynamic effects for nematic liquid crystals." *Liq. Cryst.* **26**(1999): 1267–1280.
43. T. CARLSSON, F. M. LESLIE, & J. S. LAVERTY, "Flow properties of biaxial nematic liquid crystals." *Mol. Cryst. Liq. Cryst.* **210**(1992): 95–127.
44. H. B. G. CASIMIR, "On Onsager's principle of microscopic reversibility." *Rev. Mod. Phys.* **17**(1945): 343–350.
45. V. M. CASTILLO & W. G. HOOVER, "Comment on 'Maximum of the local entropy production becomes minimal in stationary processes.' " *Phys. Rev. Lett.* **81**(1998): 5700.
46. P. E. CLADIS & H. R. BRAND, "Hedgehog-antihedgehog pair annihilation to a static soliton." *Physica A* **326**(2003): 322–332.
47. P. E. CLADIS, W. VAN SAARLOOS, P. L. FINN, & A. R. KORTAN, "Dynamics of line defects in nematic liquid crystals." *Phys. Rev. Lett.* **58**(1987): 222–225.
48. B. D. COLEMAN & V. J. MIZEL, "Existence of caloric equations of state in thermodynamics." *The Journal of Chemical Physics* **40**(1964): 1116–1125.
49. B. D. COLEMAN & W. NOLL, "The thermodynamics of elastic materials with heat conduction and viscosity." *Arch. Rational Mech. Anal.* **13**(1963): 167–178.
50. B. D. COLEMAN & C. TRUESDELL, "On the reciprocal relations of Onsager." *J. Chem. Phys.* **33**(1960): 28–31.
51. P. CONSTANTIN, I. KEVREKIDIS, & E. S. TITI, "Asymptotic states of a Smoluchowski equation." *Arch. Rational Mech. Anal.* **174**(2004): 365–384.
52. P. CONSTANTIN, I. KEVREKIDIS, & E. S. TITI, "Remarks on a Smoluchowski equation." *Discrete Contin. Dynam. Syst.* **11**(2004): 101–112.
53. E. COSSERAT & F. COSSERAT, "Sur la mécanique générale." *C. R. Acad. Sci. Paris* **145**(1907): 1139–1142.
54. E. COSSERAT & F. COSSERAT, *Théorie des Corps Déformables*. Hermann, Paris, 1909. Reprinted in 1991 by the Cornell University Library.
55. M. A. COTTER, "Generalized van der Waals theory of nematic liquid crystals: an alternative formulation." *J. Chem. Phys.* **66**(1977): 4710–4711.
56. T. A. DAVIS & E. C. GARTLAND, JR., "Finite element analysis of the Landau-de Gennes minimization problem for liquid crystals." *SIAM J. Num. Anal.* **35**(1998): 336–362.
57. P. G. DE GENNES, "Phenomenology of short-range-order effects in the isotropic phase of nematic materials." *Phys. Lett.* **30A**(1969): 454–455.
58. P. G. DE GENNES, "Short range order effects in the isotropic phase of nematics and cholesterics." *Mol. Cryst. Liq. Cryst.* **12**(1971): 193–214.

59. P. G. DE GENNES & J. PROST, *The Physics of Liquid Crystals*. Clarendon Press, Oxford, 2nd edn., 1993.
60. E. DE GIORGI, *Selected Papers*. Springer, Berlin, 2006. Papers selected, annotated, and, in part, translated by L. AMBROSIO, G. DAL MASO, M. FORTI, M. MIRANDA, and S. SPAGNOLO.
61. E. DE GIORGI, F. COLOMBINI, & L. PICCININI, *Frontiere Orientate di Misura Minima e Questioni Collegate*. Quaderni della Classe di Scienze. Scuola Normale Superiore, Pisa, 1972.
62. S. R. DE GROOT & P. MAZUR, *Non-Equilibrium Thermodynamics*. North Holland, Amsterdam, 1963.
63. G. DE MATTEIS, F. BISI, & E. G. VIRGA, "Constrained stability for biaxial nematic phases." *Continuum Mech. Themodyn.* **19**(2007): 1–23.
64. G. DE MATTEIS, S. ROMANO, & E. G. VIRGA, "Bifurcation analysis and computer simulation of biaxial liquid crystals." *Phys. Rev. E* **72**(2005): 041706.
65. G. DE MATTEIS, A. M. SONNET, & E. G. VIRGA, "Landau theory for biaxial nematic liquid crystals with two order parameter tensors." *Continuum Mech. Themodyn.* **20**(2008): 347–374.
66. G. DE MATTEIS & E. G. VIRGA, "Tricritical points in biaxial liquid crystal phases." *Phys. Rev. E* **71**(2005): 061703.
67. G. DE MATTEIS & E. G. VIRGA, "Director libration in nematoacoustics." *Phys. Rev. E* **83**(2011): 011703.
68. M. DEGIOVANNI, A. MARZOCCHI, & A. MUSESTI, "Cauchy fluxes associated with tensor fields having divergence measure." *Arch. Rational Mech. Anal.* **147**(1999): 197–223.
69. M. DEGIOVANNI, A. MARZOCCHI, & A. MUSESTI, "Edge-force densities and second-order powers." *Ann. Mat. Pura Appl.* **185**(2006): 81–103.
70. W. H. DEJEU, "On the role of spherical symmetry in the Maier-Saupe theory." *Mol. Cryst. Liq. Cryst.* **292**(1997): 13–24.
71. G. DEL PIERO, "Some properties of the set of fourth-order tensors, with application to elasticity." *J. Elast.* **9**(1979): 245–261.
72. G. DEL PIERO, "A class of fit regions and a universe of shapes for continuum mechanics." *J. Elast.* **70**(2003): 175–195.
73. F. DELL'ISOLA & P. SEPPECHER, "Edge contact forces and quasi-balanced power." *Meccanica* **32**(1997): 33–52.
74. C. DENNISTON, "Disclination dynamics in nematic liquid crystals." *Phys. Rev. B* **54**(1996): 6272–6275.
75. A. DEQUIDT & P. OSWALD, "Lehmann effect in compensated cholesteric liquid crystals." *EPL* **80**(2007): 26001.
76. S. DHAKAL & J. V. SELINGER, "Chirality and biaxiality in cholesteric liquid crystals." *Phys. Rev. E* **83**(2011): 020702.
77. A. C. DIOGO & A. F. MARTINS, "Order parameter and temperature dependence of the hydrodynamic viscosities of nematic liquid crystals." *J. Physique* **43**(1982): 779–786.
78. J. DION & A. D. JACOB, "A new hypothesis on ultrasonic interaction with a nematic liquid crystal." *Appl. Phys. Lett.* **31**(1977): 490–493.
79. J.-L. DION, "The orienting action of ultrasound on liquid crystals related to the theorem of minimum entropy production." *J. Appl. Phys.* **50**(1979): 2965–2966.
80. M. DOI, "Molecular dynamics and rheological properties of concentrated solutions of rodlike polymers in isotropic liquids and liquid crystals." *J. Polym. Sci. Polym. Phys.* **19**(1981): 229–243.

81. E. DUBOIS-VIOLETTE & B. PANSU, "Frustration and related topology of blue phases." *Mol. Cryst. Liq. Cryst.* **165**(1988): 151–182.
82. C. ECKART, "The electrodynamics of material media." *Phys. Rev.* **54**(1938): 920–923.
83. C. ECKART, "The thermodynamics of irreversible processes. I. The simple fluid." *Phys. Rev.* **58**(1940): 267–269.
84. C. ECKART, "The thermodynamics of irreversible processes. II. Fluid mixtures." *Phys. Rev.* **58**(1940): 269–275.
85. C. ECKART, "The thermodynamics of irreversible processes. III. Relativistic theory of the simple fluid." *Phys. Rev.* **58**(1940): 919–924.
86. C. ECKART, "The thermodynamics of irreversible processes. IV. The theory of elasticity and anelasticity." *Phys. Rev.* **73**(1948): 373–382.
87. C. ECKART, "Vortices and streams caused by sound waves." *Phys. Rev.* **73**(1948): 68–76.
88. C. ECKART, "Variation principles of hydrodynamics." *Phys. Fluids* **3**(1960): 421–427.
89. C. P. ELDREDGE, H. T. HEATH, B. LINDER, & R. A. KROMHOUT, "Role of dispersion potential anisotropy in the theory of nematic liquid crystals. I. The mean-field potential." *J. Chem. Phys.* **92**(1990): 6225–6234.
90. J. ERICKSEN, "Anisotropic fluids." *Arch. Rational Mech. Anal.* **4**(1959/60): 231–237. 10.1007/BF00281389.
91. J. L. ERICKSEN, "Conservation laws for liquid crystals." *Trans. Soc. Rheol.* **5**(1961): 23–34.
92. J. L. ERICKSEN, "Hydrostatic theory of liquid crystals." *Arch. Rational Mech. Anal.* **9**(1962): 371–378.
93. J. L. ERICKSEN, "Nilpotent energies in liquid crystal theory." *Arch. Rational Mech. Anal.* **10**(1962): 189–196. 10.1007/BF00281186.
94. J. L. ERICKSEN, "Inequalities in liquid crystal theory." *Phys. Fluids* **9**(1966): 1205–1207.
95. J. L. ERICKSEN, "Propagation of weak waves in liquid crystals of nematic type." *J. Acoust. Soc. Amer.* **44**(1968): 444–446.
96. J. L. ERICKSEN, "Twist waves in liquid crystals." *Quart. J. Mech. Appl. Math.* **21**(1968): 463–465.
97. J. L. ERICKSEN, "Continuum theory of liquid crystals of nematic type." *Mol. Cryst. Liq. Cryst.* **7**(1969): 153–164.
98. J. L. ERICKSEN, "On equations of motion for liquid crystals." *Quart. J. Mech. Appl. Math.* **29**(1976): 203–208.
99. J. L. ERICKSEN, "Liquid crystals with variable degree of orientation." *Arch. Rational Mech. Anal.* **113**(1991): 97–120.
100. A. C. ERINGEN, "Simple microfluids." *Int. J. Eng. Sci.* **2**(1964): 205–217.
101. A. C. ERINGEN, "Theory of micropolar fluids." *J. Math. Mech.* **15**(1966): 909–923.
102. A. C. ERINGEN, "Theory of micropolar fluids." *J. Math. Mech.* **16**(1966): 1–18.
103. A. C. ERINGEN & E. S. SUHUBI, "Nonlienar theory of simple micro-elastic solids–I." *Int. J. Eng. Sci.* **2**(1964): 189–203.
104. I. FATKULLIN & V. SLASTIKOV, "Critical points of the Onsager functional on a sphere." *Nonlinearity* **18**(2005): 2565–2580.
105. B. A. FINLAYSON, "Existence of variational principles for the Navier-Stokes equation." *Phys. Fluids* **15**(1972): 963–967.
106. D. FORSTER, T. C. LUBENSKY, P. C. MARTIN, J. SWIFT, & P. S. PERSHAN, "Hydrodynamics of liquid crystals." *Phys. Rev. Lett.* **26**(1971): 1016–1019.
107. S. FORTE & M. VIANELLO, "On surface stresses and edge forces." *Rend. Mat. Appl.* **8**(1988): 409–426.

108. R. L. FOSDICK & E. G. VIRGA, "A variational proof of the stress theorem of Cauchy." *Arch. Rational Mech. Anal.* **105**(1989): 95–103.
109. F. C. FRANK, "On the theory of liquid crystals." *Discuss. Faraday Soc.* **25**(1958): 19–28.
110. M. J. FREISER, "Ordered states of a nematic liquid." *Phys. Rev. Lett.* **24**(1970): 1041–1043.
111. M. J. FREISER, "Successive transitions in a nematic liquid." *Mol. Cryst. Liq. Cryst.* **14**(1971): 165–182.
112. E. FRIED & M. E. GURTIN, "Tractions, balances, and boundary conditions for non-simple materials with application to liquid flow at small-length scales." *Arch. Rational Mech. Anal.* **182**(2006): 513–554.
113. Y. GALERNE, "Comment on 'Thermotropic biaxial nematic liquid crystals.' " *Phys. Rev. Lett.* **96**(2006): 219803.
114. E. C. GARTLAND JR. & S. MKADDEM, "Instability of radial hedgehog configurations in nematic liquid crystals under Landau de Gennes free-energy models." *Phys. Rev. E* **59**(1999): 563–567.
115. E. C. GARTLAND, JR., A. M. SONNET, & E. G. VIRGA, "Elastic forces on nematic point defects." *Continuum Mech. Thermodyn.* **14**(2002): 307–319.
116. E. C. GARTLAND, JR., & E. G. VIRGA, "Minimum principle for indefinite mean-field free energies." *Arch. Rational Mech. Anal.* **196**(2010): 143–189.
117. W. M. GELBART, "Molecular theory of nematic liquid crystals." *J. Phys. Chem.* **86**(1982): 4298–4307.
118. W. M. GELBART & B. BARBOY, "A van der Waals picture of the isotropic-nematic liquid crystal phase transition." *Acc. Chem. Res.* **13**(1980): 290–296.
119. W. M. GELBART & B. A. BARON, "Generalized van der Waals theory of the isotropic-nematic phase transition." *J. Chem. Phys.* **66**(1977): 207–213.
120. P. GLANSDORFF & I. PRIGOGINE, "On a general evolution criterion in macroscopic physics." *Physica* **30**(1964): 351–374.
121. P. GLANSDORFF & I. PRIGOGINE, *Thermodynamic Theory of Structure, Stability and Fluctuations*. Wiley Interscience, London, 1971.
122. E. GOVERS & G. VERTOGEN, "Elastic continuum theory of biaxial nematics." *Phys. Rev. A* **30**(1984): 1998–2000.
123. E. GOVERS & G. VERTOGEN, "Fluid dynamics of biaxial nematics." *Physica A* **133A**(1985): 337–344.
124. V. A. GREANYA, A. P. MALANOSKI, B. T. WESLOWSKI, M. S. SPECTOR, & J. V. SELINGER, "Dynamics of the acousto-optic effect in a nematic liquid crystal." *Liq. Cryst.* **32**(2005): 933–941.
125. V. A. GREANYA, A. P. MALANOSKI, B. T. WESLOWSKI, M. S. SPECTOR, & J. V. SELINGER, "Dynamics of the acousto-optic effect in a nematic liquid crystal." *Liq. Cryst.* **32**(2005): 933–941.
126. V. A. GREANYA, M. S. SPECTOR, J. V. SELINGER, B. T. WESLOWSKI, & R. SHASHIDHAR, "Acousto-optic response of nematic liquid crystals." *J. Appl. Phys.* **94**(2003): 7571–7575.
127. A. E. GREEN & R. S. RIVLIN, "Simple force and stress multipoles." *Arch. Rational Mech. Anal.* **16**(1964): 325–253.
128. R. B. GRIFFITHS, "A proof that the free energy of a spin system is extensive." *J. Math. Phys.* **5**(1964): 1215–1222.
129. N. GUICCIARDINI, *Reading the Principia*. Cambridge University Press, Cambridge, 1999.

130. M. E. GURTIN, *An Introduction to Continuum Mechanics*. Academic Press, San Diego, 1981.
131. M. E. GURTIN, E. FRIED, & L. ANAND, *The Mechanics and Thermodynamics of Continua*. Cambridge University Press, Cambridge, 2010.
132. M. E. GURTIN & L. C. MARTINS, "Cauchy's theorem in classical physics." *Arch. Rational Mech. Anal.* **60**(1976): 305–324.
133. M. E. GURTIN, W. O. WILLIAMS, & W. P. ZIEMER, "Geometric measure theory and the axioms of continuum thermodynamics." *Arch. Rational Mech. Anal.* **92**(1986): 1–22.
134. W. HAMIDZADA & S. V. LETCHER, "Ultrasonic switching of a bistable liquid crystal cell." *Appl. Phys. Lett.* **42**(1983): 785–787.
135. R. HARDT & D. KINDERLEHRER, "Mathematical questions of liquid crystal theory." In J.L. ERICKSEN & D. KINDERLEHRER (Eds.), *Theory and Applications of Liquid Crystals*, vol. 5 of *The IMA Volumes in Mathematics and Its Applications*, 151–184. Springer-Verlag, New York, 1987.
136. R. HARDT, D. KINDERLEHRER, & F.-H. LIN, "Existence and partial regularity of static liquid crystal configurations." *Comm. Math. Phys.* **105**(1986): 547–570. 10.1007/BF01238933.
137. A. B. HARRIS, R. D. KAMIEN, & T. C. LUBENSKY, "Molecular chirality and chiral parameters." *Rev. Mod. Phys.* **71**(1999): 1745–1757.
138. W. HELFRICH, "Orienting action of sound on nematic liquid crystals." *Phys. Rev. Lett.* **29**(1972): 1583–1586.
139. H. HERBA & A. DRZYMAŁA, "Anisotropic attenuation of acoustic waves in nematic liquid crystals." *Liq. Cryst.* **8**(1990): 819–823.
140. J. W. HERIVEL, "A general variational principle for dissipative systems." *Proc. Roy. Irish Acad. A* **56**(1953/1954): 37–44.
141. J. W. HERIVEL, "The derivation of the equations of motion of an ideal fluid by Hamilton's principle." *Math. Proc. Cambridge Phil. Soc.* **51**(1955): 344–349.
142. S. HESS, "Irreversible thermodynamics of nonequilibrium alignment phenomena in molecular liquids and in liquid crystals." *Z. Naturforsch.* **30a**(1975): 728–738 & 1224–1232.
143. S. HESS, "Pre- and post-transitional behavior of the flow alignment and flow-induced phase transition in liquid crystals." *Z. Naturforsch.* **31a**(1976): 1507–1513.
144. S. HESS, "Transport phenomena in anisotropic fluids and liquid crystals." *J. Non-Equilib. Thermodyn.* **11**(1986): 175–193.
145. S. HESS & I. PARDOWITZ, "On the unified theory for non-equilibrium phenomena in the isotropic and nematic phases of a liquid crystal; spatially inhomogeneous alignment." *Z. Naturforsch.* **36a**(1981): 554–558.
146. H. HETTEMA, *Quantum Chemistry: Classic Scientific Papers*. World Scientific, Singapore, 2000.
147. W. G. HOOVER, "Note on 'Comment on "A check of Prigogine's theorem of minimum entropy production in a rod in a nonequilibrium stationary state," ' by Irena Danielewicz-Ferchmin and A. Ryszard Ferchmin" [Am. J. Phys. **68** (10), 962–965 (2000)], by Peter Palffy-Muhoray [Am. J. Phys. **69** (7), 825–826 (2001)]." *Am. J. Phys.* **70**(2002): 452–454.
148. R. A. HORN & C. R. JOHNSON, *Matrix Analysis*. Cambridge University Press, Cambridge, 1990.
149. H.-W. HUANG, "Hydrodynamics of nematic liquid crystals." *Phys. Rev. Lett.* **26**(1971): 1525–1527.
150. H. IMURA & K. OKANO, "Temperature dependence of the viscosity coefficients of liquid crystals." *Japan J. Appl. Phys.* **11**(1972): 1440–1445.

151. H. IMURA & K. OKANO, "Friction coefficient for a moving disclination in a nematic liquid crystal." *Phys. Lett. A* **42**(1973): 403–404.
152. G. JAUMANN, "Geschlossenes System physikalischer und chemischer Differentialgesetze." *Sitzgsber. Akad. Wiss. Wien (IIa)* **120**(1911): 385–530.
153. J. T. JENKINS, "Cholesteric energies." *J. Fluid Mech.* **45**(1971): 465–475.
154. D. JOU, J. CASAS-VÁZQUEZ, & G. LEBON, *Extended Irreversible Thermodynamics*. Springer, New York, 4th edn., 2010.
155. O. A. KAPUSTINA, "Ultrasonic properties." In D. DEMUS, G. W. GOODBY, G. W. GRAY, H.-W. SPEISS, & V. VILL (Eds.), *Physical Properties of Liquid Crystals*, 447–466. Wiley-VCH, Weinheim, 1999.
156. O. A. KAPUSTINA, "Liquid crystal acoustics: A modern view of the problem." *Crystallogr. Rep.* **49**(2004): 680–692. Translated from Kristallografiya, Vol. 49, No. 4, 2004, pp. 759–772.
157. O. A. KAPUSTINA, "Threshold structural transition in nematics in homogeneous ultrasonic fields." *JETP Lett.* **82**(2005): 586–589. Translated from Pis'ma v Zhurnal Éksperimental'noĭ i Teoreticheskoĭ Fiziki, Vol. 82, No. 9, 2005, pp. 664–667.
158. O. A. KAPUSTINA, "Threshold orientation transition in nematic liquid crystals under ultrasonic action." *Acoust. Phys.* **52**(2006): 413–417. Translated from Akusticheskiĭ Zhurnal, Vol. 52, No. 4, 2006, pp. 485–489.
159. O. A. KAPUSTINA, "On the mechanism of the effect of ultrasound on a nematic liquid crystal at oblique incidence." *Acoust. Phys.* **54**(2008): 778–782. Translated from Akusticheskiĭ Zhurnal, Vol. 54, No. 6, 2008, pp. 900–904.
160. O. A. KAPUSTINA, "Ultrasound-initiated structural transformations in liquid crystals (a review)." *Acoust. Phys.* **54**(2008): 180–196. Translated from Akusticheskiĭ Zhurnal, Vol. 54, No. 2, 2008, pp. 219–236.
161. J. KATRIEL, G. F. KVENTSEL, G. R. LUCKHURST, & T. J. SLUCKIN, "Free energies in the Landau and molecular field approaches." *Liq. Cryst.* **1**(1986): 337–355.
162. E. I. KATS, V. LEBEDEV, & S. V. MALININ, "Disclination motion in liquid crystalline films." *J. Exp. Theor. Phys.* **95**(2002): 714–727. Translated from Zhurnal Éksperimental'noĭ i Teoreticheskoĭ Fiziki, Vol. 122, No. 4, 2002, pp. 824–839.
163. O. D. KELLOGG, *Foundations of Potential Theory*. Dover, New York, 1953.
164. K. A. KEMP & S. V. LETCHER, "Ultrasonic determination of anisotropic shear and bulk viscosities in nematic liquid crystals." *Phys. Rev. Lett.* **27**(1971): 1634–1636.
165. U. D. KINI, "Isothermal hydrodynamics of orthorhombic nematics." *Mol. Cryst. Liq. Cryst.* **108**(1984): 71–91.
166. M. KLÉMAN, "Defect densities in directional media, mainly liquid crystals." *Philos. Mag.* **27**(1972): 1057–1072.
167. M. KLÉMAN, *Points, Lines and Walls*. Wiley, Chichester, 1983.
168. M. KLÉMAN & O. D. LAVRENTOVICH, *Soft Matter Physics: An Introduction*. Springer, New York, 2003.
169. M. KLÉMAN & O. D. LAVRENTOVICH, "Topological point defects in nematic liquid crystals." *Philos. Mag.* **86**(2006): 4117–4137.
170. D. J. KORTEWEG, "Sur la forme que prennent les équations du mouvement des fluides si l'on tient compte des forces capillaires causées par des variations de densité considérables mais continues et sur la théorie de la capillarité dans l'hypothèse d'une variation continue de la densité." *Arch. Néerl. Sci. Ex. Nat.* **6**(1901): 1–24.
171. E. N. KOZHEVNIKOV, "Deformation of a homeotropic nematic liquid crystal layer at oblique incidence of an ultrasonic wave." *Acoust. Phys.* **51**(2005): 688–694. Translated from Akusticheskiĭ Zhurnal, Vol. 51, No. 6, 2005, pp. 795–801.

172. S. KRALJ, E. G. VIRGA, & S. ŽUMER, "Biaxial torus around nematic point defects." *Phys. Rev. E* **60**(1999): 1858–1866.
173. C. LANCZOS, *The Variational Principles of Mechanics*. Dover Publications, New York, 4th edn., 1986. Unabridged and unaltered republication of the work first published by the University of Toronto Press.
174. L. D. LANDAU & E. M. LIFSHITZ, *Fluid Mechanics*. Pergamon Press, Oxford, 1959. Vol. 6 of Course of Theoretical Physics.
175. L. D. LANDAU & E. M. LIFSHITZ, *Electrodynamics of continuous media*. Elsevier Butterworth-Heinemann, Oxford, 2nd edn., 1984. Vol. 8 of Course of Theoretical Physics.
176. J. G. LEATHEM, "On the force exerted on a magnetic particle by a varying electric field." *Proc. Roy. Soc. London A* **89**(1913): 31–35.
177. J. D. LEE & A. C. ERINGEN, "Wave propagation in nematic liquid crystals." *J. Chem. Phys.* **54**(1971): 5027–5034.
178. O. LEHMANN, "Structur, System und magnetisches Verhalten flüssiger Krystalle und deren Mischbarkeit mit festen." *Ann. Phys.* **307**(1900): 649–705.
179. F. M. LESLIE, "Some constitutive equations for anisotropic fluids." *Quart. J. Mech. Appl. Math.* **19**(1966): 357–370.
180. F. M. LESLIE, "Some constitutive equations for liquid crystals." *Arch. Rational Mech. Anal.* **28**(1968): 265–283.
181. F. M. LESLIE, "Thermal effects in cholesteric liquid crystals." *Proc. Roy. Soc. London A* **307**(1968): 359–372.
182. F. M. LESLIE, "Thermo-mechanical effects in cholesteric liquid crystals." *J. Non-Equilib. Thermodyn.* **11**(1986): 23–34.
183. F. M. LESLIE, "Some topics in equilibrium theory of liquid crystals." In J.L. ERICKSEN & D. KINDERLEHRER (Eds.), *Theory and Applications of Liquid Crystals*, vol. 5 of *The IMA Volumes in Mathematics and Its Applications*, 211–234. Springer-Verlag, New York, 1987.
184. F. M. LESLIE, "Theory of flow phenomena in nematic liquid crystals." In J.L. ERICKSEN & D. KINDERLEHRER (Eds.), *Theory and Applications of Liquid Crystals*, vol. 5 of *The IMA Volumes in Mathematics and Its Applications*, 235–254. Springer-Verlag, New York, 1987.
185. F. M. LESLIE, "Continuum theory for nematic liquid crystals." *Continuum Mech. Thermodyn.* **4**(1992): 167–175.
186. F. M. LESLIE, J. S. LAVERTY, & T. CARLSSON, "Continuum theory for biaxial nematic liquid crystals." *Quart. J. Mech. Appl. Math.* **45**(1992): 595–606.
187. E. D. LIEBERMAN, J. D. LEE, & F. C. MOON, "Anisotropic ultrasonic wave propagation in a nematic liquid crystal placed in a magnetic field." *Appl. Phys. Lett.* **18**(1971): 280–281.
188. J. LIGHTHILL, "Acoustic streaming." *J. Sound Vib.* **61**(1978): 391–418.
189. J. LIGHTHILL, *Waves in Fluids*. Cambridge University Press, Cambridge, 1978.
190. H. LIU, H. ZHANG, & P. ZHANG, "Axial symmetry and classification of stationary solutions of Doi-Onsager equation on the sphere with Maier-Saupe potential." *Comm. Math. Sci.* **3**(2005): 201–218.
191. A. S. LODGE, "On the use of convected coordinate systems in the mechanics of continuous media." *Math. Proc. Cambridge Phil. Soc.* **47**(1951): 575–584.
192. F. LONDON, "Über einige Eigenschaften und Anwendungen der Molekularkräfte." *Z. Phys. Chem. B* **11**(1930): 222–251. Translated into English in [146], pp. 400–422.
193. F. LONDON, "On centers of van der Waals attraction." *J. Phys. Chem.* **46**(1942): 305–316.

194. L. LONGA, D. MONSELESAN, & H. R. TREBIN, "An extension of the Landau-Ginzburg-de Gennes theory for liquid crystals." *Liq. Cryst.* **2**(1987): 769–796.
195. L. LONGA, G. PAJĄK, & T. WYDRO, "Stability of biaxial nematic phase for systems with variable molecular shape anisotropy." *Phys. Rev. E* **76**(2007): 011703.
196. A. E. LORD, JR., & M. M. LABES, "Anisotropic ultrasonic properties of a nematic liquid crystal." *Phys. Rev. Lett.* **25**(1970): 570–572.
197. G. R. LUCKHURST, "Biaxial nematic liquid crystals: fact or fiction?" *Thin Solid Films* **393**(2001): 40–52.
198. G. R. LUCKHURST, "Liquid crystals – a missing phase found at last?" *Nature (London)* **430**(2004): 413–414.
199. G. R. LUCKHURST, "V-shaped molecules: New contenders for the biaxial nematic phase." *Angew. Chem., Int. Ed.* **44**(2005): 2834–2836.
200. G. R. LUCKHURST, "On the creation of director disorder in nematic liquid crystals." *Thin Solid Films* **509**(2006): 36–48.
201. G. R. LUCKHURST, "Biaxial nematics composed of flexible molecules: a molecular-field theory." *Liq. Cryst.* **36**(2009): 1295–1308.
202. G. R. LUCKHURST & S. ROMANO, "Computer simulation studies of anisotropic systems: II. Uniaxial and biaxial nematics formed by non-cylindrically symmetric molecules." *Mol. Phys.* **40**(1980): 129–139.
203. G. R. LUCKHURST & C. ZANNONI, "Why is the Maier-Saupe theory of nematic liquid crystals so successful?" *Nature (London)* **267**(1977): 412–414.
204. E. H. MACMILLAN, "On the hydrodynamics of biaxial nematic liquid crystals." *Arch. Rat. Mech. Anal.* **117**(1992): 193–239 & 241–294.
205. L. A. MADSEN, T. J. DINGEMANS, M. NAKATA, & E. T. SAMULSKI, "Thermotropic biaxial nematic liquid crystals." *Phys. Rev. Lett.* **92**(2004): 145505.
206. L. A. MADSEN, T. J. DINGEMANS, M. NAKATA, & E. T. SAMULSKI, "Madsen et al. Reply." *Phys. Rev. Lett.* **96**(2006): 219804.
207. P. L. MAFFETTONE, A. M. SONNET, & E. G. VIRGA, "Shear-induced biaxiality in nematic polymers." *J. Non-Newtonian Fluid Mech.* **90**(2000): 283–297.
208. W. MAIER & A. SAUPE, "Eine einfache molekulare Theorie des nematischen kristallinflüssigen Zustandes." *Z. Naturforsch.* **13a**(1958): 564–566. Translated into English in [305], pp. 381–385.
209. W. MAIER & A. SAUPE, "Eine einfache molekular-statistische Theorie der nematischen kristallinflüssigen Phase. Teil I." *Z. Naturforsch.* **14a**(1959): 882–889.
210. W. MAIER & A. SAUPE, "Eine einfache molekular-statistische Theorie der nematischen kristallinflüssigen Phase. Teil II." *Z. Naturforsch.* **15a**(1960): 287–292.
211. H. MAILER, K. L. LIKINS, T. R. TAYLOR, & J. L. FERGASON, "Effect of ultrasound on a nematic liquid crystal." *Appl. Phys. Lett.* **18**(1971): 105–107.
212. A. MAJUMDAR, "Equilibrium order parameters of nematic liquid crystals in the Landau-de Gennes theory." *Eur. J. Appl. Math.* **21**(2010): 181–203.
213. A. P. MALANOSKI, V. A. GREANYA, B. T. WESLOWSKI, M. S. SPECTOR, J. V. SELINGER, & R. SHASHIDHAR, "Theory of the acoustic realignment of nematic liquid crystals." *Phys. Rev. E* **69**(2004): 021705.
214. A. P. MALANOSKI, V. A. GREANYA, B. T. WESLOWSKI, M. S. SPECTOR, J. V. SELINGER, & R. SHASHIDHAR, "Theory of the acoustic realignment of nematic liquid crystals." *Phys. Rev. E* **69**(2004): 021705.
215. H. MARGENAU, "Van der Waals forces." *Rev. Mod. Phys.* **11**(1939): 1–35.
216. P. MARTINOTY & S. CANDAU, "Determination of viscosity coeffients of a nematic liquid crystal using a shear waves reflectance tecnique." *Mol. Cryst. Liq. Cryst.* **14**(1971): 243–271.

217. R. McWEENY, *Symmetry: An Introduction to Group Theory and Its Applications*. Dover, Mineola, New York, 2002. Reproduction of 1963 publication by Pergamon Press Ltd., Oxford, England.
218. J. MEIXNER, "Zur thermodynamik der irreversiblen Prozesse in Gasen mit chemisch reagierenden, dissoziierenden und anregbaren Komponenten." *Ann. Physik* **435**(1943): 244–270.
219. K. MERKEL, A. KOCOT, J. K. VIJ, R. KORLACKI, G. H. MEHL, & T. MEYER, "Thermotropic biaxial nematic phase in liquid crystalline organo-siloxane tetrapodes." *Phys. Rev. Lett.* **93**(2004): 237801.
220. C. B. MILLIKAN, "On the steady motion of viscous, incompressible fluids; with particular reference to a variation principle." *Phil. Mag* **7**(1929): 641–662.
221. K. MIYANO & J. KETTERSON, "Ultrasonic study of liquid crystals." *Phys. Rev. A* **12**(1975): 615–635.
222. S. D. MOBBS, "Variational principles for perfect and dissipative fluid flows." *Proc. Roy. Soc. London A* **381**(1982): 457–468.
223. S. E. MONROE, JR.,, G. C. WETSEL, JR.,, M. R. WOODARD, & B. A. LOWRY, "Ultrasonic investigation of the nematic-isotropic phase transition in MBBA." *J. Chem. Phys.* **63**(1975): 5139–5144.
224. B. M. MULDER, "The excluded volume of hard sphero-zonotopes." *Mol. Phys.* **103**(2005): 1411–1424.
225. M. E. MULLEN, B. LÜTHI, & M. J. STEPHEN, "Sound velocity in a nematic liquid crystal." *Phys. Rev. Lett.* **28**(1972): 799–801.
226. I. MÜLLER, "On the entropy inequality." *Arch. Rational Mech. Anal.* **26**(1967): 118–141.
227. I. MÜLLER, *A History of Thermodynamics*. Springer-Verlag, Berlin, 2007.
228. I. MÜLLER & T. RUGGERI, *Rational Extended Thermodynamics*. Springer, New York, 2nd edn., 1998.
229. I. MÜLLER & W. WEISS, *Entropy and Energy*. Springer-Verlag, Berlin, 2005.
230. F. R. N. NABARRO, "Singular lines and singular points of ferromagnetic spin systems and nematic liquid crystals." *J. Phys. France* **33**(1972): 1089–1098.
231. G. G. NATALE & D. E. COMMINS, "Temperature dependence of anisotropic-ultrasonic propagation in a nematic liquid crystal." *Phys. Rev. Lett.* **28**(1972): 1439–1441.
232. K. NEUPANE, S. W. KANG, S. SHARMA, D. CARNEY, T. MEYER, G. H. MEHL, D. W. ALLENDER, S. KUMAR, & S. SPRUNT, "Dynamic light scattering study of biaxial ordering in a thermotropic liquid crystal." *Phys. Rev. Lett.* **97**(2006): 207802.
233. I. NEWTON, *The Mathematical Papers of Isaac Newton*, vol. 1–8. Cambridge University Press, Cambridge, 1967–1981. Edited by D. T. WHITESIDE with the assistance in publication by M. A. HOSKIN and A. PRAG.
234. W. NOLL, "On the continuity of the solid and fluid states." *J. Rational Mech. Anal.* **4**(1955): 3–81.
235. W. NOLL, "A mathematical theory of the mechanical behavior of continuous media." *Arch. Rational Mech. Anal.* **2**(1958/59): 197–226.
236. W. NOLL, "The foundations of classical mechanics in the light of recent advances in continuum mechanics." In P. S. LEON HENKIN & A. TARSKI (Eds.), *The Axiomatic Method*, vol. 27 of *Studies in Logic and the Foundations of Mathematics*, 266–281. North-Holland, 1959. Reprinted in [242].
237. W. NOLL, "La mécanique classique, basée sur an axiome d'objectivité." In *La Méthode Axiomatique dans les Mécaniques Classiques et Nouvelles*, Colloque International à Paris, 1959, 47–56. Gauthier-Villars, Paris, 1963. Reprinted in [242].

238. W. NOLL, "Euclidean geometry and Minkowskian chronometry." *Amer. Math. Monthly* **71**(1964): 129–144.
239. W. NOLL, "Proof of the maximality of the orthogonal group in the unimodular group." *Arch. Rational Mech. Anal.* **18**(1965): 100–102. 10.1007/BF00282255.
240. W. NOLL, "The foundations of mechanics." In G. GRIOLI (Ed.), *Lectures given at the Centro Internazionale Matematico Estivo (C.I.M.E.) Bressanone (Bolzano), Italy, May 31–June 9, 1965*, CIME Summer Schools, 159–200. Rome, 1966.
241. W. NOLL, "Lectures on the foundations of continuum mechanics and thermodynamics." *Arch. Rational Mech. Anal.* **52**(1973): 62–92.
242. W. NOLL, *The Foundations of Mechanics and Themodynamics*. Springer-Verlag, New York, 1974.
243. W. NOLL, "Continuum mechanics and geometric integration theory." In F. LAWVERE & S. SCHNAUEL (Eds.), *Categories in Continuum Physics*, vol. 1174 of *Springer Lecture Notes in Mathematics*, 17–29. Springer, Berlin, 1986.
244. W. NOLL, *Five Contributions to Natutal Philosophy*. Pittsburgh, 2004. Available at `http://www.math.cmu.edu/~wn0g/noll/`.
245. W. NOLL, "A frame-free formulation of elasticity." *J. Elast.* **83**(2006): 291–307.
246. W. NOLL, "On the past and future of Natural Philosophy." *J. Elast.* **84**(2006): 1–11.
247. W. NOLL & B. SEGUIN, "Monoids, Boolean algebras, materially ordered sets." *Int. J. Pure Appl. Math.* **37**(2007): 187–202.
248. W. NOLL & B. SEGUIN, "Basic concepts of thermomechanics." *J. Elast.* **101**(2010): 121–151.
249. W. NOLL & E. G. VIRGA, "Fit regions and functions of bounded variations." *Arch. Rational Mech. Anal.* **102**(1988): 1–21.
250. W. NOLL & E. G. VIRGA, "On edge interactions and surface tension." *Arch. Rational Mech. Anal.* **111**(1990): 1–31.
251. W. NYBORG, "Acoustic streaming." In W. P. MASON (Ed.), *Physical Acoustics*, vol. II B, 265–331. Academic Press, New York, 1968.
252. J. G. OLDROYD, "On the formulation of rheological equations of state." *Proc. Roy. Soc. London A* **200**(1950): 523–541.
253. J. G. OLDROYD, "Non-Newtonian effects in steady motion of some idealized elastico-viscous liquids." *Proc. Roy. Soc. London A* **245**(1958): 278–297.
254. P. D. OLMSTED & P. GOLDBART, "Theory of the nonequilibrium phase transition for nematic liquid crystals under shear flow." *Phys. Rev. A* **41**(1990): 4578–4581.
255. P. D. OLMSTED & P. M. GOLDBART, "Isotropic-nematic transition in shear flow: State selection, coexistence, phase transitions, and critical behavior." *Phys. Rev. A* **46**(1992): 4966–4993.
256. L. ONSAGER, "Reciprocal relations in irreversible processes. I." *Phys. Rev.* **37**(1931): 405–426.
257. L. ONSAGER, "Reciprocal relations in irreversible processes. II." *Phys. Rev.* **38**(1931): 2265–2279.
258. L. ONSAGER, "The effects of shape on the interaction of colloidal particles." *Ann. N.Y. Acad. Sci.* **51**(1949): 627–659. Reprinted in [305], pp. 625–657.
259. C. W. OSEEN, "The theory of liquid crystals." *Trans. Faraday Soc.* **29**(1933): 883–899.
260. M. A. OSIPOV & A. S. SHUMOVSKII, "Violation of the principle of minimality of the free energy in the self-consistent field approximation for some models of ferroelectrics (in Russian)." *Teor. Mat. Fiz.* **46**(1981): 125–131. *Theor. Math. Phys.* **46**, 83–87 (1981) (English transl.).
261. P. OSWALD & J. IGNÉS-MULLOL, "Backflow-induced asymmetric collapse of disclination lines in liquid crystals." *Phys. Rev. Lett.* **95**(2005): 027801.

262. P. PALFFY-MUHORAY, "Comment on 'A check of Prigogine's theorem of minimum entropy production in a rod in a nonequilibrium stationary state' by Irena Danielewicz-Ferchmin and A. Ryszard Ferchmin [Am. J. Phys. **68** (10), 962–965 (2000)]." *Am. J. Phys.* **69**(2001): 825–826.
263. P. PALFFY-MUHORAY, "The single particle potential in mean-field theory." *Am. J. Phys.* **70**(2002): 433–437.
264. A. PARGELLIS, N. TUROK, & B. YURKE, "Monopole-antimonopole annihilation in a nematic liquid crystal." *Phys. Rev. Lett.* **67**(1991): 1570–1573.
265. O. PARODI, "Stress tensor for a nematic liquid crystal." *J. Phys. France* **31**(1970): 581–584.
266. R. PENROSE, *The Road to Reality*. Vintage, London, 2005.
267. C. PEREIRA BORGMEYER & S. HESS, "Unified description of the flow alignment and viscosity in the isotropic and nematic phases of liquid crystals." *J. Non-Equilib. Thermodyn.* **20**(1995): 359–384.
268. J. PERLO, L. E. AGUIRRE, J. REVELLI, & E. ANOARDO, "On the acoustic-director interaction in the smectic A phase." *Chem. Phys. Lett.* **450**(2007): 170–174.
269. L. M. PISMEN & B. Y. RUBINSTEIN, "Motion of interacting point defects in nematics." *Phys. Rev. Lett.* **69**(1992): 96–99.
270. H. PLEINER, "Dynamics of a disclination point in smectic-C and -C^* liquid-crystal films." *Phys. Rev. A* **37**(1988): 3986–3992.
271. P. PODIO-GUIDUGLI, "Inertia and invariance." *Ann. Mat. Pura Appl.* **172**(1997): 103–124.
272. P. PODIO-GUIDUGLI, "La scelta delle interazioni inerziali nei continui con microstruttura." *Rend. Mat. Acc. Lincei* **14**(2003): 319–326.
273. P. PODIO-GUIDUGLI & E. G. VIRGA, "Transversely isotropic elasticity tensors." *Proc. R. Soc. London Ser. A* **411**(1987): 85–93.
274. H. POINCARÉ, "Sur le problème des trois corps et les équations de la dynamique." *Acta Mathematica* **13**(1890): A3–A270.
275. R. G. PRIEST & T. C. LUBENSKY, "Biaxial model of cholesteric liquid crystals." *Phys. Rev. A* **9**(1974): 893–898.
276. I. PRIGOGINE, *Étude thermodynamique des phénomènes irréversibles*. Desoer, Liège, 1947.
277. I. PRIGOGINE, *Introduction to Thermodynamics of Irreversible Processes*. Interscience, New York, 1961.
278. I. PRIGOGINE & P. MAZUR, "Sur deux formulations de l'hydrodynamique et le problème de l'hélium liquide II." *Physica* **17**(1951): 661–679.
279. J. PROST, "Comments on the thermomechanical coupling in cholesteric liquid crystals." *Solid State Comm.* **11**(1972): 183–184.
280. T. QIAN & P. SHENG, "Generalized hydrodynamic equations for nematic liquid crystals." *Phys. Rev. E* **58**(1998): 7475–7485.
281. R. S. RIVLIN, "Further remarks on the stress-deformation relations for isotropic materials." *J. Rational Mech. Anal.* **4**(1955): 681–702.
282. G. RODNAY & R. SEGEV, "Cauchy's flux theorem in light of geometric integration theory." *J. Elast.* **71**(2003): 183–203.
283. R. ROSSO & E. G. VIRGA, "Quadrupolar projection of excluded-volume interactions in biaxial nematic liquid crystals." *Phys. Rev. E* **74**(2006): 021712.
284. H. L. ROYDEN, *Real Analysis*. Macmillan, New York, 2nd edn., 1968.
285. G. RYSKIN & M. KREMENETSKY, "Drag force on a line defect moving through an otherwise undisturbed field: Disclination line in a nematic liquid crystal." *Phys. Rev. Lett.* **67**(1991): 1574–1577.

286. A. SAUPE, "Disclinations and properties of the director field in nematic and cholesteric liquid crystals." *Mol. Cryst. Liq. Cryst.* **21**(1973): 211–238.
287. A. SAUPE, "Elastic and flow properties of biaxial nematics." *J. Chem. Phys.* **75**(1981): 5118–5124.
288. N. SCHOPOHL & T. J. SLUCKIN, "Defect core structure in nematic liquid crystals." *Phys. Rev. Lett.* **59**(1987): 2582–2584.
289. H. SCHRÖDER, "Cholesteric structures and the role of phase biaxiality." In W. HELFRICH & G. HEPPKE (Eds.), *Liquid Crystals of One- and Two-Dimensional Order*. Springer, Berlin, Heidelberg, New York, 1980, 196–204.
290. E. SCHRÖDINGER, *Statistical Thermodynamics*. Dover, New York, 1989. Reprint of the 2nd edition (1952) of the work first published in 1946 by the Cambridge University Press, Cambridge, with the following subtitle: *A Course of Seminar Lectures Delivered in January-March 1944, at the School of Theoretical Physics, Dublin Institute for Advanced Studies.*
291. R. SEGEV, "The geometry of Cauchy's fluxes." *Arch. Rational Mech. Anal.* **154**(2000): 183–198.
292. R. SEGEV, "A correction of an inconsistency in my paper 'Cauchy's theorem on manifold.' " *J. Elast.* **63**(2001): 55–59.
293. R. SEGEV & G. RODNAY, "Cauchy's theorem on manifolds." *J. Elast.* **56**(1999): 129–144.
294. J. V. SELINGER, M. S. SPECTOR, V. A. GREANYA, B. T. WESLOWSKI, D. K. SHENOY, & R. SHASHIDHAR, "Acoustic realignment of nematic liquid crystals." *Phys. Rev. E* **66**(2002): 051708.
295. B. SENYUK, Y.-K. KIM, L. TORTORA, S.-T. SHIN, S. V. SHIYANOVSKII, & O. D. LAVRENTOVICH, "Surface alignment, anchoring transitions, optical properties and topological defects in nematic bent-core materials C7 and C12." *Mol. Cryst. Liq. Cryst.* **540**(2011): 20–41.
296. B. SENYUK, H. WONDERLY, M. MATHEWS, Q. LI, S. V. SHIYANOVSKII, & O. D. LAVRENTOVICH, "Surface alignment, anchoring transitions, optical properties, and topological defects in the nematic phase of thermotropic bent-core liquid crystal A131." *Phys. Rev. E* **82**(2010): 041711.
297. J. SERRIN, "Mathematical principles of classical fluid mechanics." In S. FLÜGGE & C. TRUESDELL (Eds.), *Encyclopedia of Physics*, vol. VIII/1, 125–263. Springer-Verlag, Berlin, 1959.
298. K. SEVERING & K. SAALWÄCHTER, "Biaxial nematic phase in a thermotropic liquid-crystalline side-chain polymer." *Phys. Rev. Lett.* **92**(2004): 125501.
299. K. SEVERING, E. STIBAL-FISCHER, A. HASENHINDL, H. FINKELMANN, & K. SAALWÄCHTER, "Phase biaxiality in nematic liquid crystalline side-chain polymers of various chemical constitutions." *J. Phys. Chem. B* **110**(2006): 15680–15688.
300. M. ŠILHAVÝ, "The existence of the flux vector and the divergence theorem for general cauchy fluxes." *Arch. Rational Mech. Anal.* **90**(1985): 195–212.
301. M. ŠILHAVÝ, "Cauchy's stress theorem and tensor fields with divergences in L^p." *Arch. Rational Mech. Anal.* **116**(1991): 223–255.
302. A. P. SINGH & A. D. REY, "Theory and simulation of extensional flow-induced biaxiality in discotic mesophases." *J. Phys. II France* **5**(1995): 1321–1348.
303. G. S. SINGH & B. KUMAR, "Molecular fluids and liquid crystals in convex-body coordinate systems." *Ann. Phys. (NY)* **294**(2001): 24–47.
304. S. SINGH, "Phase transitions in liquid crystals." *Phys. Rep.* **324**(2000): 107–269.
305. T. J. SLUCKIN, D. A. DUNMUR, & H. STEGEMEYER, *Crystals That Flow*. Taylor & Francis, London, New York, 2004.

306. G. F. SMITH, "On isotropic integrity bases." *Arch. Rat. Mech. Anal.* **18**(1965): 282–292.
307. G. F. SMITH & R. S. RIVLIN, "The anisotropic tensors." *Quart. Appl. Math.* **15**(1957): 308–314.
308. A. SONNET, A. KILIAN, & S. HESS, "Alignment tensor versus director: Description of defects in nematic liquid crystals." *Phys. Rev. E* **52**(1995): 718–722.
309. A. M. SONNET, "Viscous forces on nematic defects." *Continuum Mech. Thermodyn.* **17**(2005): 287–295.
310. A. M. SONNET, P. L. MAFFETTONE, & E. G. VIRGA, "Continuum theory for nematic liquid crystals with tensorial order." *J. Non-Newtonian Fluid Mech.* **119**(2004): 51–59.
311. A. M. SONNET & E. G. VIRGA, "Dynamics of dissipative ordered fluids." *Phys. Rev. E* **64**(2001): 031705.
312. A. M. SONNET & E. G. VIRGA, "Fluids with dissipative microstructure." In P. PODIO-GUIDUGLI & M. BROCATO (Eds.), *Rational Continua, Classical and New*, 169–181. Springer, Milano, Berlin, Heidelberg, 2002.
313. A. M. SONNET & E. G. VIRGA, "Steric effects in dispersion forces interactions." *Phys. Rev. E* **77**(2008): 031704.
314. A. M. SONNET & E. G. VIRGA, "Flow and reorientation in the dynamics of nematic defects." *Liq. Cryst.* **36**(2009): 1185–1192.
315. A. M. SONNET, E. G. VIRGA, & G. E. DURAND, "Dielectric shape dispersion and biaxial transitions in nematic liquid crystals." *Phys. Rev. E* **67**(2003): 061701.
316. J. SPANIER & K. B. OLDHAM, *An Atlas of Functions*. Hemisphere, Washington, 1987.
317. C. SRIPAIPAN, C. F. HAYES, & G. T. FANG, "Ultrasonically-induced optical effect in a nematic liquid crystal." *Phys. Rev. A* **15**(1977): 1297–1303.
318. S. STALLINGA & G. VERTOGEN, "Theory of orientational elasticity." *Phys. Rev. E* **49**(1994): 1483–1494.
319. I. W. STEWART, *The Static and Dynamic Continuum Theory of Liquid Crystals*. Taylor & Francis, London, 2004.
320. A. J. STONE, *The Theory of Intermolecular Forces*, vol. 32 of *The International Series of Monographs on Chemistry*. Clarendon Press, Oxford, 1996.
321. J. P. STRALEY, "Ordered phases of a liquid of biaxial particles." *Phys. Rev. A* **10**(1974): 1881–1887.
322. H. STRUCHTRUP & W. WEISS, "Maximum of the local entropy production becomes minimal in stationary processes." *Phys. Rev. Lett.* **80**(1998): 5048–5051.
323. J. W. STRUTT (LORD RAYLEIGH), "Some general theorems relating to vibrations." *Proc. London Math. Soc.* **4**(1873): 357–368.
324. J. W. STRUTT (LORD RAYLEIGH), "Scientific Papers." Cambridge University Press, Teddington, England, 1883.
325. J. W. STRUTT (LORD RAYLEIGH), "On the circulation of air observed in Kundt's tubes, and on some allied acoustical problems." *Phil. Trans. Royal Soc. London* **175**(1884): 1–21.
326. J. W. STRUTT (LORD RAYLEIGH), *Theory of Sound*, vol. 2. Macmillan, London, 1896. Republished in [328].
327. J. W. STRUTT (LORD RAYLEIGH), "On the motion of a viscous fluid." *Phil. Mag.* **26**(1913): 776–786.
328. J. W. STRUTT (LORD RAYLEIGH), *Theory of Sound*, vol. 2. Dover, New York, 1945. An account of [325] can be found starting from p. 333.
329. M. STRUWE, *Variational Methods: Applications to Nonlinear Partial Differential Equations and Hamiltonian Systems*. Springer-Verlag, Berlin, 3rd edn., 2000.
330. D. SVENŠEK & S. ŽUMER, "Hydrodynamics of pair-annihilating disclination lines in nematic liquid crystals." *Phys. Rev. E* **66**(2002): 021712.

331. J. J. THOMSON, *Applications of Dynamics to Physics and Chemistry*. Macmillan, London, 1888.
332. W. THOMSON (LORD KELVIN), "On a universal tendency in nature to the dissipation of mechanical energy." *Philos. Mag.* **4**(1852): 304–306. Also published in the Proceedings of the Royal Society of Edinburgh for April 19, 1852.
333. W. THOMSON (LORD KELVIN), "The kinetic theory of energy dissipation." *Proc. Roy. Soc. Edinb.* **8**(1874): 325–334.
334. W. THOMSON (LORD KELVIN), "Kinetic theory of the dissipation of energy." *Nature* **9**(1874): 441–444.
335. R. C. TOLMAN, *Statistical Mechanics with Applications to Physics and Chemistry*. Chemical Catalog Company, New York, 1927. Also available at `http://ia700202.us.archive.org/15/items/statisticalmecha00tolm/statisticalmecha00tolm.pdf`.
336. G. TÓTH, C. DENNISTON, & J. M. YEOMANS, "Hydrodynamics of topological defects in nematic liquid crystals." *Phys. Rev. Lett.* **88**(2002): 105504.
337. R. A. TOUPIN, "World invariant kinematics." *Arch. Rational Mech. Anal.* **1**(1957): 181–211.
338. R. A. TOUPIN, "Elastic materials with couple-stresses." *Arch. Rational Mech. Anal.* **11**(1962): 385–414.
339. R. A. TOUPIN, "Theory of elasticity with couple-stress." *Arch. Rational Mech. Anal.* **17**(1964): 85–112.
340. C. TRUESDELL, "A new definition of a fluid. II. The Maxwellian fluid." *J. Math. Pures Appl.* **30**(1951): 111–158.
341. C. TRUESDELL, "The mechanical foundations of elasticity and fluid dynamics." *J. Rational Mech. Anal.* **1**(1952): 125–300.
342. C. TRUESDELL, *Six Lectures on Modern Natutal Philosophy*. Springer-Verlag, Berlin, 1966.
343. C. TRUESDELL, *Rational Thermodynamics*. McGraw-Hill, New York, 1969. The second edition was published as [344].
344. C. TRUESDELL, *Rational Thermodynamics*. Springer, New York, 1984.
345. C. TRUESDELL & W. NOLL, *The Non-Linear Field Theories of Mechanics*. Springer-Verlag, Berlin, 2nd edn., 1992. See [346] for an annotated edition.
346. C. TRUESDELL & W. NOLL, *The Non-Linear Field Theories of Mechanics*. Springer-Verlag, Berlin, 3rd edn., 2004. Edited by S. S. ANTMAN.
347. C. TRUESDELL & K. R. RAJAGOPAL, *An Introduction to the Mechanics of Fluids*. Birkhäuser, Boston, 2000.
348. C. TRUESDELL & R. TOUPIN, "The classical field theories." In S. FLÜGGE (Ed.), *Encyclopedia of Physics*, vol. III/1. Springer-Verlag, Berlin, 1960.
349. C. A. TRUESDELL, *A First Course in Rational Continuum Mechanics*, vol. 1. Academic Press, San Diego, 2nd edn., 1991.
350. A. UNSÖLD, "Quantentheorie des Wasserstoffmolekülions und der Born-Landéschen Abstoßungskräfte." *Z. Phys.* **43**(1927): 563–574.
351. G. VERTOGEN, "The equations of motion for nematics." *Z. Naturforsch.* **38a**(1983): 1273–1275.
352. G. VERTOGEN & W. H. DE JEU, *Thermotropic Liquid Crystals, Fundamentals*. Springer Series in Chemical Physics 45. Springer, Berlin Heidelberg New York, 1988.
353. E. G. VIRGA, *Variational Theories for Liquid Crystals*. Chapman & Hall, London, 1994.
354. E. G. VIRGA, "Variational theory for nematoacoustics." *Phys. Rev. E* **80**(2009): 031705.

355. G. E. VOLOVIK & O. D. LAVRENTOVICH, "Topological dynamics of defects: Boojums in nematic drops (in Russian)." *Zhurn. Eksp. Teor. Fiz.* **85**(1983): 1997–2010. *Sov. Phys. JETP* **58**, 1159–1167 (1983) (English transl.).
356. A. I. VOL'PERT & S. I. HUDJAEV, *Analysis in Classes of Discontinuous Functions and Equations of Mathematical Physics*. Martinus Nijhoff Publishers, Dordrecht, 1985.
357. C.-C. WANG, "A new representation theorem for isotropic functions: an answer to Professor G. F. Smith's criticism of my paper on representations for isotropic functions. Part I. Scalar-valued functions." *Arch. Rational Mech. Anal.* **36**(1970): 165–197.
358. C. C. WANG, "A new representation theorem for isotropic functions: An answer to Professor G. F. Smith's criticism of my papers on representations for isotropic functions. Part I. Vector-valued isotropic functions, symmetric tensor-valued isotropic functions, and skew-symmetric tensor-valued isotropic functions." *Arch, Rational Mech. and Anal.* **36**(1970): 198–223.
359. M. WARNER, "Interaction energies in nematogens." *J. Chem. Phys.* **73**(1980): 5874–5883.
360. P. WESTERVELT, "The theory of steady rotational flow generated by a sound field." *J. Acoust. Soc. Am.* **25**(1953): 60–67.
361. G. C. WETSEL, JR.,, R. S. SPEER, B. A. LOWRY, & M. R. WOODARD, "Ultrasonic investigation of the nematic-isotropic phase transition in MBBA." *J. Appl. Phys.* **43**(1972): 1495–1497.
362. E. T. WHITTAKER, *A Treatise on the Analytical Dynamics of Particles and Rigid Bodies*. Cambridge University Press, Cambridge, 4th edn., 1937.
363. Wolfram Research, Inc., Champaign, Illinois, *Mathematica, Version 6.0*, 2007.
364. A. WULF, "Short-range correlations and the effective orientational energy in liquid crystals." *J. Chem. Phys.* **67**(1977): 2254–2266.
365. L. J. YU & A. SAUPE, "Observation of a biaxial nematic phase in potassium laurate-1-decanol-water mixtures." *Phys. Rev. Lett.* **45**(1980): 1000–1003.
366. S. ZAREMBA, "Le principe des mouvements relatifs et les équations de la mécanique physique." *Bull. Int. Acad. Sci. Cracovie* (1903): 614–621.
367. S. ZAREMBA, "Sur une forme perfectionée de la théorie de la relaxion." *Bull. Int. Acad. Sci. Cracovie* (1903): 594–614.
368. S. ZAREMBA, *Sur une Conception Nouvelle des Forces Intérieures dans un Fluide en Mouvement*, vol. 82 of *Mém. Sci. Math.* Gauthier-Villars, Paris, 1937.
369. I. ZGURA, R. MOLDOVAN, T. BEICA, & S. FRUNZA, "Temperature dependence of the density of some liquid crystals in the alkyl cyanobiphenyl series." *Cryst. Res. Technol.* **44**(2009): 883–888.
370. Z.-D. ZHANG, Y.-J. ZHANG, & Z.-L. SUN, "Two-particle cluster theory for biaxial nematic phases based on a recently proposed interaction potential." *Chin. Phys. Lett.* **23**(2006): 3025–3028.
371. X. ZHENG, W. IGLESIAS, & P. PALFFY-MUHORAY, "Distance of closest approach of two arbitrary hard ellipsoids." *Phys. Rev. E* **79**(2009): 057702.
372. X. ZHENG & P. PALFFY-MUHORAY, "Distance of closest approach of two arbitrary hard ellipses in two dimensions." *Phys. Rev. E* **75**(2007): 061709.
373. H. ZHOU, H. WANG, M. G. FOREST, & Q. WANG, "A new proof on axisymmetric equilibria of a three-dimensional Smoluchowski equation." *Nonlinearity* **18**(2005): 2815–2825.
374. W. P. ZIEMER, *Weakly Differentiable Functions*, vol. 120 of *Graduate Texts in Mathematics*. Springer-Verlag, New York, 1989.

Index